JAN '93 P9-DHJ-421

AIR CONDITIONING PRINCIPLES AND SYSTEMS

An Energy Approach

A view of the concentrating and tracking solar collectors for the 100,000 square foot corporate headquarters of Honeywell, Inc., in Minneapolis. The collectors serve a solar heating and cooling system that provides over 50% of the building's yearly heating requirements, more than 80% of the cooling, and all of the hot water. (Honeywell, Inc.)

AIR CONDITIONING PRINCIPLES AND SYSTEMS

An Energy Approach

SECOND EDITION

Edward G. Pita

Environmental Control Technology
New York City Technical College
The City University of New York

PRENTICE HALL, Englewood Cliffs, New Jersey 07632

Library of Congress Cataloging-in-Publication Data:

Pita, Edward G.
  Air conditioning principles and systems: an energy approach
Edward G. Pita.—2nd ed.
      p.   cm.
  Bibliography: p.
  Includes index.
  ISBN 0-471-86506-0
  1. Air conditioning. 2. Buildings—Energy
conservation. I. Title.
TH7687.P446 1989   88-21884
697.9'3—dc19

 © 1981, 1989 by Prentice-Hall Inc.
A Division of Simon & Schuster
Englewood Cliffs, NJ 07632

All rights reserved. No part of this book may be
reproduced, in any form or by any means,
without permission in writing from the publisher.

Printed in the United States of America

10 9 8 7 6 5 4 3

ISBN 0-13-018151-X

Prentice-Hall International (UK) Limited, *London*
Prentice-Hall of Australia Pty. Limited, *Sydney*
Prentice-Hall Canada Inc., *Toronto*
Prentice-Hall Hispanoamericana, S.A., *Mexico*
Prentice-Hall of India Private Limited, *New Delhi*
Prentice-Hall of Japan, Inc., *Tokyo*
Simon & Schuster Asia Pte. Ltd., *Singapore*
Editora Prentice-Hall do Brasil, Ltda., *Rio de Janeiro*

Preface

This second edition of *Air Conditioning Principles and Systems* has been greatly revised. Chapter 4, on boilers and furnaces, has been expanded. Information on computer programs has been added to reflect current industry practice. An explanation of passive solar heating has been added. Additional problems have been added to the end of each chapter. The second edition also features the addition of two new elements: learning objectives and review questions.

This book is a fundamental text in heating, ventilating, and air conditioning. Students in air conditioning and refrigeration courses or programs in colleges or technical institutes, as well as practicing designers, contractors, operating engineers, and service technicians will find it useful in their studies or as a reference.

The book is designed for a two-semester course. Additional work may be added if the instructor wishes to emphasize the projects suggested.

In the text discussion the fundamental principles of air conditioning are developed and are followed by a description of equipment and systems. The relationships of theory as applied to both the design of new systems and the troubleshooting of existing ones are emphasized. This is enhanced by many illustrative examples and problems dealing with real situations.

Throughout the book, energy utilization and conservation are stressed. Energy codes and standards are described, and each topic is examined from an energy conservation viewpoint, an approach that is essential for all future work in the air conditioning field. A chapter is devoted to solar heating and cooling.

Following an overview of the scope of air conditioning, there is a review of physical principles. Heating and cooling load calculations are explained in a thorough yet understandable manner. The latest methods (now required in most states) to account for heat storage are used. Load calculation forms are furnished to aid the student. The subject of psychrometrics is presented in considerable detail, recognizing that it is at the heart of understanding air conditioning processes.

Air conditioning and refrigeration equipment and systems are covered thoroughly. Equipment construction and selection are described. Included in the discussion are reheat, dual duct, multizone, hydronic, solar heating, and combination systems. The presentation of refrigeration includes an explanation of absorption systems and heat pumps.

There is a chapter covering instrumentation and balancing and another explaining the fundamentals of automatic controls. Of special importance is the chapter devoted to energy utilization and conservation in design, installation, and operation of air conditioning systems.

Two example projects in the design of a heating and cooling system are worked out in detail. Similar projects are suggested as learning experiences. These should be of value to those who are interested in installation, operation, and service as well as design, because they require the student to analyze how the system functions.

For their aid and insights during the writing of a book that must serve the requirements of the air conditioning industry in the late twentieth century, I offer my grateful thanks to the following individuals. They are Professors Samuel Lomask; Alfred Finger; Eugene Stamper; William Lauffer, of Prince George's Community College; David L. Ernst, of Texas Technical University; and Larry Meeker, of Mt. San Antonio Community College; and also Marianne, Henry, and Dorothy Pita.

Many organizations and companies have kindly furnished material. Of particular importance is the American Society of Heating, Refrigerating and Air-Conditioning Engineers, which has played a leadership role in an industry that is devoted to the comfort and health of people.

EDWARD G. PITA

Contents

12. AIR CONDITIONING SYSTEMS AND EQUIPMENT   319

13. REFRIGERATION SYSTEMS AND EQUIPMENT   343

Vapor Compression Refrigeration System   344

15. ENERGY UTILIZATION AND CONSERVATION

16. INSTRUMENTATION, TESTING, AND BALANCING

An Air Conditioning Fable

It was a typical record-breaking July heat wave and the humidity felt like a Turkish bath. Suddenly the air conditioning system in the gigantic Acme Towers office building stopped operating. Within minutes temperatures in the offices reached 95 F. The building did not have operable windows that could be opened to relieve the oppressive heat. Computers broke down, employees started to leave, and tenants threatened lawsuits for damages.

The building operating staff became frantic. No one knew what to do. Finally one person said, "Listen, there's a fellow named Joe Schlepper who knows an awful lot about air conditioning and refrigeration, so why don't we call him?" In desperation the chief engineer agreed.

A few minutes later Joe Schlepper entered the building machine room, walked around, and looked at the complex installation capable of delivering 8000 tons of refrigeration, muttered "hmm," took out a small hammer and tapped a valve. Immediately the whole plant started functioning and soon conditions in the building were comfortable again.

The building manager thanked Joe, and asked him what the bill was. The answer was "$2005." "What!" the manager exclaimed. "$2005 for tapping a valve?"

"The bill for tapping the valve is $5," Joe answered, "the $2000 is for knowing which valve to tap."

The Scope and Uses of Air Conditioning

*F*or prehistoric people, open fires were the primary means of warming their dwellings; shade and cool water were probably their only relief from heat. No significant improvements in humankind's condition were made for millions of years. The fireplaces in the castles of medieval Europe were hardly an improvement — they only heated the area immediately around them. Paintings from those times show that the kings and queens wore furs and gloves indoors in winter!

There were a few exceptions to this lack of progress. The ancient Romans had remarkably good radiant heating in some buildings, which was achieved by warming air and then circulating it in hollow floors or walls. In the dry climate of the Middle East, people hung wet mats in front of open doorways and achieved a crude form of evaporative air cooling. In Europe Leonardo da Vinci designed a large evaporative cooler (Figure 1.1).

The development of effective heating, ventilating, and air conditioning (HVAC), however, was begun scarcely 100 years ago. Central heating systems were developed in the nineteenth century, and summer air conditioning using mechanical refrigeration has grown only in the last 50 years. Yet by 1985, HVAC systems in the United States had reached a total installed value of about $25 billion yearly, with approximately $10 billion in equipment sales.

Figure 1.1. Ventilator and cooling unit invented by Leonardo da Vinci in the fifteenth century. This air conditioning unit was for the boudoir of Beatrice d'Este, wife of da Vinci's patron, the Duke of Milan. The great wheel, a full story high, stood outside the palace wall and was turned by water power — sometimes assisted by slaves. Valves opened and closed automatically, drawing air into the drum, where it was washed and forced out through the hollow shaft and piped into the room. (Courtesy: IBM Corporation.)

OBJECTIVES

A study of this chapter will enable you to:

1. List the environmental conditions that an air conditioning system may control.
2. Describe where air conditioning is used.
3. Sketch the arrangement of the main components of an all-air air conditioning system.
4. Sketch the arrangement of the main components of a hydronic heating and cooling system.
5. Describe the internal environmental conditions that provide adequate human comfort.
6. Describe the business structure of the HVAC industry, including job opportunities.
7. Describe the organization of the building design team and the construction team.

1.1. SCOPE OF AIR CONDITIONING

To the average person, air conditioning simply means "the cooling of air." For our purposes, this definition is neither sufficiently useful nor accurate, so we will use the following definition instead:

> *Air conditioning* is the process of treating air in an internal environment to establish and maintain required standards of temperature, humidity, cleanliness, and motion.

Let us investigate how each of these conditions is controlled:

1. *Temperature.* Air temperature is controlled by heating or cooling* the air.
2. *Humidity.* Air humidity, the water vapor content of the air, is controlled by adding or removing water vapor from the air (*humidification* or *dehumidification*).
3. *Cleanliness.* Air cleanliness, or air *quality*, is controlled by either *filtration*, the removal of undesirable contaminants using filters or other devices, or by *ventilation*, the introduction of outside air into the space which dilutes the concentration of contaminants. Often both filtration and ventilation are used in an installation.
4. *Motion.* Air motion refers to air velocity and to where the air is distributed. It is controlled by appropriate air distributing equipment.

Sound control can be considered an auxiliary function of an air conditioning system, even though the system itself may be the cause of the problem. The air conditioning equipment may produce excessive noise, requiring additional sound attenuating (reducing) devices as part of the equipment.

The definition of air conditioning given here is not meant to imply that every HVAC system regulates all of the conditions described. A **hot water** or **steam heating** system, consisting of a

---

*"Cooling" technically means the *removal* of heat, in contrast to heating, the *addition* of heat.

boiler, piping, and radiation devices (and perhaps a pump) only controls air temperature, and that is only during the heating season. These types of systems are common in many individual homes (residences), apartment houses, and industrial buildings.

A **warm air** system, consisting of a furnace, ducts, and air outlet registers, also controls air temperature in winter only. However, by the addition of a humidifier in the ducts it may also control humidity in winter. Warm air systems are also popular in residences.

Some residences have combination air heating and cooling equipment that provides control of temperature and humidity in both winter and summer. Some degree of control of air quality and motion is provided in air type heating and cooling systems.

Air conditioning systems used for newer commercial and institutional buildings and luxury apartment houses usually provide year-round control of most or all of the air conditions described. For this reason it is becoming increasingly popular to call complete HVAC systems *environmental control systems*.

Applications Most air conditioning systems are used for either human comfort or for process control. From life experiences and feelings we already know that air conditioning enhances our comfort. Certain ranges of air temperature, humidity, cleanliness, and motion are comfortable, others are not.

Air conditioning is also used to provide conditions that some processes require. For example, textile, printing, and photographic processing facilities, as well as computer rooms and medical facilities, require certain air temperatures and humidity for successful operation.

1.2. COMPONENTS OF AIR CONDITIONING SYSTEMS

Heat always travels from a warmer to a cooler area (see Section 2.9). In winter, there is a con-

Figure 1.2. View of Lower Manhattan skyline with the World Trade Center Twin Towers, which have 49,000 tons of refrigeration, enough to air condition a city of 100,000 people. (Courtesy: The Port Authority of New York and New Jersey.)

tinual heat loss from within a building to the outdoors. If the air in the building is to be maintained at a comfortable temperature, heat must be continually supplied to the air in the rooms. The equipment that furnishes the heat required is called a *heating system*.

In summer, heat continually enters the building from the outside. In order to maintain the room air at a comfortable temperature, this excess heat must be continually removed from the room. The equipment that removes this heat is called a *cooling system*.

An air conditioning system may provide heating, cooling, or both. Its size and complexity may range from a single *space heater* or *window unit* for a small room to a huge system for a building complex, such as the World Trade Center (Figure 1.2), yet the basic principles are the same. Most heating and cooling systems have at a minimum the following basic components:

1. A *heating source* that adds heat to a fluid (usually air, water, or steam).

2. A *cooling source* that removes heat from a fluid.

3. A *distribution system* (a network of ducts or piping) to carry the air, water, or steam to the rooms to be heated or cooled.

4. Equipment (fans or pumps) for moving the air or water.

5. Devices (e.g., radiation) for transferring heat between the fluid and room.

We will start with a brief introduction into the function and arrangement of these major components. These and other components including automatic controls, safety devices, valves, dampers, insulation, and sound and vibration reduction devices will be discussed in more detail in later chapters of the book.

Air conditioning systems that use water as the heating or cooling fluid are called **hydronic sys-**tems; those that use air are called **all-air systems.** A system which uses both air and water is called a **combination system.**

1.3. ALL-WATER (HYDRONIC) AIR CONDITIONING SYSTEMS

A typical hydronic heating system is shown in Figure 1.3. Water is heated at the *heat source* (1), usually a hot water boiler. The heated water is circulated by a *pump* (2) and travels to each room through *piping* (3) and enters a *terminal unit* (4). The room air is heated by bringing it into contact with the terminal unit. Since the water loses its heat to the rooms, it must return to the heat source to be reheated.

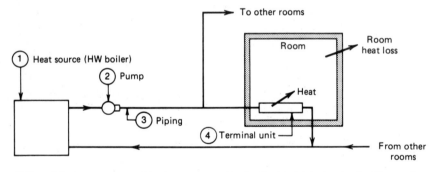

Figure 1.3. Arrangement of basic components of a (hydronic) hot water heating system.

Figure 1.4. Arrangement of basic components of a (hydronic) chilled water cooling system.

Figure 1.5. Arrangement of basic components of a hydronic heating and cooling system.

If steam is used in a heating system, the components work in the same manner with the exception that a pump is not necessary to move the steam; the pressure of the steam accomplishes this. However, when the steam cools at the terminal unit, it condenses into water and may require a condensate pump to return the water to the boiler.

A hydronic cooling system (Figure 1.4) functions in a similar manner to a hydronic heating system. Water is cooled in refrigeration equipment called a *water chiller* (1). The chilled water is circulated by a *pump* (2) and travels to each room through *piping* (3) and enters a *terminal unit* (4). The warmer room air loses its heat to the cold water in the terminal unit. Since the water is now warmed, it must return to the water chiller to be recooled.

As the reader may have guessed, hydronic systems are popular for HVAC systems that

Figure 1.6. Arrangement of basic components of an all-air heating and cooling system (many other arrangements are possible).

require both heating and cooling. This is because it is possible to use the same piping system for both by connecting a hot water boiler and water chiller in parallel (Figure 1.5), using each when needed.

1.4. ALL-AIR AIR CONDITIONING SYSTEMS

All-air systems use air to heat or cool rooms. They may also have the added capability of controlling humidity and furnishing outdoor ventilation, which hydronic systems cannot do.

A typical all-air heating and cooling system is shown in Figure 1.6. Air is heated at the heat source (1), such as a furnace. (It may also be a coil circulating hot water, or steam, heated by a remote boiler.) The heated air is circulated by a *fan* (2) and travels to each room through *supply air ducts* (3). The supply air enters the room through outlets called *air diffusers* or *registers* (4) that are designed to provide proper air distribution in the room. When the warmed supply air

Figure 1.7. Arrangement of components of all-year air conditioning equipment for a private residence (refrigeration condenser separate).

enters the room, the room is heated.

In summer, air is cooled by flowing over a cooling source (5), usually a coil of tubing containing a fluid cooled by refrigeration equipment

Figure 1.8. Rooftop-type unitary air conditioning equipment. (Courtesy: McQuay Group, McQuay-Perfex, Inc.)

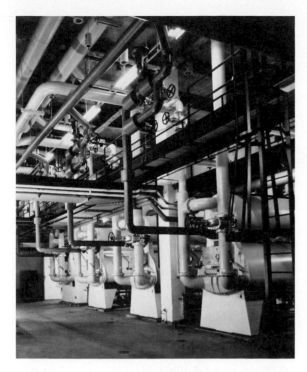

Figure 1.9. Mechanical equipment room of large central station air conditioning system, showing absorption refrigeration machines. (Courtesy: Syska & Hennessy, Inc., Engineers.)

(see Chapter 13). When the cooled supply air enters the room, the room is cooled.

Because a room's size is fixed, the same volume of air that enters the room must also exit. This is usually accomplished with *return air ducts* (6). The air is then heated or cooled again, and recirculated. An *outdoor air intake* duct (7) may be provided for introducing fresh outdoor air for increased air quality. Similarly, the same volume of air must be *exhausted* (8). Provisions may be made for cleaning the air with air filters (9), and for humidifying the air (10).

An example of all-air system equipment is shown in Figure 1.7. This arrangement is convenient for residential and light commercial air conditioning.

Combination Systems It is frequently desirable to combine water and air systems. For example, a hydronic system in a central plant might generate hot or chilled water which is then circulated to heating or cooling coils in large all-air systems in other parts of the building or even to a number of buildings.

Unitary and Central Air Conditioning Systems A *unitary* or *package* air conditioning system uses equipment where all or most of the basic components have been assembled in the factory (e.g., room air conditioner). An example of all-air unitary equipment mounted on a roof (a "rooftop" unit), like those used in supermarkets, is shown in Figure 1.8.

A *central* or *built-up* air conditioning system uses equipment centrally located in mechanical equipment rooms. Each piece of equipment is installed separately and connected on the job, rather than manufactured as a package. Figure 1.9 shows a portion of the equipment of a central system. This subject will be discussed in more detail in Chapter 12.

1.5. HUMAN COMFORT

Since the purpose of air conditioning systems is to provide a comfortable indoor environment, the system designer and operator should understand the factors that affect comfort.

Body Heat Loss The human body creates heat when it metabolizes (oxidizes) food. This body heat is continually lost to its cooler surroundings. The factor that determines whether one feels hot or cold is the *rate of body heat loss*. When the rate of heat loss is within certain limits, a comfortable feeling ensues. If the rate of heat loss is too great, cold is felt; if the rate is too low, one feels hot.

The processes by which the body loses heat to the surroundings are: *convection*, *radiation*, and *evaporation*.

In *convection*, the air immediately around the body receives heat from the body. The warmed air continually moves away, by rising naturally through the cooler air around it, or by being blown away, and is replaced by more air which in turn receives heat.

In *radiation*, body heat is transmitted through space directly to nearby objects (e.g., walls) which are at a lower temperature than the body; this is why it can be uncomfortable to sit near a window or wall in cold weather, even in a warm room. However, heating sources that are warmer than the body can radiate heat toward the body, creating a feeling of warmth even at a low surrounding air temperature; this is why one feels warm in front of a fire even on a cold day. Some restaurants now have glass-enclosed sidewalk cafés with radiant heating panels that keep the customers comfortable in winter even though the café temperature is only about 50 F (10 C).

The body is also cooled by *evaporation*: water on the skin (perspiration) which has absorbed heat from the body, evaporates into the surrounding air, taking the heat with it.

The rate of body heat loss is affected by five conditions:

1. Air temperature
2. Air humidity
3. Air motion
4. Temperature of surrounding objects
5. Clothing

The system designer and operator can control comfort primarily by adjusting three of these conditions: temperature, humidity, and air motion. How are they adjusted to improve comfort?

The indoor air temperature may be raised to decrease body heat loss (winter) or lowered to increase body heat loss (summer) by convection.

Humidity may be raised to decrease body heat loss (winter) and lowered to increase body heat loss (summer) by evaporation.

Air motion may be raised to increase body heat loss (summer) and lowered to decrease body heat loss (winter) by convection.

Occupants of the buildings, of course, have some control over their own comfort. For instance, they can control the amount of clothing that they wear, they can use local fans to increase convection and evaporative heat loss, and they can even stay away from walls and windows to keep warmer in winter.

Air Quality   Another factor, *air quality*, which affects both comfort and health, refers to the degree of purity of the air. Air quality is worsened by the presence of contaminants such as odors, smoke and dust particles, and undesirable gases. Particles can be removed by air filtration and gases removed by use of adsorbent chemicals (see Chapter 12). Contaminants such as cigarette smoke or odors can also be diluted to an acceptable level by introducing outside *ventilation* air into the building.

1.6. COMFORT STANDARDS

A set of recommended indoor conditions, called the ANSI/ASHRAE Standard 55-1981,* was developed from studies of the effects on human comfort of air temperature, humidity, motion, and clothing. Some of the results are shown in Figure 1.10. (Another measure of comfort, called the Effective Temperature (ET), has less usefulness for our purposes.)

The shaded regions in Figure 1.10, called *comfort zones*, mark those combinations of effects in which, it is believed, at least 80% of the occupants will find the environment comfortable. (*Note*: There are separate zones for winter and summer, with a slight overlap.)

---

*ANSI stands for the American National Standards Institute; ASHRAE stands for the American Society of Heating, Refrigerating, and Air-Conditioning Engineers.

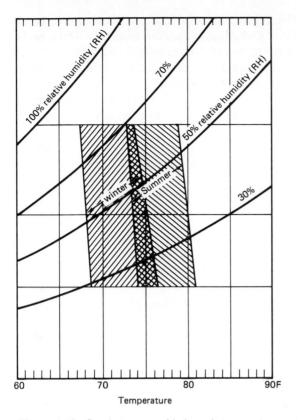

Figure 1.10. Comfort zones of indoor air temperature and humidity. These zones apply to persons clothed in typical summer or winter clothing engaged in sedentary activity. (Reprinted with permission from the *1985 Fundamentals ASHRAE HANDBOOK*.)

However, there are some restrictions. They are:

1. The comfort zones apply only to sedentary or slightly active persons.
2. The comfort zones apply only to summer clothing of light slacks and a short sleeve shirt, or equivalent (0.5 clo);* and winter clothing of heavy slacks, long sleeve shirt, and sweater or jacket, or equivalent (0.9 clo).

---

*The *clo* is a numerical unit representing a clothing ensemble's thermal insulation.

3. The comfort zones apply to air motion in the occupied zone not exceeding 30 feet per minute (FPM) in winter and 50 FPM in summer.
4. The comfort zones apply only under certain conditions of radiation between the occupant and the surrounding room surfaces.

Although these restrictions may seem to reduce the usefulness of Figure 1.10, this is not so. First, the situations specified are very common (a typical office environment). Furthermore, for changes in various conditions, the comfort zones can be adjusted to reflect these changes. The procedures for making these corrections can be found in the ANSI/ASHRAE Standard.

Applications In order to use Figure 1.10 to find if a specific set of conditions is comfortable, it is necessary to know the room air temperature and humidity. The air temperature is technically called the *dry bulb temperature* (DB). The humidity is often expressed as the *percent relative humidity* (% RH). See Chapter 7 for a complete definition of these terms.

EXAMPLE 1.1

The conditions in an office building in the summer are 77 F DB and 50% RH. The occupants are lightly clothed. Air movement in the rooms is about 30 FPM. There is negligible radiation of heat from the surroundings to the occupants. Would this be a comfortable indoor condition?

Solution. From Figure 1.10, the condition noted (the intersection of 77 F DB and 50% RH), is within the summer comfort zone, therefore most of the occupants should feel comfortable.

Air Quality Standards As mentioned earlier, air quality is maintained by filtration and venti-

TABLE 1.1. RECOMMENDED INDOOR DESIGN
CONDITIONS FOR HUMAN COMFORT

Summer	78 – 80 F DB	and 50% RH
Winter	68 – 72 F DB	> 25% RH

lation. (Air filters are discussed in Chapter 12; recommendations for air ventilation are given in Table 6.15.) The amount of outside air required depends on the type of occupancy use, and may vary from 5 to 50 cubic feet per minute (CFM) per person.

Indoor Design Conditions for Energy Conservation The comfort zones shown in Figure 1.10 leave a wide range of choices for the air conditioning system designer and operator. In recent years, in an effort to conserve energy, more specific conditions have been recommended (Table 1.1). The temperatures listed are at the low end of the comfort zone in winter and at the high end of the comfort zone in summer. These recommendations may not be a matter of choice: most states now mandate energy conserving design conditions.

The values recommended apply to general applications such as offices, residences, and public buildings, but there are exceptions. Lower indoor temperatures in winter might be used in department stores when customers are heavily clothed. Higher indoor temperatures in winter may be desirable for small children, senior citizens, and the ill. Other special applications might have different design conditions.

A minimum of 25–30% RH for comfort in winter is suggested for indoor design conditions, if humidity is to be controlled.

1.7. THE HVAC SYSTEM AS PART OF THE BUILDING CONSTRUCTION FIELD

The student who intends to work in the HVAC industry should have some understanding of how

the industry is organized and how it relates to the building construction field, of which it is part. Many of the problems faced by HVAC system designers, installers, and operators relate to construction problems.

The development of an HVAC system for a building consists of a number of steps. These are:

1. Design
2. Installation
3. Operation and regular maintenance
4. Service

We will outline who is responsible for each step, what their tasks are, and how the HVAC system relates to other building systems.

The student is strongly advised, if possible, to locate a proposed building and follow the HVAC system development through planning, installation, and operation. No textbook can substitute for this valuable learning experience.

1.8. DESIGNING THE HVAC SYSTEM

The design of a large building project is an extremely complex task. It may take months or even years and involve scores of people. The design of a private residence is much simpler and may involve as few as one or two people.

The design of an HVAC system for large projects is the responsibility of the *mechanical consulting engineers*. The electrical, structural, and plumbing systems are designed by consulting engineers specializing in their respective fields. Consulting engineers may also carry out other duties such as *cost estimating* and *field supervision* of construction. Each of these tasks is performed in cooperation with the *architects*, who carry out the overall building planning and design. An organizational flowchart of this arrangement is shown in Figure 1.11.

Coordination of the work between the architects and engineers is an important and difficult

Figure 1.11. Organizational flowchart to a building planning and design team.

task. This includes checking that the equipment and materials to be installed do not physically interfere with each other. An error in coordination can have disastrous results.

The design of an HVAC system involves determining the type of system to use, calculations of heating and cooling loads (requirements), calculations of piping and duct sizes, selection of the type and size of equipment, and planning the locations of each piece of equipment in the building. This information is shown on the building HVAC *plans* and *specifications*, which serve as instructions on how to install the system. The plans are drawings of the system. The specifications are written descriptions of materials, equipment, and so forth.

1.9. INSTALLING THE HVAC SYSTEM

The overall construction of a building is the responsibility of the *general contractor*. The general contractor is awarded a *contract* by the *owner*, which may be a real estate company, public agency, school system, or other prospec-

tive builder. The general contractor may hire *subcontractors* (mechanical, electrical, and so forth) to install each of the building's systems. The subcontractors must coordinate their work, to avoid any physical interference. The *mechanical* or *HVAC contractor* is responsible for installing the HVAC system. Figure 1.12 shows a typical organizational flowchart.

The mechanical contractor takes the consulting engineer's drawings (called *contract* or *engineering drawings*) and then prepares *shop drawings* from these. Shop drawings are larger scale, more detailed drawings of the HVAC system which will be necessary for the workers. The mechanical contractor hires the workers, which include pipefitters, sheet metal workers, insulation workers, and other skilled *building trade workers*.

The mechanical contractor also purchases all necessary HVAC equipment and materials. To do this, their employees first carry out a *take-off*, that is, they list all the equipment and materials shown on the drawings and specifications; this can be a very involved task. Overhead and labor costs must also be determined.

Figure 1.12. Organizational flowchart of a building construction team.

When the installation is complete, the mechanical contractor tests, adjusts, and balances (TAB) the HVAC system (see Chapter 16). The mechanical consulting engineer may check the installation as it proceeds and may also check the TAB work.

The Design-Build (Fast Tracking) Approach

In contrast to the procedures described, there are companies that handle all of the design and construction functions as a package: architecture, consulting engineering, and contracting. This is called the *design-build* approach. Proponents of this approach claim that construction can start and continue as plans are developed for each stage—it is not necessary to wait for engineering plans and contractor drawings, there is no delay for contractor competitive bidding, better coordination is achieved, and determination of liability is easier, since one organization is responsible for everything. The proponents claim that all of these factors result in lower cost, faster construction, and better quality.

However, proponents of construction projects using an independent architect, consulting engineers, and contractors claim that costs are kept down by competitive bidding, and quality is better because the architects and engineers have independent control over the performance of the contractors.

1.10. OPERATION, MAINTENANCE, AND SERVICE OF THE HVAC SYSTEM

When the HVAC installation is complete and after start-up and TAB, the building *operating engineering* staff takes over. Their function is to operate the system, maintaining comfortable conditions in the building while trying to minimize energy consumption, and to keep the system in proper working order.

Regular inspection and maintenance of the system is also part of the operating engineer's duties. Some routine servicing may be performed by the building operating staff, but when more complicated work is required, a *mechanical service contractor* is called in. Using instrumentation, this contractor measures the conditions and compares them with the HVAC system plans and specifications. This troubleshooting procedure leads to the cause of the problem. Proper procedures, such as repairing or replacing equipment, or adjusting its performance, are then carried out.

1.11. EMPLOYMENT IN THE HVAC INDUSTRY

It is helpful for students intending to work in some part of the HVAC industry and for those already working who wish to advance themselves to know what types of positions are available and what knowledge, responsibilities, and training are required.

One basic fact that is most important is that a fundamental knowledge of air conditioning principles is required, regardless of whether one is employed in design, installation, operation, or service, in order to succeed in the HVAC field today.

A description of the types of employers and their work, followed by a list of job titles and responsibilities will aid the student in planning his or her career.

A *mechanical consulting engineer* is a company that designs the heating, ventilating, air conditioning, and plumbing systems for buildings. They estimate costs, perform technical calculations, prepare drawings and specifications, and supervise installations. Positions include: *Project Manager*, *Designer*, *Drafter*, *Inspector*, *Computer Programmer*, and *Energy Specialist*.

A *mechanical contractor* is a company that installs the system. This includes cost estimates, preparation of drawings, and supervision of installation. Positions include: *Sales Engineer*, *Inside Representative*, *Estimator*, *Contract Man-*

ager, *Drafter*, *Purchasing Agent*, *Field Supervisor*, *Shop Technician*, and *Mechanic*.

A *service* company repairs and maintains the HVAC system. It is often a branch of a mechanical contractor. Positions include: *Sales Engineer*, *Inside Representative*, *Estimator*, *Service Manager*, *Field Service Engineer and Technician*, *Shop Technician*, and *Mechanic*.

A *manufacturer* is a company that makes HVAC equipment. This involves production, research and development, marketing, and sales. Positions include: *Sales Engineer*, *Inside Representative*, *Sales Manager*, *Application Engineer*, *Drafter*, *Purchasing Agent*, *Production Supervisor*, *Field Service Engineer and Technician*, *Shop Technician*, *Service Manager*, *Research and Development Engineer*, and *Technician*.

A *manufacturer's representative* is a company that sells HVAC equipment manufactured by another company. Their work involves sales and technical advice. Positions include *Sales Engineer*, *Inside Representative*, *Estimator*, *Sales Manager*, and *Application Engineer*.

A *building owner* may be a real estate company; a business corporation; city, state, or federal government; public authority; school system; and others. Positions within operations include: *Chief Engineer*, *Watch Engineer*, *Computer Operator*, and *Mechanic*. In addition to HVAC operating personnel, large owners may have a permanent staff which checks and supervises the work of consulting engineers and contractors that do work for them. Positions include: *Application Engineer*, *Designer*, *Drafter*, *Estimator*, *Field Supervisor*.

1.12. DESCRIPTION OF JOB RESPONSIBILITIES

Project Manager. Supervises design of project for consulting engineer. Instructs designers and drafters, checks calculations and plans, coordinates with other consultants and the architect.

Designer. Performs calculations, selects equipment, plans layout of system and specifications, supervises drafters.

Drafter. Prepares drawings with supervision. May assist in design work.

Inspector. Inspects the system installation during construction to check comformity with plans and specifications. Checks performance after testing and balancing.

Computer Programmer. Processes work to be handled by computer, such as load calculations and energy studies.

Energy Specialist. Prepares energy use analyses and conservation studies, studying varied alternatives.

Sales Engineer. Sells equipment and installation and service contracts. Furnishes technical advice to customers.

Inside Representative. Processes sales and orders by phone and correspondence. Furnishes product information and prices.

Estimator. Uses plans and specifications to determine quantity of materials, labor, and equipment for project. Prepares costs from this data.

Contract Manager. Supervises contract. Responsible for costs, time schedule, and installation.

Purchasing Agent. Orders and purchases materials and equipment. Checks technical characteristics, follows up delivery time.

Field Supervisor. Supervises installation technically. Checks comformity with drawings and specifications and resolves problems of conflict.

Mechanic. Installs and services the materials and equipment as instructed.

Service Manager. Supervises the service installation and contract.

Field Service Engineer and Technician. Determines solutions to problems (troubleshooting), obtains materials and equipment, and directs service work.

Shop Technician. Responsible for assembly or fabrication done in shop (e.g., sheet metal duct parts), or shop service and repair.

Sales Manager. Supervises the sales and marketing of a line of products for a manufacturer.

Application Engineer. Assists consulting engineer or contractor to provide technical information and aid in selection of proper equipment.

Production Supervisor. Responsible for fabrication of equipment in factory. Supervises technical work of employees, inspection, and quality.

Research and Development Engineer and Technician. Plan, develop, and test new types of equipment.

Chief Operating Engineer. Supervises the operation and maintenance of the building system. Determines method of operation for comfort and energy conservation, plans maintenance routines, and directs work of operating personnel.

Watch Engineer. Responsible for operation and maintenace of the building system under supervision.

Computer Operator. Responsible for operations of computerized building systems.

These job descriptions do not mean that a separate individual always performs each task. Often, particularly in small companies, one person may be responsible for a number of jobs if there is not enough work to employ people in each category.

The amount of education and training required for each job varies both with the type of responsibility and how complex it is. For most of the categories at least a technical institute or community college program in air conditioning and refrigeration is required, preferably equivalent to two years of study. A bachelor of engineering degree is needed for some of the categories and for improving opportunities for advancement. Further on-the-job training is also extremely valuable.

1.13. ENERGY CONSERVATION AND COMPUTERS

Energy conservation and the use of computers have become such important aspects of the air conditioning industry that they merit a special emphasis at the beginning of our studies.

The effort to conserve energy and reduce costs has revolutionized the design and operation of air conditioning systems and equipment. Perhaps the biggest change has been in the use of computers. Computer usage in the industry has spread to such an extent that many, if not most, large air conditioning installations are now designed with the aid of computers. The selection of the equipment may also be done by computers. Furthermore, virtually all large- and many medium-sized new installations are operated through computers.

Students who wish to be successful in the air conditioning field should consciously approach each task as to how it can be solved in an energy efficient manner; students should also gain some understanding of computer application and programming (see Chapter 15 for a further discussion). Emphasis on energy conservation and computer applications are placed at the appropriate places within the text.

REVIEW QUESTIONS

1. What are the two primary situations in which air conditioning is needed?
2. List the four conditions that an air conditioning system may be required to control.
3. What two methods may be used to improve air quality?

4. List the four major components of any air conditioning system. Sketch a diagram that shows their arrangement.

5. What are the major components of a hydronic heating system and a hydronic cooling system?

6. Sketch a typical all-air air conditioning system and name each component.

7. What are the indoor environmental conditions that affect human comfort?

8. Using a sketch, describe the building design team's organization and responsibilities.

9. Using a sketch, describe the building construction team's organization and responsibilities.

10. What do the terms design-build and fast tracking mean?

PROBLEMS

1.1. Sketch an environmental control system that provides heating and ventilating, but not cooling or humidity control. Label all components.

1.2. As the operating engineer of an HVAC system in a large office building, you have been instructed to raise the summer thermostat setting from 76 F to 80 F to conserve energy. Prepare a list of suggestions you might give to the building's occupants on how to minimize their decrease in comfort.

1.3. In a department store, which should be the more comfortable summer condition: 80 F DB and 40% RH or 78 F DB and 70% RH? Explain.

1.4. The conditions in an office are 70 F DB and 40% RH. Would the occupants be comfortable in winter? Would they be comfortable in summer? Explain.

1.5. The conditions in an office in summer are 75 F DB and 50% RH. Should the conditions remain as they are or should they be changed? Explain. What changes should be made, if any? Explain.

1.6. The conditions in an office in winter are 77 F DB and 10% RH. Are these conditions acceptable? Explain. What changes should be made, if any? Explain.

1.7. Select two HVAC careers that interest you. List the subjects discussed in this chapter that you think are important to learn in training for these positions. When you have completed the book, prepare a new list and compare it with this one.

Physical Principles

*T*he HVAC practitioner often encounters problems that cannot be solved without a knowledge of applied physics. In this chapter the physical principles that are useful in understanding air conditioning will be explained. (One subject in applied physics, fluid flow, will be discussed in Chapter 8.) This presentation of applied physics is not intended to substitute for a course in physics; a background in that subject will be helpful as a preparation for this book.

Generally the definitions and concepts accepted today in physics will be used here, however, in some instances terms that are in practice in the HVAC industry, even when they differ slightly in meaning, will be used. This approach will enable the student to communicate and work with others in the air conditioning field.

OBJECTIVES

A study of this chapter will enable you to:

1. Identify units and convert from one set of units to another.
2. Calculate density, specific volume, and specific gravity.
3. Express the relationship between pressure and head, and between absolute, gage, and vacuum pressure.
4. Distinguish between energy and power, and between stored energy and energy in transfer.
5. Explain the difference between temperature, heat, and enthalpy and show the relationship between temperature scales.

6. Describe and use the energy equation.

7. Identify the changes that occur when a substance changes between its liquid and vapor states.

8. Use the saturated property tables for water and the sensible and latent heat equations.

9. Make some general conclusions regarding energy conservation in HVAC.

2.1. UNITS

Concepts such as length, area, volume, temperature, pressure, density, mass, velocity, momentum, and time are called *physical characteristics*. Physical characteristics are measured by standard quantities called **units**. For instance, the foot (ft) is one of the standard units used to measure length.

For each physical characteristic, there are many different units. These units have fixed numerical relationships to each other called *equivalents* or *conversion factors*. Examples of equivalents are:

Charac- teristic	Unit Equivalents (Conversion Factors)
Length	1 ft = 12 inches (in.)
	= 0.30 meters (m)
Volume	1 ft^3 = 7.48 gallons (gal)
Time	1 minute (min) = 60 seconds (sec)
Mass	2.2 pounds (lb) = 1 kilogram (kg)

Table A.2 in the Appendix lists some useful unit equivalents. Table A.1 lists abbreviations and symbols used in this book.

2.2. CONVERSION OF UNITS

The equivalence between any two units can also be written as a ratio, by dividing both sides of the equality by either term. For instance, from Table A.2, the relation between area expressed in ft^2 or in.^{2} is 1 ft^2 = 144 in.^{2} Dividing both sides by 144 in.^{2} gives

$$\frac{1 \text{ ft}^2}{144 \text{ in.}^2} = \frac{144 \text{ in.}^2}{144 \text{ in.}^2} = 1$$

Or, dividing by 1 ft^2 gives

$$\frac{1 \text{ ft}^2}{1 \text{ ft}^2} = \frac{144 \text{ in.}^2}{1 \text{ ft}^2} = 1$$

This shows that multiplying by the ratio of equivalent units is the same as multiplying by 1. This enables us to change units, as is now explained.

This ratio arrangement is used when it is desired to change a quantity expressed in one unit in to a different unit. The procedure is carried out in the following manner:

1. Arrange the equivalency (conversion factor) between the units as a ratio, choosing that ratio that will give the results in the desired units, by canceling units that are the same in the numerator and denominator (units can be multiplied and divided in the same way as numbers).

2. Multiply the original quantity by the ratio. The result will be the correct value in the new unit.

The following example illustrates the procedure for converting units.

EXAMPLE 2.1

Some solar heating collector panels measuring 28 in. by 33 in. require insulation. The insulation is to be ordered in square feet. How much insulation would you order for each panel?

Solution. The area of the insulation for each panel is

$$\text{Area} = 28 \text{ in.} \times 33 \text{ in.} = 924 \text{ in.}^2$$

The area is not in the units needed, however. The equivalent between the known and required units is 1 ft^2 = 144 in.^{2} (Table A.2). Arranging this as a ratio, multiplying by the original quantity, and canceling units as shown:

$$\text{Area} = 924 \text{ in.}^2 \times \frac{1 \text{ ft}^2}{144 \text{ in.}^2} = 6.42 \text{ ft}^2$$

This is the amount of insulation required for each panel, expressed in ft^2.

An important point to note in this example is that there are always two possible ratios that can be used in converting units. In that case, it was either

$$\frac{1 \text{ ft}^2}{144 \text{ in.}^2} \quad \text{or} \quad \frac{144 \text{ in.}^2}{1 \text{ ft}^2}$$

Only one can be correct. Suppose the other ratio had been used. The results would be

$$\text{Area} = 924 \text{ in.}^2 \times \frac{144 \text{ in.}^2}{1 \text{ ft}^2} = 133{,}000 \frac{\text{in.}^4}{\text{ft}^2}$$

We know that this is incorrect, because the units resulting are not ft^2. Imagine what your boss would have said if you had ordered 133,000 ft^2 of insulation, when he saw it! The student should adopt the habit of always writing out the unit names when doing computations.

The procedure for changing units is the same when more than one unit is to be changed, as seen in the following example.

EXAMPLE 2.2

A U.S. manufacturer ships some air filters to Venezuela, with the note "Warning—maximum air velocity 600 ft/min." The contractor installing the filters wishes to inform the operating engineer what the maximum velocity is in meters per second (m/sec). What information should be given?

Solution. We must use the equivalency between feet and meters and that between minutes and seconds. Arranging the ratios in the form that will give the correct units in the result, and multiplying

$$\text{Velocity} = 600 \frac{\text{ft}}{\text{min}} \times \frac{1 \text{ min}}{60 \text{ sec}} \times \frac{0.30 \text{ m}}{1 \text{ ft}}$$
$$= 3.0 \frac{\text{m}}{\text{sec}}$$

Combined Conversion Factors In Example 2.2 the problem involved converting velocity from units of ft/min to m/sec. It is convenient to use combined conversion factors like this for calcula-

TABLE 2.1. PHYSICAL PROPERTIES OF SUBSTANCES

Substance	Density, lb / ft^3	Specific Heat, BTU / lb-F	Note
Water	62.4	1.0	At 32 – 60 F
Ice	57.2	0.50	
Steam	(Table A.3)	0.45	Average for water vapor in air
Air	0.075	0.24	At 70 F and 14.7 psia
Mercury	849.0		At 32 F

tions that are frequently repeated. The following example shows how one is developed.

EXAMPLE 2.3

Find the equivalence for the flow rate of water measured in units of lb/hr and gal/min (GPM), at 60 F.

Solution.   From Table 2.1 the density of water is 62.4 lb/ft^3 at 60 F. From Table A.2, 7.48 gal = 1 ft^3. Using these values, for water

$$1 \frac{\text{gal}}{\text{min}} = \left(1 \frac{\text{gal}}{\text{min}} \times 60 \frac{\text{min}}{\text{hr}} \times \frac{1 \text{ ft}^3}{7.48 \text{ gal}} \times 62.4 \frac{\text{lb}}{\text{ft}^3} \text{ water}\right) = 500 \frac{\text{lb}}{\text{hr}} \text{ water}$$

That is, 1 GPM = 500 lb/hr (for water only, at about 60 F).

2.3. U.S. AND SI UNITS

There are two systems of units used in the HVAC industry. One is called the *inch-pound*, *U.S.*, or *English* system; the other is called the *SI* or international system. SI units are part of a broader system of units called the metric system. The inch-pound system of units is generally used in the United States, whereas SI units are used in most other countries. In this book, U.S. units will be emphasized. However, SI units will be introduced in two ways: (1) some examples and tables, units will be converted between U.S. and SI units; and (2) some examples and problems will be done completely in SI units. In this way, those students who wish to become familiar with SI units may do so.

Only certain units of the metric system are standard in the SI system. The SI system of units uses only one unit of measurement for each physical characteristic. For instance, the stan-

dard SI unit of length is the meter, not the centimeter or kilometer. However, occasionally we may use metric units that are not standard SI units, because this is common practice in the HVAC industry in countries using the SI system. Table A.2 includes conversion factors for both U.S. and SI units.

2.4. MASS, FORCE, WEIGHT, DENSITY, AND SPECIFIC VOLUME

The **mass** (*m*) of an object or body is the quantity of matter it contains. The U.S. unit of mass is the pound mass. The SI unit is the kilogram (kg).

A **force** is the push or pull that one body may exert on another. The U.S. unit of force is the pound force. The SI unit is the Newton (N).

The **weight** (*w*) of a body is the force exerted on it by the gravitational pull of the earth. That is, weight is a force, not mass. Unfortunately the word *weight* is often used for mass of a body. The confusion also occurs because the word *pound* is used for both mass and force in U.S. units. However, the numerical value in pounds (lb) for the mass and weight of an object is the same, therefore no error should occur in calculations. In any case the nature of a problem indicates whether mass or weight is being considered.

Density and Specific Volume   Density (*d*) is the mass per unit of volume of a substance. **Specific volume** (*v*) is the reciprocal of density. That is

$$d = \frac{m}{\text{volume}} \qquad (2.1)$$

$$v = \frac{\text{volume}}{m} \qquad (2.2)$$

Weight density is the weight per unit volume of a substance. Although weight density and (mass) density are not the same, they are often used as such, as both are measured in lb/ft^3 in U.S. units. Density varies with temperature and pressure. Densities and other properties for some substances are shown in Table 2.1.

EXAMPLE 2.4

A contractor is going to install a cooling tower on a roof. He must inform the structural engineer how much extra weight to allow for the water in the tower basin when designing the roof. The tower basin is 15 ft by 10 ft in plan and filled with water to a depth of 1.5 ft.

Solution. The weight of water in the tank is found from Equation 2.1, after finding the volume of water. The density of water is shown in Table 2.1.

$$\text{Volume} = 15 \text{ ft} \times 10 \text{ ft} \times 1.5 \text{ ft} = 225 \text{ ft}^3$$

Solving Equation 2.1 for m

$$m = d \times \text{volume} = 62.4 \frac{lb}{ft^3} \times 225 \text{ ft}^3$$

$$= 14,000 \text{ lb}$$

Specific Gravity The *specific gravity* (*s.g.*) of a substance is defined as the ratio of its weight to the weight of an equal volume of water, at 39 F. The density of water at 39 F is 62.4 lb/ft^3, so the specific gravity is

$$s.g. = \frac{d}{d_w} = \frac{d}{62.4} \qquad (2.3)$$

where

d = density of substance, lb/ft^3
d_w = density of water at 39 F, 62.4 lb/ft^3

The value of specific gravity will change slightly with temperature, but for most calculations the values from Equation 2.3 are satisfactory.

EXAMPLE 2.5

A fuel oil has a density of 58.5 lb/ft^3. What is its specific gravity?

Solution. Using Equation 2.3

$$s.g. = \frac{d}{62.4} = \frac{58.5}{62.4} = 0.94$$

2.5. ACCURACY OF DATA

In reporting results of measurements or calculations of data, decisions must be made as to the number of *significant figures* or places of accuracy to use in numerical values. This procedure is called *rounding off*. For example, suppose the results of some calculations produced a value of 18,342 CFM for the required air supply rate to a building. This number is said to have five significant figures, because the value of the fifth digit from the left is known. The number might be used to select a fan, and then to balance the system to obtain this flow rate. However, neither the fan rating or balancing instruments can produce that accurate a value. Equipment and instrument ratings are often only accurate to within 2–5% of listed values, therefore there is no point in calculating or measuring data to an excess number of significant figures. Data in HVAC work are usually rounded off (i.e., the number of significant figures are reduced) to three or four places, and sometimes even two. If the above value is rounded off to three places, it would be reported as 18,300 CFM. Until students become familiar with good practice in rounding off values, they should use the examples of this book as a guide.

2.6. PRESSURE

Pressure (p) is defined as force (F) exerted per unit area (A). Expressed as an equation, this is

$$p = \frac{\text{force}}{\text{area}} = \frac{F}{A} \qquad (2.4)$$

If force is measured in pounds (lb) and area in square feet (ft^2), the units of pressure will be

$$p = \frac{F}{A} = \frac{\text{lb}}{\text{ft}^2}$$

If force is measured in pounds and area in square inches (in.^{2}), units of pressure will be lb/in.^{2}. The abbreviations *psf* for lb/ft^2 and *psi* for lb/in.^{2} are commonly used.

EXAMPLE 2.6

A hot water storage tank used in a solar heating system contains 3000 lb of water. The tank is 2 ft long by 3 ft wide. What is the pressure exerted on the bottom of the tank, in lb/ft^2?

Solution. A sketch of the tank is shown in Figure 2.1. Equation 2.4 will be used to find the pressure. The pressure is being exerted on an area 2 ft $\times$ 3 ft = 6 ft^2. The force acting on the bottom is the total weight of water.

$$p = \frac{F}{A} = \frac{3000 \text{ lb}}{6 \text{ ft}^2} = 500 \text{ lb/ft}^2$$

The relation between force and pressure is illustrated in Figure 2.2. A force of 3000 lb is distributed over the 2 ft $\times$ 3 ft area. The pressure is the force on each of the six 1 ft $\times$ 1 ft areas, 500 lb/ft^2.

Pressures of liquids and gases are of great importance in HVAC work. Some examples are the steam pressure in a boiler, the air pressure developed by a fan, the water pressure exerted on a valve, and the pressure exerted by the atmosphere.

Absolute, Gage, and Vacuum Pressure A space that is completely evacuated of any gas or liquid (a complete vacuum) has zero pressure. The pressure exerted by a fluid above the zero value is called its *absolute pressure* (p_{abs}). This is illustrated in Figure 2.3.

The atmospheric air above the earth exerts a pressure (p_{atm}) because of its weight. The pressure it exerts at sea level has been measured and found to be approximately 14.7 lb/in.^{2} absolute

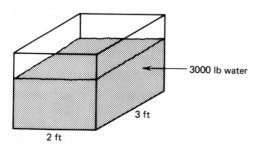

Figure 2.1. Sketch for Example 2.6.

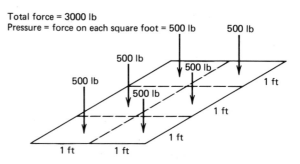

Figure 2.2. Relation between force and pressure.

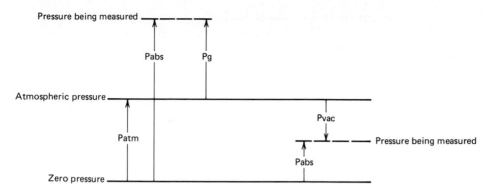

Figure 2.3. Relations of absolute, gage, and vacuum pressures.

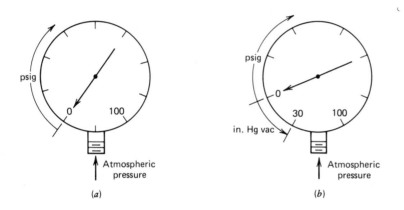

Figure 2.4. Pressure gages. (*a*) Compression gage reads gage pressure only. (*b*) Compound gage reads gage and vacuum pressure.

(psia). (This pressure decreases at higher elevations because there is less weight of air above. For example, the atmospheric pressure in Denver, Colorado is about 12.23 psia.)

Pressure measuring instruments usually measure the difference between the pressure of a fluid and the pressure of the atmosphere. The pressure measured above atmospheric pressure is called *gage pressure* (p_g). The relation between absolute, atmospheric, and gage pressures, shown in Figure 2.3, is

$$P_{abs} = P_{atm} + P_g \qquad (2.5)$$

Using gage pressure is convenient because most pressure measuring instruments are calibrated to read 0 when they are subject to atmospheric pressure. Figure 2.4(*a*) shows the dial face of a typical *compression gage*. (Pressure gages and similar instruments will be discussed in Chapter 16.)

EXAMPLE 2.7

The pressure gage connected to the discharge of a cooling tower water pump in the Trailblazers Bus Terminal in San Francisco reads 18 psi. What is the absolute water pressure at the pump discharge?

Figure 2.5. Sketch for Example 2.8.

Solution. The pressure gage reads gage pressure, 18 psig (above atmospheric). San Francisco is at sea level, so the atmospheric pressure is approximately 14.7 psia. Using Equation 2.5

$$p_{abs} = p_g + p_{atm} = 18 \text{ psi} + 14.7 \text{ psi} = 32.7 \text{ psia}$$

If a fluid exerts a pressure below atmospheric pressure, the difference from atmospheric pressure is called *vacuum pressure* (p_{vac}). The relation between absolute, atmospheric, and vacuum pressures, shown in Figure 2.3, is

$$p_{abs} = p_{atm} - p_{vac} \qquad (2.6)$$

Some gages are constructed to read both vacuum and gage pressure. This type is called a *compound gage* and is shown in Figure 2.4(*b*).

EXAMPLE 2.8

The gages on the suction and discharge lines of a compressor read 5 psiv (lb/in.² vac) and 60 psig, respectively. How much is the pressure increased by the compressor?

Solution. Referring to Figure 2.5, the pressure increase is

$$\text{pressure increase} = 60 + 5$$

$$= 65 \text{ psi}$$

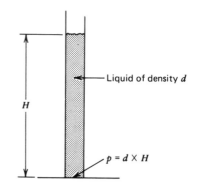

Figure 2.6. Pressure exerted by a liquid column. Pressure may be expressed as "head" (height of liquid).

2.7. PRESSURE OF A LIQUID COLUMN

A liquid exerts a pressure because of its weight, and the weight depends on the height of the column of liquid. The relation between the pressure exerted and the height, as shown in Figure 2.6, is

$$p = d \times H \qquad (2.7)$$

where

p = pressure exerted by a liquid, lb/ft²
d = density of liquid, lb/ft³
H = height of liquid, ft

Other units can be used in the equation, but these are often convenient.

Figure 2.7. Manometer reading pressures above and below atmospheric pressure. (*a*) Equal pressure on both legs. (*b*) Pressure in duct above atmospheric (gage pressure). (*c*) Pressure in duct below atmospheric (vacuum pressure).

EXAMPLE 2.9

A 300-ft vertical pipe in an office building is filled with chilled water. What is the pressure in lb/in.^{2} gage (psig) that a valve in the bottom of the line will have to withstand?

Solution. The density of water is approximately 62.4 lb/ft^3. Using Equation 2.7

$$p = d \times H$$

$$p = 62.4 \frac{\text{lb}}{\text{ft}^3} \times 300 \text{ ft} = 18,720 \frac{\text{lb}}{\text{ft}^2} \times \frac{1 \text{ ft}^2}{144 \text{ in.}^2}$$

$$= 130 \text{ psig}$$

The relation between pressure and height of a liquid is used in pressure measuring instruments that have a column of liquid. These are called *manometers*, an example of which is shown in Figure 2.7. In Figure 2.7(*a*), the pressure exerted on both legs of the manometer (atmospheric pressure) is the same, so the liquid is at the same level. In Figure 2.7(*b*) the pressure in the duct is above atmospheric. In Figure 2.7(*c*) the pressure in the duct is below atmospheric (vacuum pres-

sure), so the liquid is higher in the leg connected to the duct.

EXAMPLE 2.10

A service technician wishes to measure the pressure of air in a duct. He connects one leg of a water manometer to the duct and the other leg is exposed to the atmosphere. The difference in height of the water columns is 8 in. w.g. (inches of water gage) as shown in Figure 2.8. What is the air pressure in the duct in psig?

8 in. w.g.

Figure 2.8. Sketch for Example 2.10.

Solution. The difference in height is related to the pressure by Equation 2.7. Changing the units of H first

$$H = 8 \text{ in. w.g.} \times \frac{1 \text{ ft}}{12 \text{ in.}} = 0.667 \text{ ft w.g.}$$

$$p = d \times H$$

$$= 62.4 \frac{\text{lb}}{\text{ft}^3} \times 0.667 \text{ ft}$$

$$= 41.6 \frac{\text{lb}}{\text{ft}^2} \times \frac{1 \text{ ft}^2}{144 \text{ in.}^2} = 0.29 \text{ psig}$$

The air pressure in the duct is 0.29 psi above atmospheric pressure.

Water manometers are often used for measuring relatively small pressures, particularly when testing and balancing air systems. They are not convenient for high pressures because a very high liquid column would be needed. Manometers using mercury, a liquid with a much higher density than water, are often used for measuring higher pressures.

The *barometer* (Figure 2.9) is a special manometer used for measuring atmospheric air pressure. Mercury (Hg) is the liquid used. The tube is evacuated so that no atmospheric pressure acts on the top of the mercury column.

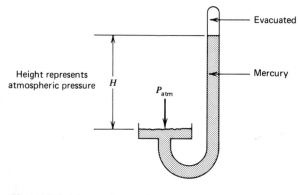

Figure labels: Evacuated, Mercury, Height represents atmospheric pressure, H, P_{atm}

Figure 2.9. Mercury barometer.

Because atmospheric pressure acts on the bottom of the mercury, the height to which the mercury column is lifted represents atmospheric pressure.

EXAMPLE 2.11

How high would the mercury column in a barometer be, in both in. Hg and mm Hg, at a location where atmospheric pressure is 14.7 psi and the temperature 32 F?

Solution. Using Equation 2.7 with proper units, noting the density of mercury (Table 2.1) is $d = 849 \text{ lb/ft}^3$ at 32 F,

$$p_{atm} = 14.7 \frac{\text{lb}}{\text{in.}^2} \frac{144 \text{ in.}^2}{1 \text{ ft}^2} = 2116.8 \frac{\text{lb}}{\text{ft}^2}$$

$$H = \frac{p}{d} = \frac{2116.8 \text{ lb/ft}^2}{849 \text{ lb/ft}^3} = 2.49 \text{ ft} \times \frac{12 \text{ in.}}{1 \text{ ft}}$$

$$= 29.92 \text{ in. Hg}$$

$$= 29.92 \text{ in. Hg} \times \frac{25.4 \text{ mm}}{1 \text{ in.}} = 760 \text{ mm Hg}$$

Head   It is often convenient to express pressure in units of head. *Head* is the equivalent of liquid column height (H) expressed in Equation 2.7. In Example 2.11, instead of stating that the pressure of the atmosphere was 14.7 psi, it could have been stated that it was 29.92 in. Hg or 760 mm Hg. In Example 2.10 the air pressure in the duct could also have been stated both ways, $p = 0.29$ psig = 8 in. w.g.

That is, there does not actually have to be a column of liquid to express any pressure in head units. Equation 2.7 can be used to convert to or from units of pressure expressed as head. Some of the equivalents for pressure expressed as head, obtained from that equation, are listed in Table A.2.

EXAMPLE 2.12

A contractor requires a pump that will have a discharge pressure of 42 psi. He looks in a manufacturer's catalog to find a suitable pump, but finds that the pump ratings are listed as "head, feet of water." What pump head should he specify in his purchase order?

Solution. Using the conversion factor equality (Table A.2) of

$$2.3 \text{ ft w.} = 1 \text{ psi}$$

$$H = 42 \text{ psi} \times \frac{2.3 \text{ ft w.}}{1 \text{ psi}} = 97 \text{ ft w.}$$

2.8. WORK, POWER, AND ENERGY

Work is the effect created by a force when it moves a body. It is expressed by the following equation

$$\text{Work} = \text{force} \times \text{distance} \quad (2.8)$$

EXAMPLE 2.13

A cooling tower weighing 6000 lb is hoisted from the street level to the roof of the Gusher Oil Co. building, 300 ft high. How much work is done in lifting it?

Solution. The force required is equal to the weight of the tower. Using Equation 2.8,

$$\text{Work} = 6000 \text{ lb} \times 300 \text{ ft} = 1,800,000 \text{ ft-lb}$$

Power is the time rate of doing work. It is expressed by the equation

$$\text{Power} = \frac{\text{work}}{\text{time}} \quad (2.9)$$

Power is usually of more direct importance than work in industrial applications; the capacity

of equipment is based on its power output or power consumption. If work is expressed in ft-lb, some units of power that would result are ft-lb/min and ft-lb/sec. More convenient units for power are the horsepower (HP) and kilowatt (KW), because the numbers resulting are not as large.

EXAMPLE 2.14

If the cooling tower in Example 2.13 is lifted by a crane in 4 minutes, what is the minimum power required?

Solution. Using Equation 2.9,

$$\text{Power} = \frac{1,800,000 \text{ ft-lb}}{4 \text{ min}} = 450,000 \text{ ft-lb/min}$$

From Table A.2, 1 HP = 33,000 ft-lb/min. Converting to HP

$$450,000 \text{ ft-lb/min} \times \frac{1 \text{ HP}}{33,000 \text{ ft-lb/min}} = 13.6 \text{ HP}$$

The actual size of the engine or motor required to hoist the cooling tower would be greater than 13.6 HP, due to friction and other losses, and to allow some excess capacity as a safety reserve.

Although it is a somewhat abstract concept, **energy** is sometimes defined as the ability to do work. For example we use the stored chemical energy in a fuel by burning it to create combustion gases at high pressure that drive the pistons of an engine and thus do work. Work is therefore one of the forms of energy.

Energy can exist in a number of forms. They can be grouped into those forms of energy that are *stored* in bodies or those forms of energy in *transfer* or *flow* between bodies. Work is one of the forms of energy in transfer between bodies. That is, one body does work on another when it moves it.

Figure 2.10. Comparison of stored energy and energy in transfer.

Energy can be stored in matter in many forms. Figure 2.10 is a diagram showing some types of *stored energy* and *energy in transfer*. At this time we will turn our attention to a form of energy in transfer or motion called heat. Some of the forms of stored energy will be discussed in Section 2.10.

2.9. HEAT AND TEMPERATURE

Heat has been described as a form of energy in transfer.

Heat is the form of energy that transfers from one body to another due to a temperature difference.

Figure 2.11 graphically describes this definition. In Figure 2.11(a) heat (Q) flows from the high temperature body, hot water, in the heating unit to the lower temperature body, the air in the room. Figure 2.11(b) shows that heat will flow from the higher temperature body, room air, to the lower temperature body, the refrigerator interior, due to the temperature difference. Note that heat can only flow naturally from a higher to a lower temperature—"downhill," so to speak, as seen in Figure 2.12. Of course if there is no temperature difference, there is no heat flow. The most common unit used for heat in the United States is the BTU (British Thermal Unit). The BTU is defined as *the quantity of heat required to raise the temperature of one pound of water one degree Fahrenheit (F) at 59 F.*

Temperature is a measure of the thermal activity in a body. This activity depends on the velocity of the molecules and other particles of which all matter is composed. It is not practical to measure temperature by measuring the velocity of molecules, however, so this definition is not of great importance in our work. Temperature is usually measured with thermometers. The most commonly used type relies on the fact that most liquids expand and contract when their temperature is raised or lowered. By creating an arbitrary scale of numbers, a temperature scale and units

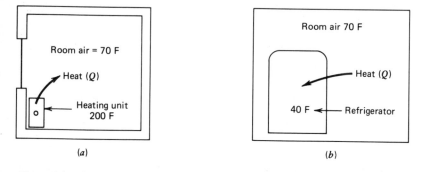

Figure 2.11. Examples of heat flow. (*a*) Heat flows from unit at higher temperature to room at lower temperature. (*b*) Heat flows from room at higher temperature to refrigerator at lower temperature.

Figure 2.12. Heat can flow only from a higher to a lower temperature. (*a*) If t_1 is greater than t_2, heat flows from *A* to *B*. (*b*) If $t_1 = t_2$, no heat flows.

Figure 2.13. Relations between temperature scales.

are developed. Other types of thermometers used in HVAC work will be discussed in Chapter 16.

The unit scale used for measuring temperature in the U.S. is the *degree Fahrenheit* (F), in which the boiling point of water is 212 F and the freezing point of water is 32 F, at atmospheric pressure. In the SI system of units the *degree Celsius* (C) is used, in which the boiling point of water is 100 C and the freezing point is 0 C, at atmospheric pressure. The relationship between these two units is therefore

$$F = 1.8\,C + 32 \qquad (2.10a)$$

$$C = \frac{F - 32}{1.8} \qquad (2.10b)$$

EXAMPLE 2.15

A room is supposed to be at a temperature of 78 F in an air conditioned building. The building maintenance engineer checks the temperature with a thermometer that has a Celsius scale. What should be the reading on the thermometer?

Solution. Using Equation 2.10*b*

$$C = \frac{F - 32}{1.8} = \frac{78 - 32}{1.8} = 25.6 \text{ C}$$

There are also two *absolute temperature* scales. These take the value 0 for the lowest temperature that can exist. They are called the *Rankine* (R) and *Kelvin* (K) temperature scales. The Rankine is used in the U.S. system, with the difference in size between each degree the same as Fahrenheit. The Kelvin is used in the SI system with the difference between each degree equal to Celsius. The relationships are

$$R = F + 460 \qquad (2.10c)$$

$$K = C + 273 \qquad (2.10d)$$

The relations between temperature scales are shown graphically in Figure 2.13.

2.10. ENTHALPY

It has been stated that energy can be classified into energy in flow and stored energy. The total energy stored in a body includes a number of types. For instance, we are well aware that a body has stored *chemical energy*, because we have experienced that this energy can be released from some substances by combustion. Other forms of stored energy are kinetic and potential energy. *Kinetic energy* is the stored energy in a body due to its motion, or velocity. *Potential energy* is the stored energy a body has due to its position, or elevation. Bodies also have additional energy due to their temperature and pressure. We are aware that a gas at a high pressure has energy (e.g., a steam boiler that bursts) and water at a high temperature can give up heat energy.

> *The stored energy of temperature and pressure is called **enthalpy** (H).*

(There is a more precise definition of enthalpy, but it serves no purpose in our discussion.) It is also common in the refrigeration industry to use the phrase *heat content* to mean the same as enthalpy. (Strictly speaking, heat is a form of energy that flows into or out of a body and enthalpy or heat content is a form of stored energy in a body.) Being a form of energy, enthalpy can also be measured in BTU. *Specific enthalpy* (*h*) is the enthalpy per unit mass of a substance, expressed in BTU per pound in U.S. units.

It is important to distinguish temperature from enthalpy (heat content). Temperature is a measure of the thermal level of a body. When heat is added to a body its temperature rises, but the total enthalpy or heat content of a body depends on the mass of a body as well as its temperature. For example, a thimbleful of molten steel at 2500 F has a much greater temperature than a large tank filled with water at 200 F but the enthalpy of the tank of water is greater. That is, there is more stored energy in the water. This is an important fact, because for many applications more heat can be obtained from the tank of water despite its lower temperature.

2.11. THE ENERGY EQUATION (FIRST LAW OF THERMODYNAMICS)

The subject we have been examining, called *thermodynamics*, is the branch of physics that deals with heat and work. The *First Law of Thermodynamics* is a principle that may be stated in various ways, for instance, "energy can neither be created nor destroyed," or "there is conservation of energy in nature." This principle is used extensively in the HVAC industry, especially when stated as an energy balance:

> *The change in total energy in a system equals the energy added to the system minus the energy removed from the system.*

The word *system* refers to any closed body or group of bodies for which the flow of energy in or out can be determined. It could be the air in a room (Figure 2.14), a boiler, a whole building, or a complete air conditioning system.

This energy balance is usually expressed as an equation, called the *Energy Equation*:

$$E_{\text{ch}} = E_{\text{in}} - E_{\text{out}} \qquad (2.11)$$

where

E_{ch} = change in stored energy in the system
E_{in} = energy added to (entering) the system
E_{out} = energy removed from (leaving) the system

Example 2.16 illustrates the use of the energy equation.

EXAMPLE 2.16

A convector in Mr. Jones office is supplying 4000 BTU/hr of heat. Heat is being transferred from the room to the outdoors at the rate of 6500 BTU/hr. What will happen in the room? What size electric heater should Mr. Jones temporarily use to solve the emergency?

Solution. We apply the Energy Equation 2.11. Figure 2.14 shows the energy (heat) added and removed

$$E_{\text{ch}} = E_{\text{in}} - E_{\text{out}}$$
$$= 4000 \ \frac{\text{BTU}}{\text{hr}} - 6500 \ \frac{\text{BTU}}{\text{hr}}$$
$$= -2500 \frac{\text{BTU}}{\text{hr}}$$

Figure 2.14. Sketch for Example 2.16.

The negative sign means the room energy is decreasing. This loss in enthalpy (heat content) will cause the room air temperature to drop, making it uncomfortable. A solution is to install a heater that will make up the heat loss which the convector does not supply, 2500 BTU/hr. There will then be no net loss of heat from the room, and the temperature will not drop. The capacity of an electric heater is normally expressed in watts (W) or kilowatts (KW) rather than BTU/hr. The heater should therefore have the following capacity.

From Table A.2

$$3410 \text{ BTU/hr} = 1000 \text{ W}$$

$$2500 \text{ BTU/hr} \times \frac{1000 \text{ W}}{3410 \text{ BTU/hr}} = 733 \text{ W}$$

The nearest size larger heater would probably be 750 W.

Example 2.16 illustrates the sign convention that will be used in the Energy Equation: An energy decrease in the system is negative, an energy increase is positive.

The example also shows that the units used for energy are interchangeable, regardless of the form of energy, whether, heat, work, or enthalpy. Example 2.17 will illustrate this, as well as a further application of the Energy Equation.

EXAMPLE 2.17

A business equipment room has 1000 watts of lighting and some small motors with a total output of 10 HP. All of the energy in the lighting and from the motors is converted into heat. What is the increase in enthalpy of the room air from these sources?

Solution. The energy added to the room air will increase its enthalpy. Applying the Energy Equation, and converting all units to BTU/hr

$$E_{ch} = E_{in} - E_{out}$$

$$E_{ch} = 1000 \text{ W} \times \frac{3.41 \text{ BTU/hr}}{1 \text{ W}}$$

$$+ 10 \text{ HP} \times \frac{2545 \text{ BTU/hr}}{1 \text{ HP}} - 0$$

$$= 28,860 \text{ BTU/hr}$$

2.12. LIQUIDS, VAPORS, AND CHANGE OF STATE

Substances can exist in three different *states* (also called *phases*)—solid, liquid, or vapor (gas). The state that a substance is in depends on its temperature and pressure. The meaning of this for liquids and vapors is best understood by describing an experiment (which the student could carry out at home to check the results), shown in Figure 2.15.

Figure 2.15(*a*) shows a pot of water at room temperature. Being open, it is subject to atmospheric pressure, 14.7 psia at sea level. At (*b*) heat (*Q*) is being added to the water, and it is noted that its temperature continually rises as heat is added. At some point in time (*c*), however, it is noted that the temperature stops rising (at 212 F) and even though more heat is added after that (*d*) the temperature does not increase for a while. What is observed now, however, is that the liquid water will gradually change into its gas or vapor state (steam). This process is called *boiling* or *vaporization*. As heat is added, no further temperature increase occurs as long as some liquid remains. At (*e*), all the water is evaporated. If more heat is added, it will be noted that the temperature (of the steam) will begin to rise again, above 212 F, as seen in (*f*). (This part of the experiment would be difficult to carry out, because the steam will escape into the room.)

The whole series of processes just described could also be carried out in reverse. Removal of heat (cooling) from the steam in Figure 2.15(*f*) lowers its temperature. When the cooling continues to (*e*) the temperature no longer drops, but the gas begins to condense to a liquid (*d*). After all of the steam is condensed (*c*) further removal of heat will result in a temperature drop of the liquid, (*b*) and (*a*).

A summary of all this information is shown in Figure 2.16 the temperature-enthalpy diagram. The line shows a change in temperature of the liquid when heat is added between 32 F and 212 F, but no change in temperature when more heat is added, until all the liquid is evaporated. When further heat is added, the temperature rises again. (Figure 2.16 also shows the change between solid and liquid state, which will be discussed later.)

Dependence of Boiling Temperature on Pressure In the experiment just described the surrounding pressure was 14.7 psia. Let us con-

Figure 2.15. Experiment showing change of state of water at atmospheric pressure (14.7 psia). (*a*) Initial condition (subcooled liquid). (*b*) Heat added, temperature increases (subcooled liquid). (*c*) Heat added, liquid reaches boiling point (saturated liquid). (*d*) Heat added, liquid changing to vapor, no temperature increase. (*e*) Heat added, all liquid vaporized (saturated vapor). (*f*) Heat added, temperature of vapor increases (superheated vapor). *Note:* Subcooled liquid is liquid below its boiling point. Saturated liquid and saturated vapor are the liquid and vapor at the boiling (condensing) point. Superheated vapor is vapor above the boiling point.

Figure 2.16. Enthalpy (heat content) change of water at 14.7 psia.

Figure 2.17. Experiment showing change of state of water at 24.9 psia. (a) Initial condition (subcooled liquid). (b) Heat added, temperature increases (subcooled liquid). (c) Heat added, temperature increases (subcooled liquid). Note that the water is not boiling at 212 F. (d) Heat added, liquid reaches boiling point, 240 F. (e) Heat added, all liquid vaporized (saturated vapor). (f) Heat added, temperature of vapor increases (superheated vapor).

duct the same experiment where the pressure is at a higher value, say 24.9 psia. Figure 2.17 represents the same heating process, or cooling if done in reverse.

When the water reaches 212 F (*c*) and more heat is added, it does not boil, but the temperature continues to rise. When the temperature reaches 240 F (*d*), however, the boiling process begins and the temperature remains constant

until the liquid has completely evaporated. This shows that the temperature at which the water boils changes with pressure. For water the boiling point is 240 F at 24.9 psia. This means that water cannot be made to boil at a temperature below 240 F if the pressure is 24.9 psia.

If the same experiment were carried out with the pressure at 6 psia, we would find that when heat was added, the boiling process would begin

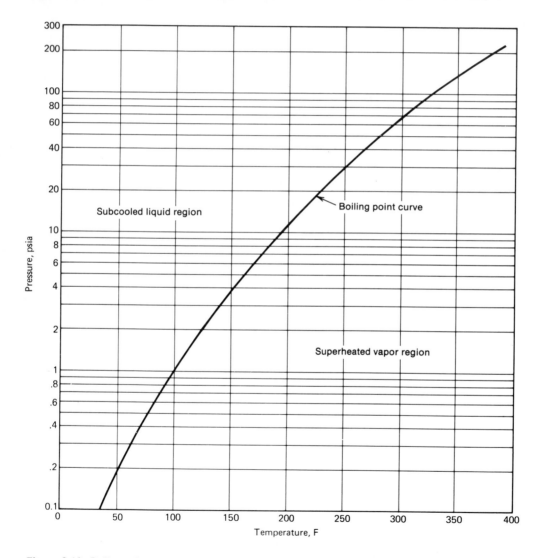

Figure 2.18. Boiling point pressure-temperature curve for water, also called the saturation
  ᴗor pressure curve.

at 170 F. These facts show that *the boiling/condensing temperature of a fluid depends on its pressure*. Figure 2.18 shows a line representing these temperature-pressure values for water; it is called the *boiling point* curve or *saturation vapor pressure* curve. Water can exist at its boiling/condensing temperature and pressure only on this line. To the left of the line it can exist only as a liquid and to the right only as a vapor. Along the line it can exist either as liquid, vapor, or as mixture of the liquid-vapor.

Figure 2.19. Slow evaporation of liquid. Escape of some molecules through surface causes a vapor pressure.

EXAMPLE 2.18

Will water exist in the liquid state, or as steam, if its temperature is 225 F and its pressure is 25 psia?

Solution. Locating the pressure-temperature ($p - t$) condition on Figure 2.18, it is found to be in the liquid region. The water is in a liquid state.

This same dependence of boiling/condensing temperature on pressure holds for all fluids, except the p-t values are different. For example, at 14.7 psia ammonia boils at −28 F, alcohol at 170 F, and copper at 4250 F.

Note that the higher the pressure on the water, the higher the boiling temperature, and the lower the pressure, the lower the temperature at which it will boil. This same relation holds for other substances.

The Molecular (Kinetic) Theory of Liquids and Gases The process of boiling and the dependence of boiling point temperature on surrounding pressure can be explained by referring to the molecular (kinetic) theory of liquids and gases. All matter is composed of particles called molecules. The molecules in a substance are constantly in motion. They are also attracted to each other by forces. The closer the molecules are to each other, the greater the attractive forces.

When a substance is in the liquid state, the molecules are closer together than when it is in its gaseous state, and therefore the attractive forces are greater. Also molecules in the gaseous state move more rapidly than molecules in the liquid state, and therefore they have more energy. This is why heat is required to boil a liquid. The heat energy is required to overcome the attractive forces holding the molecules relatively close together, so that they move further apart and change state to a gas.

The temperature of a substance is a measure of the average velocity of its molecules. The higher the average velocity, the higher the temperature. However, not all molecules move at the average velocity—some are moving faster, some slower.

Figure 2.19 shows an open vessel of water at 70 F, surrounded by air at 14.7 psia. The water is therefore in a liquid state. The average velocity of the molecules is not great enough for them to escape rapidly. However, a small fraction of molecules have velocities well above the average. If some of these molecules are near the surface they will escape. That is, there will be very slow evaporation from the surface. This leaves the remaining molecules at a slower average velocity, and therefore at a lower temperature. A slight cooling effect of the liquid has occurred as a result of the evaporation. We have all noticed this effect when alcohol is rubbed on the skin. It gradually evaporates and cools itself and the skin.

The molecules escaping from the surface of a liquid create vapor. The pressure exerted by the vapor at the surface of the liquid is called the *vapor pressure*. If the pressure exerted by a surrounding gas is above the vapor pressure, then the liquid cannot rapidly evaporate. However, if the temperature of the liquid is increased, the molecular velocity increases to a point at which the molecules break the bonds holding them together as a liquid and the liquid boils. The vapor pressure of the liquid has been increased to a value greater than the surrounding resisting pressure. Of course if the resisting pressure is higher, the temperature of the liquid must be increased further to reach the boiling point.

While the boiling process is occurring, the heat applied is breaking the bonds by overcoming forces that hold the molecules close together. It is not increasing the velocity of the molecules. That is why the temperature does not increase during boiling.

It is of importance to note what happens if the pressure exerted by a gas above a liquid is reduced to a value below the vapor pressure exerted by the liquid. In this case the liquid will suddenly boil, because the surrounding pressure is now less than the vapor pressure exerted by the liquid. The energy of the molecules is great enough to overcome the reduced resistance, and they escape rapidly. This cools the remaining liquid, because energy is removed. *Boiling has been achieved by a lowering of pressure.* This process is essential in refrigeration, as will be seen in Section 2.14.

Saturated, Subcooled, and Superheated Conditions   The pressure and temperature condition at which boiling occurs is called the saturated condition, and the boiling point is technically known as the saturation temperature and saturation pressure. As seen from the experimen-

tal description, the substance can exist as a liquid, vapor, or mixture of liquid and vapor at the saturated condition. At saturation, the liquid is called saturated liquid and the vapor is called a saturated vapor.

> *Saturated vapor is vapor at the boiling temperature, and saturated liquid is liquid at the boiling temperature. When the vapor is above its saturation temperature (boiling point) it is called a superheated vapor. When the liquid is below its saturation temperature it is called a subcooled liquid.*

Figure 2.18, a typical boiling point curve, illustrates this. Note that a substance can exist as a subcooled liquid or superheated vapor at many temperatures for a given pressure, but it can exist as a saturated liquid or vapor at only one temperature for a given pressure.

Sensible and Latent Heat   When heat added to or removed from a substance results in a temperature change then the enthalpy change in the substance is called a *sensible heat* change. When heat added to or removed from a substance results in a change in state then the enthalpy change in the substance is called a *latent heat* change. The enthalpy increase as it changes from a liquid to a vapor is called the *latent heat of vaporization*. The opposite effect, the enthalpy decrease as it changes from a vapor to a liquid is called the *latent heat of condensation*. It is equal to the latent heat of vaporization.

2.13. SATURATED PROPERTY TABLES

For various substances, saturation temperatures, corresponding pressures, and other properties at saturation conditions may be found in tables designed for that purpose. Table A.3 is a saturated property table for water and is commonly called the "Saturated Steam Table." Note that both parts of the table have the same information. If the temperature is known, use the part of the table that lists temperatures first. If the pressure is known, use the part of the table that lists pressures first. If the known value is between two listed values, interpolate to obtain the correct value. Examples 2.19–2.21 illustrate various uses of saturated property tables.

EXAMPLE 2.19

At what temperature will water boil at a pressure of 10 psia?

Solution. From Table A.3, we read the saturation temperature (boiling point) at 10 psia to be about 193 F.

EXAMPLE 2.20

Use the steam tables to determine if water is in a liquid or gas state at 300 F and 150 psia.

Solution. Using Table A.3, the saturation (boiling) temperature at 150 psia is about 358 F. The actual temperature is less, therefore the water will be in a liquid state (subcooled liquid).

EXAMPLE 2.21

The operating engineer of a hot water heating system reads the temperature and pressure at the pump suction to be 200 F and 10 psia. Should the engineer be concerned?

Solution. From Table A.3, the saturation temperature at 10 psia is below 200 F. Because the actual temperature is higher, the water will exist as steam, not liquid. Therefore the operator should be very concerned, because there is steam, not water, in the system.

2.14. REFRIGERATION

It has been stated that the boiling point of a liquid depends on the surrounding pressure. This fact can be explained by considering that all matter consists of particles (molecules) which are attracted to each other, but also have a considerable velocity energy. The pressure surrounding a liquid inhibits the escape of the molecules. If the liquid temperature increases, however, the molecules increase in velocity, and at some temperature (the boiling point) they will escape rapidly—the liquid will vaporize. If the pressure is increased, the molecules will have to reach a higher velocity—a higher temperature—to escape.

On the other hand, if the surrounding pressure is lowered enough (to the saturation point), the molecules will have enough energy to escape at a lower temperature. This is how refrigeration can be accomplished. A liquid is used that boils at a low temperature for the pressure that can be achieved. The surrounding pressure is reduced below the saturation pressure, and the liquid suddenly boils. As noted, liquids absorb heat when they boil (latent heat of vaporization). The heat absorbed from the surroundings at the low temperature is refrigeration.

Even the boiling of water can be used to achieve refrigeration, if the pressure can be lowered enough, as seen in Example 2.22.

EXAMPLE 2.22

The boiling of water is to be used to accomplish refrigeration at 50 F. To what value should the surrounding pressure be lowered?

Solution. From Table A.3, the saturation pressure of water at 50 F is 0.178 psia. If the surrounding pressure is reduced below this value, the water will boil. This requires (latent) heat. Heat will flow to the water from any surrounding body at a higher temperature, thus cooling the body.

Some absorption refrigeration machines use vaporization of water at very low pressures to achieve refrigeration. Refrigeration will be discussed in Chapter 13.

2.15. SENSIBLE AND LATENT HEAT EQUATIONS

The processes that occur in HVAC systems usually involve the addition or removal of heat from air or water. The procedures for calculating the amount of heat will be explained in this section.

Specific Heat   The *specific heat* (*c*) of a substance is defined as the amount of heat in BTUs required to change the temperature of 1 lb of the substance by 1 F. The specific heat of water is therefore 1 BTU/lb-F at 60 F, which follows from the definition of the BTU.

Sensible Heat Equation   The Energy Equation (2.11) can be applied to any substance to determine the relation between the amount of heat added or removed and the sensible heat content (enthalpy) change of the substance: Net heat added or removed = enthalpy change

$$Q = m \times c \times TC \qquad (2.12a)$$

or

$$Q = m \times c(t_2 - t_1) \qquad (2.12b)$$

where

Q = rate of heat added to or removed
　　from substance, BTU/hr
m = weight flow rate of substance, lb/hr
c = specific heat of substance, BTU/lb-F
$TC = t_2 - t_1$ = temperature change of substance, F

Equation 2.12 is called the *sensible heat equation*. The specific heat changes slightly with tem-

perature, so the equation is not exact, but it is reasonably accurate except for very large temperature changes. Approximate specific heats of a few substances are listed in Table 2.1. Examples 2.23–2.25 will illustrate the use of the sensible heat equation.

EXAMPLE 2.23

There are 5000 GPM of chilled water being circulated from the refrigeration plant to the buildings at the Interplanetary Spaceport. The water is cooled from 55 F to 43 F (Figure 2.20). What is the capacity of the refrigeration plant in BTU/hr, tons of refrigeration, and KW?

Solution.   The "capacity" of the refrigeration plant means the amount of heat it is removing from the water. First changing the water flow rate from GPM to lb/hr (Table A.2)

$$m = 5000 \text{ GPM} \times \frac{500 \text{ lb/hr}}{1 \text{ GPM}} \text{ water}$$

$$= 2,500,000 \text{ lb/hr}$$

Using Equation 2.2b, $Q = m \times c(t_2 - t_1)$

$$Q = 2,500,000 \frac{\text{lb}}{\text{hr}} \times 1 \frac{\text{BTU}}{\text{lb-F}} (55 - 43) \text{ F}$$

$$Q = 30,000,000 \text{ BTU/hr heat removal}$$

The ton of refrigeration is a unit measuring heat rate that is used often in refrigeration work. From Table A.2

$$1 \text{ ton} = 12,000 \text{ BTU/hr}$$

Figure 2.20. Sketch for Example 2.23.

Refrigeration capacity =

$$30,000,000 \text{ BTU/hr} \times \frac{1 \text{ ton}}{12,000 \text{ BTU/hr}}$$

$$= 2500 \text{ tons}$$

The capacity in KW is

$$30,000,000 \text{ BTU/hr} \times \frac{1 \text{ KW}}{3410 \text{ BTU/hr}}$$

$$= 8800 \text{ KW}$$

EXAMPLE 2.24

The fuel oil preheater for a boiler has become damaged. The oil must be heated to 180 F in order to flow readily. A spare preheater with a capacity of 100,000 BTU/hr is available. The boiler requires 10 GPM of oil. The oil is at 60 F in a storage tank. The density of the oil is 8.0 lb/gal and its specific heat is 0.5 BTU/lb-F. Is the spare preheater big enough?

Solution. First convert GPM of oil to lb/hr, and then use Equation 2.12 to find the temperature to which the oil would be heated by the spare preheater

$$m = 10 \ \frac{\text{gal}}{\text{min}} \times \frac{60 \text{ min}}{1 \text{ hr}} \times \frac{8.0 \text{ lb}}{\text{gal}}$$

$$= 4800 \text{ lb/hr}$$

$$t_2 - t_1 = \frac{Q}{m \times c} = \frac{100,000 \text{ BTU/hr}}{4800 \text{ lb/hr} \times 0.5 \text{ BTU/lb-F}}$$

$$= 42 \text{ F}$$

$$t_2 = 42 + t_1 = 42 + 60 = 102 \text{ F}$$

The preheater will heat the fuel oil to only 102 F. It will not do the job.

EXAMPLE 2.25

An electric booster heater in an air conditioning duct (Figure 2.21) has a capacity of 2 KW. The mechanical contractor is balancing the system,

Figure 2.21. Sketch for Example 2.25.

and wants to find out how much air is flowing in the duct. The contractor measures the temperature before and after the heater as 80 F and 100 F. How much air is flowing?

Solution. Convert the heater capacity to BTU/hr. Use Equation 2.12, solving for flow rate. From Table 2.1, $c = 0.24$ BTU/lb-F for air.

$$Q = 2 \text{ KW} \times \frac{3410 \text{ BTU/hr}}{1 \text{ KW}} = 6820 \text{ BTU/hr}$$

$$m = \frac{Q}{c \times TC} = \frac{6820 \text{ BTU/hr}}{0.24 \text{ BTU/lb-F} \times 20 \text{ F}}$$

$$= 1420 \text{ lb/hr}$$

Using the Saturated Property Tables to Find the Heat Added or Removed The Energy Equation can also be expressed as follows for most heating and cooling processes

$$Q = m(h_2 - h_1) \qquad (2.13)$$

where

Q = rate of heat added or removed from substance, BTU/hr

m = weight flow rate of substance, lb/hr

$h_2 - h_1$ = specific enthalpy change of substance, BTU/lb

This equation can be used instead of Equation 2.12 for the sensible heat change process, if the enthalpy is known. The property tables list enthalpy data. Example 2.26 compares the two methods.

EXAMPLE 2.26

A hot water boiler heats 10,000 lb/hr of water from 180 F to 220 F, at 30 psia. How much heat is added to the water?

Solution. The enthalpy of liquid water (h_f) is listed in Table A.3. Using Equation 2.13 at each temperature

$$Q = m(h_{f2} - h_{f1})$$
$$= 10,000 \text{ lb/hr } (188.1 - 147.9)$$
$$= 402,000 \text{ BTU/hr}$$

The result using Equation 2.12 is

$$Q = m \times c \times TC = 10,000 \frac{\text{lb}}{\text{hr}} \times 1 \frac{\text{BTU}}{\text{lb-F}} \times 40 \text{ F}$$
$$= 400,000 \text{ BTU/hr}$$

The two equations give almost identical results. Either one is acceptable.

As was done in Example 2.25, the enthalpy of the subcooled liquid is always looked up in the table at its temperature, not its pressure.

Latent Heat Equation The change in enthalpy that occurs when a substance evaporates or condenses is determined from the *latent heat equation*, found by applying the Energy Equation to the change in state

$$Q = m(h_g - h_f) = m \times h_{fg} \quad (2.14)$$

where

Q = heat added to or removed from substance, BTU/hr
m = weight flow rate, lb/hr
h_f = enthalpy of saturated liquid, BTU/lb
h_g = enthalpy of saturated vapor, BTU/lb
h_{fg} = latent heat of vaporization, BTU/lb

The latent heat of vaporization for water is shown in Table A.3. Note that it changes with temperature.

When a heating or cooling process involves both a sensible and a latent heat change to the substance, the results can be found by simply adding the two effects together.

EXAMPLE 2.27

A steam boiler generates 20,000 lb/hr of saturated steam at 20 psia. The water enters the boiler at 180 F. How much heat is required?

Solution. The enthalpy increase of the water is the sum of the sensible and latent heat change. The sensible change is to the boiling point. From Table A.3, the boiling temperature is 228 F at 20 psia. The change in sensible heat content of the liquid is

$$h_{f2} - h_{f1} = 196.2 - 147.9 = 48.3 \text{ BTU/lb}$$

The change in the latent heat content at 20 psia is

$$h_{fg} = 960.1 \text{ BTU/lb}$$

The total enthalpy change per pound is the sensible plus latent heat

$$48.3 + 960.1 = 1008.4 \text{ BTU/lb}$$

The total heat required is

$$Q = 20,000 \frac{\text{lb}}{\text{hr}} \times 1008.4 \frac{\text{BTU}}{\text{lb}}$$
$$= 20,168,000 \frac{\text{BTU}}{\text{hr}}$$

2.16. LATENT HEATS OF FUSION AND SUBLIMATION

The change of state of a substance from liquid to gas involves gaining the latent heat of vaporization. A substance in a solid state will increase in temperature when heat is added to it (sensible heat) but when a certain temperature is reached, its temperature will no longer increase when more heat is added, and the substance will begin to change state to a liquid—it will melt. If the reverse process is carried out, removal of heat from a liquid, its temperature will drop but eventually it will freeze into a solid. The heat accompanying the melting or freezing process is called the *latent heat of fusion*. For water the latent heat of fusion is 144 BTU/lb.

At very low pressure and temperature it is possible to change some substances directly from the solid to the gas state. This process is called *sublimation*. It is used in the procedure called *freeze-drying*, to prepare dried foods by first freezing them and then evaporating the ice in the food directly to vapor, at a very low pressure.

2.17. THE IDEAL (PERFECT) GAS LAWS

Under certain conditions the pressure, volume, and temperature of gases are related by an equation called the *perfect* or *ideal gas law*. Air at the temperatures and pressures in air conditioning work follow this equation. The perfect gas equation can be expressed

$$pV = mRT \qquad (2.15)$$

where

p = pressure, lb/ft² absolute
V = volume, ft³
m = weight of gas, lb
R = a gas constant
T = absolute temperature, degrees R

We will use the equation in this form in Chapter 7. By rearranging the terms in the equation, for two different conditions of the gas, 1 and 2, the following equation results

$$\frac{p_2 V_2}{T_2} = \frac{p_1 V_1}{T_1} \qquad (2.16)$$

The gas law is useful in finding changes in p, V, and T for changed conditions. If only two of

these three variables change, the equation simplifies. If the temperature is constant,

$$\frac{p_2}{p_1} = \frac{V_1}{V_2} \qquad (2.17)$$

If the volume is constant,

$$\frac{p_2}{p_1} = \frac{T_2}{T_1} \qquad (2.18)$$

If the pressure is constant,

$$\frac{V_2}{V_1} = \frac{T_2}{T_1} \qquad (2.19)$$

EXAMPLE 2.28

Compressed air required for operating the pneumatic controls in an air conditioning system is stored in a 10 ft³ tank at 150 psig. The air is used in the controls at 15 psig. What volume of air is available for the controls?

Solution. Using Equation 2.17, with absolute pressures ($p_{abs} = p_g + p_{atm}$), assuming the temperature remains constant

$$V_1 = \frac{P_2}{P_1} V_2 = \frac{164.7 \text{ psia}}{29.7 \text{ psia}} \times 10 \text{ ft}^3 = 55.5 \text{ ft}^3$$

EXAMPLE 2.29

A technician testing and balancing a system measures 5000 ft³/min (CFM) of air *entering* a heating coil at 40 F, and leaving at 120 F. However, the specifications call for measuring the volume flow rate *leaving* the coil. What is the air flow leaving the coil?

Solution. Equation 2.19 can be used, with volume flow rate instead of volume, because the same unit of time is involved. Temperatures must be in absolute units $R = (F + 460)$

$$\text{CFM}_2 = \frac{T_2}{T_1} \text{CFM}_1 = \frac{580}{500} \times 5000 = 5800 \text{ CFM}$$

The technician would now check the system design specifications. If it called for 5800 CFM, he knows that the proper amount of air is flowing, even though he measured 5000 CFM.

2.18. ENERGY UTILIZATION (SECOND LAW OF THERMODYNAMICS)

We have seen how the First Law of Thermodynamics can be used, in the form of the Energy Equation, to solve problems in HVAC work, and we will use it even further. Basically, it tells us how much energy is used for a given task (the power of a pump, the capacity of a refrigeration machine, and similar information). However, it tells us nothing about the answers to such questions as "Can I use a smaller pump, fan, or refrigeration machine?," or "How do I reduce the energy consumption of an HVAC system?"

An understanding and application of the *Second Law of Thermodynamics* will enable us to investigate problems of more efficient energy utilization. Energy conservation has become of great necessity and concern. Unfortunately, efforts in this area have sometimes been haphazard, partly due to a lack of understanding of the Second Law.

The Second Law may be expressed as an equation but it is not simple to use in energy utilization analysis. Therefore, what we will do here is state some conclusions derived from the Second Law. Throughout the book we will suggest energy conservation steps, many based on these conclusions. In Chapter 15 these ideas will be gathered together and additional ones will be discussed. The reader may wish to treat the whole subject of energy conservation at that time, or to consider each aspect as it is brought up. Some of the conclusions that can be drawn from the Second Law are:

1. Whenever heat energy is used to do work, it is never all available for a useful purpose. Some must be lost, and unavailable for the job to be performed. For instance, if we are using an engine to drive a refrigeration compressor, only part of the energy in the fuel can be used; the rest will be wasted.

2. The maximum amount of energy that can be made available in a power producing device such as an engine or turbine can be calculated. That is, we can determine the best efficiency possible, and compare it with an actual installation.

3. The minimum amount of energy required to produce a given amount of refrigeration can be calculated, and this can be compared with the actual system.

4. There are a number of physical effects that are called irreversible which cause a loss of available energy. These effects cannot be avoided, but should be reduced to a minimum. Included among them are:

   (a) *The temperature difference for heat transfer.* Greater temperature differences cause greater losses, therefore temperature differences between fluids should be kept as small as practical, as for example, in evaporators and condensers.

   (b) *Friction.* Friction causes loss of useful energy, and therefore should be minimized. For example, regular cleaning of condenser water piping will reduce the roughness of the wall. Fluid friction will be reduced, and less energy will be lost in pumping power.

   (c) *Rapid expansion.* An example of possible wasted energy that results from this is the generation of high pressure steam and then expanding it in a "flash tank" to a low pressure before using it for heating.

   (d) *Mixing.* Mixing fluids of different temperatures can result in a loss of useful energy. Mixing processes are common in HVAC systems, but should be minimized or even avoided when they cause a loss of available energy. Dual duct systems and three pipe systems are two types of air conditioning systems using mixing that can result in energy waste. They will be discussed in Chapter 12.

Any process that occurs without any of these effects is called a *reversible* process. Although a reversible process is an ideal case that is impossible to achieve, we always try to minimize irreversible effects in the interests of energy conservation.

Entropy is a physical property of substances related to energy utilization and conservation. It is defined as the ratio of the heat added to a substance to the temperature at which it is added. However, this definition is not useful here. It is important to understand that entropy is a measure of the energy that is *not available* to do work.

For any process that requires work, such as driving a refrigeration compressor, the least amount of work is required if the entropy of the fluid does not change. This is called a constant entropy, or *isentropic* process. In a constant entropy process no heat is added to or removed from the substance (*adiabatic* process) and there are no irreversible effects (e.g., friction).

A constant entropy process is an ideal reversible process that can never really occur. However, studying it gives us a goal to aim for. In any real process where work is required, the entropy increases, and we try to minimize this increase.

Practical applications of the First and Second Laws are discussed in much greater detail in later chapters, especially Chapters 3, 4, 6, 12, 13, and 15.

REVIEW QUESTIONS

1. What is a unit? What problems may arise when using units?
2. What is a conversion factor?
3. What are the advantages of the SI system of units?
4. Explain what is meant by *rounding* off.
5. Define mass, force, weight, and pressure.
6. Define density, specific volume, and specific gravity.
7. With the aid of a sketch, explain gage pressure and absolute pressure.
8. What is a compound gage?
9. What is meant by stored energy and energy in transfer? Name types of energy in each category and give an example of each.
10. What is head?
11. Define heat, temperature, and enthalpy.
12. What is the difference between work and power?
13. State the energy balance as a sentence and as an equation.
14. What are the three common states in which matter may exist?
15. Define the saturated, superheated, and subcooled conditions.
16. Explain what is meant by a sensible heat change and a latent heat change.
17. List four conditions that should be sought in HVAC systems to minimize energy use, as suggested by the Second Law of Thermodynamics.

EXAMPLE 2.30

A mechanical contractor has a choice of using copper tubing or steel piping of the same diameter in a chilled water system. Which would be the best choice to minimize energy consumption?

Solution. Copper tubing has a smoother surface and therefore less friction. Less energy will be used in the pump, according to the Second Law.

PROBLEMS

2.1. List the physical characteristic measured by each of the following units: lb/in.^{2}, hp, GPM, in. Hg, m/sec, ft^2, KW, BTU, kg/m^3, and ft^3/lb.

2.2. List the standard SI unit and a typical U.S. unit for each of the following physical characteristics: power, pressure, velocity, mass, flow rate, energy, specific volume, and density.

2.3. Change the following quantities to the new units specified:

A. 120 lb/in.^{2} to ft w.

B. 83.2 ft^3/sec to gal/min (GPM).

C. 76,500 BTU/hr to tons of refrigeration.

D. 18.2 in. Hg to lb/in.^{2}.

E. 0.91 HP to BTU/min.

2.4. Change the following quantities from the U.S. units to the SI units specified:

A. 12.6 ft^2 to m^2.

B. 629 ft^3/min (CFM) to m^3/sec.

C. 347,000 BTU/hr to KW.

D. 62.4 lb/ft^3 to kg/m^3.

2.5. Find the area in ft^2 of a window that is 4 ft 3 in. wide by 6 ft 6 in. high.

2.6. A contractor wants to lift 100 sections of steel pipe each 20 ft long. The pipe, made in Germany, is stamped "100 kg/m." How many pounds must the crane be capable of lifting?

2.7. Round off the following numbers to three significant figures:

A. 793,242.

B. 2.685.

C. 542.

D. 1.9.

E. 0.8319.

2.8. Change the following quantities to the units specified:

A. 276 gal water to lb.

B. 2760 lb/hr water to GPM.

C. 41,800 ft^3 air at 70 F and 14.7 psia to lb.

2.9. A hot water storage tank for a solar energy system measures 18 ft × 9 ft in plan. It is filled to depth of 6 ft. What is the weight of water in the tank? What is the water pressure on the bottom of the tank, in lb/in.^{2}?

2.10. What is the density of a fuel oil with a s.g. = 0.93?

2.11. The pressure gage on a boiler in Boston reads 28.7 psi. What is the absolute pressure in psi? What would the absolute pressure be if the boiler were in Denver?

2.12. The absolute pressure in the suction line to a compressor is 12.2 in. Hg. What pressure would a vacuum gage read at sea level, in in. Hg?

2.13. A vat 25 ft high is filled with Big Brew Beer that has a s.g. = 0.91. What is the pressure in psi on a value 3 ft above the bottom of the tank?

2.14. The discharge pressure of a pump is 32.6 psig. What is the pressure in ft w?

2.15. The air pressure in a tank is 3.7 ft w.g. What would be the reading on a Hg manometer attached to the tank in inches, and in mm?

2.16. Change the following temperatures to the new units specified:

A. 88 F to C.

B. 630 F to R.

C. −10 C to F.

D. 280 K to C.

E. 31 C to R.

2.17. A room receives 1200 BTU/hr of heat from solar radiation, loses 1450 BTU/hr through heat transfer to the outdoors, and has 2.2 KW of appliances operating. What is the net heat gain or loss to the room?

2.18. An electric heater is to be used to heat an enclosed porch that is losing 7900 BTU/hr to the outdoors. What size heater should be used?

2.19. Determine the state of water at the following conditions:

A. 230 F and 18 psig.

B. 180 F and 5 psia.

C. 20 psia and 400 F.

D. 0.1 psia and 50 F.

2.20. A hot water boiler heats 6400 lb/hr of water from 180 F to 220 F. How many BTU/hr of heat are required? Solve by equation and by using Table A.3.

2.21. How many tons of refrigeration are required to cool 46 GPM of milk from 80 F to 50 F? The milk has a specific heat of 0.9 BTU/lb-F and a density of 8.1 lb/gal.

2.22. A 2.5 KW electric heater in a duct is heating 1300 lb/hr of air entering at 40 F. To what temperature is the air heated?

2.23. Water enters a steam boiler at 160 F and leaves as saturated steam at 30 psig, at a flow rate of 5300 lb/hr. How much heat is the boiler supplying?

2.24. If 520 ft³ of air at atmospheric pressure at sea level is to be compressed and stored in a tank at 75 psig, what is the required volume of the tank?

2.25. An air conditioning unit takes in 15,700 CFM of outdoor air at 10 F and heats it to 120 F. How many CFM of air are leaving the unit?

2.26. What is the boiling point (saturation) temperature of water at pressures of 7.5 psia and 67.0 psia?

2.27. Is water liquid or vapor at 270°F and 50 psia?

2.28. A 24-ft.-high pipe filled with water extends from a condenser on the top floor of a building to a cooling tower on the floor above. What is the pressure exerted on the condenser in psi?

2.29. A barometer reads 705 mm Hg. What is the atmospheric pressure expressed in psi and in in. Hg?

2.30. A refrigeration unit has a cooling capacity of 327,000 BTU/hr. Express this capacity in tons of refrigeration and in KW.

2.31. A water chiller with a capacity of 150 tons of refrigeration cools 320 GPM of water entering the chiller at 52 F. At what temperature does the water leave the chiller?

CHAPTER *3*

Heating Loads

In this chapter we will discuss methods for determining the amount of heat required to keep the spaces in a building comfortable in winter. *The methods presented here will be those which are believed to be the most accurate and the most energy efficient.*

OBJECTIVES

After studying this chapter you will be able to:
1. Find *R*- and *U*-values for building components.
2. Select appropriate indoor and outdoor design conditions.
3. Calculate room and building heat transfer losses.
4. Calculate room and building infiltration and ventilation losses.
5. Determine room and building heating loads.

3.1. THE HEATING LOAD

From our own experiences we know that if the heating system in a building stops functioning in winter, the air temperature soon drops. The temperature drop occurs for two reasons: *heat transfer* from the warm inside air to the cold outside air through walls, windows, and other parts of the building envelope, and leakage of cold air through openings in the building (*infiltration*).

To counteract these heat losses, heat must be continually added to the interior of the building in order to maintain a desired air temperature. This can be shown by applying the Energy Equa-

47

tion (Chapter 2) to the air in a room or building. The energy added to the room air (E_{in}) is the heat supplied by the heating system (Q_{in}). The heat removed (E_{out}) is the heat loss (Q_{out}). The change in stored energy (E_{ch}) is the change in the room air enthalpy (H_{ch}). Substituting into Equation 2.1:

$$E_{ch} = E_{in} - E_{out}$$

we obtain

$$H_{ch} = Q_{in} - Q_{out} \qquad (3.1)$$

where

H_{ch} = change in room air enthalpy
Q_{in} = heat supplied by heating system
Q_{out} = heat losses from room air to outside

The room air temperature depends on its enthalpy. If the air enthalpy changes, its temperature changes. Since we want the room air to remain at a constant elevated temperature, the enthalpy must also be constant. Since the enthalpy does not change, $H_{ch} = 0$. Substituting into Equation 3.1, we obtain

$$0 = Q_{in} - Q_{out}$$

or

$$Q_{in} = Q_{out} \qquad (3.2)$$

Figure 3.1 illustrates the heat flow into and out of a room. Equation 3.2 tells us that if the room air enthalpy (and therefore temperature) is to be maintained at a constant desired value, *the heat supplied by the heating system must equal the heat losses from the room.* This is a valuable conclusion, because the heat losses from a building can be easily calculated and then this information can be used to determine the required capacity of the heating equipment.

The amount of heat that must be supplied to keep the building or room air at the desired temperature is called the *heating load*. The heating load must be determined because it is used in

Figure 3.1. Heat exchanges between room air and surroundings. If heat furnished to room (Q_{in}) equals heat lost from room (Q_{out}), room air temperature remains constant ($T_{ch} = 0$).

the selection of the heating equipment, piping and duct sizing, and in energy utilization studies.

Accurately determining the heating load is a fundamental step in planning a heating system. In the past, many inaccurate methods have been used to find heating loads. This is no longer acceptable because of greatly increased energy costs. Almost every state has established codes that now require precise heating and cooling load calculation methods. (Cooling load calculation methods will be discussed in Chapter 6.)

The heating load requirements for buildings result from two types of heat losses: *heat transfer* losses and *infiltration/ventilation* losses. An explanation of these will now be presented.

3.2. HEAT TRANSFER

Because building heat losses are partially a consequence of heat transfer, it is necessary to understand some basic features of this process.

It was noted previously that heat is transferred only when there is a temperature difference between two locations, and that the heat always travels from the location of higher temperature to the location of lower temperature. There are three different ways that heat transfer can occur: *conduction, convection,* and *radiation*.

Conduction *is the form of heat transfer through a body that occurs without any* *movement of the body; it is a result of molecular or electron action.*

Conduction is most familiar in heat transfer through solids—for example, when the metal body of a pot is heated on a stove the heat flows through the handle and then to your hand. Another example of conduction is heat transfer through a building wall or roof. Conduction heat transfer can also occur through liquids and gases; however, an additional form of heat transfer is more usual in fluids (convection).

Convection *is the form of heat transfer that results from gross movement of liquids or gases.*

A familiar example of convection is the air in a room heated by a unit such as a hot water convector. Heat is transferred to the air adjacent to the metal surface, increasing its temperature. This air then moves vertically upward because it is now less dense (lighter) than the surrounding cooler air. So air continually moves throughout the space (Figure 3.2). This form of convection is called *natural convection* because the fluid moves by natural gravity forces created by density differences. The less dense part of the fluid rises and the more dense (heavier) fluid drops. The rate of fluid motion created by natural convection effects is generally quite low, and therefore the resulting rate of heat transfer is small. The rate

Figure 3.3. Heat transfer by radiation from the sun to objects in a room.

of fluid motion and therefore the rate of heat transfer can be increased by using a fan for gases or a pump for liquids. This is called *forced convection.*

*Thermal **radiation** is the form of heat transfer that occurs between two separated bodies as a result of a means called electromagnetic radiation, sometimes called wave motion.*

As with all forms of heat transfer, one body must be at a higher temperature than the other. Heat transfers between the two bodies even if there is a vacuum (an absence of all matter) between them. When there is a gas between the bodies heat still transfers by radiation but usually at a lesser rate. However, the presence of an opaque solid object between the bodies will block radiation. Familiar examples of radiation are the heat our body receives when standing in front of a fire, and the heat received from the sun (Figure 3.3).

When radiation is received by a solid surface, some is *absorbed*, heating the material, and some is *reflected*. The proportion absorbed depends on the color and the roughness of the surface. Dark, rough surfaces absorb more radiant heat. Most radiation passes through transparent materials like clear glass. Color tinted glass, however, called *heat absorbing* glass, can prevent the transmission of a good part of the solar radiation.

Figure 3.2. Heat transfer by natural convection from a terminal unit (hot water convector) to room air.

3.3. RATE OF HEAT TRANSFER

The rate at which heat flows through a wall, roof, and so forth depends on three factors:

1. The temperature difference across which the heat flows
2. The area of the surface through which heat is flowing
3. The *thermal resistance* (R) of the material to heat transfer

This can be expressed by the following equation

$$Q = \frac{1}{R} \times A \times TD \qquad (3.3)$$

where

Q = heat transfer rate, BTU/hr
R = thermal resistance of material, hr-ft^2-F/ BTU
A = surface area through which heat flows, ft^2
$TD = t_H - t_L$ = temperature difference through which heat flows, from higher temperature t_H to lower temperature t_L, F

Thermal Resistance   The *thermal resistance R* of a material is its ability to resist the flow of heat through it. Equation 3.3 allows us to understand how the thermal resistance affects building heat energy losses or gains. Since R is in the denominator, high R-values mean low heat transfer (Q), and low R-values mean high heat transfer. Materials with high R-values will transfer heat at a low rate, that is, they are good thermal insulators. Building construction materi-als with a high R-value are desirable because they reduce heat losses. On the other hand, using a material with a low R-value (metal) for equipment such as a boiler is desirable because it helps to increase the rate of heat transfer from the combustion gases to the water.

The thermal resistances of various building materials are listed in Table A.4. The resistance is often written by a symbol, for example "R-6". This means $R = 6$.

EXAMPLE 3.1

A 110 ft long by 20 ft high wall is made of 4 in. common brick. The temperature on the inside surface of the wall is 65 F and on the outside surface the temperature is 25 F. What is the rate of heat transfer through the wall?

Solution.   Figure 3.4 illustrates the conditions. From Table A.4,

$$R = 0.20 \text{ hr-ft}^2\text{-F/BTU per in.} \times 4 \text{ in.}$$

$$= 0.80 \text{ hr-ft}^2\text{-F/BTU}$$

$$\text{Area of wall } A = 110 \text{ ft} \times 20 \text{ ft} = 2200 \text{ ft}^2$$
$$TD = t_H - t_L = 65 - 25 = 40 \text{ F}$$

Using Equation 3.3,

$$Q = 1/R \times A \times TD$$
$$= 1/0.80 \text{ hr-ft}^2\text{-F/BTU} \times 2200 \text{ ft}^2 \times 40 \text{ F}$$
$$= 110,000 \text{ BTU/hr}$$

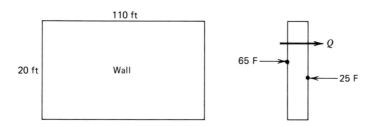

Figure 3.4. Sketch for Example 3.1.

Thermal Resistance of Surface Air Film

Heat transfer through a building material is also affected by the resistance of air next to the solid surface. In determining the rate of heat transfer in Example 3.1, the temperatures of the surfaces of the walls were used. In actual practice, it is the inside and outside air temperatures that are known. These temperatures will not be the same as that of the adjacent wall surfaces, because on each side of the wall (or roof) there is a thin film of somewhat stagnant air. This film has a thermal resistance, just as the wall itself does.

The resistance of the air film depends on the spatial orientation of the surface (vertical, horizontal, or on a slope), and on the air velocity near the surface. It is usually assumed that the air is not moving indoors and is at 15 MPH outdoors in winter. Table A.5 lists thermal resistances of surface air films. Example 3.2 illustrates the use of Table A.5.

EXAMPLE 3.2

A wall of a supermarket measures 80 ft by 18 ft. The temperature of the air in the store is 70 F and the inside surface of the wall is 60 F. What is the heat loss through the wall?

Solution. The resistance is the inside air film on a vertical surface (Figure 3.5). From Table A.5, $R = 0.68$.

$$A = 80 \times 18 = 1440 \text{ ft}^2, \text{ TD} = 70 - 60 = 10 \text{ F}$$

Figure 3.5. Sketch for Example 3.2.

Using Equation 3.3,

$$Q = 1/R \times A \times \text{TD}$$
$$= 1/0.68 \times 1440 \times 10$$
$$= 21,200 \text{ BTU/hr}$$

Conductance and Conductivity

Besides thermal resistance, *conductance* and *conductivity* are terms which are used to describe a material's ability to transfer heat.

The *thermal conductance* (C) of a material is the reciprocal of its resistance

$$C = 1/R \qquad (3.4)$$

C is measured in units of BTU/hr-ft^2-F in the U.S. system. The thermal conductance may be thought of as the ability of a material to transfer heat, the opposite meaning of resistance. The thermal conductance of the air film adjacent to a surface is also called the *film coefficient*.

The *thermal conductivity* (k) of a material is defined as its conductance per unit of thickness. The units used for conductivity are usually BTU/hr-ft^2-F per inch of thickness. Its relationship to conductance is

$$C = \frac{k}{L} \qquad (3.5)$$

where

C = conductance, BTU/hr-ft^2-F
k = conductivity, BTU/hr-ft^2-F per in. of thickness
L = thickness of material, in.

EXAMPLE 3.3

A roof has 4 in. of glass fiber insulation with a thermal conductivity $k = 0.24$ BTU/hr-ft^2-F per in. thickness. What is the thermal resistance of the insulation?

Solution. From Equation 3.5, the conductance is

$$C = \frac{k}{L}$$

$$= \frac{0.24}{4}$$

$$= 0.06 \; \frac{\text{BTU}}{\text{hr-ft}^2\text{-F}}$$

Using Equation 3.4, solving for R,

$$R = \frac{1}{C}$$

$$= \frac{1}{0.06}$$

$$= 16.7 \; \frac{\text{hr-ft}^2\text{-F}}{\text{BTU}}$$

It is not necessary to memorize the definitions just described. The concept of thermal resistance is the important one to understand, as it is being used in all new building standards and codes. Remember that for a material with high thermal resistance, heat transfer will be low, and for a material with a low resistance, heat transfer will be high.

EXAMPLE 3.4

Archie Tech has a choice of two materials for the walls of a bubble gum warehouse he is building, either 8 in. common brick or 8 in. poured concrete, with sand and gravel aggregate. He is concerned about keeping the heating bills low. He calls Art Warmer, an HVAC consultant, to ask his opinion. Which material should Art recommend?

Solution. From Table A.4
   common brick R = 0.20 per in. $\times$ 8 in. = 1.60
   concrete R = 0.08 per in. $\times$ 8 in. = 0.64

The brick, with the higher thermal resistance, will reduce heat loss. The overall (total) resistance will still be too low, however; insulation should be added to the wall.

3.4. OVERALL THERMAL RESISTANCE

The heat transfer through the walls, roof, floor, and other elements of a building is through the air film on one side, through the solid materials, and then through the air film on the other side. Furthermore, the element is often made up of layers of different materials. The *overall thermal resistance* of the combination can be found very simply by adding the individual thermal resistances as follows:

$$R_0 = R_1 + R_2 + R_3 + \text{etc.} \qquad (3.6)$$

where

$$R_0 = \text{overall thermal resistance}$$
$$R_1, \; R_2, \text{etc.} = \text{individual thermal resistance of}$$
$$\text{each component, including}$$
$$\text{air films}$$

Once the overall resistance R_0 is known, Equation 3.3 can be used to find the heat transfer, as illustrated in Example 3.5.

EXAMPLE 3.5

The exterior wall of a building is constructed of 8 in. sand and gravel aggregate concrete (not over dried), R-5 insulation, and 1/2 in. gypsum board. The wall is 72 ft long by 16 ft high. The indoor and outdoor temperatures are 70 F and −10 F. What is the heat transfer through the wall?

Solution. The wall construction is illustrated in Figure 3.6. All of the resistances are found in Tables A.4 and A.5. The thermal resistances of

Figure 3.6. Sketch for Example 3.5.

3.5. OVERALL HEAT TRANSFER COEFFICIENT

For each application, the designer can calculate the overall thermal resistance for each part of the building structure through which heat flows, but fortunately, these calculations have already been made for many different combinations of building materials. However, most tables do not list the results as overall resistance, but as overall conductance, called the *overall heat transfer coefficient* (*U*), BTU/hr-ft²-F. The relationship between R_0 and U is

$$U = \frac{1}{R_0} \qquad (3.7)$$

In terms of *U*, the heat transfer equation becomes

$$Q = U \times A \times \text{TD} \qquad (3.8)$$

where

Q = heat transfer rate, BTU/hr
U = overall heat transfer coefficient, BTU/hr-ft²-F
A = surface area through which heat flows, ft²
TD = temperature difference, F

U-values may be read directly from Tables A.6, A.7, and A.8. The following comments should be helpful in using these tables correctly:

1. Table A.6 shows the cross sections of some typical building components that you will encounter. Use it to familiarize yourself with various construction arrangements.

2. Table A.7 lists *U*-values for walls, roofs, floors/ceilings, partitions, and doors (except glass doors). If the *U*-value cannot be determined directly, it may be calculated from individual *R*-values, as explained earlier, or it may be found in HVAC engineering handbooks containing more complete tables.

the air films on the inside and outside surfaces must also be considered.

The overall thermal resistance R_0 is found by adding the individual resistances (See Equation 3.6).

Wall Item	R
Inside air film	0.68
Gypsum board	0.45
Insulation	5.0
Concrete ($R = 0.08/\text{in.} \times 8$ in.)	0.64
Outside air film	0.17
Overall resistance R_0 =	6.94

The wall area is

$$A = 72 \text{ ft} \times 16 \text{ ft} = 1152 \text{ ft}^2$$

The temperature difference is

$$\text{TD} = 70 - (-10) = 80 \text{ F}$$

Using Equation 3.3 and the overall resistance R_0, the heat transfer loss is

$$Q = \frac{1}{R_0} \times A \times \text{TD}$$

$$= \frac{1}{6.94} \times 1152 \times 80$$

$$= 13,280 \text{ BTU/hr}$$

The insulated wall and roof sections listed in Table A.7 have been chosen to illustrate those arrangements that are likely to satisfy many state energy code requirements. They also reflect some of the popular construction methods used currently.

3. Table A.8, which lists *U*-values for glass, should be used for windows and glass doors. Note that the *U*-values for glass often need adjustments, as indicated in the table. For example, the type of sash (wood, metal) affects the *U*-value. Similarly, the addition of a storm window to single glass changes the *U*-value.

   Insulated glass refers to double (two) or triple (three) panes of glass separated by air spaces in the window assembly.

4. Note that *U*-values may differ in summer and winter.

Examples 3.6, 3.7, and 3.8 show how to use Equation 3.8 and the *U*-value tables to calculate heat transfer through building components.

EXAMPLE 3.6

Find the *U*-value for the wall described in Example 3.5, using Table A.7. Compare it with the value that would be found from Example 3.5, using Equations 3.6 and 3.7.

Solution. From Table A.7, $U = 0.14$. From Example 3.5, $R_0 = 6.94$. Using Equation 3.7,

$$U = \frac{1}{R_0} = \frac{1}{6.94} = 0.14$$

The results agree.

EXAMPLE 3.7

A building 120 ft long by 40 ft wide has a flat roof constructed of 8 in. lightweight aggregate

concrete, with a finished ceiling. The inside temperature is 65 F and the outdoor temperature is 5 F. What is the heat transfer loss through the roof?

Solution. From Table A.7, $U = 0.09$ BTU/hr-ft²-F. Using Equation 3.8,

$$Q = U \times A \times TD$$

$$= 0.09 \; \frac{BTU}{hr\text{-}ft^2\text{-}F} \times 4800 \; ft^2 \times 60 \; F$$

$$= 25,900 \; BTU/hr$$

EXAMPLE 3.8

A frame wall of the bedroom of a house has the following specifications:

Wall: 12 ft by 8 ft, wood siding, wood sheathing, 2 in. of insulation with R-7 value, and inside finish

Window: 3 ft by 4 ft 6 in., single glass, metal sash.

The room temperature is 68 F and the outdoor temperature is 2 F. What is the heat transfer loss through the wall and window?

Solution. Figure 3.7 illustrates the wall. There is heat transfer through the frame part of the wall and through the window. The *U*-values from

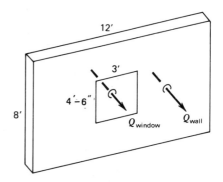

Figure 3.7. Sketch for Example 3.8.

Table A.7 and the areas are:

$$\text{Window } U = 1.10 \text{ BTU/hr-ft}^2\text{-F}$$
$$A = 3 \times 4.5 = 13.5 \text{ ft}^2$$
$$\text{Wall} \quad U = 0.09 \text{ BTU/hr-ft}^2\text{-F}$$
$$\text{Gross } A = 12 \times 8 = 96 \text{ ft}^2$$
$$\text{Net } \quad A = 96 - 13.5 = 82.5 \text{ ft}^2$$

The heat transfer is:

Wall

$$Q = 0.09 \times 82.5 \times 66 = \quad 490 \text{ BTU/hr}$$

Window

$$Q = 1.10 \times 13.5 \times 66 = \underline{\quad 980 \text{ BTU/hr}}$$
$$\text{total } Q = 1470 \text{ BTU/hr}$$

3.6. HEAT TRANSFER LOSSES: BASEMENT WALLS AND FLOORS

Equation 3.8 is also used to calculate the heat losses through basement walls and floors. However, the fact that the basement floor and part or all of the wall is underground (below grade) complicates the use of the equation.

If any part of the basement is above ground the U-values from Table A.7 are used. The inside and outside design air temperatures are also used to find the TD (see Section 3.9).

However, for the part of the structure which is below grade, the U-values and TD used in Equation 3.8 will be different. This is due to the effect of the surrounding ground on the thermal resistance and the heat flow path. Table 3.1 lists recommended U-values for below grade basement walls and floors.

To find the TD for the below grade wall or floor, the outside winter design temperature is taken to be the deep ground temperature value. This temperature varies between 40 F and 60 F in cold climates in the continental United States. When not known accurately, many designers assume it is 50 F.

Example 3.9 illustrates a calculation of heat losses from a below grade basement.

TABLE 3.1. OVERALL HEAT TRANSFER COEFFICIENT U (BTU / hr-ft^2-F) FOR BELOW GRADE BASEMENT WALLS AND FLOORS

Item	U
Floor	0.10
Wall (no insulation)[a]	0.20
Wall (1 in. insulation)	0.10

[a]A minimum of 1 in. insulation for the first 2 ft of wall depth is recommended even for uninsulated basement walls, and required now in most states.

EXAMPLE 3.9

The recreation room of a basement has a floor area of 220 ft^2 and a wall area (uninsulated) below ground of 400 ft^2. The room temperature is 70 F, and the ground temperature 50 F. What is the heat loss from the room?

Solution. Using recommended U-values,

Floor	$= 0.10 \times 220 \times 20 = \quad 440 \text{ BTU/hr}$
Wall	$Q = 0.20 \times 400 \times 20 = \underline{1600 \text{ BTU/hr}}$
	$\text{total } Q = 2040 \text{ BTU/hr}$

Remember that if part of the basement wall is above ground and part is below, the heat transfer losses from each part should be calculated separately, using the appropriate U, A, and TD values. The separate heat losses should then be added to find the total wall heat loss.

Basement Inside Temperature A number of possible conditions may exist in a basement.

1. *Basement heated (with terminal units).* The basement heat losses should be calculated using the basement inside design temperature. If the basement is partitioned, a separate calculation for each area is required.

2. *Unheated basement.* The temperature in an unheated basement (with no heat sources) will be between the design inside and design

outside temperatures. No heat loss calculation from the basement should be made.

3. *Basement heated by heat source equipment.* This situation exists when the furnace and ducts or boiler and piping are located in the basement. The following guidelines are recommended:

(a) For any area that the heat source is in or substantial hot ducts or piping pass through, assume the inside temperature is equal to that of the rest of the building, and calculate the resulting heat transfer loss.

(b) For other partitioned off areas assume an unheated basement (as in item 2).

In items 2 and 3b there will still be a heat loss from the floor above which should be added to the heat losses from those rooms. If the specific basement conditions are not known, an estimated unheated basement temperature of 50 F should be used. (A temperature considerably below this should not be permitted anyway, because of the possibility of freezing water in piping.)

3.7. HEAT TRANSFER LOSSES: FLOOR ON GROUND AND FLOOR OVER CRAWL SPACE

Special calculations also apply for the heat transfer which occurs through a concrete floor slab on grade and through a floor with a crawl space below.

Floor Over Crawl Space   If the crawl space is vented during the heating season (to prevent moisture condensation), the crawl space air temperature will equal the outside air design temperature. The heat loss through the floor area should be calculated using Equation 3.8. In cold climates, the floor should be insulated; an overall R-20 value is a typical requirement in state energy codes.

If the crawl space is used for the warm air ductwork, the heat loss calculation is more complicated. Consult the *ASHRAE Handbook* or energy codes.

Floor Slab On Grade   When a floor is on the ground, the heat loss is greatest near the outside edges (perimeter) of the building, and is proportional to the length of these edges, rather than the area of the floor (Figure 3.8). This is the only case where heat transfer is not calculated using Equation 3.8. The following equation is used instead:

$$Q = E \times L \times \text{TD} \qquad (3.9)$$

where

Q = heat transfer loss through floor on grade, BTU/hr

E = edge heat loss coefficient, BTU/hr-F per ft of edge length

L = total length of outside (exposed) edges of floor, ft

TD = design temperature difference between inside and outside air, F

Figure 3.8. Heat loss through floor of building without basement.

TABLE 3.2. EDGE HEAT LOSS COEFFICIENT, *E*, FOR FLOOR SLAB ON GRADE
(BTU / hr-F per ft of edge)

Wall Construction	Edge Insulation	Degree Days		
		3000	5400	7400
8 in. block with face brick	None	0.62	0.68	0.72
	R-5	0.48	0.50	0.56
4 in. block with face brick	None	0.80	0.84	0.93
	R-5	0.47	0.49	0.54
Metal stud with stucco	None	1.15	1.20	1.34
	R-5	0.51	0.53	0.58
Poured concrete, heated floor	None	1.84	2.12	2.73
	R-5	0.64	0.72	0.90

Table 3.2 lists values of *E* for various wall constructions, both with and without edge insulation. Insulation is often required by energy codes. The last entry applies to a heated floor slab, which is recommended in more severe winter climates. The *degree days*, a measure of the severity and length of the heating season, are higher in colder climates. The degree days for a particular locality may be found in Table A.9. Degree days are discussed in detail in Chapter 15.

EXAMPLE 3.10

A 60 ft by 30 ft (in plan) garage built with a concrete floor slab on grade is maintained at 65 F. The outdoor air temperature is 4 F. Degree days are 5200. The floor edge has R-5 insulation. The walls are 8 in. block with face brick. What is the heat loss through the floor?

Solution. Use Equation 3.9. First find TD, *E*, and *L*. TD = 65 − 4 = 61 F. From Table 3.2,

E = 0.50 for 5200 degree days. Edge length *L* = 2(60 + 20) = 160 ft.

$$Q = E \times L \times TD$$
$$= 0.50 \times 160 \times 61$$
$$= 4880 \text{ BTU/hr}$$

The edge loss method (Equation 3.9) is recommended for buildings with small floor slab areas. For buildings with large floor slab areas, the edge loss method is recommended for perimeter rooms. For the interior areas Equation 3.8 should be used, with the *U* and TD values for basements.

3.8. INFILTRATION AND VENTILATION HEAT LOSS

In addition to the heat required to offset heat transfer losses in winter, heat is also required to offset the effects from any cold air that may enter a building. The two means by which cold air may enter the building are called *infiltration* and *ventilation*. The resulting amounts of heat required are called the *infiltration heating load* and the *ventilation heating load*.

Sensible Heat Loss Effect of Infiltration Air
Infiltration occurs when outdoor air enters through building openings, due to wind pressure. The openings of most concern to us are cracks around window sashes and door edges, and open doors. Infiltration air entering a space in winter would lower the room air temperature. Therefore, heat must be furnished to the room to maintain its design air temperature. This is expressed by Equation 2.12, the Sensible Heat Equation (Section 2.15)

$$Q = m \times c \times TC \qquad (2.12)$$

To determine the sensible heat loss effect of the infiltration air, it is more convenient for us to use units of CFM (ft³/min) for the air flow rate instead of mass units of lb/hr as in Equation 2.12. After converting units (using the appropriate specific heat of air), the following equation results:

$$Q_s = 1.1 \times CFM \times TC \qquad (3.10)$$

where

Q_s = sensible heat required for infiltration or ventilation air, BTU/hr
CFM = air infiltration or ventilation rate, ft³/min
TC = temperature change between inside and outdoor air, F

Latent Heat Loss Effect of Infiltration Air
Since infiltration air is often less humid than the room air, the room air humidity may fall to an unacceptable level for comfort. If the room air humidity is to be maintained, water vapor must be added. The addition of this moisture requires heat (latent heat of vaporization of water). This is expressed by the following equation:

$$Q_L = 0.68 \times CFM \times (W_i' - W_o') \quad (3.11)$$

where

Q_L = latent heat required for infiltration or ventilation air, BTU/hr
CFM = air infiltration or ventilation rate, ft³/min.
W_i', W_o' = higher (inside) and lower (outdoor) humidity ratio in grains water/lb dry air (gr w/lb d.a.)

To sum up, Equations 3.10 and 3.11 are used to find the room air sensible and latent heat losses resulting from infiltration air. The sensible heat loss should always be calculated. If the lower room air humidity resulting from infiltration is acceptable, then the latent heat loss effect may be neglected.

TABLE 3.3. RECOMMENDED MAXIMUM ALLOWED DESIGN INFILTRATION RATES THROUGH EXTERIOR WINDOWS AND DOORS

Item	Infiltration Rate
Windows	0.75 CFM / ft crack
Doors	1.0 CFM / ft crack

For the interested student, Equations 3.10 and 3.11 are derived in Chapter 7. The humidity ratios (W') for Equation 3.11 can be read from the psychometric chart, also introduced in Chapter 7.

There are two methods used to estimate the CFM of infiltration air: the *crack method* and the *air change method*.

Crack Method The crack method assumes that a reasonably accurate estimate of the rate of air infiltration per foot of crack opening can be measured or established. Energy codes list maximum permissible infiltration rates for new construction or renovation upgrading. Table 3.3 lists typical energy code infiltration rates, based on a 25 MPH wind.

The crack lengths and areas should be determined from building architectural plans or field measurements. Example 3.11 illustrates the use of Table 3.3 and Equation 3.10 to find the infiltration loss.

EXAMPLE 3.11

The entrance door to a house is to be weather stripped to meet recommended energy conservation infiltration standards. The door is 7 ft high × 3 ft wide. The indoor and outdoor temperatures are 70 F and 10 F. What is the sensible heat loss due to infiltration?

Solution. From Table 3.3 the infiltration rate after upgrading is 1.0 CFM/ft of crack. The crack length is 2(7) + 2(3) = 20 ft (Figure 3.9).

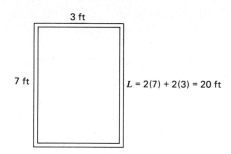

Figure 3.9. Sketch for Example 3.11.

The total infiltration rate around the edges is 1.0 CFM/ft × 20 ft = 20 CFM. Using Equation 3.10, the infiltration heat loss is

$$Q_s = 1.1 \times CFM \times TD$$
$$= 1.1 \times 20 \times (70 - 10)$$
$$= 1200 \text{ BTU/hr}$$

The quality of installation and the maintenance of windows and doors greatly affect the resultant crack infiltration. Poorly fitted windows may have up to five times the sash leakage shown in Table 3.3.

Corner Room Infiltration When the infiltration rate is calculated for a room with two adjacent exposed walls (a corner room) with door or window openings on both sides, we assume that infiltration air comes through cracks on one side only, since the wind can only come from one direction at any given time. The wind changes direction of course, but the infiltration effects cannot be additive, since they occur at different times.

If the wind comes obliquely (toward the corner), the projected crack lengths for each side are less; the overall effect is the same as if the wind came directly from one side only (using its actual crack lengths). If there are different types or sizes of openings on each side, the side that has the *greater* CFM should be used for the calculation.

TABLE 3.4. INFILTRATION RATES FOR FREQUENT DOOR USAGE

Type	ft³ per Person
Swinging door, no vestibule	900
Swinging door, vestibule	550
Revolving door	60

The procedure for combining the infiltration rates of individual rooms on opposite walls, in order to find the total building infiltration rate, will be discussed in Section 3.11.

Door Usage For buildings that have frequent door usage (e.g., department stores), the infiltration that results from door opening should be included. The rate of door usage (number of people per minute) is first determined, with the advice of the architect or owner. Some average infiltration rates are shown in Table 3.4.

For doors that are left indefinitely open, special means may be used to try to offset infiltration. *Unit heaters*, which can blow warm air directly at the opening, and *air curtains*, which direct a vertical warm air barrier across the opening, are two such methods. However, it is difficult to determine the effect on the building heating load of these methods.

Additional air infiltration may occur through a porous wall. If the walls have significant porosity, sealant coatings or other coverings may be applied to them.

In high rise buildings, a thermal stack effect may increase infiltration through existing cracks. This occurs when the warmer inside air, which rises through the building and exits out through cracks on upper stories, is replaced by colder outside air entering through cracks on lower floors. Consult the *ASHRAE Handbook* for more information.

Much publicity is being given today to reducing infiltration in existing buildings by use of weather stripping, and the sealing of cracks around frames, sillplates, wall penetrations, and other openings. There are many excellent publications available from governmental agencies and utility companies on this subject.

Air Change Method The air change method assumes that the number of air changes per hour (one *air change* is equal to one room volume) in a room due to infiltration air can be determined. This determination is based on past experience and testing. Table 3.5 lists typical values for residential construction.

In modern "tight" residential construction, sealing techniques may reduce the infiltration rate to as little as 0.2 to 0.5 air changes per hour. There is concern, however, that this reduction in natural ventilation may increase indoor pollution levels to an unhealthy level.

Using the definition of an air change, Equation 3.12 can be used to find the air infiltration rate in CFM.

$$\text{CFM} = A.C. \times \frac{V}{60} \qquad (3.12)$$

where

CFM = air infiltration rate to room, CFM
$A.C.$ = number of air changes per hour for room (Table 3.5)
V = room volume, ft^3

Example 3.12 should help to clarify the meaning and use of the air change method.

EXAMPLE 3.12

A 20 ft long × 10 ft wide × 8 ft high room in a house has a window on one side, with weather stripping. Find the infiltration rate in CFM.

Solution. From Table 3.5, $A.C. = 0.7$, $V = 20 \times 10 \times 8 = 1600$ ft^3.

$$\begin{aligned}
\text{CFM} &= A.C. \times \frac{V}{60} \\
&= 0.7 \times 1699/60 \\
&= 17 \text{ CFM}
\end{aligned}$$

Crack Method versus Air Change Method The obvious question arises as to which of these two methods should be used. There is no unqualified answer, but the following suggestions may be helpful:

1. The air change method is used primarily in residential construction estimates, but there is no reason why the crack method cannot be used if reliable data are available.
2. The crack method is generally used in nonresidential construction. Reliable data from manufacturers and quality control of installa-

TABLE 3.5. AIR CHANGES PER HOUR OCCURRING UNDER AVERAGE CONDITIONS IN RESIDENCES, EXCLUSIVE OF AIR PROVIDED FOR VENTILATION

Kind of Room	Single Glass, No Weather Strip	Storm Sash or Weather Stripped
No windows or exterior doors	0.5	0.3
Windows or exterior doors on one side	1	0.7
Windows or exterior doors on two sides	1.5	1
Windows or exterior doors on three sides	2	1.3
Entrance halls	2	1.3

tion and maintenance may provide good estimates using this method. However, it is often difficult to estimate leakage rates in older buildings because the condition of the windows is not known; but since heating load calculations for existing buildings are usually being made when upgrading for energy conservation, this should not be a problem.

Ventilation (Outside Air) Load Some outside air is usually brought into nonresidential buildings through the mechanical ventilation equipment (air handling units) in order to maintain the indoor air quality. The outside ventilation air will be an additional part of the building heating load, since the entering air is at the outdoor temperature and humidity.

Equations 3.10 and 3.11 are also used to find the ventilation heating load. However, the ventilation air is heated (and humidified, if required) in the air conditioning equipment, *before* it enters the room. Therefore, it is part of the total building heating load, but *not* the individual room heating loads. The procedures for determining the appropriate quantity of outside ventilation air are explained in Chapter 6.

Mechanical ventilation systems for buildings are often designed and installed so that fans create a slightly positive air pressure in the building. This will reduce or even prevent infiltration. It is impossible to give a single figure for the amount, since it depends on the circumstances. However, some designers limit the reduction in infiltration to 50%; in some cases, when it is felt that the building is relatively tight and pressurized, no allowance for infiltration is made; only the outside air ventilation load is included. A separate word of caution on pressurizing buildings: it is not uncommon to find that overpressurization results in doors that require great force to open or close.

Some non-residential buildings have sealed windows (no openable part). In this case, of course, crack infiltration is limited to exterior doors only.

Air distributing systems in residences often use recirculated air only, so there is no ventilation load component. Infiltration generally provides adequate fresh air. However, in modern "tight" residences, there is concern that there may be inadequate natural infiltration, resulting in long-term health problems from indoor air pollutants. Introduction of outside ventilation air should be considered in such cases.

Example 3.13 illustrates the calculation of ventilation loads.

EXAMPLE 3.13

A building with sealed windows is maintained at 72 F, with an outdoor temperature of −5 F. The mechanical ventilation system introduces 5000 CFM of outside air. What is the additional sensible heating requirement from this effect?

Solution. The inside humidity conditions are not specified as being fixed, so only the sensible heat of the ventilation air is calculated. Using Equation 3.10

$$Q_s = 1.1 \times CFM \times TC$$
$$= 1.1 \times 5000 \times 77 = 423,500 \text{ BTU/hr}$$

3.9. DESIGN CONDITIONS

Heating and cooling load calculations are generally based on indoor and outdoor *design conditions* (of temperature and humidity). Indoor design conditions are defined as those combinations of temperature and humidity which are comfortable (within the comfort zones in Table 1.1). Outdoor winter design conditions are based on relatively low expected temperatures in an average winter, using weather records. If outside air latent loads are to be calculated, it is usually assumed that the outdoor air moisture content in winter is zero.

Table A.9 lists recommended outdoor design conditions in summer and winter for various cities in the United States and other countries. The value in the column entitled 97.5% DB (dry bulb) is the outdoor design air temperature most commonly used for heating load calculations in winter. The 97.5% figure means that in an average winter, from December through February, the outdoor temperature will be above the listed design value 97.5% of the time. In other words, the outdoor temperature will be below the design value 2.5% of the time. In theory, when the outdoor temperature drops below the design temperature, the indoor temperature should also drop below the indoor design temperature because the heating system would be inadequate. Fortunately, this rarely occurs because the heating system usually has excess capacity, and there is heat storage in the building, as well as an allowable temperature differential associated with the heating system controls (see Chapter 14).

There are also 99 and 95% outdoor design temperature values (not shown) for more and less critical applications. The advantage of using the 97.5% value instead of lower outdoor design temperatures (often used by designers in the past) is that it reduces oversizing of equipment, resulting in less costly systems, the equipment is run more steadily, increasing its life span, and energy use may be reduced.

usually heavily clothed and active. If necessary, the heating system should have enough surplus capacity to raise the temperature slightly.

Unheated Space Temperature   Unheated rooms or spaces between a heated room and the outdoors will have a temperature lower than the heated room. The heat loss from the heated room through the separating partition should be calculated.

Some designers assume that the temperature of the unheated space is halfway between indoor and outdoor design conditions. If the unheated space has a large exposed glass area, it is better to assume the space is at outdoor temperature. If an unheated space is totally surrounded by heated spaces, it can be assumed to be at indoor design conditions, so the heat transfer can be neglected.

3.10. ROOM HEAT LOSS AND ROOM LOAD

The *room heat loss is the sum of all the room heat transfer and infiltration losses*.

Example 3.15 illustrates the procedure for calculating the room heat loss.

EXAMPLE 3.14

Plans are being prepared for the Big Bargain Department Store in Chicago, Illinois. What indoor and outdoor winter temperatures should be used for designing the heating system?

Solution.   From Tables 1.1 and A.9, the temperatures selected are 68 F and 2 F (using the 97.5% value).

In Example 3.14, the indoor design temperature of 68 F, or the low end of the comfort zone recommended in Table 1.1, is probably adequate because the occupants in a department store are

---

EXAMPLE 3.15

The office shown in Figure 3.10 is in a one-story building in Des Moines, Iowa. The building has a heated basement. Find the room heat loss at design conditions. Construction is as follows

Wall:	8 in. concrete, *R*-5 insulation, 1/2 in. wallboard
Window:	5 ft × 4 ft pivoted type with storm window
Roof:	2.5 in. wood deck, *R*-5.5 insulation, finished ceiling

Ceiling height: 9 ft

Figure 3.10. Floor plan for Example 3.15.

Solution. The heat losses, using the appropriate values from the design tables, are calculated below. An inside design temperature of 72 F is selected. Table A.9 lists an outdoor design temperature of −5 F. The infiltration rate of 0.75 CFM/ft is from Table 3.3.

		U	A	TD	Q
		BTU/hr-ft²-F	ft²	F	BTU/hr
Wall	Q	= 0.14	×106	× 77	= 1140
Window	Q	= 0.50	× 20	× 77	= 770
Roof	Q	= 0.10	×168	× 77	= 1290
Total heat transfer loss					− 1140
Infiltration $Q = 1.1 \times 0.75 \dfrac{\text{CFM}}{\text{ft}}$			× 18 ft	× 77	= 1140
Total room heat loss					= 4340

In calculating design heat losses, usually no credit is taken for heat gains. A solar heat gain could not be guaranteed, of course. For buildings that have steady internal gains, however, there is no reason why they should not be considered in calculating net heat losses.

Room Heating Load If Equation 3.2 is applied to the room, we find that the *heat which must be added to the room is equal to the room heat loss*, if the room is to be maintained at design conditions. The room heat required is called the *room heating load*.

3.11. THE BUILDING HEATING LOAD

In addition to calculating the individual room heating loads, the building heating load should also be determined.

The *building heating load* is the amount of heat required for the building at outdoor design conditions.

The building heating load is the sum of the building heat transfer losses, infiltration losses, and ventilation load, if any.

Building Heat Transfer Loss This is the sum of the heat transfer losses to the outdoors through the exposed walls, windows, doors, floor, and roof of the building. (It is also the sum of the room heat transfer losses, but it is preferable to calculate it directly, measuring total building areas.)

Building Infiltration Loss Although the building generally has more than one side with openings, it should be understood that infiltration air cannot enter through all sides at the same time. (A similar situation occurs with corner rooms.)

The procedures for determining the infiltration CFM when using the crack method and the air change method are different.

For the air change method, the building air infiltration is taken as *one-half the sum of the room air infiltrations*.

To estimate the building infiltration heat loss when using the crack method, two separate values are first found.

1. Find the infiltration CFM on each side of the building. Select the greatest of these values. Assume here that the greatest building infiltration occurs on the side of the building from which the wind is blowing.

2. Take one-half the sum of the infiltration CFM on all sides of the building. Assume here that

the amount of building infiltration is affected not only by the positive wind pressure, but also by the negative wind pressures on the other sides. To express this in another way, exfiltrating air on one side may increase the infiltration on another side. This occurs if there are relatively free passages for air to flow in the building. (If this is not the case, it is more difficult to estimate the infiltration, and beyond the scope of our discussion here.)

The greater value from 1 and 2 should be chosen as the estimated building infiltration CFM loss.

Example 3.16 illustrates how the building air infiltration is found.

EXAMPLE 3.16

Find the heating load of the building shown in Figure 3.11. The sum of the room heat transfer losses has already been determined as 370,000 BTU/hr, and infiltration losses are as shown.

Solution. The largest value of the infiltration heat loss on any one side of the building is seen to be 70,000 BTU/hr. However, comparing this with one-half the sum of the infiltration losses on all sides:

$$Q = \frac{160,000 \text{ BTU/hr}}{2} = 80,000 \text{ BTU/hr}$$

Figure 3.11. Sketch for Example 3.16.

This last figure is larger and should be used. The building heating load is, therefore,

$$Q = 370,000 + 80,000 = 450,000 \text{ BTU/hr}$$

If the infiltration loss from all sides had been used, the calculated building load would have been

$$Q = 370,000 + 160,000 = 530,000 \text{ BTU/hr}$$

This is significantly greater than the true heating requirements and would result in using oversized equipment.

3.12. OTHER HEAT LOSSES

Besides the two fundamental types of building heat losses, heat transfer and infiltration/ventilation, there are often other building system heat losses peculiar to an installation. Some of these losses are continual building losses, and some apply only to sizing equipment.

Duct Heat Loss   In a warm air heating system, if the ducts travel through unheated spaces (e.g., attics, basements, crawl spaces), there will be heat transfer from the air in the duct to the cooler surrounding spaces. The heat loss should be included as part of the building system heating load. Table 3.6 shows some recommended values, expressed as percents, to be added to the calculated building load. These values are suitable for residences and small commercial installations.

For larger installations, the heat transfer loss should be calculated. Ducts passing through unheated spaces should be insulated so that the *U*-value is no greater than 0.25.

Duct Leakage   There may be air leakage from supply ducts, which would constitute a heat loss if the leakage is into unconditioned spaces. The loss is not generally considered in residential design, but can be as much as 5–10% of the load in larger systems (see Chapter 6).

Piping Loss In boiler systems, the heat that is lost from hot piping running through unconditioned spaces should be calculated and added to the building load, if it is significant. However, piping is usually well insulated and heat losses are negligible. A related subject, the piping factor, will be discussed shortly.

Pickup Loss When a building is intermittently heated, or when regular nighttime temperature setback is practiced, if the equipment capacity just equals the building load the heating equipment may not be able to bring the rooms up to design temperature quickly enough. In this situation, it may be desirable to oversize the central heating equipment. Some designers allow an extra 10% for intermittently heated buildings, and 40% for a 10 F night setback in residential equipment. Of course, this additional capacity is gained at a sacrifice to cost.

In large buildings, an equipment allowance for setback is not standard practice. Many strategies are available for bringing the space temperature up to design in sufficient time. For

instance, if the weather forecast is for cold weather, the system may be started earlier. Furthermore, in large buildings multiple boilers and excess standby capacity are common.

It is normal practice in the boiler industry to specify a combined *piping and pickup factor* that can be used when sizing a heating boiler. This factor combines the piping and pickup losses described earlier. The standard piping and pickup factor varies from 15–25%, depending on boiler type and size (see Chapter 4). The actual piping and pickup losses may be greater or less than these values, as suggested here; if there are doubts, they should be calculated.

Since duct and piping losses are constant, they are added to the building load (and thus the equipment load). The pickup loss is a load on the equipment only. It affects the size of the equipment, but not the amount of heat delivered to the building.

Example 3.17 illustrates the use of Table 3.6.

EXAMPLE 3.17

A residence with a floor area of 1200 ft^2 has a design building heat loss of 36,000 BTU/hr. The

TABLE 3.6. DUCT HEAT GAIN AND LOSS ALLOWANCE

Duct Description	Cooling — Heat Gain, %	Heating — Heat Loss, %			
		Design Heat Loss BTU/(hr · ft^2)	Below 0	0 to +15	Above +15
Located in the attic or the crawl space and with duct insulation of R-5 or more	10	Less than 35	20	15	10
		35 and over	15	10	5
Unconditioned basement	5	Less than 35	20	15	10
		35 and over	10	10	5

outside winter design temperature is 10 F. The ducts for the warm air system are located in the attic and have R-5 insulation. What is the total building heating load? A clock thermostat is programmed for 10 F night setback. What is the required warm air furnace capacity?

Solution. Table 3.6 is used. Design heat loss per ft^2 = 36,000/1200 = 30 BTU/hr-ft^2. From Table 3.6, duct heat loss allowance = 15%.

Net heating load	= 36,000
Duct heat loss (0.15 × 36000)	= 5,400
Total building load	= 41,400 BTU/hr
Allowance for setback	
(0.40 × 41,400)	= 16,560
Required furnace capacity	= 57,960 BTU/hr

Domestic Hot Water Heating The heat output of a boiler is sometimes used to heat domestic hot water (hot water for kitchens, baths, and so forth) as well as for space heating. This load should then be included when sizing the boiler. Procedures for determining hot water use loads may be found in the *ASHRAE Handbooks*.

3.13. SUMMARY OF HEATING LOAD PROCEDURES

The following step-by-step instructions summarize how to calculate heating loads. The individual room data and results are recorded on a room heating load calculations form (Figure 3.12). Figure 3.13 is a building heating load calculations form. The forms are suitable for both residential and commercial estimates. Examples of completed forms are shown in Figures 3.14 and 3.15.

Room Heating Loads

1. Select appropriate indoor and outdoor design temperatures (Tables 1.1 and A.9).

2. Obtain dimensions from architectural plans. For each room, find areas of exposed windows, walls, and so forth through which there will be heat transfer.

3. Select appropriate overall heat transfer coefficients (U-values) from Tables A.7 and A.8, or calculate from R-values if necessary (Tables A.4 and A.5).

4. Calculate the heat transfer losses through all exposed surfaces in the room (Equation 3.8). Total these to find the room heat transfer loss.

  (a) For basement floors and walls below grade, use the U-values from Table 3.1 and outdoor ground temperature.

  (b) For floor slabs on grade, use Equation 3.9 and Table 3.2.

5. Find the room infiltration heat loss. Choose either the crack method or the air change method, depending on available information. For the crack method:

  (a) Use the architectural plans to find lengths of cracks and appropriate areas, according to Table 3.3.

  (b) Use Tables 3.3 and 3.4 and the information from a to find the infiltration rate.

  (c) Use Equation 3.10 to find the infiltration heat loss.

  For the air change method:

  (a) Use Table 3.5 and Equation 3.12 to find the air infiltration rate.

  (b) Use Equation 3.10 to find the infiltration heat loss.

6. Find the room heating load.

   Room heating load

$$= \text{sum of} \begin{cases} \text{room heat transfer loss} \\ + \\ \text{room infiltration loss} \end{cases}$$

Remember to treat basement areas below ground and floor slabs on grade as described previously.

Room Heating Load Calculations

Project _____ Location _____ Indoor DB _____ F

Engrs. _____ Calc. by _____ Chk. by _____ Outdoor DB _____ F

Room			
Plan Size			
Heat Transfer	U x A x TD = BTU/hr	U x A x TD = BTU/hr	U x A x TD = BTU/hr
Wall			
Window			
Door			
Roof/ceiling			
Floor			
Partition			
Subtotal			
Infiltration	(CFM) 1.1 x A x B x TC =	(CFM) 1.1 x A x B x TC =	(CFM) 1.1 x A x B x TC =
Window	1.1	1.1	1.1
Door	1.1	1.1	1.1
Subtotal			
Room Load			

Room			
Plan Size			
Heat Transfer	U x A x TD = BTU/hr	U x A x TD = BTU/hr	U x A x TD = BTU/hr
Wall			
Window			
Door			
Roof/ceiling			
Floor			
Partition			
Subtotal			
Infiltration	(CFM) 1.1 x A x B x TC =	(CFM) 1.1 x A x B x TC =	(CFM) 1.1 x A x B x TC =
Window	1.1	1.1	1.1
Door	1.1	1.1	1.1
Subtotal			
Room Load			

Infiltration CFM	Column A	Column B
Crack method	CFM per ft or crack	length of crack, ft
air change method	air changes per hr	room volume/60, ft^3

Figure 3.12. Room heating load calculation form.

Building Heating Loads

1. Calculate the heat transfer losses through all exposed surfaces. Use total wall, roof, and other areas from architectural plans, rather than adding up areas from each room.

2. Find the building infiltration heat loss. For the crack method:

(a) Find the infiltration loss on the side of the building with the greatest infiltration.

(b) Take one-half the sum of the infiltration losses on all sides.

(c) Use the greater loss of a or b.

For the air change method, use one-half the sum of the infiltration losses of all rooms.

3. Find the outdoor air ventilation load, if any.

(a) Find the CFM from Table 6.15.

(b) Calculate the ventilation sensible load, using Equation 3.10.

Building Heating Load Calculations

Project _____

Location _____ Calc. by _____

Engineers _____ Chk. by _____

	DB, F	W; gr./lb.
Indoor		
Outdoor		
Diff.		

Heat Transfer	Q =	U	x	A	x	TD	=	BTU/hr
Roof								
Walls								
Windows								
Doors								
Floor								

Subtotal

Infiltration Q_S = 1.1 x _____ CFMx _____ TC =

Ventilation Q_S = 1.1 x _____ CFMx _____ TC =

Humidific Q_L = 0.68 x _____ CFMx ___ gr./lb.

Building Net Load

Piping or duct loss _____ %

Duct leakage _____ %

Pickup factor _____ %

Domestic HW Load

Boiler or furnace gross load

Notes:

Figure 3.13. Building heating load calculation form.

Room Heating Load Calculations

Project **W. Guthrie Residence** Location **Des Moines, Iowa** Indoor DB _72_ F

Engrs. **Enersave Assoc.** Calc. by **EP 1-22-88** Chk. by **KM 2-1-88** Outdoor DB _-5_ F

Room	Bedroom No. 1				Bedroom No. 2				Bedroom No. 3			
Plan Size	9.5' x 12'+ 2'x 3'				11' x 12.5'				9.5'x 9.5'+ 2'x 4'			
Heat Transfer	U	x A	x TD	= BTU/hr	U	x A	x TD	= BTU/hr	U	x A	x TD	= BTU/hr
Wall	.09	148	77	1030	.09	164	77	1140	.09	60	77	420
Window	.60	24	77	1110	.60	24	77	1110	0.60	16	77	740
Door												
Roof/ceiling	.07	120	77	650	.07	138	77	740	.07	98	77	530
Floor												
Partition												
Subtotal				2790				2990				1690
Infiltration	1.1 x	A (CFM)	x B x TC =		1.1 x	A (CFM)	x B x TC =		1.1 x	A (CFM)	x B x TC =	
Window	1.1	.5	17	77 720	1.1	.5	17	77 720	1.1	.5	20	77 850
Door	1.1				1.1				1.1			
Subtotal				720				720				850
Room Load				3510				3710				2540

Room	Bath				Kitchen				Living Room			
Plan Size	7'x 9'+ 4'x 2'				17' x 9'				22' x 13'			
Heat Transfer	U	x A	x TD	= BTU/hr	U	x A	x TD	= BTU/hr	U	x A	x TD	= BTU/hr
Wall	.09	46	77	320	.09	82	77	570	.09	217	77	1500
Window	.60	10	77	460	.60	36	77	1660	.60	45	77	2130
Door					.64	21	77	1030	.64	21	77	1030
Roof/ceiling	.07	55	77	300	.01	153	77	820	.07	286	77	1540
Floor												
Partition												
Subtotal				1080				4080				6200
Infiltration	1.1 x	A (CFM)	x B x TC =		1.1 x	A (CFM)	x B x TC =		1.1 x	A (CFM)	x B x TC =	
Window	1.1	.5	15.5	77 660	1.1	.5	33	77 1400	1.1	.5	57	77 2410
Door	1.1				1.1	1.0	20	77 1690	1.1	1.0	20	77 1690
Subtotal				660				3090				4100
Room Load				1740				7170				10300

Infiltration CFM	Column A	Column B
Crack method	CFM per ft or crack	length of crack, ft
air change method	air changes per hr	room volume/60, ft^3

Figure 3.14. Example of room heating load calculations (see Example Project I, Chapter 17).

Project __W. Guthrie Res.__   Location _____ Indoor DB _____ F

Engrs. _____ Calc. by _____ Chk. by _____ Outdoor DB _____ F

Room	Dining Area				Basement							
Plan Size	9' x 9'				42' x 22' + 11' x 3'							
Heat Transfer	U	x A	x TD	= BTU/hr	U	x A	x TD	= BTU/hr	U	x A	x TD	= BTU/hr
Wall	.09	106	77	730	.20	670	20	2680				
					.09	220	77	1520				
Window	.60	20	77	920	1.10	48	77	4070				
Door	.64	21	77	1030								
Roof/ceiling	.07	81	77	440								
Floor					.10	957	20	1910				
Partition												
Subtotal				3120				10180				
Infiltration	1.1 x	(CFM) A x	B x	TC =	1.1 x	(CFM) A x	B x	TC =	1.1 x	(CFM) A x	B x	TC =
Window	1.1				1.1	.8	30	77	2030	1.1		
Door	1.1	1.0	20	77	1690	1.1				1.1		
Subtotal				1690				2030				
Room Load				4810				12210				

Room												
Plan Size												
Heat Transfer	U	x A	x TD	= BTU/hr	U	x A	x TD	= BTU/hr	U	x A	x TD	= BTU/hr
Wall												
Window												
Door												
Roof/ceiling												
Floor												
Partition												
Subtotal												
Infiltration	1.1 x	(CFM) A x	B x	TC =	1.1 x	(CFM) A x	B x	TC =	1.1 x	(CFM) A x	B x	TC =
Window	1.1				1.1				1.1			
Door	1.1				1.1				1.1			
Subtotal												
Room Load												

Infiltration CFM	Column A	Column B
Crack method air change method	CFM per ft or crack air changes per hr	length of crack, ft room volume/60, ft^3

Figure 3.14. Continued

70

Building Heating Load Calculations

Project _W. Guthrie Residence_

Location _Des Moines, Iowa_   Calc. by _EP 2-8-88_

Engineers _Enersave Assoc._   Chk. by _KM 2-12-88_

	DB, F	W; gr./lb.
Indoor	72	
Outdoor	-5	
Diff.	77	

Heat Transfer	Q =	U	x	A	x	TD	=	BTU/hr
Roof		.07		957		77		5160
Walls *(1st floor)*		.09		1072		77		7430
(base, below gr.)		.20		670		20		2680
(base, above gr.)		.09		220		77		1520
Windows *(1st floor)*		.60		175		77		8090
(basement)		1.10		48		77		4070
Doors		.64		63		77		3100
Floor *(basement)*		.10		957		20		1910

Subtotal — 33960

Infiltration Q_S = 1.1 x _116_ CFMx _77_ TC = 9830

Ventilation Q_S = 1.1 x _____ CFMx _____ TC =

Humidific Q_L = 0.68 x _____ CFMx _____ gr./lb.

Building Net Lead

Piping or duct loss _____ %   43790

Duct leakage _____ %

Pickup factor _40_ %   17520

Domestic HW Load

Boiler or furnace gross load   61310

Note: Calculation of infiltration CFM

	N	E	S	W
1st fl. windows	44		47	17
doors	20	20	20	
base. windows	24		24	16
CFM per side	88	20	91	33

½ total CFM   (88 + 20 + 91 + 33) x ½ = 116 CFM
Use 116 CFM

Figure 3.15. Example of building heating load calculations (see Example Project I, Chapter 17).

(c) Reduce or neglect the infiltration load if the building is sufficiently pressurized.

(d) Calculate the ventilation latent load, using Equation 3.11, if the building is humidified.

4. Find the duct or piping losses and pickup factor (Section 3.12).

5. Find the building design (net) heating load:

Building heating load

$$= \text{sum of} \begin{cases} \text{building heat transfer loss} \\ + \\ \text{infiltration loss} \\ + \\ \text{ventilation losses} \end{cases}$$

6. Find the required boiler/furnace capacity:

Boiler/furnace capacity

$$= \text{sum of} \begin{cases} \text{building heating load} \\ + \\ \text{duct or piping loss} \\ + \\ \text{pickup factor} \end{cases}$$

An example of heating load calculations for a building may be found in Example Project I in Chapter 17. The student is advised to read Sections 17.1 and 17.2 at this point for an account of some practical problems encountered in doing an actual estimate, such as what to do with odd shaped spaces like closets, stairs, and hallways. For reference here, the results of the load calculations from Chapter 17 are reproduced in Figures 3.14 and 3.15.

3.14. ENERGY CONSERVATION

Reducing the building heating load provides a major opportunity for energy conservation. Some ways this can be achieved are

1. Use ample insulation throughout the building. Building construction in the past has been scandalously wasteful of energy due to inadequate insulation. For instance, an overall roof-ceiling resistance of R-20 to R-30 is recommended for residential buildings in colder climates.

2. Use the 97.5% winter outdoor design temperature level where possible.

3. Use inside winter design temperatures that provide comfort but not excessive temperature. The practice of using temperatures as high as 75 F is often unnecessary. Consider using 68 to 72 F.

4. Be certain all windows and doors are weather stripped, and either double glass or storm windows are used, except in mild climates. Windows with a maximum of 0.50–0.75 CFM/ft leakage should be used.

5. Heat losses shall be calculated using thorough, correct procedures.

6. The building architectural design (orientation, use of glass, type of materials, and so forth) should be consistent with reducing energy consumption.

7. Follow applicable energy conservation construction standards.

REVIEW QUESTIONS

1. What is the *heating load* and what items make it up?

2. What is the *infiltration loss*?

3. What is the *ventilation load*?

4. Define resistance and conductance. What is their relationship?

5. What are the two methods for estimating infiltration?

6. How is the infiltration for a corner room found when using the air change method?

7. What is meant by the "97.5% outside design temperature?"

8. What outside temperature is used to find the heat transfer from below grade surfaces?

9. Through what part of a building is the heat transfer loss proportional to the perimeter?

PROBLEMS

3.1. A homeowner asks an energy specialist to find the heat loss from his home. On one wall, measuring 15 ft by 9 ft (without windows) the specialist measures a temperature of 66 F on the inside surface of the wall and 18 F on the outside surface. The wall has a thermal resistance of 0.35 ft^2-F-hr/BTU. What is the rate of heat loss through the wall?

3.2. An insulating material has a thermal conductivity of $k = 0.23$ BTU/hr-ft^2-F per inch. How many inches of the material should the contractor install if energy conservation specifications call for insulation with an R-12 value?

3.3. Find the overall R value and "U" factor in winter for the wall with construction as shown in Figure 3.16.

3.4. The wall in Problem 3.3 is 30 ft long by 12 ft high. The indoor temperature is 70 F. What is the heat transfer loss through the wall on a day when the outdoor temperature is −5 F.

3.5. A state energy code requires a certain wall to have an overall R-8 value. What must be the R-value of insulation added to the wall in Problem 3.3?

3.6. A roof is constructed to built-up roofing on top of a metal deck, R-5.61 insulation, with a suspended 1/2 in. acoustical tile ceiling. Calculate the R-value of the roof in winter. Compare the result with the value from Table A.5.

3.7. Find the R-value of the roof in Problem 3.6 if 2 in. more of glass fiber insulating board is added.

3.8. Calculate the heat transfer loss through a 4 ft-6 in. wide by 3 ft high wood sash window with indoor and outdoor temperatures of 68 F and 3 F, respectively.

3.9. Calculate the heat transfer loss through a 25 ft by 30 ft roof-ceiling of a house with pitched asphalt shingle roof, vented attic, R-8 insulation, and 1/2 in. acoustical tile on furring ceiling. Inside and outside temperatures are 72 F and −2 F.

3.10. What is the heat transfer loss through a 40 ft by 20 ft basement floor when the room is at 65 F and the ground temperature is 50 F?

3.11. A warehouse in Cleveland, Ohio, built on grade, is 100 ft by 40 ft in plan. The wall construction is 4 in. block with brick facing. R-5 insulation is used around the edge.

6 in. sand and gravel aggregate (not dried) concrete
R-5 insulation
1/2 in. gypsum board

Figure 3.16. Sketch for Problem 3.3.

16 ft

Office

28 ft

Figure 3.17. Floor plan for Problem 3.19.

The inside temperature is 68 F. What is the heat loss through the floor?

3.12. Find the sensible heat loss from infiltration through a casement window with a 3 ft wide by 4 ft high operable section if the infiltration rate is 0.8 CFM/ft. The room temperature is 69 F and the outdoor temperature is −8 F.

3.13. A room 15 ft by 20 ft by 10 ft has an air infiltration rate of 1.5 air changes per hour. The room is at 72 F and the outdoor temperature is 1 F. What is the heat loss from infiltration?

3.14. A building in Milwaukee, Wisconsin has 2000 ft² of single-glazed metal sash windows. The inside temperature is 72 F. Using the 97.5% recommended outdoor winter design temperature, calculate the design heat transfer loss through the windows.

3.15. To save energy it has been decided to install storm windows on the building in Problem 3.14. What is the reduction in the design heat transfer loss?

3.16. Find the total heat loss from heat transfer and infiltration through a single-glazed 5 ft

All windows 3′–6″ × 4′
Scale 1/8 in. = 1 ft–0 in.

Figure 3.18. Building plan for Problem 3.20.

wide by 4 ft high double-hung metal sash window in a building in Springfield, Illinois, maintained at 71 F. The infiltration rate is 0.6 CFM/ft. Use the 97.5% design value.

3.17. A corner room in a building in Pittsburgh, Pennsylvania, has two 3 ft wide by 4 ft high casement windows on one side with an infiltration rate of 0.7 CFM/ft, and a 7

ft by 3 ft door on the other side with an infiltration rate of 1.2 CFM/ft. The room is at 68 F. Find the design infiltration heat loss from the room, using the 97.5% value.

3.18. Find the total heat loss through the exterior wall and windows of a room at 72 F, as described:

WALL: *8 in. concrete, R-5 insulation, 1 / 2 in. gypsum wallboard, 30 ft long by 15 ft high.*

Factory building

Height = 12 ft
Windows: 8 ft W × 6 ft H
Doors: 8 ft H × 5 ft W
Scale: 1/16 in. = 1'–0"

Figure 3.19. Building plan for Problem 3.21.

WINDOWS: *(5) 4 ft wide by 4 ft high metal casement type. Infiltration rate 0.5 CFM / ft.*
LOCATION: *Salt Lake City, Utah. Use 97.5% value.*

3.19. Find the design heat loss from the room shown in Figure 3.17 on an intermediate floor in an office building in Toronto, Ontario. Use recommended energy conservation design values suggested in this chapter.

WALL: *4 in. face brick, 8 in. cinder block, 1 / 2 in. furred gypsum wallboard.*
WINDOWS: *4 ft wide by 5 ft high double hung metal sash.*
CEILING HEIGHT: *9 ft.*

3.20. Calculate the individual room heating loads and building heating load for the house shown in Figure 3.18. Use recom-

mended energy conservation design values.

LOCATION: *Hartford, Connecticut.*
WALLS: *wood siding, building paper, wood sheathing, R-4 insulation, 3 / 8 in. gypsum board interior.*
WINDOWS: *double-hung metal sash with storm windows.*
ROOF: *pitched, asphalt shingles, building paper, wood sheathing, R-8 insulation, gypsum board ceiling, attic.*
DOORS: *1-1 / 2 in. wood.*
CEILING HEIGHT: *9 ft. No basement.*

3.21. Calculate the heating load for the building shown in Figure 3.19. The building is constructed on grade.

LOCATION: *Topeka, Kansas. Factory.*
WALLS: *8 in. concrete block, 1 / 2 in. furred gypsum board.*
WINDOWS: *sealed double-glazed, metal sash.*
MECHANICAL VENTILATION: *2500 CFM.*
DOORS: *1 in. wood.*
ROOF: *4 in. lightweight concrete, finished ceiling.*

4

Furnaces and Boilers

*T*his chapter will examine boilers and furnaces, the most common heat source equipment used in air conditioning. The heat which is produced by this equipment is most often used for heating, but may also be used in cooling *(see the discussion on absorption refrigeration in Chapter 13). The heat pump, used for both heating and cooling, will be discussed in Chapter 13. Solar heaters will be discussed in Chapter 18.*

OBJECTIVES

After studying this chapter you will be able to:

1. Describe the basic features of warm air furnaces and heating boilers.
2. Explain the functions of the basic operating and safety controls for furnaces and boilers.
3. Explain the function of flame safety controls.
4. Explain draft and how it is created and controlled.
5. Describe commonly used fossil fuels, products of combustion, pollutants, and methods of pollution control.
6. Describe the basic types of gas burners and oil burners.
7. Select a warm air furnace or heating boiler.
8. Describe the energy conservation methods that are associated with the use of furnaces and boilers.

4.1. WARM AIR FURNACES

A *warm air furnace* heats by delivering warmed air to the spaces in a building. Warm air furnaces are popular in private residences and small commercial installations, since, in very small buildings, warm air systems with ductwork are often less expensive than hydronic systems. Also, if the ductwork is already installed, by using a combined heating/cooling central unit or add-on cooling unit, summer air conditioning may be easily added at a minimal cost. A third advantage of warm air systems over hydronic systems is that when nighttime temperature set-back is used, full heat can be delivered to rooms faster in the morning. The hydronic system, however, does have advantages in many applications (Chapter 5).

Components The main parts of a warm air furnace are the heat exchanger, fuel burner, air blower (fan), controls, and insulated housing cabinet. The basic components of a warm air furnace are shown in Figure 4.1. The furnace may also have a humidifier and air filter.

Figure 4.1. Components of a warm air furnace.

Furnaces may have coal, oil, gas, or wood burners, or electric heaters, as a heat source. Oil and gas burners are discussed in Section 4.7.

The construction of furnaces for residential or commercial use is similar, except that commercial furnaces have larger capacities, are structurally stronger, and may have more complex controls than residential furnaces.

Operation Circulating air enters the furnace through the return air inlet (Figure 4.1). Pushed by a blower, the air passes over the outside of the heat exchanger, which has been warmed from the inside by hot combustion gas passing through it. The heated air exits through the warm air outlet and travels through ducts to the rooms in the building.

The hot combustion gas (also called *flue gas*) inside the heat exchanger is produced by the burning of fuel. After its heat is transferred to the circulating air, the combustion gas is exhausted through a *vent* to the outdoors. The vent may be a pipe, sheet metal stack, or masonry chimney. (Heating units in which the combustion gases are discharged to the outdoors are called *vented appliances*. Some heating units, called *unvented appliances*, discharge combustion gases directly into the room where the heater is located.)

Types of Furnaces To fit in different spaces, furnaces are made in a variety of physical configurations. Figure 4.2 shows some arrangements of residential type furnaces. The *upflow* or *highboy* type is suitable for full-height basement or utility room installations with overhead ductwork. The *low-boy* type may be used if there is less headroom. The *downflow* type is practical when the supply air ductwork is under a floor or grade, or in a crawl space. The *horizontal* type is suitable for an attic; when weatherized for outdoor service, it is popular for commercial rooftop installations.

Additional heating equipment that is usually grouped with furnaces includes *space heaters*, *wall* and *floor furnaces*, *duct heaters*, and *unit heaters.* Unlike warm air furnaces, these are not designed to be connected with ductwork (except for duct heaters), but instead deliver air directly into the space to be heated.

Space heaters are usually freestanding units resting on the floor. Some space heaters have blowers; others rely solely on the convected motion of the warm air. The wall furnace and floor furnace are designed to be recessed into a wall or floor. Duct heaters are mounted in a section of duct; air flow in duct heaters is created by a separate blower-housed unit. Unit heaters are generally hung from a ceiling; they may be gas or oil fired, or may use heating coils (Chapter 5). Another version is the gas-fired *radiant* heater

which works by using the flame and hot combustion gases to heat an element to a very high temperature; the heat is then radiated directly from the element to solid objects in the space, rather than having warm air circulate.

Because incomplete combustion may cause toxic pollutants, the use of unvented appliances is often restricted by legal codes. If unvented appliances are to be used, they must have ample room ventilation and special safety shut-off devices.

Capacity and Performance Manufacturers rate heating capacity in BTU/hr at the furnace outlet (bonnet). Residential type furnaces are

Figure 4.2. Arrangements of residential warm air furnaces.

available in capacities from about 35,000–175,000 BTU/hr. Commercial furnaces are available up to about 1 million BTU/hr.

The system designer needs to know the net heat available to heat the room or building, not the gross furnace output at the bonnet. Allowances must be made for any duct or pickup losses (see Chapter 3). When no specific data are available, some designers add 15% to the net heating load to allow for these losses. It is better to calculate the actual heat losses, if possible.

In addition to the heating capacity, the CFM of air to be circulated and the duct system air static pressure loss requirements must be determined (see Chapter 8). Combination heating/cooling units generally have two-speed fans to enable them to provide more air flow in the summer.

The steady-state efficiency of warm air furnaces typically ranges from 75–80%, except for so-called high-efficiency furnaces (Section 4.12), which may have efficiencies as high as 95%.

4.2. FURNACE CONTROLS

The controls for a warm air furnace are of two types: *operating controls* and *safety* (limit) *controls*. Operating controls regulate the burner (Section 4.7) and the air circulation fan during normal operation. Safety controls, also called limit controls, stop or prevent furnace operation if safe limits are exceeded. Safety controls may sense air and combustion gas temperatures, and air and gas flow; a safety control that detects flame presence is also used (Section 4.8).

The controls used in each application depend on the equipment, type of fuel, and safety code regulations. The following sequence outlines the operation of a typical residential gas-fired furnace:

1. On a call for heat, a switch in the room thermostat closes. A pilot flame (pilot light) safety control checks for the standing (continuous) pilot flame.

2. If the flame is present, the main gas valve opens. The pilot flame ignites the main gas. The safety control will continually check for flame presence; if the flame fails later, the valve will close.

   Instead of a standing pilot flame, the pilot gas itself may be ignited on a call for heating. When the furnace is not heating, the pilot flame is off. This arrangement, called *intermittent ignition*, is used to save energy.

3. A fan control thermostat located in the circulating air fan plenum (discharge) automatically starts the fan when the air has been warmed to a comfortable level and stops it when the air is too cool. An alternate arrangement, which achieves the same result, is to use a timer to delay the fan's start until a short period after the air starts warming. Fan shutdown can also be delayed until after the valve closes, utilizing the "free" remaining heat left in the heat exchangers, thus saving energy.

   Manual control of the fan is provided to allow continual air circulation in the summer.

4. A fan limit switch thermostat (high limit) will stop the fan if the air temperature becomes dangerously high (about 200 F). A simple occurrence such as blocked air flow from dirty filters can cause this. When the temperature drops, the switch may reclose to start the fan at a lower high limit (say 175 F); alternately, a manual reset switch may be used, if preferred.

The operation of a residential oil-fired furnace is similar, except that the space thermostat starts the oil burner pump motor and activates the circuit providing an ignition spark.

Commercial furnaces need more complex control arrangements. For instance, the quantity of gas-air mixture that remains in a large furnace after a shutdown is enough to be an explosion hazard; therefore, purge cycles are incorporated into the furnace control operations. In these more complex arrangements, the sequence of events is often called a *programming control sequence*, because it consists of programmed steps. A typical programming control sequence for a larger gas-fired furnace might be:

1. On a call for heating, the space thermostat closes the heating control circuit (only if all safety controls are also closed).
2. The air circulating fan starts.
3. Air circulation is tested by a pressure switch, and if air flow is proven,
4. The combustion fan starts.
5. Combustion gas flow is tested, and if proven,
6. A timed prepurge cycle (typically 30–60 seconds) exhausts combustion gases which may remain in the furnace from the last operation.
7. The main gas valve opens and the spark ignition circuit is activated.
8. Ignition is tested (with the flame safety control).
   (a) If ignition does not occur (in about 4 seconds), the ignition circuit is deactivated, and steps 6 and 7 are repeated (prepurge cycle and attempted ignition).
   (b) If the second ignition attempt fails, the system shuts down and "locks out" (a manual reset is necessary). (During operation, the flame safety control continually checks the flame. If the flame fails or is unstable, the system shuts down.)
9. When the thermostat is satisfied, the main gas valve closes.
10. A timed postpurge cycle exhausts combustion gases from the furnace.

4.3. HEATING BOILERS

Boilers produce hot water or steam, which is then delivered through pipes to space heating equipment. A *hot water boiler* heats water to a high temperature, but does not boil it. Since it does not actually boil water, a hot water boiler would be better named a hot water *generator*. A *steam boiler*, also called a steam generator, heats water to the boiling point to make steam. Both hot water and steam boilers, having similar features, will be discussed together; any important differences will be noted.

At this point a few words about safety should be said. Of all the equipment used in HVAC systems, probably the boiler is where one should be most sensitive about safety to life. This is because there is a tremendous amount of energy packed into a boiler. In New York City in 1962 a heating boiler in a telephone company building exploded, killing 21 and injuring 95. This incident is mentioned to stress the importance of obtaining a thorough working knowledge of boilers.

Components The main parts of a boiler are the combustion chamber, burner, heat exchanger, controls, and enclosure (Figures 4.3 and 4.5).

Boilers may be classified in various ways, according to:

1. Their specific application
2. Their pressure and temperature ratings
3. Their materials of construction
4. Whether water or combustion gas is inside the tubes (watertube or firetube)
5. Whether the boiler and accessories are assembled on the job site or at the factory
6. The type of fuel used
7. How draft (airflow) is achieved

The basic features of these groupings will be explained separately. Items 6 and 7, fuels and draft, will be discussed in relation to boilers and furnaces together, since they apply to both types of equipment.

Pressure and Temperature Ratings The *American Society of Mechanical Engineers*

(ASME) has developed standards for the construction and permissible operating pressure and temperature limits for low pressure *heating boilers* which are used in the United States.

The ASME *Code for Heating Boilers* limits maximum working pressure to 15 psig for steam and 160 psig for water. Hot water temperatures are limited to 250 F.

Hot water boilers are usually manufactured for 30 psig maximum working pressure, since this is more than adequate for the vapor pressure exerted by 250 F water.

For higher temperatures and pressures, the ASME *Code for Power Boilers* applies. *Power boilers*, although so named because they are used to generate steam to be utilized in generating electric power, can be used in high temperature hot water (HTW) hydronic heating systems. Another application for high pressure boilers in HVAC systems is to use them with steam turbine-driven centrifugal refrigeration machines, because steam turbines require steam at a relatively high pressure. Low pressure heating boilers do not require the attendance of a licensed operating engineer in many locations, whereas high pressure boilers do. This may affect the choice of low or high pressure boilers for an installation. The HVAC engineer should always check that a boiler conforms to ASME codes.

Materials of Construction *Cast iron boilers* have a heat exchanger constructed of hollow cast iron sections. The water flows inside with the combustion gases outside. The sections are assembled together similar to cast iron radiators. An advantage of this construction is that when the boiler is too large to fit through the building access opening, it can be shipped in parts and assembled on site. They range from small to fairly large capacity, up to about 10 million BTU/hr. Figure 4.3 shows a small cast iron boiler complete with burner, controls, and housing.

Steel boilers have a heat exchanger constructed of steel tubes arranged in a bundle, as seen in Figures 4.4 and 4.5.

Firetube or Watertube Steel boilers can be classified as either *firetube* or *watertube* In firetube boilers (Figure 4.4) the combustion gases flow inside the tubes and the water circulates outside. In watertube boilers the water flows inside the tubes and the combustion gases outside. Firetube boilers are less expensive than watertube boilers but are less durable. Firetube boilers range from small capacities to about 20 million BTU/hr. Watertube heating boilers range from medium size to about 100 million BTU/hr.

The differences among types of firetube boilers (e.g., *locomotive*, *horizontal return tube*, and *Scotch marine* type) are in their construction (a specialized subject, and not important for our purposes.) The Scotch marine type firetube boiler

Figure 4.3. Cutaway view of small gas-fired cast iron hot water package boiler. Note draft hood and automatic flue gas damper. (*Courtesy:* Burnham Corporation — Hydronics Division.)

Figure 4.4. Steel firetube boiler arrangement.

(Figures 4.4 and 4.5) is the most popular for commercial heating service because of its compactness, low cost, and reliability.

Watertube boilers are not often used in HVAC installations. Their main application is for large steam power plants or for creating process steam to be used in industry.

Built-Up and Package Boilers A *built-up* boiler is a boiler whose components are assembled at the job site; this includes the combustion

Figure 4.5. Package firetube boiler. (*Courtesy:* Cleaver-Brooks, Inc.)

Water
level

Figure 4.6. Low water cut-off. (*Courtesy:* McDonnell & Miller ITT.)   **Figure 4.7.** Water column and gage glass.

Note: Unions or equivalent for service not shown.

Figure 4.8. Typical piping arrangement and accessories for a hot water boiler. (In small systems, pump may be located in return line.)

chamber, heat exchanger, burner, and accessories. A *package boiler* is completely assembled and tested in the factory. This procedure reduces cost, increases reliability by ensuring that components are properly matched, and decreases the contractor's field work. The small cast iron residential type boiler shown in Figure 4.3, and the Scotch marine type boiler shown in Figure 4.5 are both package boilers.

Boiler Accessories Certain accessories are needed for the proper operation, maintenance, and safety of boilers. Some accessories are optional; others are required by codes or by law.

Accessories Required for a Steam Boiler A *low water cut-off* (Figure 4.6), which senses water level in a steam boiler; it will stop burner operation if the water level falls below a safe level.

A *water column* with a *gauge glass* (Figure 4.7), which when mounted on the side of a steam boiler, allows the operator to see the water level.

A pressure gauge and thermometer mounted on or near the boiler outlet aids the operating engineer in checking performance. A typical piping arrangement with accessories for a hot water heating boiler is shown in Figure 4.8.

Accessories Required for a Hot Water Boiler A *dip tube*, which is a piece of pipe from the boiler outlet extending down below the water line, prevents air that may be trapped in the top of the boiler from getting into the water supply line.

An *expansion tank* provides space for the increased volume of water when it is heated (Chapter 11).

A *flow check valve* closes when the pump stops. Without this valve, hot water would circulate by natural convection, heating rooms even when no heat is called for.

A *make-up water connection* allows for filling the system and replenishing water losses.

Figure 4.9. Safety relief valve. (*Courtesy:* TACO, Inc., Cranston, RI.)

A *pressure reducing valve* (PRV) prevents excess pressure from being exerted on the boiler from the water make-up source.

Air control devices may be required in the water circuit, these devices divert air in the system to the expansion tank.

Accessories Required for Both Steam and Hot Water Boilers A *safety relief valve* (Figure 4.9) which opens if boiler pressure is excessive. The ASME Code specifies the type of valve which is acceptable for a particular application. This valve must be connected separately at the boiler. In the New York explosion incident mentioned earlier, one of the claims in the investigation was that the safety relief valve did not open and relieve the excessive pressure which had developed. A combustion gas connection, called a *vent* on small boilers and *breeching* on larger boilers, through which combustion gas travels from the boiler to the chimney or flue.

Thermal insulation, which when applied around the boiler, reduces heat loss. It may be field installed or applied in the factory.

A *preheater*, if heavy fuel oil is to be used; it heats the oil to a temperature at which it will flow easily.

4.4. BOILER CONTROLS

Boiler controls, like furnace controls, are of two types, operating controls and safety (limit) controls.

Operating Controls   Operating controls regulate the burner (Section 4.7) during normal operation.

In smaller units, a room thermostat starts and stops the burner in response to space conditions.

In larger units a controller sensing a condition in the boiler regulates the burner operation: in steam boilers a pressure controller is used; in hot water boilers, a temperature controller ("aquastat") is used.

When controls sensing conditions in the boiler are used, secondary operating controls are generally also used. For instance, room thermostats may be used to control the flow of hot water to the terminal units in each room.

Safety Controls   As discussed earlier, safety (limit) controls prevent or stop an unsafe condition from occurring. Conditions that boiler safety controls may check for are:

1. High steam pressure (steam boilers)

2. High hot water temperature (hot water boilers)

3. High or low fuel oil/fuel gas pressure

4. High or low fuel oil temperature

5. Low water level

6. Flame failure

Programming Control Sequence   Commercial boiler controls, like furnace controls (Section 4.2), operate automatically in a specific programmed sequence. A typical programming control sequence for a small commercial gas-fired boiler might be:

1. *Prepurging.* On a call for heat, a fan starts, purging any combustion gases that might remain in the boiler from the last operation.

2. After an automatically timed period (a few seconds), the pilot gas valve opens, an ignition transformer is energized, and the pilot flame is lit. If the pilot flame does not light, the flame safety control shuts the pilot flame gas valve.

3. When the pilot flame lights, the main gas valve opens. If the main flame does not light, the flame safety control shuts down the burner. During operation, the flame safety control continually checks the flame. If the flame fails or is unstable, the system shuts down.

4. When the thermostat is satisfied, the main fuel valve closes.

5. *Postpurging.* On shutdown, the fan continues to run for a short time after the burner stops to purge the remaining combustion gases.

4.5. BOILER AND FURNACE DRAFT

Since boilers and furnaces need a constant supply of fresh combustion air, a pressure differential, called *draft*, must be created to force air and gas through the equipment and chimney. The pressure differential must be great enough to allow the flowing air and gas to overcome the resistance from friction in the combustion chamber, heat exchanger, and flue.

The term draft is also used in an associated manner to refer to the air or gas flow itself caused by the pressure differential. It is important to keep in mind both meanings of draft when dealing with combustion problems: draft is both the pressure to move the air and gas, and the flow itself.

Draft can be created either naturally or mechanically. *Natural draft* results from the difference in densities between the hot gas in the combustion chamber and the surrounding cool air. Since the heated gas is lighter than the cool air outside, it rises through the chimney, drawing cool air into the boiler through the openings at the bottom (the chimney flue effect).

A small negative pressure (relative to the atmospheric pressure) exists in the furnace in a natural draft unit. Under normal circumstances, this pressure prevents leakage of potentially toxic combustion gas into the equipment room.

When operating correctly, the overfire (combustion chamber) draft pressure reading in a typical residential furnace will be about −0.02 in. w.g. The manufacturer will provide precise values. Note that this is an extremely small pressure. The pressure in the flue outlet will be slightly more negative, perhaps −0.04 in. w.g., to allow for the pressure drop through the furnace heat exchanger.

Natural draft provides enough combustion air for simpler equipment, but as the equipment's size and complexity increases, the resistance to airflow also increases, resulting in a need for more draft. Natural draft can be increased by using a taller chimney, but this approach has physical limits, plus problems with aesthetics, legalities, and cost.

The term *mechanical draft* is used when fans are used to develop the pressure to move the gases through the boiler/furnace and flue. Either an *induced draft* fan or a *forced draft* fan may be used, or both. An induced draft fan, located at the equipment's outlet, pulls combustion air through the equipment and discharges it into the stack. A forced draft fan, located at the equipment's air inlet, blows air through the furnace. Since the forced draft fan creates positive pressure in the boiler/furnace, care must be taken in the equipment's design and maintenance to prevent combustion gas leakage into the room.

Generally, if mechanical draft is to be used, a fan with adequate pressure is used to create both furnace and stack draft, and only a short stack is needed. However, a draft fan may be used with a tall chimney, so that the two together develop the needed draft. In this arrangement, the fan creates only the furnace draft, and the chimney flue effect handles the required stack draft.

In addition to being able to develop more pressure than natural draft, mechanical draft fans provide closer control of draft (air flow quantity) than natural draft does, as will be seen in the following discussion.

Draft Control Draft (air flow) should remain constant for a given fuel firing rate. Too low a draft supplies insufficient air so that combustion is incomplete, thereby wasting fuel; excess draft results in too much air, reducing efficiency because the excess air is being heated and then thrown away.

Changes in temperature and outside air conditions (e.g., wind) can cause changes in the draft through the equipment, adversely affecting the equipment's performance. A sudden downdraft (gas flow down the chimney) due to outside disturbances can even blow out the flame, creating a dangerous situation. Because of these potential problems, a means of maintaining constant draft is needed.

Figure 4.10. Draft hood.

For smaller units, adequate draft control can usually be achieved using a *draft hood* (also called a *draft diverter*) or *barometric damper*.

A draft hood (Figures 4.3 and 4.10) is used on vented gas-fired equipment. A momentary increase in updraft will draw more surrounding air from the room into the flue via the hood. The greater volume of air in the flue increases the resistance to air flow; this reduces the chimney draft to its previous level. A momentary decrease in updraft will be canceled by an opposite reaction in the flue. If a downdraft occurs, the draft hood diverts the air into the surrounding space rather than into the combustion chamber, where it could blow out the pilot or main flame or cause poor combustion. Note that the draft hood is a safety device as well as a means of maintaining approximately constant draft.

The barometric damper, also called a barometric draft regulator (Figure 4.11), is used to control draft in oil-fired equipment and some power gas-fired units. (A power unit has a combustion air fan in the burner.) As stack updraft increases or decreases, the damper will open or close, canceling the momentary change in draft in the same way that a draft hood does.

Figure 4.11. Barometric damper.

Because the draft hood and barometric damper are located in the vent stack, they control stack draft better than furnace draft. Consequently, the furnace draft may vary too much to maintain an efficient air/fuel ratio. This may be acceptable with small equipment, but the inefficiency is too costly for larger units. In larger equipment, an *automatically controlled outlet damper* is often used instead of a barometric damper. This device is regulated by a pressure sensor in the furnace and adjusts automatically to maintain the proper furnace draft.

Furnaces or boilers that have power burners (i.e., with burner draft fans) may use one of the draft regulating devices discussed, or may rely only on the burner fan to control draft.

Newer higher efficiency residential units often use a mechanical draft combustion air fan and an *automatic vent damper* (Figure 4.3). (The vent damper can also be installed in existing systems.) The fan is needed to overcome the greater resistance of the larger heat exchanger (Section 4.12); the fan also improves draft control. The vent damper in the flue is not modulated, but instead closes automatically to prevent air flow through the stack when the system is shut down. This may reduce energy losses; for example, if the boiler/furnace is in a heated space, a closed damper prevents the loss of warm air from the building. However, another problem arises. When the vent damper closes, the chimney cools down for long periods and water vapor may condense on start-up, eventually causing corrosion; special chimney or vent materials may be required to prevent this occurrence.

4.6. FUELS AND COMBUSTION

The three major fossil fuels used in boilers and furnaces are gas, oil, and coal. Gas and oil have largely replaced coal as the fuels for space heating, due to their greater ease of handling, cleanliness, and reduced pollutant products. Coal is still used in many large power plant boilers because of its lower cost.

Gas *Natural gas* is the most commonly used fuel gas. It is delivered from gas wells to the consumer's equipment via a pipeline. Since no storage is necessary, gas is more convenient than oil or coal. Its primary combustible component is methane (CH_4); the by-products of its complete combustion are carbon dioxide (CO_2) and water.

The *liquified petroleum* (LP) gases, propane and butane, are stored in containers as liquids, but are used in their gaseous form. They are convenient in isolated locations where piped natural gas is not available.

Manufactured fuel gases are also available, often as by-products of other industrial processes, or from decomposing garbage (biogas).

Oil Fuel oils are available in different grades (Numbers 1, 2, 4, 5, and 6). The lower number grades are lighter in density, have a lower viscosity, and slightly lower heating values. Use of No. 1 oil (kerosene) is limited to small space heaters. No. 2 oil is generally used in residential and small commercial furnaces and boilers. No. 6 and often No. 5 oils require preheating before they become fluid enough to be used. This limits their usage to larger installations with auxiliary

heating equipment. Their advantage is that they cost less than No. 2 oil.

Since oil must be stored in tanks, it is less convenient than natural gas. In smaller quantities, oil may be stored indoors; for larger installations, oil storage tanks are buried underground or placed outdoors.

Fuel Choice The system designer should consider availability, cost, convenience, and pollution effects of the various fuels. Coal-fired burners may regain popularity as a means of conserving scarce oil and gas. Wood-fired units are growing in use, particularly in those areas where wood is abundant. Waste-fired boilers, which use garbage, save depletable fuel resources and aid in garbage disposal problems. A few cities are already successfully using large waste-fired boilers.

For residences, the popular choices are No. 2 heating oil and natural gas. The cost of each varies with location, market conditions, and legislated price controls. In some cases gas is cheaper, in others, oil.

Combustion The chemical reaction in which a fuel combines with oxygen in the air and releases heat is called *combustion*, or more commonly, burning. The amount of air required for complete combustion is called the *theoretical air*

TABLE 4.1. FUEL COMBUSTION DATA

Fuel	Theoretical Air / Fuel Ratio	Percent CO_2 in Combustion Gas Quantity of Air Supplied			Heating Value
		Theoretical	20% Excess	40% Excess	
Natural gas	9.6 ft^3 / ft^3	12.1	9.9	8.4	1000 BTU / ft^3
No. 2 fuel oil	1410 ft^3 / gal	15.0	12.3	10.5	140,000 BTU / gal
No. 6 fuel oil	1520 ft^3 / gal	16.5	13.6	11.6	153,000 BTU / gal
Bituminous coal	940 ft^3 / lb	18.2	15.1	12.9	13,000 BTU / lb

NOTES. Air / fuel ratios based on air density of 0.075 lb / ft^3, are by volume. Values are approximate, as the composition of fuels varies.

quantity and the resulting ratio of air to fuel is called the *theoretical air/fuel ratio*.

Because it is not practical or economical to construct equipment to mix air and fuel perfectly, if a boiler or furnace is furnished with the exact theoretical air quanity, combustion will not be complete. The result is unburned fuel and a waste of energy. Furthermore, incomplete combustion produces carbon monoxide (CO), a toxic pollutant.

To prevent this problem, *excess air* (air above the theoretical quantity) is always furnished. However, the efficiency of a boiler/furnace is maximized by using the minimum excess air needed for complete combustion. If too much excess air is used, efficiency is reduced, because

the excess air is being heated and then thrown away. In practice, many installations are operated with huge oversupplies of excess air resulting in a tremendous energy waste.

The minimum amount of excess air actually needed for complete combustion to occur depends on the type of fuel and the construction of the heating device, and may vary from 5–50% above 100% theoretical air. Manufacturers furnish data on recommended air/fuel ratios for their equipment. Table 4.1 and Figure 4.12 show theoretical air/fuel ratios, CO_2 content in the combustion gases for different excess air quantities, and fuel heating values.

Examples 4.1 and 4.2 illustrate how the heating specialist can use measurement of percentage

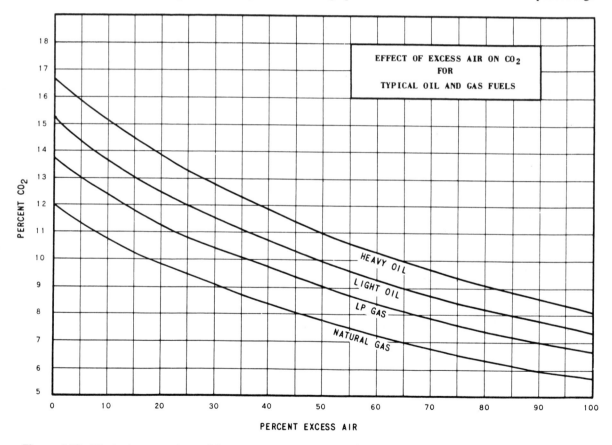

Figure 4.12. Effect of excess air on CO_2 percentage in flue gas. (*Courtesy:* Dunham Bush /Iron Fireman.)

of CO_2 in the stack gas to determine the necessary amount of combustion air and the percent excess air used. After such a test, the technician may adjust the air/fuel ratio if the air is excessive, in accordance with the manufacturer's data.

EXAMPLE 4.1

A boiler firing 15 gal/hr of No. 2 oil requires 20% excess air for complete combustion. How many CFM of combustion air should be supplied?

Solution. Using the air/fuel ratio data from Table 4.1, and noting that the actual air quantity should be 1.2 times the theoretical (20% excess),

$$CFM = 1410 \ \frac{ft^3}{gal} \times 15 \ \frac{gal}{hr} \times \frac{1 \ hr}{60 \ min} \times 1.2$$

$$= 423 \ CFM$$

EXAMPLE 4.2

A technician measures 8% CO_2 in the combustion gas of a natural gas-fired boiler that requires 15% excess air for complete combustion. Is the air/fuel ratio satisfactory?

Solution. From Figure 4.12, 45% excess air is being used. The answer is a loud NO! A great deal of excess hot gas is going up the stack.

Measuring Efficiency The *thermal efficiency* of a boiler or furnace is defined at the ratio of useful heat output to heat input, expressed as a percentage (see Section 4.10). The heat input is the amount of heat released from complete combustion of the fuel. Two measurements that can be used to determine thermal efficiency are *stack* (flue) *gas temperature* and *percent* CO_2.

The stack gas temperature should be relatively low, because this indicates that the heat exchanger has transferred more heat to the working

fluid, and less heat is wasted in the stack gas. Stack gas temperatures typically range from 300–600 F. Stack gas temperatures below about 300 F have traditionally not been desired, because of possible water vapor condensation and the resulting corrosion of the chimney or equipment. In some newer equipment, however, lower stack gas temperatures are being used to save energy (see Section 4.12).

Percent CO_2 in the stack gas should be relatively high, because this indicates that combustion is complete. The amount of CO_2 in the combustion gas can be measured with flue gas analysis equipment (Chapter 16).

Using measurements of the stack gas temperature and percent CO_2 in the flue gas, the heating specialist can determine the thermal efficiency of the equipment from Table 4.2. After the results are compared to the manufacturer's data, adjustments may be performed to improve the unit's efficiency. This testing should be performed regularly. (*Note:* Table 4.2 uses the difference between the temperatures of the stack gas and of the boiler room.)

EXAMPLE 4.3

In carrying out an energy study, readings on a boiler burning No. 2 oil are taken. The flue gas analyzer reads 10% CO_2. The stack gas temperature is 570 F and the boiler room temperature is 70 F. What is the thermal efficiency of the boiler?

Solution. The difference between the stack gas and room temperatures is $570 - 70 = 500$ F. Using Table 4.2, at 500 F and 10% CO_2, for No. 2 oil, the heat loss to the flue gas is 21.0%. The thermal efficiency is $100.0 - 21.0 = 79.0\%$.

EXAMPLE 4.4

A soot blower is used to clean off the heating surfaces of the boiler in Example 4.3. The stack

TABLE 4.2. EFFECT OF FLUE GAS TEMPERATURE AND CO_2 ON HEAT LOST
(%) TO FLUE GAS

	% CO_2	DIFFERENCE BETWEEN FLUE GAS AND ROOM TEMPERATURE IN DEGREES F.										
		300	350	400	450	500	550	600	650	700	750	800
NATURAL GAS	4.0	25.1	27.7	30.4	33.1	35.8	38.3	40.9	43.5	46.2	48.8	
	4.5	23.6	25.9	28.3	30.7	33.0	35.4	37.8	40.1	42.6	44.8	47.2
	5.0	22.2	24.4	26.8	28.7	30.9	33.0	35.7	37.3	39.7	41.8	43.8
	5.5	21.2	23.4	25.2	27.3	29.2	31.3	33.2	35.3	37.3	39.2	41.0
	6.0	20.4	22.3	24.1	25.8	27.8	29.6	31.5	33.3	35.2	36.8	38.8
Fuel Analysis	6.5	19.8	21.4	23.2	24.8	26.5	28.3	30.0	31.7	33.5	34.6	36.8
1120 BTU/cu.ft.	7.0	19.1	20.7	22.3	23.9	25.5	27.1	28.8	30.4	32.0	33.8	35.3
% by Volume	7.5	18.5	20.0	21.5	23.0	24.6	26.1	27.7	29.1	30.8	32.2	33.8
CH4 79.9	8.0	18.0	19.5	20.9	22.4	23.8	25.2	26.7	28.1	29.5	31.0	32.4
C2H6 17.3	8.5	17.6	19.0	20.4	21.7	23.1	24.5	25.8	27.1	28.6	29.9	31.3
CO2 0.3	9.0	17.2	18.5	19.9	20.1	22.5	23.8	25.2	26.4	27.8	29.0	30.3
N2 2.5	9.5	16.9	18.1	19.5	20.7	21.9	23.1	24.4	25.6	26.9	28.2	29.4
	10.0	16.6	17.8	19.0	20.2	21.4	22.6	23.8	25.0	26.2	27.4	28.6
	11.0	16.1	17.1	18.4	19.4	20.5	21.6	22.8	23.9	25.0	26.2	27.2
NO. 2 FUEL OIL	5.0	22.7	25.4	28.2	30.9	33.8	36.3	39.2	42.0	44.7	47.4	50.1
	5.5	21.3	23.8	26.3	28.9	31.4	34.0	36.4	39.0	41.7	44.0	46.5
	6.0	20.4	22.5	24.9	27.2	29.5	31.9	34.2	36.4	38.9	41.0	43.5
	6.5	19.3	21.3	23.6	25.7	27.8	30.1	32.3	34.4	36.5	38.7	40.8
	7.0	18.4	20.5	22.4	24.5	26.5	28.5	30.5	32.6	34.6	36.5	38.6
Fuel Analysis	7.5	17.7	19.6	21.3	23.4	25.2	27.1	29.0	30.9	32.9	34.8	36.5
BTU 19750/lb.	8.0	17.1	18.9	20.7	22.4	24.2	26.0	27.8	29.6	31.5	33.2	35.0
% by Weight	8.5	16.5	18.2	20.0	21.6	23.3	24.9	26.7	28.4	30.1	31.8	33.5
C 86.1	9.0	16.0	17.6	19.3	20.9	22.4	24.1	25.7	27.3	28.9	30.5	32.1
H 13.6	9.5	15.7	17.1	18.6	20.2	21.7	23.2	24.8	26.3	27.9	29.4	31.0
O 0.2	10.0	15.2	16.6	18.1	19.6	21.0	22.5	24.0	25.4	27.0	28.3	29.9
N 0.1	11.0	14.5	15.8	17.2	18.5	20.0	21.2	22.6	23.9	25.3	26.7	28.0
	12.0	13.9	15.1	16.4	17.6	18.9	20.2	21.4	22.6	24.0	25.2	26.5
	13.0	13.4	14.5	15.8	16.9	18.1	19.3	20.5	22.5	22.8	24.0	25.2
	14.0	13.0	14.0	15.3	16.3	17.4	18.5	19.7	20.8	21.8	22.9	24.1
NO. 6 FUEL OIL	5.0	22.8	25.9	29.0	32.0	35.3	38.1	41.0	44.3	47.5	50.1	53.6
	5.5	21.3	24.0	26.9	29.7	32.5	35.2	37.9	40.7	43.8	46.1	49.1
	6.0	20.0	22.5	25.2	27.7	30.3	32.9	35.3	37.9	40.5	43.0	45.8
	6.5	18.9	21.3	23.7	26.0	28.5	30.9	33.4	35.6	38.0	40.2	42.8
	7.0	17.9	20.1	22.4	24.6	26.8	28.8	31.2	33.4	35.8	37.9	40.1
Fuel Analysis	7.5	17.3	19.2	21.4	23.3	25.4	27.5	29.6	31.6	34.9	35.9	37.9
BTU 18,150/lb.	8.0	16.4	18.4	20.4	22.3	24.2	26.2	28.2	30.2	32.1	34.1	36.0
% by Weight	8.5	15.7	17.6	19.6	21.3	23.3	25.1	26.8	28.6	30.8	32.6	34.2
C 89.36	9.0	15.3	17.0	18.8	20.5	22.3	24.0	25.7	27.5	29.4	31.1	32.9
H 9.30	9.5	14.7	16.3	18.1	19.7	21.4	23.1	24.7	26.4	28.1	29.8	31.2
S 0.90	10.0	14.4	15.8	17.5	19.0	20.6	22.2	23.8	25.4	27.0	28.7	30.1
N 0.20	11.0	13.5	15.0	16.5	17.9	19.4	20.9	22.3	23.8	25.2	26.8	28.1
O 0.19	12.0	12.8	14.1	15.6	16.9	18.3	19.6	21.0	22.4	23.8	25.0	26.4
Ash 0.05	13.0	12.3	13.5	14.8	16.0	17.3	18.6	19.8	21.1	22.4	23.8	24.9
	14.0	11.8	13.0	14.3	15.4	16.6	17.7	18.8	20.1	21.2	22.5	23.7

(*Courtesy:* Dunham Bush / Iron Fireman)

gas temperature now reads 470 F. What is the thermal efficiency?

Solution. The difference between the stack gas and room temperatures is now $470 - 70 = 400$ F. From Table 4.2, the heat loss is 18.1%. The thermal efficiency is $100.0 - 18.1 = 81.9\%$. A significant fuel savings (about 3%) has been accomplished.

Pollution Control Control of possible air pollutants from combustion has become an important concern to the public. Heating system specialists should have some awareness of how this affects their work.

The main combustible constituents of gas, oil, and coal are hydrocarbons (compounds of hydrogen and carbon). When burned completely, hydrocarbons leave the by-products carbon dioxide (CO_2) and water.

Incomplete combustion of fossil fuels produces carbon monoxide (CO) gas, a colorless and odorless, but extremely toxic air pollutant. The presence of carbon monoxide indicates that not all of the heat has been obtained from the fuel, that is, a waste of energy. The cause is usually insufficient combustion air, which can often be easily remedied by adjusting dampers or other devices in the boiler or furnace to increase furnace draft. (If sufficient openings from the equipment room to the outside are not provided, or if these openings are closed, this will also starve the boiler or furnace of adequate air.)

A small amount of sulfur is present in most fuel oils and coals that when burned, releases sulfur dioxide (SO_2) into the atmosphere. Sulfur dioxide, a harmful air pollutant, is known to be a contributor to respiratory illnesses, and is often responsible for acid rain, which damages the ecology of lakes and forests. Consequently, many localities now limit the use of fuel oils to those with a low sulfur content, or in very large boilers, require the use of sulfur cleaning devices.

Smoke, another air pollutant, may result from incomplete combustion of coal or oil, a cold furnace on start-up, and related problems. Since combustion is incomplete, the presence of smoke indicates inefficient operation. Appearing as a dense black cloud, smoke contributes to respiratory problems and property damage. Smoke can usually be eliminated by adjustment of the air/fuel ratio and/or other combustion controls.

4.7. GAS AND OIL BURNERS

The *fuel burner* is a device for delivering the fuel and part or all of the combustion air to the furnace/boiler. It also helps to mix the fuel and air, and when fuel oil is used, breaks up the liquid oil into a spray of small droplets.

Gas Burners *Atmospheric gas burners* and *power gas burners* differ in how the air and fuel gas are delivered to the combustion chamber. In the atmospheric burner (Figure 4.13), the flow of fuel through a Venturi (a nozzle shaped tube) draws part of the combustion air (called *primary air*) through an opening, into the gas stream; the air/gas mixture then goes into the combustion chamber. The amount of primary air may be varied by using adjustable shutters or dampers at the opening. The remainder of the combustion air (called *secondary air*) is drawn by natural draft directly into the combustion chamber around the burner head ports (openings). (This arrangement is also called a *premix type burner*, since some of the air and gas mix before entering the burner ports.) In some burners, mixing of the fuel gas and primary air is enhanced by vanes or other devices.

Various burner head arrangements are available, each designed to match the furnace characteristics and size. *Inshot* burners have ports located to deliver the gas horizontally; *upshot* burners (Figure 4.13) deliver the gas vertically upward. The burner may have one or more ports in a pipe; multiple pipes can be arranged in parallel. Another version has a narrow slot (ribbon). A ring burner has the ports arranged around a ring-shaped pipe (like a kitchen stove).

Power gas burners use fans to deliver the air. The fan creates turbulence to promote air and gas mixing. The fan may be designed for complete forced draft (to overcome furnace and stack draft loss), or only to overcome furnace draft loss, relying on induced draft for the stack.

Gas Burner Ignition   Fuel ignition in a gas burner may be achieved by a *standing pilot*, an *intermittent pilot*, or *direct spark ignition*. A standing pilot is a continuously burning small gas flame. When the main gas fuel enters the combustion chamber, the pilot flame ignites it. Since the pilot flame remains burning when the

burner is off, there is a perpetual waste of heat up the stack.

To save energy, intermittent pilot ignition may be used. The pilot is lit by a spark, only on a call for heating. The pilot flame then lights the main fuel. The pilot and main gas are both shut off when the heating requirement is satisfied. Intermittent pilot ignition is also useful on rooftop equipment, where a standing pilot may be blown out by wind.

Some gas fuel equipment uses direct spark ignition. There is no pilot. On a call for heating, the fuel gas is fed into the combustion chamber and a spark ignites the gas directly. Descriptions

Figure 4.13. Atmospheric gas burner. Burner head is upshot type.

Figure 4.14. Gas manifold valves and burner arrangements. (*Courtesy:* North American Heating & Air Conditioning Wholesalers Association, Home Study Institute Division.)

of typical ignition procedures were given in Sections 4.2 and 4.4.

Many states have banned standing pilots on new gas-fired equipment, as an energy conservation measure. Although not usually legally required, existing equipment may often be easily converted to intermittent pilot ignition.

Typical gas manifold connections to a burner with a pilot flame (Figure 4.14) include a manual shut-off valve or cock, a gas pressure regulator, a pilot safety control, the electrically controlled main valve, and the pilot piping, burner and sensor. On newer residential equipment, a single *combination valve* (Figure 4.15), serves the functions of gas cock, pressure regulator, pilot safety control, and main gas valve.

Oil Burners   An oil burner mixes fuel oil and combustion air and delivers the mixture to the combustion chamber.

Except for one type, the *vaporizing pot burner*, all oil burners have the additional function of mechanically assisting in vaporizing the fuel oil. Vaporizing the oil is necessary since oil will not burn in its liquid state.

The vaporizing pot burner is basically a bowl filled with fuel oil. The oil at the surface vapor-

izes naturally due to its vapor pressure, aided by the turbulence and heat from the combustion gas. Because the vaporization is slow and difficult to control, it is used mainly in smaller space heaters burning No. 1 oil, which vaporizes more rapidly than heavier oils.

The other types of oil burners help vaporize oil by breaking it up into very small droplets. This process, called *atomizing*, increases the oil's surface area, causing it to vaporize faster. Each burner type has an oil pump, a combustion air fan, and an ignition system. They differ considerably from each other in the means of atomizing the oil and introducing it into the combustion chamber.

Steam atomizing or *air atomizing* burners use steam or air under pressure to cause mixing, turbulence, and atomizing. The *horizontal rotary cup burner* (Figure 4.16) has a rotating cup which throws the oil into the air stream, causing atomization; these types of burners are used with larger equipment, and are suitable for both heavy and light oils. The *gun burner*, or *mechanical pressure atomizing burner* (Figure 4.17), atomizes oil by pumping it under high pressure through a small diameter nozzle. A combustion air fan introduces air through a tube surrounding the

Figure 4.15. Combination gas valve. (*Courtesy:* North American Heating & Air Conditioning Wholesalers Association, Home Study Institute Division.)

nozzle. Deflector vanes at the burner outlet cause mixing and proper distribution of the oil-air mixture. The gun burner, which usually uses No. 2 oil, is used in both residential and smaller commercial oil-fired boilers and furnaces. *Retention head* gun burners, a recent development, have a head (outlet) that is designed to improve mixing and combustion, resulting in higher combustion efficiency.

Steam or air atomizing burners and rotary cup burners have a relatively high *turn-down ratio*, which is the ratio of maximum to minimum fuel oil flow that the burner can handle. A high ratio is desirable because the boiler/furnace can operate at a low capacity when necessary, thereby saving fuel. Gun burners have a relatively low turn-down ratio, which limits their ability to operate at part capacity.

Combination Burners   A combination burner can burn both oil and gas. Essentially it has the components of both burner types in the burner housing. It is useful when there is the possibility of a shortage of one of the two fuels, or where prices may change so that the relative costs of the fuels reverse.

Oil Burner Ignition   Ignition of the oil-air mixture in gun burners is done with a high voltage electric spark. Ignition of rotary cup and steam or air atomizing burners is usually done with a gas pilot flame, or a spark ignited pilot oil burner.

Figure 4.16. Rotary oil cup burner.

Gas and Oil Burner Firing Rate Control

Burner capacity (firing rate) may be controlled by a space thermostat, which is used with warm air furnaces and residential boilers; or by a steam pressure or water temperature controller, which is often used with commercial boilers. Other controls are available, too. The methods of burner capacity control are: *On/Off*, *High/Low/Off*, and *modulation*.

An On/Off control simply starts and stops the burner; this limits the boiler/furnace to full capacity operation only. A High/Low/Off (or High/Medium/Low/Off) control provides more flexibility in operating capacity, allowing the furnace/boiler to be more closely matched to load variations. Although considerable heat is wasted in starting up the heating equipment and when it is in the off cycle, use of these control methods is common in smaller equipment because of their low cost, and since loss of efficiency is often not critical.

To achieve maximum efficiency in larger units, full modulation of burner capacity, by varying fuel and air flow rates over the entire turn-down range, is standard. Fuel valves and air dampers are automatically modulated by the temperature or pressure controllers. A common method for doing this is to interlock the fuel valve and air damper after the initial adjustments have been made to provide the most efficient air-fuel ratio. However, the most desirable air-fuel ratio may change with temperature, fuel, humidity, and load. If the extra initial expense is warranted, an air-fuel ratio control system should be used. This control senses oxygen and combustion gas directly, and then adjusts the fuel valve and dampers for the most efficient combustion under all conditions.

4.8. FLAME SAFETY CONTROLS

A safety control that deserves special consideration is the *flame safety control*, because it ensures safe burner operation. This control shuts off the fuel supply if the fuel does not ignite, or if the

Figure 4.17. High pressure atomizing ("gun type") burner. (*Courtesy:* Air Conditioning Contractors of America (ACCA).)

Figure 4.18. Photo cell type flame safety control sensor ("cad cell" element). (*Courtesy:* Air Conditioning Contractors of America (ACCA).)

flame fails during operation. If fuel were to continue to enter the furnace and not be burned, a serious explosion hazard would quickly arise.

The flame safety control consists of a flame sensing element and a means of relaying its signal to start or stop fuel flow. The sensor can detect one of three possible effects of the flame: temperature (heat), flame electrical conductivity, or radiation (either visible light or infrared or ultraviolet radiation).

Gas-fired residential furnaces and boilers with a standing pilot use a heat sensing *thermocouple* for flame detection. The thermocouple consists of two wires of different metals that at a high

Figure 4.19. Stack switch flame safety control. (*Courtesy:* Air Conditioning Contractors of America (ACCA).)

temperature create a very small voltage. The thermocouple is placed in the pilot flame. If the flame fails, the control circuit will be deenergized and the gas valve will close.

Because it reacts slowly, the thermocouple is not a satisfactory flame safety control for larger heating equipment nor for equipment that uses intermittent pilot or direct ignition. Since it may take the main gas valve 30–40 seconds to close, enough gas will collect in the combustion chamber so that when the spark ignites the mixture on a call for heat, an explosion may occur.

The *flame rod* is a suitable flame safety control for gas ignition systems, since it closes the gas valve quickly (1–3 seconds). It consists of two electrodes placed at the flame location. Since a flame conducts electricity, the flame's presence completes a circuit opening the gas valve. Flame failure will open the circuit causing the valve to close. The flame rod is not used with oil-fired systems because the flame temperature is too high for the electrodes.

Oil-fired residential units use either a *photo cell* or a *stack switch* for flame detection. The photo cell (Figure 4.18) is a light sensitive device whose conductivity increases in the presence of strong light radiation. It is pointed at the flame location and allows the burner to operate only when the cell conducts. (The photo cell is not used with a gas flame because the intensity of light from a gas flame is too low.) The photo cell for small boilers and furnaces is often made of cadmium sulfide, and is then given the name "cad cell."

The stack switch (Figure 4.19) is a heat sensitive device placed in the stack to sense gas temperature. The stack switch is often a bimetal type thermostat. The bimetal closes a circuit on high temperature, permitting burner operation. The photo cell has replaced the stack switch as flame safety control in newer equipment both because of its faster response and its direct sensing of flame presence.

Flame safety controls for larger commercial equipment must react more quickly because fuel enters at a greater rate, reaching an explosive concentration sooner. Radiation sensing safety controls are ideal, since they react very fast. The three types of radiation flame sensors used in commercial equipment are: photo, infrared, and ultraviolet cells. In each case, the type of radiation from the flame changes an electrical property of the detector, which is then used in a relay to start or stop fuel flow.

4.9. BOILER APPLICATIONS

Hot water heating boilers generate hot water that is used directly in hydronic heating systems. Steam heating boilers generate steam that may be used directly in steam heating systems or the steam may be used to heat the hot water with a heat exchanger called a *converter*. In one circuit

Figure 4.20. Arrangement of steam boiler and hot water convertors in hydronic heating system for high rise building.

steam from the boiler flows and in the other the water to be heated. An obvious question arises as to why a hot water boiler should not be used, considering the additional expense and complication of the converter. The answer is that low pressure hot water boilers usually are designed for a maximum pressure of 30 psig. This corresponds to a head of $2.3 \times 30 = 69$ ft water. If a hot water boiler were installed in a basement or lower floor with more than 69 ft of height of water piping above it, it would be subject to an unsafe pressure. Very high buildings may be split into zones to prevent excess pressure on equipment. Figure 4.20 shows this arrangement. Another solution is to put the boiler on the roof.

The hot water or steam generated in boilers may be used for space heating, space cooling, or heating of service (domestic) hot water. It may seem strange that hot water or steam can be used for cooling, but absorption refrigeration machines require heat to produce refrigeration (Chapter 13). Often the boiler used for space heating is also used to heat service hot water. It is possible to obtain small boilers with a service hot water heating coil furnished internally. With large boilers a separate heat exchanger is usually specified.

4.10. BOILER RATING AND SELECTION

Manufacturers present rating data in tables, from which the proper boiler can be selected for a given application. Boilers are rated by their heat output in BTU/hr. It is also desirable to specify the temperature and flow rate of steam or hot water required. Two other units beside the BTU/hr have been used in specifying boiler capacity, boiler horsepower, and equivalent direct radiation (EDR). The use of these units is disappearing. They are confusing and sometimes misleading, and it is recommended that they be avoided. If necessary, their conversion equalities can be found in tables.

It is possible to produce increased capacity output from a boiler by firing it at very high fuel rates and by other variations in operating procedures. It is also possible to produce increased output by sacrificing other characteristics, such as increasing the draft loss. These practices may shorten the life of the boiler. For these and further reasons, standards have been adopted on procedures for testing and rating boilers. These standards may also recommend characteristics such as required amount of heating surface, furnace volume, fuel firing rates, and draft loss. The *Hydronics Institute*, an independent industry organization, has established recommended standards, called I-B-R ratings. It is suggested that the HVAC engineer check if a boiler has been tested and rated in accordance with I-B-R standards before selection. This is usually stated in the manufacturer's catalog. The *American Gas Association* (AGA) also recommends standards for gas-fired boilers.

Piping and Pickup Loss   The heat output of a boiler is used to deliver the building heating load. However, the actual output capacity of the boiler must be greater than the building heating load, because of two factors:

1. There is a constant loss of heat through hot piping to surrounding areas, some of which is not useful heating (piping in shafts). Insulation will reduce, but not eliminate, this loss. This is called the *piping loss*.
2. There is an additional heat loss when starting up a cold system. Before the boiler can deliver heat to the building, all the piping, water, and equipment of the heating system itself must be heated. This is called the *pickup loss*.

Because of these losses the boiler must have excess capacity so that the remaining capacity can handle the building load. This is expressed as follows:

Boiler gross output

= (boiler net output + piping loss

+ pickup loss)

RATINGS* FOR NATURAL AND PROPANE GASES

MODEL NUMBER†	A.G.A. RATINGS FOR WATER		NET I-B-R RATINGS	
	WATER INPUT (Btuh)	GROSS OUTPUT (Btuh)	WATER (Btuh)	WATER (SQ. FT.)
GG-75	75,000	60,000	52,200	350
GG-100	100,000	80,000	69,600	460
GG-125	125,000	100,000	87,000	580
GG-150	150,000	120,000	104,300	690
GG-175	175,000	140,000	121,700	810
GG-200	200,000	160,000	139,100	930
GG-225	225,000	180,900	156,500	1040

MODEL NUMBER†	A.G.A. RATINGS FOR WATER		NET I-B-R RATINGS	
	WATER INPUT (Btuh)	GROSS OUTPUT (Btuh)	WATER (Btuh)	WATER (SQ. FT.)
GG-250	250,000	200,000	173,900	1160
GG-275	275,000	220,000	191,300	1280
GG-300	300,000	240,000	208,700	1390
GG-325	325,000	260,000	226,100	1510
GG-350	350,000	280,000	243,500	1620
GG-375	375,000	300,000	260,900	1740

*Net ratings for baseboard radiation are based on a piping and pick-up allowance of 1.15 (hot water). Slant/Fin should be consulted before selecting a boiler for installation, having unusual piping and pick-up requirements. For elevations above 2,000 feet, reduce ratings by 4% for each 1,000 feet above sea level.

†Add suffix "S" for standard water boiler, "P" for packaged water boiler, Type of gas: After model number, specify gas by name: "Natural," or "Propane".

DIMENSIONS

FRONT

RIGHT SIDE HOT WATER BOILER

MODELS	GG-75	GG-100	GG-125	GG-150	GG-175	GG-200	GG-225	GG-250	GG-275	GG-300	GG-325	GG-350	GG-375
A JACKET WIDTH	13⅝₆	13⅝₆	13⅝₆	16⅝₆	16⅝₆	19⅝₆	19⅝₆	23⅝₆	23⅝₆	26¹¹₆	26¹¹₆	30⅝₆	30⅝₆
B DRAFT HOOD HEIGHT	46⅜	53¼	53⅛	53¼	53¼	57½	57½	56⅝	56⅝	56⅝	56⅝	63⅝	63⅝
C FLUE COLLAR DIAMETER	5	6	6	6	6	7	7	8	8	8	8	9	9
D JACKET TOP TO DRAFT HOOD	6¾	13	13	13	13	16	16	14	14	14	14	20	20
E CIRCULATOR RETURN FLANGE	1¼	1¼	1¼	1¼	1¼	1¼	1¼	1½	1½	1½	1½	1½	1½
F GAS CONNECTION NAT./PROP.	½ / ½	½ / ½	½ / ½	½ / ½	½ / ½	½ / ½	¾ / ½	¾ / ½	¾ / ½	¾ / ½	¾ / ½	¾ / ¾	¾ / ¾

Crates for all models are 30" wide, 38" high; depth 30" (GG-75 thru GG-225), 38" (GG-250 thru GG-375)

EQUIPMENT

BASIC WATER BOILER (SUFFIX S) includes pre-assembled heat exchanger with built-in air eliminator and insulated extended jacket; base; flue collector; gas burners, gas orifices and manifold assembly; combination gas valve including manual shut-off, pressure regulator (not on propane), pilot adj. and 100% shut-off, automatic pilot-thermocouple safety; hi limit control; altitude, pressure and temperature gauge; pressure relief valve (ASME); draft hood.

PACKAGED WATER BOILER (SUFFIX P) includes all equipment listed for model S, plus 2-way combination control (hi limit and circulator relay), instead of hi limit control; 1¼" circulator; 24 volt thermostat; drain cock.

OPTIONAL EQUIPMENT: Deluxe room thermostat, 24 volt; millivolt (self energized) controls; combination gas valve, combination limit controls; propane conversion kit; air package consisting of diaphragm expansion tank, fill and pressure reducing valve, automatic air vent; base for combustible floor.

Figure 4.21. Ratings for small gas-fired hot water boilers. (*Courtesy:* the Slant / Fin Corporation.)

The *net output* is the useful load, which is the sum of the requirements for space heating and service hot water, if any. The *gross output* is the actual heat output of the boiler. The piping and pickup losses could be calculated, but the practice is usually to make a standard allowance to be added to the net output. Manufacturers' rating tables list both a gross and net output to allow for piping and pickup loss, based on I-B-R recommendations. For an unusual case the actual heat loss from the piping should be calculated. Most buildings are heated continuously and therefore do not require much allowance for a pickup loss. In a building where the heating system is shut down regularly, an additional pickup allowance should be added.

A piping and pickup allowance of 15% has been established as an I-B-R standard for automatic fired hot water boilers. The piping allowance for steam boilers varies. Values can be found in Hydronics Institute publications.

EXAMPLE 4.5

A building has a net space heating load of 370,000 BTU/hr and a service hot water load of 32,000 BTU/hr. The building is heated intermittently. A piping allowance of 15% and pickup of 10% are required. What should the gross output be?

Solution. From the definitions given,

Boiler net output	= 370,000 + 32,000
	= 402,000 BTU/hr
Piping allowance	= 0.15 × 402,000
	= 60,000 BTU/hr
Pickup allowance	= 0.10 × 402,000
	= 40,000 BTU/hr
Required boiler gross output =	502,000 BTU/hr

The required heat input to a boiler will be greater than the gross output because some of the heat available in the fuel will be lost. Most of this loss is wasted in the hot combustion gases (in Chapter 15 we will see how some of this energy can be recovered). The efficiency is defined as

$$\text{Efficiency} = \frac{\text{gross heat output}}{\text{heat input}} \times 100\%$$

Modern automatic-fired boilers have efficiencies of 75–80% at full load when new or well maintained. Poor maintenance and operating procedures often result in boilers actually operating at 50–60% efficiency, or even less, a scandalous waste of expensive energy these days. Ratings and dimensions for a group of small cast iron, gas-fired hot water boilers are shown in Figure 4.21.

EXAMPLE 4.6

Select a gas-fired hot water boiler for the Moneybags Mansion. The heating load is 220,000 BTU/hr.

Solution. The net output (rating) of the boiler must be at least equal to the building heating load. From the ratings in Figure 4.21, a Model GG-325 is the smallest boiler that will do the job, with a net I-B-R rating of 226,100 BTU/hr. Note that the gross output is 260,000 BTU/hr, which includes a 15% piping and pickup allowance.

EXAMPLE 4.7

For a Model GG-200 boiler, determine the full load efficiency and ft^3/hr (CFH) of natural gas consumed at full load.

Solution. From the definition of efficiency,

$$\text{Efficiency} = \frac{\text{gross heat output}}{\text{heat input}} \times 100\%$$

$$= \frac{160,000}{200,000} \times 100 = 80\%$$

The heating value of natural gas is about 1000 BTU/ft^3 (Table 4.1), therefore the amount of gas required is

$$\text{CFH of gas} = 200{,}000 \ \frac{BTU}{hr} \times \frac{1 \ ft^3}{1000 \ BTU}$$

$$= 200 \ CFH$$

As we discussed in Chapter 3, nighttime temperature setback in residential applications requires a large pickup allowance (40% for a 10 F setback). This special pickup allowance is not usually necessary when sizing boilers for commercial installations. There are two reasons for this. First, standby or reserve boiler capacity is often provided by using two or more boilers. Typically, the excess capacity is from 25–100%, which normally covers the breakdown or maintenance of one boiler. The excess capacity also covers large pickup requirements. Second, the capable operating engineer knows how to operate the heating system to ensure that in the morning, temperatures are brought to comfortable levels before the building is occupied.

4.11. BOILER INSTALLATION

Each manufacturer furnishes specific instructions for the installation of the boiler when it is shipped to the job. We will not attempt to repeat these detailed instructions, but will instead list some procedures that are generally useful.

1. Allow ample size openings and passages into the boiler room for the boiler. The architect must be informed of the dimensions needed so that he or she can provide them. If an existing building that requires a new boiler does not have adequate openings for a tubular boiler, a sectional cast iron boiler may be the solution.
2. On high-rise buildings, consider a penthouse location for a gas-fired boiler. This eliminates the need for a flue running the whole height of the building.
3. Provide ample space on all sides of the boiler for maintence. Allow adequate distance in front of the boiler for tube cleaning and removal.
4. Locate the boiler as close to the flue as possible. Install the breeching to the flue without offsets.
5. Provide sufficient openings to the outdoors for both combustion air and ventilation air. Fixed grilles in walls or doors are one method. This is an extremely important point. If the openings are not adequate, the boiler may be starved of sufficient air for combustion, resulting in the production of toxic carbon monoxide.

4.12. HIGH EFFICIENCY BOILERS AND FURNACES

The increase in fuel costs in recent years has led to the development of so-called high efficiency boilers and furnaces for residential use. In "conventionally" designed equipment, a substantial part of the available heat from combustion is lost because the combustion gases are discharged to the stack at a relatively high temperature (e.g., 400–500 F). Efficiency can be increased by transferring more heat from the combustion gases.

By increasing the heat exchanger surface area, and designing their shapes so that the gases take a longer path through them, high efficiency furnaces utilize more of the heat in the combustion gases. This, however, results in a greater resistance to flow, possibly requiring the use of combustion gas fans; in conventional residential units, natural draft alone is adequate.

Some high efficiency units reduce the combustion gas temperature to about 300 F before discharging it to the stack. This is still well above the condensation point of the water vapor in the air. Others reduce the temperature to about 110 F, condensing the water vapor. Efficiency in this instance is higher not only because a greater part of the sensible heat of the gases is used, but also because the latent heat of condensation of the water vapor is used.

The sections of the heat exchangers that handle the lower temperature gases are made of stainless steel or other corrosion resistant materials. Stack vents and drains made of plastic pipe are used if the temperature goes below the condensation point. Some units have a sealed combustion system. Air is brought directly from outdoors through a pipe to the boiler/furnace, and gases are discharged outdoors through plastic pipes; this is called direct venting. No draft hood is used.

Figure 4.22. Pulse combustion type high efficiency hot water boiler. (*Courtesy:* Hydrotherm, Inc.)

Along with the efficiency improvements described, another type of high efficiency hot water boiler uses *pulse combustion* (Figure 4.22). In this unit the fuel gas and air are introduced into the combustion chamber in very small quantities intermittently. Each charge is ignited by the heat from the previous combustion.

High efficiency units are generally equipped for intermittent ignition rather than for using a standing pilot flame. They are also often designed to close off the vent flue when not operating, saving the losses that would result if cold air would be drawn through the unit. This further increases their efficiency. Of course, conventional boilers can also be retrofitted with these last two energy saving features.

A term that expresses the efficiency of a boiler or furnace is the Annual Fuel Utilization Efficiency (AFUE). The AFUE evaluates the seasonal efficiency of the unit, rather than just the peak operating efficiency in steady conditions. The AFUE for conventional residential units is about 60%. High efficiency units that operate above the water vapor condensing temperature have AFUEs of about 80–85%. Those that cool the stack gases down to 100 F or so have AFUEs of 90–95%.

Energy saving practices such as intermittent ignition and the closing off of stack vents with dampers on shut down are used with large boilers regularly. Additional recovery of stack gas heat can be achieved by using a separate stack gas heat exchanger. This has proven cost effective by retrofitting older installations.

4.13. ENERGY CONSERVATION

Some methods of conserving energy with boilers and furnaces are:

1. Adjust the air-fuel ratio so that excess air is the minimum recommended for the equipment.

2. Clean all heat transfer surfaces regularly (boilertubes, heat exchangers).

3. Do not use oversized boilers or furnaces. Efficiency is less at part loads.

4. For larger projects install multiple boilers. At part loads boilers will then operate closer to full capacity, where efficiency is higher.

5. Consider the use of heat exchange devices to use some of the waste heat in the hot flue gases (Chapter 15).

6. Use proper water treatment methods. Consult a specialist for advice.

7. Clean burner nozzles regularly.

8. Consider installation of a solar heating system.

9. Install an automatic damper in the flue. This device closes when the combustion unit is not operating, thereby reducing extra infiltration air that would go up the stack.

10. Install a flame retention type of oil burner in residential equipment. This type of burner uses less excess air and results in better heat transfer than other types.

11. Use intermittent ignition, rather than a standing pilot flame. (Care must be taken that this or automatic dampers do not result in water vapor condensation, which might cause corrosion.)

12. Consider the use of high efficiency boilers or furnaces.

REVIEW QUESTIONS

1. List the major components of a warm air furnace.

2. List four physical arrangements of warm air furnaces and where they would be located in a residence.

3. Name four types of warm air heating devices.

4. Describe the basic operating and safety controls for a warm air furnace.

5. Describe a typical programming control sequence for the operation of a warm air furnace.

6. List the major components of a hot water boiler and a steam boiler. List and explain the purpose of common boiler accessories.

7. Sketch a typical hot water boiler piping arrangement, listing essential components. What are piping loss and pickup loss? What are gross and net boiler output?

8. What are the pressure and temperature ratings for low pressure boilers?

9. Explain the difference between a firetube and watertube boiler.

10. Describe the basic operating and safety controls for hot water and steam boilers.

11. Describe a typical programming control sequence for the operation of a boiler.

12. What are the two meanings of draft?

13. Describe the different methods of achieving draft.

14. How is draft usually controlled in small gas- and oil-fired equipment? How is draft controlled in larger boilers?

15. List the major fuels used in heating plants and their relative advantages and disadvantages. List the by-products of the combustion of these fuels. What pollutants may result from incomplete combustion of these fuels?

16. Explain the terms theoretical air and excess air.

17. Describe the basic types of gas and oil burners.

18. Describe the methods of burner firing rate control.

19. List the types of flame safety controls. How do these controls work?

20. What problem may arise if stack gas temperature is too low?

21. What is the meaning of AFUE?

22. List five possible ways of increasing heating equipment efficiency.

PROBLEMS

4.1. A residence has a net heating load of 120,000 BTU/hr. Select a natural gas-fired hot water boiler, assuming a standard piping and pickup allowance. Determine the full load efficiency of the boiler.

4.2. If the boiler in Problem 4.1 is operating at full load efficiency, how much gas would be required at full load?

4.3. A building has a net heating load of 175,000 BTU/hr. The piping heat loss is 25,000 BTU/hr and the pickup loss is 30,000 BTU/hr. Select a natural gas-fired hot water boiler for this application.

4.4. A hot water boiler has a design heat input of 800,000 BTU/hr and a full load efficiency of 78%. It is to be used in a building with a piping and pickup loss of 100,000 BTU/hr. What is the maximum heating load the boiler can handle?

4.5. A boiler is using 1.3 GPM of No. 2 fuel oil. It has an efficiency of 72% and the piping and pickup loss is 22%. What is gross output and net output of the boiler?

4.6. A furnace burning natural gas is designed to operate with 30% excess air. What percentage reading of CO_2 will indicate proper operation? If the difference between the flue gas temperature and room temperature is 400 F, what would be the furnace efficiency at design conditions (neglecting other losses)?

4.7. A boiler uses No. 2 fuel oil. Using a combustion gas analyzer, a technician measures 12% CO_2 in the stack gases. What is the percent excess air? Flue gas temperature reads 520 F. Room temperature is 70 F. What is the boiler efficiency (neglecting other losses)?

4.8. For the boiler referred to in Problem 4.7, a stack heat exchanger is installed, reducing flue gas temperature to 370 F. What is the approximate boiler efficiency?

5

Hydronic Piping Systems and Terminal Units

The piping that is used to circulate hot or chilled water for air conditioning is called a hydronic piping system. The terminal units are the heat exchangers that transfer the heat be- tween the water and the spaces to be heated or cooled. In this chapter we will examine types of hydronic piping arrangements and terminal units.

OBJECTIVES

After studying this chapter you will be able to:
1. Identify the types of hydronic piping system arrangements and describe their features.
2. Identify the types of hydronic terminal units and describe their features.
3. Select baseboard radiation.
4. Lay out a hydronic system and determine its temperatures and flow rates.

5.1. PIPING ARRANGEMENTS

The connections between the piping and the terminal units may be made in any of these four basic ways:

1. Series loop
2. One-pipe main
3. Two-pipe direct return
4. Two-pipe reverse return

5.2. SERIES LOOP

A diagram of a *series loop* arrangement is shown in Figure 5.1. It is so named because all of the units are in a series, and one loop is formed.

Note that the entire water supply flows through each terminal unit, and then returns to the generator and pump. Because all of the water flows through each unit, and the units cannot be isolated from each other, the series loop has several disadvantages:

1. The maintenance or repair of any terminal unit requires shutdown of the entire system.
2. Separate capacity control of each unit by changing its water flow rate or temperature is not possible. (Control is possible by use of air dampers, however.)
3. The number of units is limited. Since the water temperature continually decreases as it gives up heat in each unit in series, the water temperature in later units may be too low for heating.

These disadvantages can be partially remedied by arranging the piping in two or more split series loops, as shown in Figure 5.2. This creates two or more zones which can be controlled separately.

The series loop arrangement is simple and inexpensive. It is limited to small, low-budget applications such as residences.

5.3. ONE-PIPE MAIN

A diagram of a *one-pipe main* arrangement is shown in Figure 5.3. As in the series loop, there is one main pipe through which the water flows,

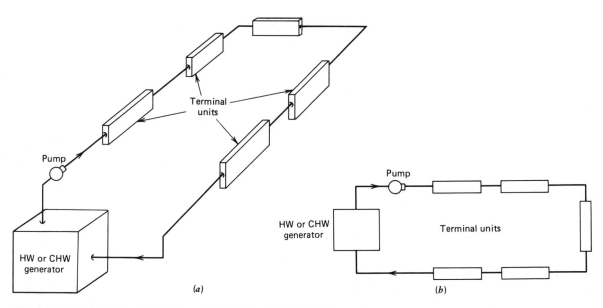

Figure 5.1. Series loop piping system. (*a*) Isometric. (*b*) Schematic.

but instead of being in series with this main, each terminal unit is connected by a supply and a return branch pipe to the main. By locating valves in the branch lines, each unit can be separately controlled and serviced. As in the series loop, if there are too many units, the water going to the later units may be too cool to heat the rooms adequately.

Flowing water seeks the path of least resistance. Consequently, water circulating in the main tends to flow through the straight run of the tee fitting at each supply branch, thus starving the terminal unit. To overcome this problem,

special *diverting tees* (Figure 5.4) are used at each supply branch take-off, directing some of the water to the branch. Additionally, if the terminal unit is below the main, a special tee is also needed at the return branch to prevent backflow.

5.4. TWO-PIPE DIRECT RETURN

To get the water temperature supplied to each terminal unit to be equal, the two-pipe arrangement is used. Figure 5.5 shows a *two-pipe direct*

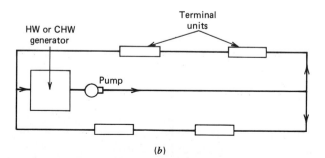

Figure 5.2. Split series loop piping system. (*a*) Isometric. (*b*) Schematic.

Figure 5.3. One-pipe main piping system. (*a*) Isometric. (*b*) Schematic.

return system. There are two mains, one for supply water and one for return. Each terminal unit is fed by an individual supply branch; a return branch carries the water back to the return main. In this manner, all units receive water directly from the source.

The total system flow rate (GPM) is split up among the terminal units, according to the design.

Although its cost is higher than one-pipe main and series loop arrangements, the two-pipe system allows each terminal unit to be separately controlled and serviced, and because the supply water temperature to each unit is the same, it can be used on any size installation. All larger systems use two-pipe arrangements.

Figure 5.4. Diverting tee used in one-pipe main systems.

The two-pipe arrangement in Figure 5.5 is called direct return because the return main is routed to bring the water back to the source by the shortest path. However, this creates a problem. Note in Figure 5.5 that the path the water takes from the pump to the first unit and back is shorter than the unit furthest away. Since flowing water prefers the path with least resistance, there will be too much water going to the units nearest the pump and too little to the units furthest from the pump. To overcome this problem, balancing valves can be installed in every branch, but the balancing process is difficult and requires considerable expense. The problem is largely solved with a *reverse return*.

5.5. TWO-PIPE REVERSE RETURN

The balancing problem in the direct return arrangement would be overcome if the circuit length to each terminal unit and back was made approximately the same. This is accomplished by piping the return main in a *reverse return* arrangement, as shown in Figure 5.6. Note that the path length for the water is about the same regardless of which unit it passes through. With this arrangement, it is a relatively simple process to balance the flow rates.

The relative costs of the direct return and reverse return piping arrangements depend on the building shape and location of terminal units. In some cases the costs are not significantly different; in others, the reverse return piping may be more expensive.

It may seem from the prior discussion on balancing that the two-pipe reverse return system would always be chosen over the direct return. In some situations, however, it may not be difficult to balance a direct return system. These are:

1. If the terminals are all far from the pump and grouped near each other, there may be little difference between the length of each path.

2. A very high resistance in the terminal units may make fluid flow through them approximately equal.

3. It is possible to make the fluid resistance in each circuit approximately equal in a direct return system by using smaller diameter piping in the closer branches. This depends on the piping layout, but it may then cause other problems (see Chapter 9).

In each case the planner must examine the layout before making a choice.

The two-pipe and one-pipe main arrangements can be split into two or more systems (if this is useful), as was shown with the series loop arrangement.

Two-pipe arrangements are almost always used for chilled water distribution to terminal cooling units. The water temperature to units far from the chiller would be too high for adequate cooling with the series loop or one-pipe main arrangement.

5.6. COMBINATION ARRANGEMENTS

It is sometimes useful to combine some of the four basic piping arrangements, taking advan-

Figure 5.5. Two-pipe direct return system.

tage of the best features of each. Figure 5.7 shows an example of a combined two-pipe reverse return with a group of units on each floor in series. This might be chosen for a high-rise building where separate control of each unit on a floor is not needed, yet flow balance will be simple and costs reduced as compared with a complete reverse return.

5.7. THREE-PIPE SYSTEM

The supply main in the two-pipe arrangements can be furnished with either chilled or hot water for cooling or heating, if the system is connected to both a water chiller and hot water boiler. However, only one can be used at any given time. In modern buildings heating is often required in some rooms and cooling in others at the same time. An example might be on a cool day with solar radiation on one side of the building only.

Simultaneous heating or cooling can be made available by use of a *three-pipe* arrangement (Figure 5.8). There are two supply mains, one circulating chilled water, the other hot water. Three-way control valves in the branch to each terminal unit will determine whether the unit

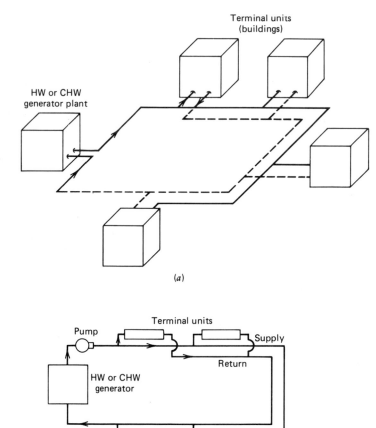

(a)

(b)

Figure 5.6. Two-pipe reverse return system. (a) Isometric — two-pipe reverse return to a number of buildings. (b) Schematic.

Figure 5.7. Combination reverse return (riser) and series loop system.

receives hot or chilled water. The return main receives the water from each unit. The connections to units can be made either direct or reverse return.

Because the return main mixes hot and chilled water, the three-pipe system can waste energy. The chilled water is warmed and the hot water is cooled in the mixing process, which results in extra heating and cooling in the boiler and chiller. This problem can be minimized by careful design, but it should be recognized.

5.8. FOUR-PIPE SYSTEM

The *four-pipe* system is actually two separate two-pipe systems, one for chilled water and one for hot water, and therefore no mixing occurs. This is an ideal arrangement, but of course it is expensive.

5.9. HYDRONIC TERMINAL UNITS

The *terminal units* are heat exchangers that transfer the heat between the room air and the

circulating water. Generally, the type of units used for heating and cooling are different from each other. The following will be discussed here:

Heating

1. *Radiators*
2. *Convectors*
3. *Baseboard*
4. *Fin-tube*
5. *Radiant panels* (heating and cooling)
6. *Unit heaters*

Cooling

1. *Fan-coil units* (heating and cooling)
2. *Induction units* (heating and cooling)

Radiators, convectors, baseboard, and fin-tube are collectively called *radiation*. This is a misleading name because they transfer heat to the room largely by natural convection. The air adjacent to the unit is warmed and rises naturally, creating a natural circulation.

All types of radiation should be located along exposed walls and under windows. In this location heat is suplied where the heat loss is great-

Figure 5.8. Three-pipe system.

Figure 5.9. Correct and incorrect location of radiation.

est, and cold downdrafts are prevented. Figure 5.9 shows good and poor locations of radiation.

Except for radiators and some convectors, the heating or cooling element of all hydronic terminal units is usually made of finned tubing. The fins increase the heat transfer. The material may be steel pipes with steel fins, or copper tube with either aluminum or copper fins.

5.10. RADIATORS

This type of radiation is constructed of hollow cast iron sections through which the hot water flows (Figure 5.10). Once very popular, radiators are not used much in new construction. The other types of radiation are often less expensive, require less space, and are more attractive looking.

5.11. CONVECTORS

Convectors have a finned tube or small cast iron heating element enclosed by a sheet metal cabinet (Figure 5.11). Room air enters through an opening in the bottom and leaves through an outlet grille at the top.

Convectors are available in varied arrangements to suit the architectural needs of the building. *Flush* units are mounted against the wall, whereas *recessed* units are recessed into an opening provided in the wall (Figure 5.12). Recessed

Figure 5.10. Cast iron radiator.

units have the advantage of not taking up floor space.

Free standing units rest on the floor whereas *wall hung* units are off the floor and are supported by the wall (Figure 5.13). Wall hung units allow easier floor and carpet cleaning.

Figure 5.11. Convector (free standing type).

Figure 5.12. Recessed convector.

Free standing   Wall hung

Figure 5.13. Free standing and wall hung convector.

Flush type units are available with the outlet grille on top, at the top front, or with a sloping top (Figure 5.14). The sloping top prevents placement of objects or sitting on the cabinet.

Convectors are used in rooms, vestibules, and stairwells. They are available in standard lengths and heights.

5.12. BASEBOARD

This type of radiation is located close to the floor in front of the architectural baseboard strip. It consists of a finned tube heating element with a sheet metal cover open at the bottom and with a slotted opening in the top (Figure 5.15).

The cover is often installed the whole length of the wall for a neater appearance, even when the heating element is not required for the whole length. Baseboard radiation is very popular in residences because it is inexpensive and unobtrusive. Tubing diameter is small, usually 1/2 or 3/4 in. The cover and fins are thin and therefore will not withstand heavy abuse.

5.13. FIN-TUBE

This type of radiation is similar to baseboard. The heating element is usually made of larger tubing (3/4 to 2 in.) and both the element and cover are heavier and stronger than that used for baseboard (Figure 5.16).

Covers are available with flat or sloping tops and varied quality of appearance. The capacity of fin-tube is greater per foot of length than baseboard because of larger fins and larger pipe. Tubing may be stacked more than one row high to increase output. (However, capacity does not increase proportionally with number of rows.) Fin-tube is widely used in commercial and indus-

Top outlet   Sloping top outlet   Top front outlet

Figure 5.14. Outlet arrangements.

Figure 5.15. Baseboard radiation. (*Courtesy:* Slant / Fin Corporation.)

Figure 5.16. Fin-tube radiation. (*Courtesy:* Vulcan Radiator Company.)

trial applications where radiation is desired along exposed walls.

The capacity of convectors, baseboard, and fin-tube can all be manually controlled by dampers located at the air outlet.

5.14. RADIANT PANELS

A radiant panel system has tubing installed in walls, floors, or ceiling, and extending over all or a considerable part of the surface. Both heating and cooling systems are available. Ceiling panels are used for cooling, so that the cooled air drops and circulates through the room. Because the heating or cooling source is spread out, radiant panel systems produce uniform temperatures and

comfortable air motion. It is an ideal system, but it can be very expensive.

5.15. UNIT HEATERS

The unit heater differs from the previous types of terminal units in having a fan that forces the air through the unit at a greater rate than would be achieved by natural convection. The heating element is finned tubing, arranged in coils to achieve a more compact arrangement. As a result, unit heaters have a high heating capacity for a given physical size. Two kinds of unit heaters will be discussed here.

Propeller Unit Heaters This type of unit heater is available in two versions—horizontal or vertical discharge. Each has a finned-tube coil heating element, propeller fan, motor, and casing (Figure 5.17).

The *horizontal blow* heater is usually mounted at 7–10 ft elevations. It has adjustable outlet dampers to control air direct. Air is directed toward work spaces or door openings. It is often

Makes 3 Times the output on baseboard heat

Horizontal propellor unit heater
(*a*)

Vertical down-blow propellor unit heater
(*b*)

Figure 5.17. Horizontal and vertical propeller unit heaters. (*a*) Horizontal propeller unit heater. (*b*) Vertical down-blow propeller unit heater.

used at loading platforms, vestibules, garage doors—wherever doors may be opened frequently and "spot" heating is needed.

The *vertical down-blow* unit heater is suitable for heating spaces with high ceilings and large floor areas. The units are mounted at high elevations. Adjustable outlet diffusers are available so that the amount of floor area heated can be varied. These units are often used in factories and warehouses.

Propeller fan unit heaters are generally limited to industrial applications or service areas of commercial buildings because they are unsightly, bulky, and noisy.

Cabinet Unit Heaters This type has a finned-tube heating element arranged as a ser-

Figure 5.18. Cabinet unit heater — floor mounted.

pentine coil, small centrifugal fans, an air filter, and a cabinet enclosure. In outward appearance, it looks like a convector (Figure 5.18).

The cabinet unit heater is often used where a convector would be suitable, but where the required heat output is larger, as in vestibules. It can also be mounted flat against a ceiling when this is architecturally desirable, because the outlet grilles will direct the air in the desired direction (Figure 5.19).

Cabinet unit heaters can be used in commercial applications because they have a pleasing appearance and are relatively quiet. They are sometimes called fan-coil units but we will use this name for a unit quite similar in construction that is used for either heating or cooling.

Figure 5.19. Cabinet unit heater — ceiling mounted.

Figure 5.20. Fan-coil unit.

Figure 5.21. Induction unit.

5.16. FAN-COIL UNITS

This terminal unit has a serpentine-shaped finned-tube heat exchange element, small centrifugal fans, air filter, and cabinet (Figure 5.20). It is basically the same as a cabinet unit heater.

The fan-coil unit can be used for both cooling and heating by supplying either chilled or hot water to the coil. The coil required for cooling is usually larger than that necessary for heating.

Fan-coil units are often constructed with an opening and damper in the rear to bring in outdoor ventilation air directly. The unit is then located along an exposed wall that has an opening for the fresh air connection. Fan-coil units may also be mounted horizontally at ceiling levels.

5.17. INDUCTION UNITS

This type of terminal unit is supplied with both chilled (or hot) water and *primary air* from a central air handling unit. A finned-tube serpentine coil is used. The primary air supply enters a small plenum and then is forced out vertically through small nozzles at a higher velocity. This creates a low pressure and therefore an induction effect—that is, room air is induced into the unit through the opening in the bottom. The room air is cooled as it flows across the chilled water coil.

Then room air (*secondary air*) and primary air mix and exit through the outlet grille in the cabinet enclosure at the top of the unit (Figure 5.21).

The induction unit, as seen from this illustration, is used with a combination air-water air conditioning system. The unit and system have many advantages, which will be discussed in Chapter 12. It is extremely popular in high-rise office buildings. (There are other variations of the induction unit, however, they will not be described here.)

Water heating and cooling coils that are installed in air handling units will be considered in Chapter 12.

5.18. SYSTEM WATER TEMPERATURES AND FLOW RATES

Hydronic heating systems are classified into temperature categories as follows:

LTW (low temperature hot water)—temperature below 250 F.

MTW (medium temperature hot water)—250–350 F.

HTW (high temperature hot water)—350–450 F.

These categories are important because different types of boilers and equipment are required for each. For example, as the water temperature increases, the boiler pressure must be increased to prevent the water from evaporating. Consequently, equipment of greater strength is required to handle the increased pressure.

In theory, a high supply water temperature is preferred because the terminal units may be made smaller. A high temperature drop is desirable because less water is required, allowing smaller pumps and piping to be used, and pump energy consumption is reduced.

However, there are reasons for limiting the water temperature. As mentioned earlier, lower water temperature/pressures do not require the heavy and expensive equipment that higher water temperatures do. In addition, the lower water temperature lessens the severity of a burn from accidental contact, a greater possibility with children, for instance.

The usual practice in designing LTW systems is to select a supply water temperature between 180–240 F, and a system temperature drop between 10–40 F. For private residences, supply water temperatures of 180–210 F and a temperature drop of 20 F or less are often satisfactory, since the room heating loads are small. The designer should consider supply temperatures up to 240 F and temperature drops up to 40 F for commercial applications.

In HTW systems, much greater temperature drops are often chosen (up to 100 F) to reduce pipe sizes and power use. HTW systems are used in very large projects.

Hydronic cooling systems using chilled water (CHW) do not have temperature categories. The suply temperature required in CHW systems depends on the dehumidification needed (Chapter 7), and usually ranges from 40–50 F. The system temperature rise usually ranges from 5–15 F. Here also, high temperature rises should be considered when planning in order to reduce energy consumption. Manufacturers often suggest desir-

able temperature ranges for their heating and cooling equipment.

The relationship between water temperature, flow rate, and heat gain or loss was shown previously by Equation 2.12

$$Q = m \times c \times \text{TC}$$

Because $c = 1.0$ BTU/lb-F for water, this becomes

$$Q = m \times \text{TC} \qquad (5.1)$$

where

Q = heat gain or loss of water, BTU/hr
m = flow rate of water, lb/hr
$\text{TC} = t_1 - t_2$ = temperature change of water, F

A more convenient form of the equation is to express the flow rate in GPM. Because (approximately) 1 GPM = 500 lb/hr of water, the equation becomes

$$Q = 500 \times \text{GPM} \times \text{TC} \qquad (5.2)$$

where Q, TC are as before and

$$\text{GPM} = \text{flow rate, gal/min}$$

Although the conversion factor of 1 GPM = 500 lb/hr is correct only at cold water temperatures, it may be used with insignificant error to 250 F.

EXAMPLE 5.1

A hydronic heating system is to be installed in the Square Tire Company factory. The building heating load is 8 million BTU/hr. A system water supply temperature of 240 F and return temperature of 200 F is chosen. What is the required system flow rate in GPM?

Solution. The flow rate is found by using Equation 5.2:

$$GPM = \frac{Q}{500 \times TC} = \frac{8,000,000}{500 \times (240 - 200)}$$

$$= 400 \ GPM$$

EXAMPLE 5.2

A water chiller with a capacity of 30 tons of refrigeration cools 80 GPM of water entering at 54 F. What is the temperature of the water leaving the chiller?

Solution. Changing units of cooling capacity,

$$Q = 30 \ tons \times \frac{12,000 \ BTU/hr}{1 \ ton}$$

$$= 360,000 \ BTU/hr$$

Using Equation 5.2,

$$TC = t_1 - t_2 = \frac{Q}{500 \times GPM} = \frac{360,000}{500 \times 80} = 9 \ F$$

Solving for t_2,

$$t_2 = t_1 - TC = 54 - 9 = 45 \ F$$

5.19. SELECTION OF TERMINAL UNITS

The rating (capacity) of terminal units is measured and reported by manufacturers in their equipment catalogs. Standard testing procedures for measuring ratings have been established. The designer and installer should check that any unit being considered has been tested according to a standard rating procedure, such as that of the *Hydronics Institute*.

The manufacturer's catalog ratings are used to select the terminal units required. Table 5.1 shows the ratings for a typical baseboard radiation. The heating capacity is listed in BTU/hr per foot of length. From this the required length of baseboard can be chosen. Note that the capacity depends on both the flow rate and average water temperature in the unit. The capacity also depends on the entering air temperature. Most ratings are listed for 65 F air entering the unit. For a room maintained at 68–70 F, no correction for the ratings at 65 F is usually necessary.

EXAMPLE 5.3

Select the length of 1/2 in. baseboard required for a room with a heat loss of 12,000 BTU/hr.

TABLE 5.1. RATINGS FOR TYPICAL BASEBOARD RADIATION

Nominal Tube Size in.	Flow Rate GPM	Velocity FPS	Hot Water Ratings, BTU / hr per foot Length at Following Average Water Temperatures, F							
			170	180	190	200	210	220	230	240
	1	0.6	510	580	640	710	770	840	910	970
	2	1.2	520	590	650	730	790	860	930	990
3/4	3	1.8	530	600	670	740	800	870	950	1010
	4	2.4	540	610	680	750	810	890	960	1030
	1	1.2	550	620	680	750	820	880	950	1020
	2	2.4	560	630	700	770	840	900	970	1040
1/2	3	3.6	570	640	710	780	850	920	990	1060
	4	4.8	580	660	720	790	870	930	1000	1080

NOTES: Tubing is copper with 2 1/8 × 2 5/16 in. aluminum fins, 55 fins per foot. Height of unit with enclosure is 8 in. Ratings are based on air entering at 65 F. For flow rates over 4 GPM use 4 GPM ratings. Type M tubing.

The unit has 2 GPM of water flowing through it at an entering temperature of 218 F.

Solution. Table 5.1 will be used to select the unit. The average water temperature in the unit must first be determined. To find this, Equation 5.2 will be used. The heating capacity of the unit must equal the room heat loss, according to the Energy Equation.

Using equation 5.2,

$$TC = \frac{Q}{500 \times GPM} = \frac{12,000}{500 \times 2} = 12\ F$$

Leaving $t = 218 - 12 = 206\ F$

Average $t = \dfrac{218 + 206}{2} = 212\ F$

Using Table 5.1, at a flow rate of 2 GPM and an average water temperature of 210 F (the next lowest temperature rating listed is used to be certain that the unit has adequate capacity), the rated capacity is listed as 840 BTU/hr per foot of length. The length required for a capacity of 12,000 BTU/hr is therefore

$$Length = \frac{12,000\ BTU/hr}{840\ BTU/hr\ per\ ft} = 14.3\ ft\ (use\ 15\ ft)$$

The contractor would order 15 ft of the radiation, rather than deal with the fractional amount. This also provides a little extra capacity.

The manufacturer should be consulted before using flow rates greatly outside the range shown in their tables. Generally, however, a good guideline is to use flow rates between values that result in water velocities between 1 and 5 ft/sec. Velocities above 5 ft/sec inn occupied areas may result in objectionable noise and velocities below 1 ft/sec may not be enough to carry dirt particles through the unit. Table 5.2 lists water velocities for different flow rates.

EXAMPLE 5.4

A contractor is about to install a hydronic system and notes that the engineer's specifications call for 1/2 in. diameter tubing with a flow rate of 7 GPM. What should the contractor do?

Solution. This flow rate results in a very high velocity, as seen in Table 5.2, and would probably be very noisy. The contractor should call the engineer and discuss possible changes in the design.

Rating tables and selection procedures are similar for other types of radiation, and will not be included here. Catalogs can be obtained from manufacturers.

5.20. SYSTEM DESIGN PROCEDURE

In the previous sections we discussed piping arrangements, water temperatures and flow rates, and selection of terminal units. It is often difficult for the student to put all this information together in planning a system. The following procedures should be helpful:

1. Choose the types of terminal units best suited for the application (Sections 5.9–5.19).
2. Choose the type of piping arrangement best suited for the application (Sections 5.1–5.8).

TABLE 5.2. WATER VELOCITIES, FT / SEC, IN TYPE L TUBING

Tube Diameter	Flow Rate, GPM				
	1	2	4	6	8
1 / 2 in.	1.4	2.7	5.5	8.5	
3 / 4 in.	0.6	1.3	2.6	4.0	5.5

3. Prepare a diagrammatic sketch of the piping system and the terminal units connected together.

4. Select a trial value of the system water temperature change (Section 5.20) and calculate the system water flow rate required to handle the building load (Equation 5.2).

5. Check to see if this is a satisfactory flow rate. Velocities through units should be within recommended values (Section 5.21). Also check manufacturer's recommendations on flow rates.

    (a) For a series loop system the flow rate through every unit is the same, of course.

    (b) For a one pipe main system the flow rates through each unit may be arbitrarily selected within recommended values, but cannot of course be greater than the total flow rate.

    (c) For a two pipe system the total system flow rate is distributed among all the units. The flow rate through each unit may be arbitrarily chosen within the manufacturer's recommended values, but of course the sum of the flow rates through each of the units must equal the total system flow rate.

6. If the flow rate is not satisfactory according to this check, a new trial value of the system water temperature change is taken, and a new flow rate calculated. After a little experience a designer can usually select an appropriate temperature change on the first or second trial.

7. Calculate the water temperature change through each unit, based on the required capacity of each unit (Equation 5.2).

8. Choose a suitable system supply temperature (Section 5.20). Determine the water temperature entering and leaving each unit. It is helpful to record all flow rates and temperatures on the piping sketch.

9. Select the terminal units from the manufacturer's catalog. Prepare a table showing all the information collected.

Two examples will illustrate this procedure.

EXAMPLE 5.5

The S. O. Smith residence has a design heating load of 68,000 BTU/hr. A series loop hydronic heating system will be used, with 3/4 in. diameter baseboard radiation. A sketch of the piping system and units is shown in Figure 5.22. Table 5.3 lists the required heating capacity of each unit.

Determine appropriate water temperatures, flow rates, and select the terminal heating units.

Figure 5.22. Sketch for Example 5.5.

Solution. The system design procedure recommended will be followed. (The results are noted in Table 5.3.)

1 – 3. For this small house a series loop baseboard system has been chosen.

4. Try a system temperature drop of 10 F. Using Equation 5.2 to find the resulting flow rate

$$\text{GPM} = \frac{Q}{500 \times \text{TC}} = \frac{68{,}000}{500 \times 10} = 13.6 \text{ GPM}$$

5. This flow rate will result in a greatly excessive velocity (Table 5.2).

6. Try a temperature drop of 30 F.

$$\text{GPM} = \frac{68{,}000}{500 \times 30} = 4.5 \text{ GPM}$$

This is a satisfactory flow rate.

7. The first unit has a required capacity of 9200 BTU/hr. The temperature change is

$$\text{TC} = \frac{9200}{500 \times 4.5} = 4 \text{ F}$$

8. A supply temperature of 220 F is chosen. The return temperature is then $220 - 30 = 190$ F. A higher supply temperature, say 240 F, might have been chosen for a small saving in the amount of radiation required. Practical cost analysis of a job would show whether this was worthwhile. The temperature entering the first unit is 220 F. The leaving temperature is $220 - 4 = 216$ F. The same procedure is used to find the remaining temperatures.

9. The required length of baseboard is now determined. The average water temperature in the first unit is

$$\text{Average } t = \frac{220 + 216}{2} = 218 \text{ F}$$

From Table 5.1, using the rating at 4 GPM and 220 F the heat output is 890 BTU per foot of length. The length required is therefore

$$\text{Length} = \frac{9200}{890} = 10.3 \text{ ft (use 11 ft)}$$

The choice of 11 ft of radiation instead of 10.3 ft will make up for selecting it at 220 F instead of 218 F. For more accuracy, values can be interpolated between listed temperatures. The results for the other units are shown in Table 5.3. The student should check these. (The sum of the unit capacities is slightly greater than the building load, as explained in Chapter 3, due to infiltration. This results in a calculated return temperature slightly less than originally chosen which, however, will not seriously affect the accuracy of the selection of the terminal units.)

If the radiation selected above is excessively long, the problem might be resolved by raising the water supply temperature. Another solution is to use radiation that has a greater output per unit length.

TABLE 5.3. RESULTS OF TERMINAL UNIT SELECTION FOR EXAMPLE 5.5

Unit	Capacity, BTU / hr	Flow Rate, GPM	t_{in}, F	t_{out}, F	t_{ave}, F	Length, ft
1	9,200	4.5	220	216	218	11
2	12,400	4.5	216	211	213	16
3	8,700	4.5	211	207	209	11
4	9,600	4.5	207	203	205	13
5	6,300	4.5	203	200	201	9
6	13,600	4.5	200	194	197	20
7	11,800	4.5	194	189	191	18

TABLE 5.4. RESULTS OF EXAMPLE 5.6

Unit	Total[a] Capacity, BTU / hr	Flow Rate, GPM	t_{in}, F	t_{out}, F
1	52,000	7.3	44	58
2	41,000	7.3	44	55
3	53,000	7.3	44	59
4	47,000	7.3	44	57
5	27,000	7.3	44	51

[a]In selecting cooling units the sensible and latent capacities must also be determined. These are not shown here. Furthermore, the building load will not generally be as great as the sum of the room loads, as will be seen in Chapter 6.

EXAMPLE 5.6

Determine the chilled water temperatures and flow rates for the two pipe system shown in Figure 5.23. The terminal units are fan-coil units. Capacities are listed in Table 5.4. The building is a small group of medical offices, with a cooling load of 220,000 BTU/hr.

Solution. The design procedures recommended will be followed. All results are listed in Table 5.4.

1 – 3. After a study of the building plans, it has been decided to use fan-coil units and a two pipe reverse return system. The fan-coil units will fit nicely under the window in each office. The building shape is such that there may be unbalanced flow if a direct return layout is used.

4. Try a system temperature rise of 12 F. Using Equation 5.2, the flow rate is

$$GPM = \frac{Q}{500 \times TC} = \frac{220,000}{500 \times 12}$$
$$= 36.7 \text{ GPM}$$

The flow rate is arbitrarily distributed equally among all five units, giving 36.7/5 = 7.3 GPM each.

5 – 6. Referring to a manufacturer's catalog of fan-coil units (not shown here) it is noted that their coil has a 3/4 in. nominal diameter tubing. From Table 5.2, the velocity will be satisfactory.

7. The temperature change for the first unit is

$$TC = \frac{52,000}{500 \times 7.3} = 14 \text{ F}$$

The temperature change for the other units is found in the same manner.

8. The system supply water temperature chosen depends on the selection of the refrigeration equipment and costs. Assume a CHW supply temperature of 44 F has been found satisfactory. The temperature leaving the first unit is then 44 + 14 = 58 F. The same procedure is carried out for each unit.

Figure 5.23. Sketch for Example 5.6.

9. From the above information the terminal units can be selected from the manufacturer's catalog. We will not describe that process here. Each manufacturer has slightly different procedures, which are described in their catalogs.

In the previous example an alternate procedure could have been used—to assume that every unit has exactly the same water temperature rise, and then to calculate the required flow rate. This procedure is just as acceptable as those chosen, but remember that the flow rates found should be checked to see if they are within recommended values. The student should work out this solution as a learning exercise.

REVIEW QUESTIONS

1. List the four basic types of hydronic system piping arrangements.

2. Sketch the arrangements for the four types of hydronic piping system arrangements.

3. List the applications and the advantages and disadvantages of the four types of hydronic piping system arrangements.

4. What application do three- and four-pipe systems have?

5. What undesirable feature does a three-pipe system have?

6. List the types of hydronic terminal units used for heating and/or for cooling. Describe a suitable application for each.

7. What are the basic parts of a unit heater?

8. List the types of unit heaters and one application for each type.

PROBLEMS

5.1. The operating engineer wants to check the capacity of a refrigeration water chiller. The instruments show 240 GPM of water flowing through the chiller, entering at 52 F and leaving at 40 F. What is the chiller's capacity in tons?

5.2. A building has a heating load of 630,000 BTU/hr. A hydronic heating system is used, supplying 40 GPM of water at 240 F. What is the return water temperature?

5.3. A fan-coil unit is to be used to cool a room with a cooling load of 12,000 BTU/hr. If the water temperature rise in the unit is 14 F, what is the flow rate in GPM?

5.4. The flow rate through a convector is 4.5 GPM. Water enters the unit at 220 F and leaves at 208 F. What is the heat output of the convector?

5.5. In Figure 5.24 terminal unit A has a heat output of 9300 BTU/hr and unit B of 8100 BTU/hr. What is the water temperature leaving unit A and unit B?

5.6. In Figure 5.25 terminal units A, B, and C have cooling capacities of 14,000 BTU/hr, 7,200 BTU/hr, and 12,700 BTU/hr, respectively. Determine the water tempera-

Figure 5.24. Sketch for Problem 5.5.

Figure 5.25. Sketch for Problem 5.6.

tures and flow rates at points (1), (2), and (3). The flow rate through each unit is 3 GPM.

5.7. What is the heating capacity of a 7 ft length of 1/2 in. baseboard radiation of the type listed in Table 5.1, with a flow rate of 2 GPM and an average water temperature of 200 F?

5.8. A room has a heating load of 9600 BTU/hr. Find the required length of 3/4 in. baseboard radiation to heat the room, of the type shown in Table 5.1, if the baseboard is supplied with 3 GPM of water at 235 F.

5.9. Select the terminal units for the residence described in Example 5.5, using 1/2 in. baseboard, a supply temperature of 230 F, and a temperature drop of 35 F.

5.10. Select baseboard radiation for the residence described in Example 5.5 using a split series loop piping system, and suitable temperatures and flow rates.

5.11. Find the flow rate in each section of pipe for the hydronic cooling system of Example 5.6.

5.12. Using a system temperature rise of 10 F for Example 5.6 and equal flow rates to each unit, calculate the temperature rise in each unit and the flow rate in each section of pipe.

5.13. Assuming a system temperature rise of 14 F for Example 5.6 and the same temperature rise in each unit, calculate the flow rate through each unit.

5.14. Lay out a hydronic piping system and terminal units for the house shown in Problem 3.20. (Use the type of system and terminal units as assigned by your instructor, or select your own. Use the heating loads calculated previously, or those specified by the instructor.)

5.15. Lay out a hydronic piping system and terminal units for the building shown in Problem 3.21. (Use the type of system and terminal units as assigned by your instructor, or select your own. Use the heating loads calculated previously, or those specified by the instructor.)

CHAPTER 6

Cooling Load Calculations

*T*he objective of this chapter is to learn how to determine the amount of cooling required to keep the rooms in a building comfortable in summer. In Chapter 3 we learned how to calculate the winter heating requirements of a building. The procedures described for calculating cooling needs are similar but involve additional items that make the subject more complicated.

OBJECTIVES

After studying this chapter you will be able to:

1. Calculate the heat gains to a space.
2. Select appropriate design conditions for cooling.
3. Determine peak load conditions.
4. Find required ventilation rates.
5. Perform a commercial cooling load analysis.
6. Perform a residential cooling load analysis.

6.1. THE COOLING LOAD

The interior of a building gains heat from a number of sources. If the temperature and humidity of the air in rooms are to be maintained at a comfortable level, heat must be extracted to offset these heat gains. The net amount of heat that is removed is called the *cooling load*. In this chapter we will examine each of the components of the heat gain and how they are calculated, and then how the cooling load is determined. The cooling load must be determined because it is the basis for selection of the proper size cooling

Figure 6.1. Heat flow diagram showing building heat gain, heat storage, and cooling load.

equipment, piping, and ducts. It is also used to analyze energy use and conservation.

Calculation of the cooling load is based on the heat transfer principles discussed in Chapter 3. In that chapter these principles were used to calculate the heat losses and heating load. The cooling load is similar but opposite in direction, in that it results from heat gains to the rooms. The same heat transfer principles are used for finding heat gains, and for this reason the reader should review Chapter 3 before studying the material in this chapter.

The methods for calculating cooling loads that will be used here are essentially those recommended by *ASHRAE* (1985 *Fundamentals Handbook*).

These calculation procedures described are considerably more accurate than past methods. The improved accuracy often leads to selection of smaller size equipment and more efficient use of energy. Many new building standards and codes are requiring that calculations be carried out using these or similar methods.

The procedures that will be explained apply to commercial and industrial buildings. Modifications that are used for residential buildings will be explained later.

6.2. THE HEAT STORAGE EFFECT

In the explanation of heating load calculations it was shown that the room heating load was simply the sum of the room heat losses (taking into account the proper rate of infiltration). The relationship between the heat gains and cooling load is more complex, as can be seen in Figure 6.1.

The gross room *heat gain* is the rate at which heat is being received in the room at any time. This heat gain is made up of components from many sources (solar radiation, lighting, conduction and convection, people, equipment, infiltration). All the heat received from these sources usually does not go immediately to heating the room air. Some of the heat, especially radiant energy (from sun, lights, people) is absorbed by the room materials, both the structure and the furnishings. This is called the *heat storage* effect —heat is absorbed and stored in the building materials. As a result the *net* heat gain to the room air from the heat sources is often less than the *gross* heat gains. The room *cooling load*, the rate of heat that must be removed from the room to maintain the design conditions, is the sum of these net heat gains.

It is very important to account for the heat storage effect because it may result in considerably reduced actual loads. The storage effect can also be considered as a heat time lag. That is, some of the heat received by the room is delayed in time from reaching the room air. Eventually the temperature of the building materials will rise enough so that they will give off heat to the room air. Usually, however, the storage continues past the peak load time of day, and the net effect is a reduction of peak loads, as seen in Figure 6.2. The tourist who has been inside the great cathedrals of Italy on a hot summer day

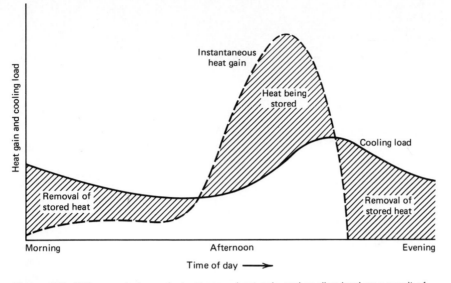

Figure 6.2. Difference between instantaneous heat gain and cooling load as a result of heat storage effect.

notices how cool it is inside. The walls, sometimes 6 to 8 ft thick, have a huge heat storage capacity and time lag.

The methods that we will use account for the storage effect and time lag in a direct and simple manner. There are two other related effects, described later, which are treated separately—*heat transfer to the surroundings* and *temperature swing.*

6.3. ROOM HEAT GAINS

The components that contribute to room heat gain consist of the following (Figure 6.3):

1. Conduction through exterior walls, roof, and glass
2. Conduction through interior partitions, ceilings, and floors
3. Solar radiation through glass
4. Lighting
5. People
6. Equipment
7. Heat from infiltration of outside air through openings

It is convenient to arrange these heat gains into two groups—those from external sources outside the room, and those internally generated. From the earlier description it is seen that items 1 through 3 are *external heat gains*, and items 4 through 6 are *internal heat gains*. Infiltration can be considered as a separate class.

It is also convenient to arrange the heat gains into a different set of two groups: *sensible* and *latent* heat gains. Sensible heat gains result in increasing the air temperature and latent heat gains are due to addition of water vapor, thus increasing humidity. Items 1 through 4 are solely sensible gains. Items 5 and 7 are part sensible and part latent, and item 6 can fall in either category or both, depending on the type of equipment. As will be noted in the study of psychrometrics (Chapter 7), it is necessary to separate the sensible and latent gains because the

Figure 6.3. Room heat gain components, Q.

6.4. CONDUCTION THROUGH EXTERIOR STRUCTURE

The conduction heat gains through the exterior roof, walls, and glass are each found from the following equation:

$$Q = U \times A \times CLTD_c \qquad (6.1)$$

where

> Q = net room conduction heat gain through roof, wall, or glass, BTU/hr
> U = overall heat transfer coefficient for roof, wall, or glass, BTU/hr-ft²-F
> A = area of roof, wall, or glass, ft²
> $CLTD_c$ = cooling load temperature difference, F

The area of each component is found from building plans. The U-values, discussed in Chapter 3, are found in Tables 6.1, 6.3, A.7, A.8 or they may be calculated from R-values. If calculated note that air film R-values may be different in summer than winter, because wind velocities change.

The CLTD is a temperature difference that accounts for the heat storage effect. Tables 6.1 and 6.2 list CLTD values for a few different roof and wall constructions. Table 6.3 describes wall constructions for Table 6.2.

The values in the tables are based on an indoor design temperature of 78 F and an outdoor daily average temperature of 85 F, and with dark-colored walls and roofs, for July 21 at 40°N latitude.

The CLTD values read from Tables 6.1 and 6.2 must be corrected as follows

$$CLTD_c = \big[(CLTD + LM) \times K + (78 - t_R)$$
$$+ (t_0 - 85)\big] \times f \quad (6.2)$$

where

> $CLTD_c$ = corrected value of CLTD, F
> $CLTD$ = temperature from Table 6.1 or 6.2, F
> LM = correction for latitude and month from Table 6.4
> K = correction for color of surface
> > K = 1.0 for dark color or industrial areas
> > K = 0.5 for light-color roof in rural areas
> > K = 0.65 for light-color wall in rural areas

TABLE 6.1. COOLING LOAD TEMPERATURE DIFFERENCES (CLTD) FOR CALCULATING COOLING LOAD FROM FLAT ROOFS, 1, F

Roof No	Description of Construction	Weight lb/ft²	U-value Btu/(h·ft²·°F)	1	2	3	4	5	6	7	8	9	10	11	12	13	14	15	16	17	18	19	20	21	22	23	24
														Without Suspended Ceiling												
1	Steel sheet with 1-in. (or 2-in.) insulation	7 (8)	0.213 (0.124)	1	-2	-3	-3	-5	-3	6	19	34	49	61	71	78	79	77	70	59	45	30	18	12	8	5	3
2	1-in. wood with 1-in. insulation	8	0.170	6	3	0	-1	-3	-3	-2	4	14	27	39	52	62	70	74	74	70	62	51	38	28	20	14	9
3	4-in. l.w concrete	18	0.213	9	5	2	0	-2	-3	-3	1	9	20	32	44	55	64	70	73	71	66	57	45	34	25	18	13
4	2-in. h.w. concrete with 1-in. (or 2-in.) insulation	29	0.206 (0.122)	12	8	5	3	0	-1	-1	3	11	20	30	41	51	59	65	66	66	62	54	45	36	29	22	17
5	1-in. wood with 2-in. insulation	19	0.109	3	0	-3	-4	-5	-7	-6	-3	5	16	27	39	49	57	63	64	62	57	48	37	26	18	11	7
6	6-in. l.w. concrete	24	0.158	22	17	13	9	6	3	1	1	3	7	15	23	33	43	51	58	62	64	62	57	50	42	35	28
7	2.5-in. wood with 1-in. insulation	13	0.130	29	24	20	16	13	10	7	6	6	9	13	20	27	34	42	48	53	55	56	54	49	44	39	34
8	8-in. l.w. concrete	31	0.126	35	30	26	22	18	14	11	9	7	7	9	13	19	25	33	39	46	50	53	54	53	49	45	40
9	4-in. h.w. concrete with 1-in. (or 2-in.) insulation	52 (52)	0.200 (0.120)	25	22	18	15	12	9	8	8	10	14	20	26	33	40	46	50	53	53	52	48	43	38	34	30
10	2.5-in. wood with 2-in. insulation	13	0.093	30	26	23	19	16	13	10	9	8	9	13	17	23	29	36	41	46	49	51	50	47	43	39	35
11	Roof terrace system	75	0.106	34	31	28	25	22	19	16	14	13	13	15	18	22	26	31	36	40	44	45	46	45	43	40	37
12	6-in. h.w. concrete with 1-in. (or 2-in.) insulation	75 (75)	0.192 (0.117)	31	28	25	22	20	17	15	14	14	16	18	22	26	31	36	40	43	45	45	44	42	40	37	34
13	4-in. wood with 1-in. (or 2-in) insulation	17 (18)	0.106 (0.078)	38	36	33	30	28	25	22	20	18	17	16	17	18	21	24	28	32	36	39	41	43	43	42	40
														With Suspended Ceiling												
1	Steel Sheet with 1-in. (or 2-in.) insulation	9 (10)	0.134 (0.092)	2	0	-2	-3	-4	-4	-1	9	23	37	50	62	71	77	78	74	67	56	42	28	18	12	8	5
2	1-in. wood with 1-in. insulation	10	0.115	20	15	11	8	5	3	2	3	7	13	21	30	40	48	55	60	62	61	58	51	44	37	30	25
3	4-in. l.w. concrete	20	0.134	19	14	10	7	4	2	0	0	4	10	19	29	39	48	56	62	65	64	61	54	46	38	30	24
4	2-in. h.w. concrete with 1-in. insulation	30	0.131	28	25	23	20	17	15	13	13	14	16	20	25	30	35	39	43	46	47	46	44	41	38	35	32
5	1-in. wood with 2-in. insulation	10	0.083	25	20	16	13	10	7	5	5	7	12	18	25	33	41	48	53	57	57	56	52	46	40	34	29
6	6-in. l.w. concrete	26	0.109	32	28	23	19	16	13	10	8	7	8	11	16	22	29	36	42	48	52	54	54	51	47	42	37
7	2.5-in. wood with 1-in. insulation	15	0.096	34	31	29	26	23	21	18	16	15	15	16	18	21	25	30	34	38	41	43	44	44	42	40	37
8	8-in. l.w. concrete	33	0.093	39	36	33	29	26	23	20	18	15	14	14	15	17	20	25	29	34	38	42	45	46	45	44	42
9	4-in. h.w. concrete with 1-in. (or 2-in.) insulation	53 (54)	0.128 (0.090)	30	29	27	26	24	22	21	20	20	21	22	24	27	29	32	34	36	38	38	38	37	36	34	33
10	2.5-in. wood with 2-in. insulation	15	0.072	35	33	30	28	26	24	22	20	18	18	18	20	22	25	28	32	35	38	40	41	41	40	39	37
11	Roof terrace system	77	0.082	30	29	28	27	26	25	24	23	22	22	22	23	23	25	26	28	29	31	32	33	33	33	33	32
12	6-in. h.w. concrete with 1-in. (or 2-in) insulation	77 (77)	0.125 (0.088)	29	28	27	26	25	24	23	22	21	21	22	23	25	26	28	30	32	33	34	34	34	33	32	31
13	4-in. wood with 1-in. (or 2-in.) insulation	19 (20)	0.082 (0.064)	35	34	33	32	31	29	27	26	24	23	22	21	22	22	24	25	27	30	32	34	35	36	37	36

Reprinted with permission from the *1985 Fundamentals, ASHRAE Handbook & Product Directory*.

t_R = room temperature, F
t_0 = average outside design temperature, F
f = correction for ceiling ventilation (for roof only)
  $f = 0.75$ for attic fan; use $f = 1.0$ otherwise

The hours shown in all tables are Solar Time, approximately equal to Standard Time. Add one hour to find Daylight Savings Time.

It is recommended that $K = 1.0$ to be used in any location where colors may darken from pollution or other causes.

TABLE 6.2. COOLING LOAD TEMPERATURE DIFFERENCES (CLTD) FOR CALCULATING LOAD FROM SUNLIT WALLS, 1, F

North Latitude Wall Facing	1	2	3	4	5	6	7	8	9	10	11	12	13	14	15	16	17	18	19	20	21	22	23	24	Hr of Maximum CLTD	Minimum CLTD	Maximum CLTD	Difference CLTD
Group A Walls																												
N	14	14	14	13	13	13	12	12	11	11	10	10	10	10	10	10	11	11	12	12	13	13	14	14	2	10	14	4
NE	19	19	19	18	17	17	16	15	15	15	15	15	16	16	17	18	18	18	19	19	20	20	20	20	22	15	20	5
E	24	24	23	23	22	21	20	19	19	18	19	19	20	21	22	23	24	24	25	25	25	25	25	25	22	18	25	7
SE	24	23	23	22	21	20	20	19	18	18	18	18	18	19	20	21	22	23	23	24	24	24	24	24	22	18	24	6
S	20	20	19	19	18	18	17	16	15	14	14	14	14	15	16	17	18	19	19	20	20	20	20	20	23	14	20	6
SW	25	25	25	24	24	23	22	21	20	19	19	18	17	17	17	17	18	19	20	22	23	24	25	25	24	17	25	8
W	27	27	26	26	25	24	24	23	22	21	20	19	19	18	18	18	18	19	20	22	23	25	26	26	1	18	27	9
NW	21	21	21	20	20	19	19	18	17	16	16	15	15	14	14	15	15	16	17	18	19	20	21		1	14	21	7
Group B Walls																												
N	15	14	14	13	12	11	11	10	9	9	9	8	9	9	9	10	11	12	13	14	14	15	15	15	24	8	15	7
NE	19	18	17	16	15	14	13	12	12	13	14	15	16	17	18	19	19	20	20	21	21	21	20	20	21	12	21	9
E	23	22	21	20	18	17	16	15	15	15	17	19	21	22	24	25	26	26	27	27	26	26	25	24	20	15	27	12
SE	23	22	21	20	18	17	16	15	14	14	15	16	18	20	21	23	25	26	26	26	26	25	24		21	14	26	12
S	21	20	19	18	17	15	14	13	12	11	11	11	11	12	14	15	17	19	20	21	22	22	22	21	23	11	22	11
SW	27	26	25	24	22	21	19	18	16	15	14	13	13	14	15	17	20	22	25	27	28	28	28		24	13	28	15
W	29	28	27	26	24	23	21	19	18	17	16	15	14	14	14	15	17	19	22	25	27	29	29	30	24	14	30	16
NW	23	22	21	20	19	18	17	15	14	13	12	12	12	11	12	12	13	15	17	19	21	22	23	23	24	11	23	12
Group C Walls																												
N	15	14	13	12	11	10	9	8	8	7	7	8	8	9	10	12	13	14	15	16	17	17	17	16	22	7	17	10
NE	19	17	16	14	13	11	10	10	11	13	15	17	19	20	21	22	22	23	23	23	22	21	20		20	10	23	13
E	22	21	19	17	15	14	12	12	14	16	19	22	25	27	29	30	30	30	29	28	27	26	24		18	12	30	18
SE	22	21	19	17	15	14	12	12	12	13	16	19	22	24	26	28	29	29	29	29	28	27	26	24	19	12	29	17
S	21	19	18	16	15	13	12	10	9	9	9	10	11	14	17	20	22	24	25	26	25	25	24	22	20	9	26	17
SW	29	27	25	22	20	18	16	15	13	12	11	11	11	13	15	18	22	26	29	32	33	33	32	31	21	11	33	22
W	31	29	27	25	22	20	18	16	14	13	12	12	12	13	14	16	20	24	29	32	35	35	35	33	22	12	35	23
NW	25	23	21	20	18	16	14	13	11	10	10	10	10	11	12	13	15	18	22	25	27	27	27	26	22	10	27	17
Group D Walls																												
N	15	13	12	10	9	7	6	6	6	7	8	10	12	13	15	17	18	19	19	19	18	16			21	6	19	13
NE	17	15	13	11	10	8	7	8	10	14	17	20	22	23	23	24	24	25	24	23	22	20	18		19	7	25	18
E	19	17	15	13	11	9	8	9	12	17	22	27	30	32	33	33	32	32	31	30	28	26	24	22	16	8	33	25
SE	20	17	15	13	11	10	8	8	10	13	17	22	26	29	31	32	32	32	31	30	28	26	24	22	17	8	32	24
S	19	17	15	13	11	9	8	7	6	6	7	9	12	16	20	24	27	29	29	29	27	26	24	22	19	6	29	23
SW	28	25	22	19	16	14	12	10	9	8	8	10	12	16	21	27	32	36	38	38	37	34	31		21	8	38	30
W	31	27	24	21	18	15	13	11	10	9	9	9	10	11	14	18	24	30	36	40	41	40	38	34	21	9	41	32
NW	25	22	19	17	14	12	10	9	8	7	7	8	9	10	12	14	18	22	27	31	32	32	30	27	22	7	32	25
Group E Walls																												
N	12	10	8	7	5	4	3	4	5	6	9	11	13	15	17	19	20	21	23	20	18	16	14		20	3	22	19
NE	13	11	9	7	6	4	5	9	15	20	24	25	25	26	26	26	26	25	24	22	19	17	15		16	4	26	22
E	14	12	10	8	6	5	6	11	18	26	33	36	37	36	34	33	32	30	28	25	22	20	17		13	5	38	33
SE	15	12	10	8	7	5	5	8	12	19	25	31	35	37	37	36	34	33	31	28	26	23	20	17	15	5	37	32
S	15	12	10	8	7	5	4	3	4	5	9	13	19	24	29	32	34	33	31	29	26	23	20	17	17	3	34	31
SW	22	18	15	12	10	8	6	5	5	6	7	9	12	18	24	32	38	43	45	44	40	35	30	26	19	5	45	40
W	25	21	17	14	11	9	7	6	6	6	7	9	11	14	20	27	36	43	49	49	45	40	34	29	20	6	49	43
NW	20	17	14	11	9	7	6	5	5	5	6	8	10	13	16	20	26	32	37	38	36	32	28	24	20	5	38	33
Group F Walls																												
N	8	6	5	3	2	1	2	4	6	7	9	11	14	17	19	21	22	23	24	23	20	16	13	11	19	1	24	23
NE	9	7	5	3	2	1	5	14	23	28	30	29	28	27	27	27	26	24	22	19	16	13	11		11	1	30	29
E	10	7	6	4	3	2	6	17	28	38	44	45	43	39	36	34	32	30	27	24	21	17	15	12	12	2	45	43
SE	10	7	6	4	3	2	4	10	19	28	36	41	43	42	39	36	34	31	28	25	21	18	15	12	13	2	43	41
S	10	8	6	4	3	2	1	1	3	7	13	20	27	34	38	39	38	35	31	26	22	18	15	12	16	1	39	38
SW	15	11	9	6	5	3	2	2	4	5	8	11	17	26	35	44	50	53	52	45	37	28	23	18	18	2	53	51
W	17	13	10	7	5	4	3	3	4	6	8	11	14	20	28	39	49	57	60	54	43	34	27	21	19	3	60	57
NW	14	10	8	6	4	3	2	2	3	5	8	10	13	15	21	27	35	42	46	43	35	28	22	18	19	2	46	44
Group G Walls																												
N	3	2	1	0	-1	2	7	8	9	12	15	18	21	23	24	24	25	26	22	15	11	9	7	5	18	-1	26	27
NE	3	2	1	0	-1	9	27	36	39	35	30	26	26	27	27	26	25	22	18	14	11	9	7	5	9	-1	39	40
E	4	2	1	0	-1	11	31	47	54	55	50	40	33	31	30	29	27	24	19	15	12	10	8	6	10	-1	55	56
SE	4	2	1	0	-1	5	18	32	42	49	51	48	42	36	32	30	27	24	19	15	12	10	8	6	11	-1	51	52
S	4	2	1	0	0	1	5	12	22	31	39	45	46	43	37	31	25	20	15	12	10	8	5		14	-1	46	47
SW	5	4	3	1	0	0	2	5	8	12	16	26	38	50	59	63	61	52	37	24	17	13	10	8	16	0	63	63
W	6	5	3	2	1	1	2	5	8	11	15	19	27	41	56	67	72	67	48	29	20	15	11	8	17	1	72	71
NW	5	3	2	1	0	0	2	5	8	11	15	18	21	27	37	47	55	55	41	25	17	13	10	7	18	0	55	55

Reprinted with permission from the *1985 Fundamentals, ASHRAE Handbook & Product Directory*.

TABLE 6.3. WALL CONSTRUCTION GROUP DESCRIPTION

Group No.	Description of Construction	Weight (lb/ft²)	U-Value Btu/(hr·ft²·F)	Heat Capacity Btu/(ft²·F)
4-in. Face Brick+(*Brick*)				
C	Air Space+4-in. Face Brick	83	0.358	18.3
D	4-in. Common Brick	90	0.415	18.4
C	1-in. Insulation or Air space+4-in. Common Brick	90	0.174-0.301	18.4
B	2-in. Insulation+4-in. Common Brick	88	0.111	18.5
B	8-in. Common Brick	130	0.302	26.4
A	Insulation or Air space+8-in. Common Brick	130	0.154-0.243	26.4
4-in. Face Brick+(*H.W. Concrete*)				
C	Air Space+2-in. Concrete	94	0.350	19.7
B	2-in. Insulation+4-in. concrete	97	0.116	19.8
A	Air Space or Insulation+8-in. or more Concrete	143-190	0.110-0.112	29.1-38.4
4-in. Face Brick+(*L.W. or H.W. Concrete Block*)				
E	4-in. Block	62	0.319	12.9
D	Air Space or Insulation+4-in. Block	62	0.153-0.246	12.9
D	8-in. Block	70	0.274	15.1
C	Air Space or 1-in. Insulation+6-in. or 8-in. Block	73-89	0.221-0.275	15.5-18.5
B	2-in. Insulation+8-in. Block	89	0.096-0.107	15.5-18.6
4-in Face Brick+(*Clay Tile*)				
D	4-in. Tile	71	0.381	15.1
D	Air Space+4-in. Tile	71	0.281	15.1
C	Insulation+4-in. Tile	71	0.169	15.1
C	8-in. Tile	96	0.275	19.7
B	Air Space or 1-in. Insulation+8-in. Tile	96	0.142-0.221	19.7
A	2-in. Insulation+8-in. Tile	97	0.097	19.8
H.W. Concrete Wall+(*Finish*)				
E	4-in. Concrete	63	0.585	12.5
D	4-in. Concrete+1-in. or 2-in. Insulation	63	0.119-0.200	12.5
C	2-in. Insulation+4-in. Concrete	63	0.119	12.7
C	8-in. Concrete	109	0.490	21.9
B	8-in. Concrete+1-in. or 2-in. Insulation	110	0.115-0.187	22.0
A	2-in. Insulation+8-in. Concrete	110	0.115	21.9
E	12-in. Concrete	156	0.421	31.2
A	12-in. Concrete+Insulation	156	0.113	31.3
L.W. and H.W. Concrete Block+(*Finish*)				
F	4-in. Block+Air Space/Insulation	29-36	0.161-0.263	5.7-7.2
E	2-in. Insulation+4-in. Block	29-37	0.105-0.114	5.8-7.3
E	8-in. Block	41-57	0.294-0.402	6.3-11.3
D	8-in. Block+Air Space/Insulation	41-57	0.149-0.173	8.3-11.3
Clay *Tile*+(*Finish*)				
F	4-in. Tile	39	0.419	7.8
F	4-in. Tile+Air space	39	0.303	7.8
E	4-in. Tile+1-in. Insulation	39	0.175	7.9
D	2-in. Insulation+4-in. Tile	40	0.110	7.9
D	8-in. Tile	63	0.296	12.5
C	8-in. Tile+Air Space/1-in. Insulation	63	0.151-0.231	12.6
B	2-in. Insulation+8-in. Tile	63	0.099	12.6
Metal Curtain Wall				
G	With/without Air Space+1-in./2-in./3-in. Insulation	5-6	0.091-0.230	0.7
Frame Wall				
G	1-in. to 3-in. Insulation	16	0.081-0.178	3.2

Reprinted with permission from the *1985 Fundamentals, ASHRAE Handbook & Product Directory*.

TABLE 6.4. CLTD CORRECTION FOR LATITUDE AND MONTH APPLIED TO WALLS AND ROOFS, NORTH LATITUDES, 1, F

Lat.	Month	N	NNE NNW	NE NW	ENE WNW	E W	ESE WSW	SE SW	SSE SSW	S	HOR
0	Dec	-3	-5	-5	-5	-2	0	3	6	9	-1
	Jan/Nov	-3	-5	-4	-4	-1	0	2	4	7	-1
	Feb/Oct	-3	-2	-2	-2	-1	-1	0	-1	0	0
	Mar/Sept	-3	0	1	-1	-1	-3	-3	-5	-8	0
	Apr/Aug	5	4	3	0	-2	-5	-6	-8	-8	-2
	May/Jul	10	7	5	0	-3	-7	-8	-9	-8	-4
	Jun	12	9	5	0	-3	-7	-9	-10	-8	-5
8	Dec	-4	-6	-6	-6	-3	0	4	8	12	-5
	Jan/Nov	-3	-5	-6	-5	-2	0	3	6	10	-4
	Feb/Oct	-3	-4	-3	-3	-1	-1	1	2	4	-1
	Mar/Sept	-3	-2	-1	-1	-1	-2	-2	-3	-4	0
	Apr/Aug	2	2	2	0	-1	-4	-5	-7	-7	-1
	May/Jul	7	5	4	0	-2	-5	-7	-9	-7	-2
	Jun	9	6	4	0	-2	-6	-8	-9	-7	-2
16	Dec	-4	-6	-8	-8	-4	-1	4	9	13	-9
	Jan/Nov	-4	-6	-7	-7	-4	-1	4	8	12	-7
	Feb/Oct	-3	-5	-5	-4	-2	0	2	5	7	-4
	Mar/Sept	-3	-3	-2	-2	-1	-1	0	0	0	-1
	Apr/Aug	-1	0	-1	-1	-1	-3	-3	-5	-6	0
	May/Jul	4	3	3	0	-1	-4	-5	-7	-7	0
	Jun	6	4	4	1	-1	-4	-6	-8	-7	0
24	Dec	-5	-7	-9	-10	-7	-3	3	9	13	-13
	Jan/Nov	-4	-6	-8	-9	-6	-3	3	9	13	-11
	Feb/Oct	-4	-5	-6	-6	-3	-1	3	7	10	-7
	Mar/Sept	-3	-4	-3	-3	-1	-1	1	2	4	-3
	Apr/Aug	-2	-1	0	-1	-1	-2	-1	-2	-3	0
	May/Jul	1	2	2	0	0	-3	-3	-5	-6	1
	Jun	3	3	3	1	0	-3	-4	-6	-6	1
32	Dec	-5	-7	-10	-11	-8	-5	2	9	12	-17
	Jan/Nov	-5	-7	-9	-11	-8	-4	2	9	12	-15
	Feb/Oct	-4	-6	-7	-8	-4	-2	4	8	11	-10
	Mar/Sep	-3	-4	-4	-4	-2	-1	3	5	7	-5
	Apr/Aug	-2	-2	-1	-2	0	-1	0	1	1	-1
	May/Jul	1	1	1	0	0	-1	-1	-3	-3	1
	Jun	1	2	2	1	0	-2	-2	-4	-4	2
40	Dec	-6	-8	-10	-13	-10	-7	0	7	10	-21
	Jan/Nov	-5	-7	-10	-12	-9	-6	1	8	11	-19
	Feb/Oct	-5	-7	-8	-9	-6	-3	3	8	12	-14
	Mar/Sep	-4	-5	-5	-6	-3	-1	4	7	10	-8
	Apr/Aug	-2	-3	-2	-2	0	0	2	3	4	-3
	May/Jul	0	0	0	0	0	0	0	0	1	1
	Jun	1	1	1	0	1	0	0	-1	-1	2
48	Dec	-6	-8	-11	-14	-13	-10	-3	2	6	-25
	Jan/Nov	-6	-8	-11	-13	-11	-8	-1	5	8	-24
	Feb/Oct	-5	-7	-10	-11	-8	-5	1	8	11	-18
	Mar/Sep	-4	-6	-6	-7	-4	-1	4	8	11	-11
	Apr/Aug	-3	-3	-3	-3	-1	0	4	6	7	-5
	May/Jul	0	-1	0	0	1	1	3	3	4	0
	Jun	1	1	2	1	2	1	2	2	3	2
56	Dec	-7	-9	-12	-16	-16	-14	-9	-5	-3	-28
	Jan/Nov	-6	-8	-11	-15	-14	-12	-6	-1	2	-27
	Feb/Oct	-6	-8	-10	-12	-10	-7	0	6	9	-22
	Mar/Sep	-5	-6	-7	-8	-5	-2	4	8	12	-15
	Apr/Aug	-3	-4	-4	-4	-1	1	5	7	9	-8
	May/Jul	0	0	0	0	2	2	5	6	7	-2
	Jun	2	1	2	1	3	3	4	5	6	1

Reprinted with permission from the 1979 *ASHRAE Load Calculations Manual.*

EXAMPLE 6.1

A 30 ft × 40 ft roof of a building in Washington, D.C., is constructed of 4 in. h.w. (heavy weight) concrete with 1 in. insulation and a suspended ceiling. The inside design condition is 78 F. Determine the net heat gain to the building through the roof at 2 PM Solar Time on July 21.

Solution. From Table 6.1 the roof is type No. 9, and at 2 PM (14 hrs), CLTD = 29 F.

Correction for *LM* = 1 F for July. (From Table 6.4, use HOR column. D.C. is at 38°N latitude, From Table A.6. Use 40°N.)

$$K = 1.0 \text{ (assume roof is dark)}$$
$$f = 1.0 \text{ (no ceiling ventilation)}$$

Find average outside summer design temperature as follows: From Table A.9, summer design DB temperature = 91 F and range = 18 F. Therefore

$$t_0 = \frac{(91 - 18)}{2} = 83 \text{ F}$$

Find CLTD$_c$:

$$\text{CLTD}_c = ([\text{CLTD} + \text{LM}]) \times K + [78 - t_R]$$
$$+ ([t_0 - 85]) \times f$$
$$= ([29 + 1] \times 1 + [78 - 78]$$
$$+ [83 - 85]) \times 1$$
$$= 28 \text{ F}$$
$$U = 0.13 \text{ (Table 6.1)}$$
$$A = 30 \text{ ft} \times 40 \text{ ft} = 1200 \text{ ft}^2$$

Find conduction heat gain (Equation 6.1)

$$Q = U \times A \times \text{CLTD}_c$$
$$= 0.13 \times 1200 \times 28$$
$$= 4370 \text{ BTU/hr}$$

EXAMPLE 6.2

A south-facing masonry wall of a building in Pittsburgh, Pa., has a net area of 2200 ft². The wall is constructed of 4 in. face brick, 4 in. concrete cinder block and 3/8 in. gypsum board interior. The inside design temperature is 76 F. Determine the net heat gain through the wall at 4 PM Solar Time on June 21.

Solution. The wall is in Group E (Table 6.3). From Table 6.2, CLTD = 32 F. From Table 6.4 LM = −1 F. From Table A.6, the mean daily range is 19 F. Therefore, the average outside design temperature is

$$t_0 = 88 - \tfrac{19}{2} = 78 \text{ F}$$

Using Equation 6.2

$$\text{CLTD}_c = ([\text{CLTD} + \text{LM}] \times K + [78 - t_R]$$
$$+ [t_0 - 85])$$
$$= (32 - 1) \times 1 + (78 - 76)$$
$$+ (78 - 85)$$
$$= 26 \text{ F}$$

From Table 6.3, U = 0.32. Using Equation 6.1,

$$Q = U \times A \times \text{CLTD}_c = 0.32 \times 2200 \times 26$$
$$= 18,300 \text{ BTU/hr}$$

For those familiar with past methods for calculating loads, note that the CLTD is used instead of the Total Equivalent Temperature Difference (TETD). The TETD does not completely account for the heat storage effect.

Table 6.5 lists CLTD values for glass. The corrected CLTD is found as follows

$$\text{CLTD}_c = \text{CLTD} + (78 - t_R) + (t_0 - 85) \tag{6.3}$$

These corrections have the same meaning as with walls and roofs.

Values in Table 6.5 may be interpolated between the listed hours.

TABLE 6.5. COOLING LOAD TEMPERATURE DIFFERENCES
FOR CONDUCTION THROUGH GLASS

Hour	2	4	6	8	10	12	14	16	18	20	22	24
CLTD, F	0	−2	−2	0	4	9	13	14	12	8	4	2

Reprinted with permission from the *1985 Fundamentals, ASHRAE Handbook & Product Directory*.

EXAMPLE 6.3

A room has 130 ft^2 of exterior single glass with no interior shading. The inside design condition is 78 F and the outdoor daily average temperature is 88 F. Determine the net conduction heat gain through the glass at 12 noon Solar Time in summer.

Solution. From Table 6.5, CLTD = 9 F. Correcting this by Equation 6.3,

$$CLTD_c = CLTD + (78 - t_R) + (t_0 - 85)$$
$$= 9 + 0 + 3 = 12 \text{ F}$$

From Table A.8, U = 1.04 BTU/hr-ft^2-F. Using Equation 6.1,

$$Q = U \times A \times CLTD = 1.04 \times 130 \times 12$$
$$= 1620 \text{ BTU/hr}$$

6.5. CONDUCTION THROUGH INTERIOR STRUCTURE

The heat that flows from interior unconditioned spaces to the conditioned spaced through partitions, floors, and ceilings can be found from Equation 3.5

$$Q = U \times A \times TD \qquad (3.5)$$

where

Q = heat transfer rate through partition, floor, or ceiling, BTU/hr
U = overall heat transfer coefficient for partition, floor, or ceiling, BTU/hr-ft^2-F
A = area of partition, floor, or ceiling, ft^2
TD = temperature difference between unconditioned and conditioned space, F

If the temperature of the unconditioned space is not known, an approximation often used is to assume that it' is at 5 F less than the outdoor temperature. Spaces with heat sources, such as boiler rooms, may be at a much higher temperature.

6.6. SOLAR RADIATION THROUGH GLASS

Radiant energy from the sun passes through transparent materials such as glass and becomes a heat gain to the room. Its value varies with time, orientation, shading, and storage effect. The net heat gain can be found from the following equation:

$$Q = SHGF \times A \times SC \times CLF \qquad (6.4)$$

where

Q = net solar radiation heat gain through glass, BTU/hr
SHGF = maximum solar heat gain factor, BTU/hr-ft^2
A = area of glass, ft^2
SC = shading coefficient
CLF = cooling load factor for glass

The maximum solar heat gain factor (SHGF) is the maximum solar heat gain through 1/8 in. single clear glass at a given month, orientation, and latitude. Values are shown in Table 6.6 for the twenty-first day of each month.

TABLE 6.6. MAXIMUM SOLAR HEAT GAIN FACTORS FOR GLASS, BTU/HR-FT^2, NORTH LATITUDES

0 Deg

	N	NNE/NNW	NE/NW	ENE/WNW	E/W	ESE/WSW	SE/SW	SSE/SSW	S	HOR
Jan.	34	34	88	177	234	254	235	182	118	296
Feb.	36	39	132	205	245	247	210	141	67	306
Mar.	38	87	170	223	242	223	170	87	38	303
Apr.	71	134	193	224	221	184	118	38	37	284
May	113	164	203	218	201	154	80	37	37	265
June	129	173	206	212	191	140	66	37	37	255
July	115	164	201	213	195	149	77	38	38	260
Aug.	75	134	187	216	212	175	112	39	38	276
Sep.	40	84	163	213	231	213	163	84	40	293
Oct.	37	40	129	199	236	238	202	135	66	299
Nov.	35	35	88	175	230	250	230	179	117	293
Dec.	34	34	71	164	226	253	240	196	138	288

16 Deg

	N	NNE/NNW	NE/NW	ENE/WNW	E/W	ESE/WSW	SE/SW	SSE/SSW	S	HOR
Jan.	30	30	55	147	210	244	251	223	199	248
Feb.	33	33	96	180	231	247	233	188	154	275
Mar.	35	53	140	205	239	235	197	138	93	291
Apr.	39	99	172	216	227	204	150	77	45	289
May	52	132	189	218	215	179	115	45	41	282
June	66	142	194	217	207	167	99	41	41	277
July	55	132	187	214	210	174	111	44	42	277
Aug.	41	100	168	209	219	196	143	74	46	282
Sep.	36	50	134	196	227	224	191	134	93	282
Oct.	33	33	95	174	223	237	225	183	150	270
Nov.	30	30	55	145	206	241	247	220	196	246
Dec.	29	29	41	132	198	241	254	233	212	234

8 Deg

	N	NNE/NNW	NE/NW	ENE/WNW	E/W	ESE/WSW	SE/SW	SSE/SSW	S	HOR
Jan.	32	32	71	163	224	250	242	203	162	275
Feb.	34	34	114	193	239	248	219	165	110	294
Mar.	37	67	156	215	241	230	184	110	55	300
Apr.	44	117	184	221	225	195	134	53	39	289
May	74	146	198	220	209	167	97	39	38	277
June	90	155	200	217	200	141	82	39	39	269
July	77	145	195	215	204	162	93	40	39	272
Aug.	47	117	179	214	216	186	128	51	41	282
Sep.	38	66	149	205	230	219	176	107	56	290
Oct.	35	35	112	187	231	239	211	160	108	288
Nov.	33	33	71	161	220	245	233	200	160	273
Dec.	31	31	55	149	215	246	247	215	179	265

24 Deg

	N	NNE/NNW	NE/NW	ENE/WNW	E/W	ESE/WSW	SE/SW	SSE/SSW	S	HOR
Jan.	27	27	41	128	190	240	253	241	227	214
Feb.	30	30	80	165	220	244	243	213	192	249
Mar.	34	45	124	195	234	237	214	168	137	275
Apr.	37	88	159	209	228	212	169	107	75	283
May	43	117	178	214	218	190	132	67	46	282
June	55	127	184	214	212	179	117	55	43	279
July	45	116	176	210	213	185	129	65	46	278
Aug.	38	87	156	203	220	204	162	103	72	277
Sep.	35	42	119	185	222	225	206	163	134	266
Oct.	31	31	79	159	211	237	235	207	187	244
Nov.	27	27	42	126	187	236	249	237	224	213
Dec.	26	26	29	112	180	234	247	247	237	199

EXAMPLE 6.4

What is the maximum solar heat gain factor through the windows on the Southwest side of a building located at 32°N latitude on September twenty-first?

Solution. From Table 6.6 SHGF = 218 BTU/hr-ft^2.

The SHGF gives maximum heat gain values only for the type of glass noted and without any shading devices. To account for heat gains with different fenestration arrangements, the shading coefficient SC is introduced. The SC is the pro-portional amount of maximum heat gain through arrangements other than 1/8 in. single clear glass. Table 6.7 lists some values of SC.

EXAMPLE 6.5

What is the value of SC to be applied to the solar heat gain for 1/8 in. single clear glass and medium colored inside venetian blinds?

Solution. From Table 6.7, SC = 0.64.

The cooling load factor CLF accounts for the storage of part of the solar heat gain. Values of CLF to be applied to the solar heat gain calcula-

TABLE 6.6 *(Continued).*

32 Deg

	N (Shade)	NNE/ NNW	NE/ NW	ENE/ WNW	E/ W	ESE/ WSW	SE/ SW	SSE/ SSW	S	HOR
Jan.	24	24	29	105	175	229	249	250	246	176
Feb.	27	27	65	149	205	242	248	232	221	217
Mar.	32	37	107	183	227	237	227	195	176	252
Apr.	36	80	146	200	227	219	187	141	115	271
May	38	111	170	208	220	199	155	99	74	277
June	44	122	176	208	214	189	139	83	60	276
July	40	111	167	204	215	194	150	96	72	273
Aug.	37	79	141	195	219	210	181	136	111	265
Sep.	33	35	103	173	215	227	218	189	171	244
Oct.	28	28	63	143	195	234	239	225	215	213
Nov.	24	24	29	103	173	225	245	246	243	175
Dec.	22	22	22	84	162	218	246	252	252	158

48 Deg

	N (Shade)	NNE/ NNW	NE/ NW	ENE/ WNW	E/ W	ESE/ WSW	SE/ SW	SSE/ SSW	S	HOR
Jan.	15	15	15	53	118	175	216	239	245	85
Feb.	20	20	36	103	168	216	242	249	250	138
Mar.	26	26	80	154	204	234	239	232	228	188
Apr.	31	61	132	180	219	225	215	194	186	226
May	35	97	158	200	218	214	192	163	150	247
June	46	110	165	204	215	206	180	148	134	252
July	37	96	156	196	214	209	187	158	146	244
Aug.	33	61	128	174	211	216	208	188	180	223
Sep.	27	27	72	144	191	223	228	223	220	182
Oct.	21	21	35	96	161	207	233	241	242	136
Nov.	15	15	15	52	115	172	212	234	240	85
Dec.	13	13	13	36	91	156	195	225	233	65

40 Deg

	N (Shade)	NNE/ NNW	NE/ NW	ENE/ WNW	E/ W	ESE/ WSW	SE/ SW	SSE/ SSW	S	HOR
Jan.	20	20	20	74	154	205	241	252	254	133
Feb.	24	24	50	129	186	234	246	244	241	180
Mar.	29	29	93	169	218	238	236	216	206	223
Apr.	34	71	140	190	224	223	203	170	154	252
May	37	102	165	202	220	208	175	133	113	265
June	48	113	172	205	216	199	161	116	95	267
July	38	102	163	198	216	203	170	129	109	262
Aug.	35	71	135	185	216	214	196	165	149	247
Sep.	30	30	87	160	203	227	226	209	200	215
Oct.	25	25	49	123	180	225	238	236	234	177
Nov.	20	20	20	73	151	201	237	248	250	132
Dec.	18	18	18	60	135	188	232	249	253	113

56 Deg

	N (Shade)	NNE/ NNW	NE/ NW	ENE/ WNW	E/ W	ESE/ WSW	SE/ SW	SSE/ SSW	S	HOR
Jan.	10	10	10	21	74	126	169	194	205	40
Feb.	16	16	21	71	139	184	223	239	244	91
Mar.	22	22	65	136	185	224	238	241	241	149
Apr.	28	58	123	173	211	223	223	213	210	195
May	36	99	149	195	215	206	187	181	192	222
June	53	111	160	199	213	213	196	174	168	231
July	37	98	147	192	211	214	201	183	177	221
Aug.	30	56	119	165	203	216	215	206	203	193
Sep.	23	23	58	126	171	211	227	230	231	144
Oct.	16	16	20	68	132	176	213	229	234	91
Nov.	10	10	10	21	72	122	165	190	200	40
Dec.	7	7	7	7	47	92	135	159	171	23

tion are shown in Table 6.8. Note that there are separate listings for Light (L), Medium (M), and Heavy (H) construction, as described. The value are also dependent on whether or not there are internal shading devices.

EXAMPLE 6.6

A building wall facing southwest has a window area of 240 ft². The glass is $\frac{1}{4}$ in. single clear glass with light-colored interior venetian blinds. The building is of medium construction, and is located at 40°N latitude. Find the net heat gain through the window due to solar radiation in August at 3 PM Solar Time.

Solution. Equation 6.4 will be used. From the appropriate tables the terms in the equation are selected:

$$Q = \text{SHGF} \times A \times \text{SC} \times \text{CLF}$$
$$= 196 \times 240 \times 0.55 \times 0.83$$
$$= 21{,}470 \text{ BTU/hr}$$

The area of glass is that portion of the area which receives *direct* solar radiation. Outside projections or adjacent buildings may cut off radiation, shading part or all of the glass. The effect of overhead projections can be found from Table 6.9. The values in the table are the feet of shade on the wall caused per foot of extension of overhang.

TABLE 6.7. SHADING COEFFICIENTS FOR GLASS WITHOUT OR WITH INTERIOR SHADING BY VENETIAN BLINDS OR ROLLER SHADES

	Type of Glass	Nominal Thickness Each Light [a]	Solar Trans. [b]	No Interior Shading $h_o = 4.0$	Type of Interior Shading				
					Venetian Blinds		Roller Shades		
							Opaque		Translucent
					Medium	Light	Dark	Light	Light
SINGLE GLASS	Single								
	Clear	3/32 to 1/4	0.87-0.80	1.00					
	Clear	1/4 to 1/2	0.80-0.71	0.94					
	Clear	3/8	0.72	0.90	0.64	0.55	0.59	0.25	0.39
	Clear	1/2	0.67	0.87					
	Clear Pattern	1/8 to 9/32	0.87-0.79	0.83					
	Heat Absorbing Pattern	1/8		0.83					
	Heat Absorbing [c]	3/16 to 1/4	0.46	0.69					
	Heat Absorbing Pattern	3/16 to 1/4		0.69	0.57	0.53	0.45	0.30	0.36
	Tinted	1/8 to 7/32	0.59-0.45	0.69					
	Heat Absorbing or Pattern		0.44-0.30	0.60	0.54	0.52	0.40	0.28	0.32
	Heat Absorbing [c]	3/8	0.34	0.60					
	Heat Absorbing or Pattern	1/2	0.44-0.30 0.24	0.53	0.42	0.40	0.36	0.28	0.31
	Reflective Coated Glass			0.30	0.25	0.23			
				0.40	0.33	0.29			
				0.50	0.42	0.38			
				0.60	0.50	0.44			
INSULATING GLASS	Double [d]								
	Clear Out Clear In	3/32, 1/8	0.71 [a]	0.88	0.57	0.51	0.60	0.25	0.37
	Clear Out Clear In	1/4	0.61 [a]	0.81					
	Heat Absorbing Out Clear In	1/4	0.36 [a]	0.55	0.39	0.36	0.40	0.22	0.30
	Reflective Coated Glass			0.20	0.19	0.18			
				0.30	0.27	0.26			
				0.40	0.34	0.33			
	Triple								
	Clear	1/4		0.71					
	Clear	1/8		0.80					

Reprinted with permission from the *1985 Fundamentals, ASHRAE Handbook & Product Directory*.

EXAMPLE 6.7

A building at 32°N latitude has a wall facing west with a 4 ft overhang, and a 5 ft wide by 6 ft high window whose top is 1 ft below the overhang. How much of the glass receives direct solar radiation at 3 PM?

Solution. Table 6.9 will be used. Figure 6.4 shows the arrangement. The vertical proportion of shade, from Table 6.9, is 0.97. The total vertical distance the shade extends down is therefore

$$L = 0.97 \times 4 = 3.9 \text{ ft}$$

The height of shade on the window is $3.9 - 1 = 2.9$ ft, and the unshaded height is $6 - 2.9 = 3.1$ ft. The unshaded area of window is

$$A = 3.1 \times 5 = 15.5 \text{ ft}^2$$

When glass is shaded from outside obstructions it still receives some indirect solar radiation

TABLE 6.8. COOLING LOAD FACTORS FOR GLASS WITHOUT INTERIOR SHADING (INCLUDES REFLECTIVE AND HEAT ABSORBING GLASS)

N. Latitude Fenestration Facing	Room Construction	1	2	3	4	5	6	7	8	9	10	11	12	13	14	15	16	17	18	19	20	21	22	23	24
N	L	0.17	0.14	0.11	0.09	0.08	0.33	0.42	0.48	0.56	0.63	0.71	0.76	0.80	0.82	0.82	0.79	0.80	0.84	0.61	0.48	0.38	0.31	0.25	0.20
	M	0.23	0.20	0.18	0.16	0.14	0.34	0.41	0.46	0.52	0.59	0.65	0.70	0.73	0.75	0.76	0.74	0.75	0.79	0.61	0.50	0.42	0.36	0.31	0.27
	H	0.25	0.23	0.21	0.20	0.19	0.38	0.45	0.50	0.55	0.60	0.65	0.69	0.72	0.73	0.72	0.70	0.70	0.74	0.57	0.46	0.39	0.34	0.31	0.28
NE	L	0.04	0.04	0.03	0.02	0.02	0.23	0.41	0.51	0.51	0.45	0.39	0.36	0.33	0.31	0.28	0.26	0.23	0.19	0.15	0.12	0.10	0.08	0.06	0.05
	M	0.07	0.06	0.06	0.05	0.04	0.21	0.36	0.44	0.45	0.40	0.36	0.33	0.31	0.30	0.28	0.26	0.23	0.21	0.17	0.15	0.13	0.11	0.09	0.08
	H	0.09	0.08	0.08	0.07	0.07	0.23	0.37	0.44	0.44	0.39	0.34	0.31	0.29	0.27	0.26	0.24	0.22	0.20	0.16	0.14	0.13	0.12	0.11	0.10
E	L	0.04	0.04	0.03	0.02	0.02	0.19	0.37	0.51	0.57	0.57	0.51	0.42	0.36	0.32	0.29	0.25	0.22	0.19	0.14	0.12	0.09	0.08	0.06	0.05
	M	0.07	0.06	0.06	0.05	0.04	0.18	0.33	0.44	0.50	0.51	0.45	0.39	0.35	0.32	0.29	0.26	0.23	0.21	0.17	0.15	0.13	0.11	0.10	0.08
	H	0.09	0.09	0.08	0.08	0.07	0.21	0.34	0.45	0.50	0.49	0.43	0.36	0.32	0.29	0.26	0.24	0.22	0.19	0.17	0.15	0.13	0.12	0.11	0.10
SE	L	0.05	0.04	0.04	0.03	0.02	0.13	0.28	0.43	0.55	0.62	0.63	0.57	0.48	0.42	0.37	0.33	0.28	0.24	0.19	0.15	0.12	0.10	0.08	0.07
	M	0.09	0.08	0.07	0.06	0.05	0.14	0.26	0.38	0.48	0.54	0.55	0.51	0.45	0.40	0.36	0.33	0.29	0.25	0.21	0.18	0.16	0.14	0.12	0.10
	H	0.11	0.10	0.10	0.09	0.08	0.17	0.28	0.40	0.49	0.53	0.53	0.48	0.41	0.36	0.33	0.30	0.27	0.24	0.20	0.18	0.16	0.14	0.13	0.12
S	L	0.08	0.07	0.05	0.04	0.04	0.06	0.09	0.14	0.22	0.34	0.48	0.59	0.65	0.65	0.59	0.50	0.43	0.36	0.28	0.22	0.18	0.15	0.12	0.10
	M	0.12	0.11	0.09	0.08	0.07	0.08	0.11	0.14	0.21	0.31	0.42	0.52	0.57	0.58	0.53	0.47	0.41	0.36	0.29	0.25	0.21	0.18	0.16	0.14
	H	0.13	0.12	0.12	0.11	0.10	0.12	0.14	0.17	0.24	0.33	0.43	0.51	0.56	0.55	0.50	0.43	0.38	0.32	0.26	0.22	0.20	0.18	0.16	0.15
SW	L	0.12	0.10	0.08	0.06	0.05	0.06	0.08	0.10	0.12	0.14	0.16	0.24	0.36	0.49	0.60	0.66	0.66	0.58	0.43	0.33	0.27	0.22	0.18	0.14
	M	0.15	0.13	0.12	0.10	0.09	0.09	0.10	0.12	0.13	0.15	0.17	0.23	0.33	0.44	0.53	0.58	0.59	0.53	0.41	0.33	0.28	0.24	0.21	0.18
	H	0.15	0.14	0.13	0.12	0.11	0.12	0.13	0.14	0.16	0.17	0.19	0.25	0.34	0.44	0.52	0.56	0.56	0.49	0.37	0.30	0.25	0.21	0.19	0.17
W	L	0.12	0.10	0.08	0.07	0.05	0.06	0.07	0.08	0.10	0.11	0.13	0.14	0.20	0.32	0.45	0.57	0.64	0.61	0.44	0.34	0.27	0.22	0.18	0.14
	M	0.15	0.13	0.11	0.10	0.09	0.09	0.09	0.10	0.11	0.12	0.13	0.14	0.19	0.29	0.40	0.50	0.56	0.55	0.41	0.33	0.27	0.23	0.20	0.17
	H	0.14	0.13	0.12	0.11	0.10	0.11	0.12	0.13	0.14	0.14	0.15	0.16	0.21	0.30	0.40	0.49	0.54	0.52	0.38	0.30	0.24	0.21	0.18	0.16
NW	L	0.11	0.09	0.08	0.06	0.05	0.06	0.08	0.10	0.12	0.14	0.16	0.17	0.19	0.23	0.33	0.47	0.59	0.60	0.43	0.33	0.26	0.21	0.17	0.14
	M	0.14	0.12	0.11	0.09	0.08	0.09	0.10	0.11	0.13	0.14	0.16	0.17	0.18	0.21	0.30	0.42	0.51	0.53	0.39	0.32	0.26	0.22	0.19	0.16
	H	0.14	0.12	0.11	0.11	0.10	0.11	0.12	0.13	0.15	0.16	0.18	0.19	0.19	0.22	0.30	0.41	0.50	0.51	0.36	0.29	0.23	0.20	0.17	0.15
HOR	L	0.11	0.09	0.07	0.06	0.05	0.07	0.14	0.24	0.36	0.48	0.58	0.66	0.72	0.74	0.73	0.67	0.59	0.47	0.37	0.30	0.24	0.19	0.16	0.13
	M	0.16	0.14	0.12	0.11	0.09	0.11	0.16	0.24	0.43	0.43	0.52	0.59	0.64	0.67	0.66	0.62	0.55	0.47	0.38	0.32	0.28	0.24	0.21	0.18
	H	0.17	0.16	0.15	0.14	0.13	0.15	0.20	0.27	0.45	0.45	0.52	0.59	0.62	0.64	0.62	0.58	0.51	0.42	0.35	0.29	0.26	0.23	0.21	0.19

Solar Time, hr

L = Light construction: frame exterior wall, 2-in. concrete floor slab, approximately 30 lb of material/square feet of floor area.
M = Medium construction: 4-in. concrete exterior wall, 4-in. concrete floor slab, approximately 70 lb of building material/square feet of floor area.
H = Heavy construction: 6-in. concrete exterior wall, 6-in. concrete floor slab, approximately 130 lb of building materials/square feet of floor area.

TABLE 6.8 (Continued).

N. Latitude Fenestration Facing	Room Construction	1	2	3	4	5	6	7	8	9	10	11	12	13	14	15	16	17	18	19	20	21	22	23	24
N	L	0.07	0.05	0.04	0.04	0.05	0.70	0.65	0.65	0.74	0.81	0.87	0.91	0.91	0.88	0.84	0.77	0.80	0.92	0.27	0.19	0.15	0.12	0.10	0.08
	M	0.08	0.07	0.06	0.06	0.07	0.73	0.66	0.65	0.73	0.80	0.86	0.89	0.89	0.86	0.82	0.75	0.78	0.91	0.24	0.18	0.15	0.13	0.11	0.09
	H	0.09	0.09	0.08	0.07	0.09	0.75	0.67	0.66	0.74	0.80	0.86	0.89	0.88	0.85	0.80	0.73	0.76	0.88	0.23	0.17	0.14	0.13	0.11	0.10
NE	L	0.02	0.01	0.01	0.01	0.02	0.55	0.76	0.75	0.60	0.39	0.31	0.28	0.27	0.25	0.23	0.20	0.16	0.12	0.06	0.05	0.04	0.03	0.02	0.02
	M	0.03	0.02	0.02	0.02	0.02	0.56	0.76	0.74	0.58	0.37	0.29	0.27	0.26	0.24	0.22	0.20	0.16	0.12	0.06	0.05	0.04	0.04	0.03	0.03
	H	0.03	0.03	0.03	0.03	0.04	0.57	0.77	0.74	0.58	0.36	0.28	0.26	0.25	0.23	0.21	0.19	0.16	0.11	0.06	0.05	0.05	0.04	0.04	0.04
E	L	0.02	0.01	0.01	0.01	0.01	0.45	0.71	0.80	0.77	0.64	0.43	0.29	0.25	0.23	0.20	0.17	0.14	0.10	0.06	0.05	0.04	0.03	0.02	0.02
	M	0.03	0.02	0.02	0.02	0.02	0.47	0.72	0.80	0.76	0.62	0.41	0.27	0.24	0.22	0.20	0.17	0.14	0.11	0.06	0.05	0.04	0.04	0.03	0.03
	H	0.04	0.03	0.03	0.03	0.03	0.48	0.72	0.80	0.75	0.61	0.40	0.25	0.22	0.21	0.19	0.16	0.14	0.10	0.06	0.05	0.05	0.04	0.04	0.04
SE	L	0.02	0.02	0.01	0.01	0.01	0.29	0.56	0.74	0.82	0.81	0.70	0.52	0.35	0.30	0.26	0.22	0.18	0.13	0.08	0.06	0.05	0.04	0.03	0.03
	M	0.03	0.03	0.02	0.02	0.02	0.30	0.56	0.74	0.81	0.79	0.68	0.49	0.33	0.28	0.25	0.22	0.18	0.13	0.08	0.07	0.06	0.05	0.04	0.04
	H	0.04	0.04	0.04	0.03	0.04	0.31	0.57	0.74	0.81	0.79	0.67	0.48	0.31	0.27	0.23	0.20	0.17	0.13	0.07	0.07	0.06	0.05	0.05	0.05
S	L	0.03	0.03	0.02	0.02	0.02	0.08	0.15	0.22	0.37	0.58	0.75	0.84	0.82	0.71	0.53	0.37	0.29	0.20	0.11	0.09	0.07	0.06	0.05	0.04
	M	0.04	0.04	0.03	0.03	0.03	0.09	0.16	0.22	0.38	0.58	0.75	0.83	0.80	0.68	0.50	0.35	0.27	0.19	0.11	0.09	0.08	0.07	0.06	0.05
	H	0.05	0.05	0.04	0.04	0.04	0.11	0.17	0.24	0.39	0.59	0.75	0.82	0.79	0.67	0.49	0.33	0.26	0.18	0.10	0.08	0.07	0.06	0.06	0.05
SW	L	0.05	0.04	0.03	0.02	0.02	0.06	0.10	0.13	0.16	0.18	0.22	0.38	0.59	0.76	0.84	0.83	0.72	0.48	0.18	0.13	0.11	0.08	0.07	0.06
	M	0.06	0.05	0.04	0.04	0.03	0.07	0.11	0.14	0.16	0.19	0.22	0.38	0.59	0.75	0.83	0.81	0.69	0.45	0.15	0.12	0.10	0.08	0.07	0.06
	H	0.06	0.05	0.05	0.04	0.04	0.08	0.12	0.15	0.18	0.20	0.23	0.39	0.59	0.75	0.82	0.80	0.68	0.43	0.14	0.11	0.09	0.08	0.07	0.06
W	L	0.05	0.04	0.03	0.02	0.02	0.05	0.08	0.11	0.13	0.14	0.15	0.17	0.30	0.53	0.72	0.83	0.83	0.63	0.19	0.14	0.11	0.08	0.07	0.06
	M	0.05	0.05	0.04	0.04	0.03	0.06	0.09	0.11	0.13	0.15	0.16	0.17	0.31	0.53	0.72	0.82	0.81	0.61	0.16	0.12	0.10	0.08	0.07	0.06
	H	0.05	0.05	0.04	0.04	0.04	0.07	0.10	0.12	0.14	0.16	0.17	0.18	0.31	0.54	0.71	0.81	0.80	0.59	0.15	0.11	0.09	0.07	0.06	0.06
NW	L	0.04	0.04	0.03	0.02	0.02	0.06	0.10	0.13	0.16	0.19	0.20	0.21	0.22	0.30	0.52	0.73	0.83	0.71	0.19	0.13	0.10	0.08	0.07	0.05
	M	0.05	0.04	0.04	0.03	0.03	0.07	0.11	0.14	0.17	0.19	0.20	0.21	0.22	0.30	0.52	0.73	0.82	0.69	0.16	0.12	0.09	0.08	0.07	0.06
	H	0.05	0.04	0.04	0.04	0.04	0.08	0.12	0.15	0.18	0.20	0.21	0.22	0.23	0.30	0.52	0.73	0.81	0.67	0.15	0.11	0.08	0.07	0.06	0.05
HOR	L	0.04	0.03	0.03	0.02	0.02	0.10	0.26	0.43	0.59	0.72	0.81	0.87	0.87	0.83	0.74	0.60	0.44	0.27	0.15	0.12	0.09	0.08	0.06	0.05
	M	0.06	0.05	0.04	0.04	0.03	0.12	0.27	0.44	0.59	0.72	0.81	0.85	0.85	0.81	0.71	0.58	0.42	0.25	0.14	0.12	0.10	0.08	0.07	0.06
	H	0.06	0.06	0.06	0.05	0.05	0.13	0.29	0.45	0.60	0.72	0.81	0.85	0.84	0.79	0.70	0.56	0.40	0.23	0.13	0.11	0.09	0.08	0.08	0.07

L = Light construction: frame exterior wall, 2-in. concrete floor slab, approximately 30 lb of material/sq ft of floor area.
M = Medium construction: 4-in. concrete exterior wall, 4-in. concrete floor slab, approximately 70 lb of building material/sq ft of floor area.
H = Heavy construction: 6-in. concrete exterior wall, concrete floor slab, approximately 130 lb of building materials/sq ft of floor area.
Reprinted with permission from the 1985 ASHRAE Load Calculation Manual.

TABLE 6.9. SHADING FROM OVERHEAD PROJECTIONS

Latitude	24°				32°				40°				48°				56°			
St'd. Time	9 am	Noon	3 pm	6 pm	9 am	Noon	3 pm	6 pm	9 am	Noon	3 pm	6 pm	9 am	Noon	3 pm	6 pm	9 am	Noon	3 pm	6 pm
(Facing) N	--	--	--	.58	--	--	--	.63	--	--	--	.83	--	--	--	1.37	--	--	--	1.61
NE	1.89	--	--	--	2.17	--	--	--	2.13	--	--	--	3.03	--	--	--	3.45	--	--	--
E	1.00	--	--	--	.97	--	--	--	.89	--	--	--	.83	--	--	--	.74	--	--	--
SE	.93	4.55	--	--	1.00	3.33	--	--	.86	2.33	--	--	.73	1.67	--	--	.61	1.33	--	--
S	4.35	3.57	4.35	--	2.63	2.38	2.63	--	1.85	1.59	1.85	--	1.33	1.19	1.33	--	1.08	.93	1.08	--
SW	--	4.55	.93	--	--	3.33	1.00	--	--	2.33	.86	--	--	1.67	.73	--	--	1.33	.61	--
W	--	--	1.00	*	--	--	.97	*	--	--	.89	*	--	--	.83	*	--	--	.74	*
NW	--	--	1.89	*	--	--	2.17	*	--	--	2.13	*	--	--	3.03	*	--	--	3.45	*

Reprinted with permission from the *1985 Fundamentals, ASHRAE Handbook & Product Directory*.

NOTES. Values apply from April to September.

*Shading not effective.

—Completely shaded.

Figure 6.4. Sketch for Example 6.7.

from the sky. This is called *diffuse* radiation. The solar heat gain is calculated in the same manner as it is with the glass receiving direct solar radiation, except that the SHGF is much less. The values in the tables for the north facing orientation can be used with reasonable accuracy for shaded glass for latitudes greater than 24°N. For lower latitudes consult the *ASHRAE Handbook*.

EXAMPLE 6.8

A room with a wall facing east at 40°N latitude has a total window glass area of 80 ft^2. The building is of heavyweight construction. The glass is $\frac{1}{4}$ in. single heat-absorbing glass with no interior shading device. At 10 AM in June an adjacent building shades 30 ft^2 of the window. What is the solar heat gain through the window at that time?

Solution. Equation 6.4 will be used. The externally shaded and unshaded portions of the glass must be handled separately, however, because they receive different radiation. For the part

receiving direct radiation

$$Q = 216 \times 50 \times 0.69 \times 0.49 = 3650 \text{ BTU/hr}$$

For the part receiving only diffuse radiation, using the SHGF for north orientation

$$Q = 48 \times 30 \times 0.69 \times 0.49 = 490 \text{ BTU/hr}$$

The total solar heat gain through the window is

$$Q = 3650 + 490 = 4140 \text{ BTU/hr}$$

6.7. DESIGN CONDITIONS

The cooling load calculations are usually based on inside and outdoor *design conditions* of temperature and humidity. The inside conditions are those that provide satisfactory comfort. Table 1.1 lists some suggested values. The outdoor summer design conditions are based on reasonable maximums, using weather records. Table A.6 lists conditions for some cities. These values will normally be equaled or exceeded 2.5% of the summertime hours. The DB (dry bulb temperature) and coincident WB (wet bulb temperature) occurring at the same time are listed together, and should be used as the *corresponding* design values. The separate design WB listed does not usually occur at the same time as the DB listed and therefore should not be used in load calculations. Previous weather data tables showed this value, which if used, would result in too large a design cooling load. The separate WB value may be needed, however, in selecting a cooling tower or for special applications. Definitions of DB and WB are given in Chapter 7.

The 2.5% values will provide a reasonable compromise between comfort and cost for most applications. For different requirements 1 and 5% values can be found in the ASHRAE Handbook tables.

The table also lists latitudes and mean daily DB temperature ranges.

June to September can be used as months for summer outdoor design temperatures in the northern hemisphere. Occasionally, maximum cooling loads occur in other months due to solar radiation, and therefore it is necessary to know the expected design conditions at those times. Table 6.10 lists appropriate values.

6.8. LIGHTING

The equation for determining heat gain from lighting is

$$Q = 3.4 \times W \times \text{BF} \times \text{CLF} \qquad (6.5)$$

where

Q = net heat gain from lighting, BTU/hr
W = lighting capacity, watts
BF = ballast factor
CLF = cooling load factor for lighting

The term W is the rated capacity of the lights in use, expressed in watts. In many applications all of the lighting is on at all times, but if it is not, the actual amount should be used. The value 3.4 converts watts to BTU/hr.

The factor BF accounts for heat losses in the ballast in fluorescent lamps, or other special losses. A typical value of BF is 1.25 for fluorescent lighting but should be checked if in doubt. For incandescent lighting there is no extra loss, and BF = 1.0.

The factor CLF accounts for storage of part of the lighting heat gain. The storage effect depends on how long the lights and cooling system are operating, as well as the building construction, type of lighting fixture, and ventilation rate.

No storage effect can be allowed for any of the following conditions:

1. Cooling system operates only during occupied hours

2. Cooling system operates more than 16 hr

TABLE 6.10. COOLING DESIGN DRY BULB AND MEAN COINCIDENT WET BULB

CITY	LAT DEG.MIN.		LONG DEG.MIN.		ELEV	DESIGN DB (2.5%)								DESIGN COINCIDENT WB (2.5%)							
						JAN	FEB	MAR	APR	MAY	OCT	NOV	DEC	JAN	FEB	MAR	APR	MAY	OCT	NOV	DEC
Birmingham, Al	33	34	86	45	630	67	72	76	83	91	84	74	68	62	63	64	67	71	69	64	60
Yuma, Az	32	40	114	36	206	76	82	87	96	99	99	85	75	54	57	58	61	63	66	58	54
Little Rock, Ar	34	44	92	14	265	68	70	76	83	89	84	74	67	63	61	64	67	72	71	65	61
Arcata, Ca	40	59	124	6	217	61	60	59	59	62	66	63	59	53	54	53	54	57	58	56	55
Bishop, Ca	37	22	118	22	4112	60	66	72	81	88	85	72	64	43	45	48	52	57	55	49	45
Los Angeles, Ca	33	56	118	23	122	72	71	72	74	74	80	79	74	54	55	55	61	61	63	56	53
San Diego, Ca	32	44	117	10	37	69	69	70	73	72	79	74	71	53	56	56	60	60	64	55	54
Colorado Springs	38	49	104	42	6170	56	61	63	73	79	77	64	59	39	43	43	47	52	50	44	40
Wilmington, De	39	40	75	36	78	50	55	63	78	84	78	65	55	48	52	54	64	70	69	60	51
Jacksonville, Fl	30	25	81	39	24	75	80	83	86	93	86	81	75	66	68	68	69	73	74	69	67
Augusta, Ga	33	22	81	58	182	70	73	78	85	91	84	77	71	62	63	66	68	73	71	65	63
Boise, Id	43	34	116	13	2857	51	53	62	72	83	74	59	48	44	46	49	53	62	55	48	43
Chicago-OHare, Il	41	59	87	54	667	40	49	58	74	83	78	64	50	45	47	52	63	68	66	56	49
Fort Wayne, In	41	0	85	12	828	47	52	60	76	82	78	63	53	47	50	53	62	68	63	58	51
Indianapolis, In	39	44	86	16	793	53	57	64	78	82	79	66	55	52	54	57	64	68	65	59	53
Des Moines, Ia	41	32	93	39	963	42	49	59	76	84	79	63	52	39	44	49	62	69	64	55	48
Dodge City, Ks	37	46	99	58	2592	58	64	72	82	89	83	68	59	46	49	51	58	64	60	52	45
Covington, Ky	39	4	84	40	888	55	61	65	79	84	79	68	58	53	56	55	64	70	64	59	53
Louisville, Ky	38	11	85	44	488	58	63	69	82	88	81	71	61	56	57	59	66	72	67	61	57
Lake Charles, La	30	1	93	9	32	72	75	78	84	89	88	79	74	67	68	69	70	75	74	73	68
New Orleans, La	29	59	90	15	20	74	77	79	84	89	86	79	74	69	70	70	72	74	75	72	69
Portland, Me	43	39	70	19	61	41	44	49	62	77	69	58	48	33	43	44	53	64	60	55	47
Battle Creek, Mi	42	18	85	14	939	49	48	64	72	86	77	62	49	50	45	57	59	66	64	55	49
Minneapolis, Mn	44	53	93	15	838	36	42	52	73	83	76	57	43	34	39	45	58	65	62	52	43
Jackson, Ms	32	20	90	13	332	70	74	78	84	89	87	77	70	64	65	67	70	73	72	68	63
Kansas City, Mo	39	7	94	35	750	54	60	69	81	87	83	69	58	49	51	55	65	71	67	57	52
Springfield, Mo	37	14	93	23	1270	59	62	72	80	84	83	69	59	54	54	59	65	70	66	60	53
Billings, Mt	45	48	108	32	3583	51	54	62	70	80	77	59	52	41	44	47	52	59	56	47	41
North Platte, Ne	41	8	100	42	2787	52	58	64	77	83	80	64	54	42	45	49	56	63	57	49	43
Tonopah, Ne	38	4	117	8	5422	53	58	65	75	81	78	64	55	40	43	45	49	54	52	46	41
Albuquerque, NM	35	3	106	37	5314	55	63	69	79	87	79	64	56	42	45	47	51	56	55	46	42
Albany, NY	42	45	73	48	277	43	47	54	73	81	73	60	49	42	45	49	59	67	65	58	46
Greensboro, NC	36	5	79	57	891	63	64	70	83	87	81	71	63	57	56	59	65	70	68	59	58
Bismarck, ND	46	46	100	45	1660	41	42	56	70	82	77	53	43	37	39	46	53	62	57	43	38
Akron-Canton, Oh	40	55	81	26	1236	49	53	60	74	79	75	64	54	48	50	53	63	67	62	57	52
Toledo, Oh	41	36	83	48	692	44	48	59	76	84	77	64	52	42	46	53	63	69	63	58	51
Tulsa, Ok	36	11	95	54	674	63	69	75	83	88	86	73	63	57	58	60	66	73	69	61	53
Medford, Or	42	23	122	52	1329	55	59	66	75	84	78	63	53	48	50	53	58	64	60	52	49
Portland, Or	45	36	122	36	24	54	57	60	69	79	73	59	54	50	51	51	56	63	59	54	53
Pittsburgh, Pa	40	30	80	13	1151	49	53	63	78	82	77	64	55	46	50	53	63	68	64	57	51
Sioux Falls, SD	43	34	96	44	1422	41	45	57	75	84	78	58	47	37	41	48	58	65	60	49	43
Bristol, Tn	36	30	82	21	1566	59	64	69	81	86	79	70	60	54	55	58	64	70	66	61	54
Amarillo, Tx	35	14	101	46	3700	63	69	75	84	90	84	70	65	46	50	51	56	61	59	50	47
Midland, Tx	31	56	102	12	2858	72	74	81	88	94	88	77	71	53	54	55	59	65	63	56	52
Wichita Falls, Tx	33	59	98	31	1039	66	74	82	88	93	89	76	67	56	58	61	66	71	69	60	54
Cedar City, Ut	37	42	113	6	5616	53	56	63	73	81	78	63	54	41	44	46	50	55	53	46	42
Burlington, Vt	44	28	73	9	331	39	40	49	68	79	70	58	46	38	38	44	56	66	61	55	44
Blackstone, Va	37	4	77	58	438	65	64	71	83	87	82	72	65	59	56	61	67	72	70	61	60
Roanoke, Va	37	19	79	58	1174	60	63	69	82	87	81	69	62	54	52	56	63	69	66	56	55
Everett, Wa	47	54	122	17	596	50	54	61	64	72	64	54	53	47	49	49	55	61	58	52	49
Charleston, WV	38	22	81	36	989	62	64	70	83	86	80	72	64	55	55	56	63	67	66	60	55
Huntington, WV	38	25	82	27	565	65	65	72	83	87	84	71	63	58	56	60	66	70	69	60	57
Green Bay, Wi	44	29	88	8	699	36	39	48	70	77	69	58	42	36	39	44	57	65	60	54	41
Madison, Wi	43	8	89	20	866	38	44	53	73	81	74	61	44	37	42	47	60	68	62	55	43
Cheyenne, Wy	41	9	104	49	6144	51	56	58	69	76	74	60	54	37	41	41	48	52	50	43	41

Reprinted with permission from the *1979 Load Calculation Manual*.

3. Temperature of the space is allowed to rise during nonoccupied hours (temperature swing)

These conditions cover so many possible situations that the author believes heat storage effects for lighting should be used with extreme caution. Building use patterns often change and may be unpredictable. Energy conservation operating techniques may also result in one of the conditions discussed earlier, even though not planned for originally.

For these reasons the CLF tables for lighting are not presented here. For those cases where they are applicable, they may be found in the ASHRAE *Fundamentals Volume*. Otherwise use a value of CLF = 1.0.

EXAMPLE 6.9

A room has eight 40 W fluorescent lighting fixtures in use. The cooling system operates only during occupied hours. What is the heat gain from the lighting?

Solution. A value of BF = 1.25 for the ballast heat will be assumed. CLF = 1.0 for the operating conditions. Using Equation 6.5,

$$Q = 3.4 \times W \times BF \times CLF$$
$$= 3.4 \times 320 \times 1.25 \times 1.0 = 1360 \text{ BTU/hr}$$

6.9. PEOPLE

The heat gain from people is composed of two parts, sensible heat and the latent heat resulting from perspiration. Some of the sensible heat may be absorbed by the heat storage effect, but not the latent heat. The equations for sensible and latent heat gains from people are

$$Q_s = q_s \times n \times CLF \qquad (6.6)$$
$$Q_l = q_l \times n \qquad (6.7)$$

where

Q_s, Q_l = sensible and latent heat gains
q_s, q_l = sensible and latent heat gains per person
n = number of people
CLF = cooling load factor for people

The rate of heat gain from people depends on their physical activity. Table 6.11 lists values for some typical activities. The rates are suitable for a 78 F DB room temperature. Values vary slightly at other temperatures as noted.

The heat storage effect factor CLF applies to the sensible heat gain from people. If the air conditioning system is shut down at night, however, no storage should be included, and CLF = 1.0. Table 6.12 lists values of CLF for people.

EXAMPLE 6.10

What is the heat gain from 240 people in a movie theater at 78 F DB?

Solution. Equations 6.6 and 6.7 will be used. Because the air conditioning system of a theater is normally shut down at night, no storage effect is included in calculating the sensible heat gain.

$$Q_s = 210 \times 240 \times 1.0 = 50,400 \text{ BTU/hr}$$
$$Q_l = 140 \times 240 = 33,600 \text{ BTU/hr}$$
$$\text{Total } Q = 84,000 \text{ BTU/hr}$$

6.10. EQUIPMENT

The heat gain from equipment may sometimes be found directly from the manufacturer or the nameplate data, allowing for intermittent use. Some equipment produces both sensible and latent heat. Some values of heat output for typical appliances are shown in Table 6.13. Note that when hoods are used to exhaust the hot air, the heat output to the space is less. The CLF factors (not shown) apply if the system operates 24 hours.

TABLE 6.13. HEAT GAIN FROM APPLIANCES, BTU/HR

TYPE OF APPLIANCE	ELECTRIC				GAS				STEAM			
	Without Hood			Hood²	Without Hood			Hood²	Without Hood			Hood²
	Sensible	Latent	Total	All Sensible	Sensible	Latent	Total	All Sensible	Sensible	Latent	Total	All Sensible
Broiler-Griddle 31 in. x 20 in. x 18 in.					11,700	6,300	18,000	3,600				
Coffee brewer/warmer												
per burner	770	230	1,000	340	1,750	750	2,500	500				
per warmer	230	70	300	90								
Coffee urn:												
3 gallon	2,550	850	3,400	1,000	3,500	1,500	5,000	1,000	2,180	1,120	3,300	1,000
5 gallon	3,850	1,250	5,100	1,600	5,250	2,250	7,500	1,500	3,300	1,700	5,000	1,600
8 gallon (twin)	5,200	1,600	6,800	2,100	7,000	3,000	10,000	2,000	4,350	2,250	6,600	2,100
Deep fat fryer:												
15# fat	2,800	6,600	9,400	3,000	7,500	7,500	15,000	3,000				
21# fat	4,100	9,600	13,700	4,300								
Dry food warmer per sq ft top	320	80	400	130	560	140	700	140				
Griddle, frying per sq ft top	3,000	1,600	4,600	1,500	4,900	2,600	7,500	1,500				
Hot plate (two heating units)					5,300	3,600	8,900	2,800				
Short order stove (open grates) per burner					3,200	1,800	5,000	1,000				
Steam table per sq ft top					750	500	1,250	250	500	325	825	260
Toaster:												
Continuous—												
360 slices per hour	1,960	1,740	3,700	1,200	3,600	2,400	6,000	1,200				
720 slices per hour	2,700	2,400	5,100	1,600	6,000	4,000	10,000	2,000				
Pop-up (4 slice)	2,230	1,970	4,200	1,300								
Waffle iron 18 in. x 20 in. x 13 in. (2 grids)	1,680	1,120	2,800	900								
Hair dryer:												
Blower type	2,300	400	2,700									
Helmet type	1,870	330	2,200									
Lab burners:												
Bunsen					1,680	420	2,100					
Fishtail					2,800	700	3,500					
Meeker					3,360	840	4,200					
Neon sign per foot of tube	60		60									
Sterilizer	650	1,200	1,850									
Vending machines:												
Hot drink			1,200									
Cold drink			625									

Reprinted with permission from the *1972 Fundamentals, ASHRAE Handbook & Product Directory.*

EXAMPLE 6.11

Mary's Coffee Shop has the following electrical equipment operating in the air conditioned area, without hoods:

1   coffee brewer/warmer (one burner each)
1   toaster (four slices)
1   waffle iron

What is the heat gain from this equipment?

Solution.   Using Table 6.13,

	Sensible	Latent
Coffee brewer	770 BTU / hr	230 BTU / hr
Coffee warmer	230	70
Toaster	2230	1970
Waffle iron	1680	1120

Q_s = 4910 BTU / hr Q_l = 3390 BTU / hr

Total Q = 8300 BTU / hr

The heat output from motors and the equipment driven by them results from the conversion of the electrical energy to heat. The proportion of heat generated that is gained by the air conditioned space depends on whether the motor and driven load are both in the space or only one of them is. Table 6.14 lists heat outputs for each condition.

EXAMPLE 6.12

A hotel with 150 rooms has a fan-coil air conditioning unit in each room, with a $\frac{1}{6}$ HP motor. What is the heat gain to the building from the units?

Solution.   Both the motor and fan are in the conditioned spaces. From Table 6.14, the heat gain is

$$Q = 710 \text{ BTU/hr} \times 150 = 106,500 \text{ BTU/hr}$$

TABLE 6.14. HEAT OUTPUT FROM MOTOR-DRIVEN EQUIPMENT, BTU / HR

Motor Horsepower	Location of Equipment with Respect to Air Stream or Conditioned Space		
	Motor and Driven Machine In	Motor Out Driven Machine In	Motor In Driven Machine Out
1/8	580	320	260
1/6	710	430	280
1/4	1,000	640	360
1/3	1,290	850	440
1/2	1,820	1,280	540
3/4	2,680	1,930	750
1	3,220	2,540	680
1-1/2	4,770	3,820	950
2	6,380	5,100	1,280
3	9,450	7,650	1,800
5	15,600	12,800	2,800
7-1/2	22,500	19,100	3,400
10	30,000	25,500	4,500
15	44,500	38,200	6,300
20	58,500	51,000	7,500
25	72,400	63,600	8,800

6.11. INFILTRATION

Infiltration of air through cracks around windows or doors results in both a sensible and latent heat gain to the rooms. Procedures and equations for calculating infiltration heat losses were explained in detail in Chapter 3. The same procedure is used for calculating heat gains.

Most summer air conditioning systems have *mechanical ventilation* using some outside air, which reduces or eliminates infiltration by creating a positive air pressure within the building. In this case the ventilation air is not a load on the room, but is a load on the central cooling equipment. Many modern buildings have sealed windows and therefore have no infiltration loss, except for entrances.

6.12. HEAT TRANSFER TO SURROUNDINGS

Some of the sensible heat gain to the room is transferred through the structure to the surroundings and never appears as part of the room load. This is a separate effect from heat storage. This heat loss should be calculated and a correction made to the room sensible heat gains from conduction, solar radiation, lights, people, and equipment heat gains. The following equations are used to determine the correction for heat transfer to the surroundings

$$F_c = 1 - 0.02K \qquad (6.8)$$

$$K = (U_w A_w + U_g A_g)/L \qquad (6.9)$$

where

F_c = multiplier to correct each room sensible heat gain
K = unit length conductance, BTU/hr-ft-F
L = length of exterior wall, ft
U_w, U_g = heat transfer coefficients, w = wall, g = glass, BTU/hr-ft^2-F
A_w, A_g = area of wall, glass, ft^2

Windows 4′ × 5′
Ceiling height 9′
U_{glass} = 1.0
U_{wall} = 0.20

Figure 6.5. Sketch for Example 6.13.

EXAMPLE 6.13

For the room plan shown in Figure 6.5 find the correction factor F_c for heat transfer to the surroundings.

Solution. Equations 6.8 and 6.9 will be used.

$$K = \frac{U_w A_w + U_g A_g}{L}$$

$$= \frac{0.20 \times 176 + 1.0 \times 40}{24} = 3.13$$

$$F_c = 1 - 0.02K = 1 - 0.02(3.13) = 0.94$$

The sensible heat gains from conduction, solar radiation, lights, people, and equipment would each be multiplied by 0.94. That is, these loads would be reduced by 6%.

6.13. ROOM COOLING LOAD

The *room cooling load* is the rate at which heat must be removed from the space, and is the sum of each of the corrected net room heat gains. We will use the abbreviations RSHG, RLHG, and

RTHG for the corrected room sensible, latent, and total heat gains, the RTHG being the same as the room cooling load.

When calculating cooling loads, prepared forms are useful for organizing and reporting data and results. A Commercial Cooling-Load Calculation form is shown in the Appendix and in Figure 6.7.

6.14. PEAK COOLING LOADS

We have learned how to calculate the cooling loads but not how to determine their maximum values. Because the air conditioning system must be sized to handle *peak* periods, we must know how to find them.

The external heat gain components vary in intensity with time of day and time of year because of changing solar radiation as the orientation of the sun changes and because of outdoor temperature changes. This results in a change in the total room heat gain. Sometimes it is immediately apparent by inspecting the tables at what time the peak load occurs, but often calculations are required at a few different times. Some general guidelines can be offered to simplify this task. From the tables we can note the following:

1. For west-facing glass maximum total heat gains occur in mid-summer in the afternoon.
2. For east-facing glass maximum solar heat gains occur in early or mid-summer in the morning.
3. For south-facing glass maximum solar heat gains occur in the fall or winter in early afternoon.
4. For southwest-facing glass maximum solar heat gains occur in the fall in the afternoon.
5. For roofs maximum heat gains occur in the summer in the afternoon or evening.
6. For walls maximum heat gains occur in the summer in the afternoon or evening.

These generalizations can be used to localize approximate times of room peak loads. For instance, we might expect a south-facing room with a very large window area to have a peak load in early afternoon in the fall—not in the summer! If the room had a small glass area, however, the wall and glass heat conduction might dominate and the peak load time would be a summer afternoon. Once the appropriate day and time are located a few calculations will determine the exact time and value of the peak load.

EXAMPLE 6.14

A room facing east in the Shelton Motel in St. Louis, Mo., has a 60 ft^2 window and a 40 ft^2 wall area. The glass is $\frac{1}{4}$ in. single clear glass with light-colored interior venetian blinds. The wall is a metal curtain wall with a *U*-value of 0.14. Building construction is lightweight. Find the time and value of room peak cooling load. The room is a 78 F DB.

Solution. The glass area in the room is large enough compared with the wall so that it may be used to determine the peak load time. From Table 6.6 for 40°N latitude the peak SHGF is in April (224 BTU/hr-ft^2). However, there will be a large conduction heat loss through both glass and wall in the morning at that time of year. In August the SHGF = 216, almost as large as in April, so the total heat gain will be maximum in August. Referring to the tables the CLF for the glass is maximum at 8 AM and the CLTD for the wall is maximum at 10 AM. It appears as if 8, 9, or 10 AM are the possible peak times for the room cooling load. We are assuming that the room is not on the top floor, otherwise the roof load might affect the peak time.

Proceeding to check the possibilities:

at 8 AM
Solar, glass $Q = 216 \times 60 \times 0.55 \times 0.80 = 5700$
Conduction, glass $Q = 1.041 \times 60 \times 0 = \quad 0$
Conduction, wall $Q = 0.14 \times 40 \times 47 = \quad 260$
Total $= \overline{5960}$ BTU/hr

at 9 AM

    Solar, glass $Q = 216 \times 60 \times 0.55 \times 0.77 = 5490$

    Conduction, glass $Q = 1.04 \times 60 \times 2 =$ 120

    Conduction, wall $Q = 0.14 \times 40 \times 54 =$ 300

                                               Total = $\overline{5910}$ BTU/hr

at 10 AM

    Solar, glass $Q = 216 \times 60 \times 0.55 \times 0.64 = 4560$

    Conduction, glass $Q = 1.04 \times 60 \times 4 =$ 250

    Conduction, wall $Q = 0.14 \times 40 \times 55 =$ 310

                                               Total = $\overline{5120}$ BTU/hr

The peak load for this room is at 8 AM in August. Even though the conduction heat gain through the glass and wall increases later in the morning, the solar gain is large enough to dominate. On the other hand, if the window were small, the peak time would be later.

Another point that needs comment here is the possibility of peak load in April, mentioned earlier. But the early morning outside temperature in April would result in a considerable conduction heat loss from the room, and the net gain would probably be less. If in doubt, however, the calculation should be made.

Each building must be analyzed in a similar way to determine time of room peak loads so that the proper room load is calculated.

6.15. BUILDING PEAK LOAD

The value of each room peak load is used to determine the design cooling rate to each room. If these maximum heat gains to each room were added, however, the total would be greater than the peak cooling load required for the whole building, because they do not occur at the same time. Therefore, the designer must also determine the time of year and time of day at which the building cooling load is at a peak, and then calculate it. A reasoning and investigation similar to that carried out in finding room peak loads is used. From our previous discussion and a

study of the tables, the following guidelines emerge:

1. For buildings that are approximately square-shaped in plan with similar construction on all four walls, the peak load is usually in late afternoon in summer. This is because the outside temperature is highest then, and there is no special influence of solar radiation on one side of a building.

2. For buildings with a long south or southwest exposure having large glass areas the peak load may occur in the fall, around mid-day, because radiation is highest then. This case requires careful analysis.

3. For one-story buildings with very large roof areas the peak load usually occurs in the afternoon in summer.

These suggestions must be verified in each case because there are so many variations in building orientation and construction. Once the peak load time is determined the total building heat gains can be calculated.

6.16. REFRIGERATION OR COOLING LOAD

After the building peak sensible and latent heat gains are determined, the next step is to find the refrigeration load, which is also called the cooling load. This is the load on the cooling equipment. The refrigeration load will be greater than the building load because there are heat gains to the air conditioning system from outside the air conditioned rooms. These gains may be from:

1. Ventilation (outside air)

2. Heat gains to ducts

3. Heat produced by the air conditioning system fans and pumps

4. Air leakage from ducts

6.17. VENTILATION

Some outside air is generally brought into a building for health and comfort reasons. The sensible and latent heat of this air is greater than that of the room air, so it becomes part of the cooling load. The excess heat is usually removed in the cooling equipment, however, so it is part of the refrigeration load but not the room load.

The equations for determining the sensible and latent cooling loads from ventilation air, explained in Chapters 3 and 7, are

$$Q_s = 1.1 \times \text{CFM} \times \text{TC} \qquad (3.7)$$
$$Q_l = 0.68 \times \text{CFM} \times (W_0' - W_i') \qquad (3.8)$$

where

Q_s, Q_l = sensible and latent cooling loads from ventilation air, BTU/hr
CFM = air ventilation rate, ft^3/min
TC = temperature change between outdoor and inside air, F
W_0', W_i' = outdoor and inside humidity ratio, gr w./lb d.a.

The total heat Q_t removed from the ventilation air is $Q_t = Q_s + Q_l$.

Suggested outdoor air ventilation rates and typical occupancy levels for some applications are listed in Table 6.15. Note that there are minimum and recommended values. The minimum would be used when economy is primary and the recommended when a better air quality is required.

EXAMPLE 6.15

The Stellar Dome enclosed athletic stadium seats 40,000 people. The space design conditions are 80 F and 50% RH, and outdoor design conditions 94 F DB and 74 F WB. What is the refrigeration load due to ventilation?

Solution. Equations 3.7 and 3.8 will be used. Relatively low ventilation requirements will be used because economy is essential and the occu-

pancy time is brief. Table 6.15 lists 10–20 CFM per person for auditoriums with smoking. We will choose the minimum, 10 CFM per person.

$$Q_s = 1.1 \times \text{CFM} \times \text{TC}$$
$$= 1.1 \times 10 \times 40,000 \times 14$$
$$= 6,160,000 \text{ BTU/hr}$$

The humidity ratios at the inside and outdoor conditions are 77.0 and 95.0 gr.w./lb d.a. (see Chapter 7).

$$Q_l = 0.68 \times \text{CFM} \times (W_0' - W_i')$$
$$= 0.68 \times 10 \times 40,000 \times (95.0 - 77.0)$$
$$= 4,900,000 \text{ BTU/hr}$$
$$Q_t = 6,160,000 + 4,900,000$$
$$= 11,060\,000 \text{ BTU/hr}$$
$$\times \frac{1 \text{ ton}}{12,000 \text{ BTU/hr}} = 922 \text{ tons}$$

If the peak load does not occur at the time of the day that the outdoor temperature is at a maximum, a correction must be made to the outdoor temperature used for calculating ventilation and infiltration. Table 6.16 lists this correction

6.18. HEAT GAIN TO DUCTS

The conditioned air flowing through ducts will gain heat from the surroundings. If the duct passes through conditioned spaces, the heat gain results in a useful cooling effect, but for the ducts passing through unconditioned spaces it is a loss of sensible heat that must be added to the RSHG. The heat gain can be calculated from the heat transfer Equation 3.5

$$Q = U \times A \times \text{TD} \qquad (3.5)$$

where

Q = duct heat gain, BTU/hr
U = overall coefficient of heat transfer, BTU/hr

TABLE 6.15. VENTILATION REQUIREMENTS FOR OCCUPANTS

	Estimated Persons per 1000 ft² Floor Area	Required Ventilation. Air per Person	
		Minimum CFM	Recommended CFM
RESIDENTIAL			
Single unit dwellings			
Living areas, bedrooms	5	5	7–10
Kitchens, baths, toilet rooms	—	20	30–50
Multiple unit dwellings			
Living area, bedrooms	7	5	7–10
Kitchens, baths, toilet rooms	—	20	30–50
COMMERCIAL			
Public rest rooms	100	15	20–25
Merchandising			
Sales floors (basement and ground floors)	30	7	10–15
Sales floor (upper floors)	20	7	10–15
Dining rooms	70	10	15–20
Kitchens	20	30	35
Cafeterias	100	30	35
Hotels, Motels			
Bedrooms	5	7	10–15
Living rooms	20	10	15–20
Baths, toilets	—	20	30–50
Beauty shops	50	25	30–35
Barber shops	25	7	10–15
Parking garages	—	1.5	2–3
Theaters			
Lobbies	150	20	25–30
Auditoriums (no smoking)	150	5	5–10
Auditoriums (smoking permitted)	150	10	10–20
Bowling alleys (seating area)	70	15	20–25
Gymnasiums and arenas			
Playing floors	70	20	25–30
Locker rooms	20	30	40–50
Spectator areas	150	20	25–30
Swimming pools	25	15	20–25
Offices			
General office space	10	15	15–25
Conference rooms	60	25	30–40
INSTITUTIONAL			
Schools			
Classrooms	50	10	10–15
Auditoriums	150	5	5–7.5
Gymnasiums	70	20	25–30
Libraries	20	7	10–12
Locker rooms	20	30	40–50
Hospitals			
Single, dual bedrooms	15	10	15–20
Wards	20	10	15–20
Operating rooms, delivery rooms	—	20	—

TABLE 6.16. DECREASE FROM PEAK DESIGN OUTDOOR DB TEMPERATURE, F

| Daily Range, F | \multicolumn{24}{c}{hour} |
	1	2	3	4	5	6	7	8	9	10	11	12	13	14	15	16	17	18	19	20	21	22	23	24
10	9	9	10	10	10	10	9	8	7	6	4	2	1	0	0	0	1	2	3	5	6	7	8	8
15	13	14	14	15	15	15	14	13	11	8	6	3	2	0	0	0	2	3	5	7	9	10	11	12
20	17	18	19	20	20	20	19	17	14	11	8	5	2	1	0	1	2	4	7	9	12	14	15	16
25	22	23	24	25	25	25	23	21	18	14	10	6	3	1	0	1	3	5	9	12	15	17	19	21
30	26	28	29	30	30	30	28	25	21	17	12	7	3	1	0	1	3	6	10	14	17	20	23	25
35	30	33	34	35	35	35	33	29	25	20	14	8	4	1	0	1	4	7	12	16	20	24	27	29

Reprinted with permission from the *1979 ASHRAE Load Calculation Manual*.

A = duct surface area, ft^2
TD = temperature difference between air in duct and surrounding air, F

It is recommended that cold air ducts passing through unconditioned areas be insulated to at least an overall value of R-4 (U = 0.25).

EXAMPLE 6.16

A 36 in. × 12 in. duct, 50 ft long, carrying air at 60 F, runs through a space at 90 F. The duct is insulated to an overall U = 0.25. What is the heat gain to the air in the duct?

Solution. The surface area of the duct is

$$A = \left(2 \times 36 \text{ in.} \times \frac{1 \text{ ft}}{12 \text{ in.}} + 2 \times 12 \text{ in.} \right.$$
$$\left. \times \frac{1 \text{ ft}}{12 \text{ in.}}\right) \times 50 \text{ ft} = 400 \text{ ft}^2$$

Using Equation 3.5,

$$Q = U \times A \times \text{TD} = 0.25 \times 400 \times (90 - 60)$$
$$= 3000 \text{ BTU/hr}$$

If there is significant heat gain to return air ducts, it should also be calculated, but it is only added to the refrigeration load, not the RSHG.

Although the heat gain to supply ducts in conditioned spaces is not wasted, care should be taken that it does not affect the distribution of cooling. If there is a long run of duct with a number of outlets, the heat gains in the first sections of duct might be enough so that the air temperature at the last outlets is too high. In this case it might be useful to insulate the duct even though it is in the conditioned area (see Chapter 10).

For residential and small commercial buildings, Table 3.6 can be used to estimate duct heat gains.

6.19. FAN AND PUMP HEAT

Some of the energy from the system fans and pumps is converted into heat through friction and other effects, and becomes part of the heat gain that should be added to the refrigeration load. For a draw-through fan arrangement (downstream from the cooling coil) the heat is added to the RSHG, whereas for a blow-through arrangement (upstream from the coil) the heat gets added to the refrigeration load only. An approximate allowance for fan heat can be made as follows:

For 1 in. w.g. pressure add 2.5% to RSHG

For 2 in. w.g. pressure add 5% to RSHG

For 4 in. w.g. pressure add 10% to RSHG

The heat from the chilled water pump on small systems is generally small and may be neglected, but for large systems it may range from 1–2% of sensible heat, and should be added to the refrigeration load.

6.20. DUCT AIR LEAKAGE

Duct systems will leak air at joints. Unfortunately, many systems have unnecessarily high air leakage due to sloppy installation. However, a careful job should limit duct leakage to 5% of the total CFM. If ducts are outside the conditioned space, the effect of leakage must be added to the system RSHG and RTHG. If the air leaks into the conditioned space, then it does useful cooling, but care should be taken that it is not distributed to the wrong location.

6.21. TEMPERATURE SWING

The cooling load calculations described here are based on maintaining a constant space temperature. By allowing the room temperature to rise above the thermostat set point—called *temperature swing* the load can be reduced still further. Calculation of this effect requires a detailed hourly analysis of loads and will not be discussed here. The use of computers is helpful in this task. Because room temperatures are being raised to conserve energy, the author recommends against correcting for temperature swing, except in special cases.

For very large projects the designer can also often reduce the calculated peak load by considering *diversity.* This refers to the fact that heat gains from people and lighting may shift with time. In a large educational building or group of buildings, for example, at certain times the students may shift from classrooms to a large auditorium. It would be incorrect and wasteful to assume both are used at the same time when calculating the refrigeration load. Typical diver-

sity factors may vary from 0.70–0.90 for people and lighting in large buildings. Each case must be analyzed individually.

6.22. SUPPLY AIR CONDITIONS

After the sensible and latent heat gains are calculated, the required supply air conditions (flow rate, temperature, and humidity) necessary to satisfy room conditions are determined. This subject is covered in Chapter 7.

6.23. SUMMARY OF COMMERCIAL COOLING LOAD CALCULATION PROCEDURES

The steps in determining commercial cooling loads can be summarized as follows.

1. Select indoor and outdoor design conditions from Tables 1.1 and A.9.
2. Use architectural plans to measure dimensions of all surfaces through which there will be external heat gains, for each room.
3. Calculate areas of all these surfaces.
4. Select heat transfer coefficient U-values for each element from Tables A.7 or A.8, or calculate from individual R-values.
5. Determine time of day and month of peak load for each room by calculating external heat gains at times that they are expected to be a maximum. Search Tables 6.1, 6.2, and 6.6 to find maximum values. Often calculations at a few different times will be required, but the suggestions in Section 6.14 should be helpful.
6. Calculate each room peak RSHG and RLHG, using the values for the external heat gains determined above and by calculating and adding the internal heat gains from people, lights, and equipment. The ar-

COMMERCIAL COOLING LOAD CALCULATIONS

Project *Superb Supermarket*   Room *Building, peak*   Engr. *Energy Association*

Location *Indianapolis, Ind.*   Lat. *39 N*   Calc. by *EP 5/5/87*   Check *VL 5/8/87*

		DB F	WB F	RH %	W gr/lb	
Design Conditions	Outdoor	90	74		101	Daily Range *22* F. Ave. *79 F*
	Room	78		50	72	Day *July 21* Time *5 PM*

Conduction	Dir.	Color	U	A, ft² Gross	A, ft² Net	CLTD, F Table	CLTD, F Corr.	F_c	RSHG BTU/hr
Glass	W		1.04		830	13	7	.91	5500
	W		1.04		42	13	7	.91	280
	E		1.04		42	13	7	.91	280
Wall Group B	N	D	.13		840	11	5	.91	500
	S	D	.13		840	17	12	.91	1190
	E	D	.13	1260	1176	26	20	.91	2780
	W	D	.13	1260	388	17	11	.91	500
Roof/ceiling		D	.09		5400	71	66	.91	29,190
Floor			.35		5400		12	.91	20,640
Partition									
Door	E		.18		42	27	22	.91	150

$K = (U_w A_w + U_g A_g)/L = (.13 \times 3244 + 1.04 \times 914)/300 = 4.58$

$F_c = 1 - 0.02K = 1 - 0.02(4.58) = 0.91$

$CFM = \dfrac{RSHG}{1.1\,(T_R - T_{SA})} =$

$RSHR = RSHG/RTHG =$

Solar	Dir.	Sh.	SHGF	A	SC	CLF	F_c		
Glass	W	no	216	830	.69	.56	.91	63,040	
	W	no	216	42	1.0	.56	.91	4620	
	E	no	216	42	1.0	.23	.91	1900	

Lights *16,200*   W × 3.41 × *1.2*   BF × *1.0*   CLF × *.91* F_c

People *315*   SHG × *60*   n × *1.0*   CLF × *.91* F_c

325   LHG × *60*   n

Equipment _____

Infiltration   1.1 × _____ CFM × _____ TC

0.68 × _____ CFM × _____ gr/lb

SA duct gain

SA duct leakage ___*0*___ %

SA fan gain (draw through) ___*0*___ %

Room Heat Gain

Ventilation 1.1 × *600* CFM × *12* TC

0.68 × *600* CFM × *29* gr/lb

SA fan gain (blow through) *2.5* %

Pump gain

RA duct gain

RA fan gain ___*0*___ %

Cooling Load

Tons

	RLHG BTU/hr	RTHG BTU/hr
60,320		
17,200		
	19,500	
Subtotal		
208,090	19,500	
7920		
	11,830	
5200		
221,210	31,330	252,510
	21.0	

Figure 6.6. Cooling load calculations for Example 6.17.

chitect or building owner will furnish the data needed for the calculations.

If there is infiltration this must be added to the room load.

7. Find the time of building peak load using a similar search process as in item 5, and the suggestions in Section 6.15.

8. Calculate the building RSHG and RLHG at peak time, adding all external and internal gains and infiltration, if any. Add supply duct heat gain (Section 6.18), duct heat leakage (Section 6.19), and draw-through supply fan heat gain (Section 6.19), if significant. Correct for heat transfer to surroundings (Section 6.12) and diversity (Section 6.20).

9. Find the refrigeration load by adding the ventilation load to the building heat gains. Add blow-through fan, return air fan, and pump heat gains, if significant.

10. Calculate required supply air conditions (Chapter 7).

Example 6.17 will illustrate the procedures. The data and results are tabulated on a Commer-cial Cooling Load Calculation form (Figure 6.6), which should be carefully studied in relation to the explanations in the example.

EXAMPLE 6.17

The Superb Supermarket shown in Figure 6.7, is located in Indianapolis, Indiana. It is a one-story building with a basement used for storage. Construction and conditions are as follows:

Roof is 4 in. l.w. concrete slab, R-5 insulation, gypsum board ceiling, $U = 0.09$ BTU/hr-ft^2-F

Floor is 4 in. concrete slab, $U = 0.35$

Walls are 4 in. face brick, 4 in. common brick, R-4 insulation, $\frac{1}{2}$ in. gypsum wallboard, $U = 0.13$

Front window is $\frac{1}{4}$ in. single heat absorbing glass, 10 ft high

Doors are $\frac{1}{4}$ in. single clear glass

Receiving door is hollow steel with insulation

Occupancy is 60 people

Figure 6.7. Sketch for Example 6.17.

Lighting is 3 watts per square foot of floor area, fluorescent fixtures

Outdoor ventilation rate is 10 CFM per person

Store is open from 10 AM to 8 PM (9 AM to 7 PM Standard Time)

Determine the required cooling load.

Solution. The procedures recommended previously will be followed (Section 6.23).

1. Indoor and outdoor design conditions are 78 F DB/50% RH and 90 F DB/74 WB. Latitude is 39°N. Daily temperature range is 22 F. Inside and outdoor humidity ratios are 72 and 101 gr w./lb d.a.

2. Dimensions are shown on the plan.

3. Areas are calculated and recorded on the form.

4. U-values found from tables are listed on the form.

5 – 7. The building has only one room. The construction and orientation indicates that the roof and West glass will determine the peak load time. Peak glass load is in both July and August, therefore either will be peak time of year. Peak CLF for the glass is at 5 PM. Peak TD for the roof is at 4 PM. The roof and glass load will be calculated to determine the peak. Construction is *M* weight.

At 4 PM

Outdoor average temperature = 90 − 22/2 = 79 F
$CLTD_c$ = 73 + 1 + (79 − 85) = 68 F

Roof Q = 0.09 × 5400 × 68	=	33,050 BTU/hr
Window solar Q = 216 × 830		
×0.69 × 0.50	−	61,850 BTU/hr
Sum	=	94,900 BTU/hr

At 5 PM

Roof Q = 0.09 × 5400 × 66	=	32,080 BTU/hr
Window solar Q = 216 × 830		
×0.69 × 0.56	=	69,270 BTU/hr
Sum	=	101,350 BTU/hr

The peak load time is 5 PM. Conduction

through the glass has not reduced appreciably (Table 6.5), so this does not affect the results. Besides, wall loads will be higher at 5 PM than at 4 PM.

8. Individual heat gains items are calculated and recorded on the form. Students should see if they obtain the same values from the appropriate tables. The basement is assumed to be at the outside temperature. Walls are in Group B.

   No storage effect for people or lighting is taken because the system is shut down when the store closes, and does not operate until shortly before the store opens.

   No infiltration is included. Ventilation air is assumed to prevent any significant infiltration, because the doors are not used heavily.

   The supply duct is exposed in store, therefore heat gain and leakage are useful cooling and do not add to the load.

   The effect of heat transfer to the surroundings is calculated from Equations 6.8 and 6.9. No diversity is considered.

$$L = 300 \text{ ft}$$

$$K = \frac{(U_w A_w + U_g A_g)}{L}$$

$$= \frac{(0.13 \times 3244 + 1.04 \times 914)}{300} = 4.58$$

$$F = 1 - 0.02K = 0.91$$

9. Ventilation loads are calculated and shown on the form. Return air fan gain is negligible, and there is no pump. A blow through unit will be used, and it is assumed the fan gain is 2.5%. Refrigeration load is the sum of all loads.

10. It is assumed supply air is at 55 F, from which the supply air CFM is calculated. The RSHR concept is explained in Chapter 7.

Another cooling load calculation example will be carried out as part of Example Project II (Chapter 17).

Residential Cooling Loads

The procedures described previously are used for calculating cooling loads for commercial and industrial buildings. The procedures for determining cooling loads for residences are based on the same heat transfer principles, but are simplified somewhat. There are a number of reasons for this. Residential air conditioning equipment and controls usually do not have refined provisions for zoning, humidity control, and part load operation. Homes are often conditioned 24 hours a day. These factors all lead to a simplification of load calculations. A standardized procedure which has been developed and adopted by the industry will be explained here. An understanding of the material discussed earlier in this chapter is necessary before studying the following.

Only sensible heat gains are calculated. An allowance is made for latent gains, and lighting loads are neglected. Approximations are used for people and infiltration loads. The procedure does not require determination of peak time of load or of heat storage effect, this being included in the data.

6.24. HEAT GAIN THROUGH STRUCTURE

The heat gains through walls, roof, ceiling, and floor are each calculated by use of the following equation

$$Q = U \times A \times \text{ETD} \qquad (6.10)$$

where

Q = sensible heat gain, BTU/hr
U = overall heat transfer coefficient, BTU/hr-ft^2-F
A = area, ft^2
ETD = equivalent temperature difference, F

TABLE 6.17. DESIGN EQUIVALENT TEMPERATURE DIFFERENCES, F

Design Temperature, deg F	85		90			95			100		105	110
Daily Temperature Range [a]	L	M	L	M	H	L	M	H	M	H	H	H
WALLS AND DOORS												
1. Frame and veneer-on-frame	17.6	13.6	22.6	18.6	13.6	27.6	23.6	18.6	28.6	23.6	28.6	33.6
2. Masonry walls, 8-in. block or brick	10.3	6.3	15.3	11.3	6.3	20.3	16.3	11.3	21.3	16.3	21.3	26.3
3. Partitions, frame	9.0	5.0	14.0	10.0	5.0	19.0	15.0	10.0	20.0	15.0	20.0	25.0
masonry	2.5	0	7.5	3.5	0	12.5	8.5	3.5	13.5	8.5	13.5	18.5
4. Wood doors	17.6	13.6	22.6	18.6	13.6	27.6	23.6	18.6	28.6	23.6	28.6	33.6
CEILINGS AND ROOFS [b]												
1. Ceilings under naturally vented attic or vented flat roof—dark	38.0	34.0	43.0	39.0	34.0	48.0	44.0	39.0	49.0	44.0	49.0	54.0
—light	30.0	26.0	35.0	31.0	26.0	40.0	36.0	31.0	41.0	36.0	41.0	46.0
2. Built-up roof, no ceiling—dark	38.0	34.0	43.0	39.0	34.0	48.0	44.0	39.0	49.0	44.0	49.0	54.0
—light	30.0	26.0	35.0	31.0	26.0	40.0	36.0	31.0	41.0	36.0	41.0	46.0
3. Ceilings under unconditioned rooms	9.0	5.0	14.0	10.0	5.0	19.0	15.0	10.0	20.0	15.0	20.0	25.0
FLOORS												
1. Over unconditioned rooms	9.0	5.0	14.0	10.0	5.0	19.0	15.0	10.0	20.0	15.0	20.0	25.0
2. Over basement, enclosed crawl space or concrete slab on ground	0	0	0	0	0	0	0	0	0	0	0	0
3. Over open crawl space	9.0	5.0	14.0	10.0	5.0	19.0	15.0	10.0	20.0	15.0	20.0	25.0

[a] Daily Temperature Range
    L (Low) Calculation Value: 12 deg F.   M (Medium) Calculation Value: 20 deg F.   H (High) Calculation Value: 30 deg F.
    Applicable Range: Less than 15 deg F.   Applicable Range: 15 to 25 deg F.   Applicable Range: More than 25 deg F.
[b] Ceiling and Roofs: For roofs in shade, 18-hr average = 11 deg temperature differential. At 90 deg F design and medium daily range, equivalent temperature differential for light-colored roof equals 11 + (0.71)(39 − 11) = 31 deg F.

Reprinted with permission from the *1985 Fundamentals, ASHRAE Handbook & Product Directory*.

The equivalent temperature difference accounts for both conduction and the effect of radiation. Values of the ETD are listed in Table 6.17. The *L*, *M*, *H* outdoor daily temperature ranges are shown in the footnotes to the table, and can be found from Table A.6. The tables are based on an inside temperature of 75 F. For other temperatures, the values of the ETD should be corrected by 1 F for each 1 F difference from the values shown. Outdoor temperatures can be interpolated between listed values.

EXAMPLE 6.18

A home has a ceiling area of 900 ft². The roof is dark colored with a vented attic and ceiling. The inside design condition is 78 F; outdoor design condition is 90 F. The outdoor daily temperature range is 20 F. The combined roof-ceiling U-factor is 0.23. Find the roof heat gain.

Solution. Equation 6.10 will be used. The ETD will be found from Table 6.17, correcting it for the inside design temperature of 78 F. The outdoor temperature range falls in the *M* class.

$$ETD = 39 - (78 - 75) = 36 \text{ F}$$
$$Q = 0.23 \times 900 \times 36 = 7450 \text{ BTU/hr}$$

6.25. HEAT GAIN THROUGH WINDOWS

The CLF (cooling load factor) gains through windows are listed in Table 6.18. These values include both solar radiation and conduction through the glass. Outdoor temperatures can be interpolated. The heat gain is found from Equation 6.11

$$Q = A \times CLF \tag{6.11}$$

where

Q = heat gain through window, BTU/hr
A = area of glass, ft²
CLF = cooling load factor, BTU/hr-ft²

EXAMPLE 6.19

A residence has 80 ft² of regular single glass on the southwest side, with awnings. The outdoor design temperature is 95 F. Find the total heat gain through the glass.

Solution. From Table 6.18 the CLF is 29 BTU/hr-ft², and the window heat gain is, using Equation 6.11

$$Q = 29 \text{ BTU/hr-ft}^2 \times 80 \text{ ft}^2 = 2370 \text{ BTU/hr}$$

If the glass is shaded by permanent outside overhangs, the calculation is carried out differently. First, the extent of shading is determined. This can be done with the aid of Table 6.19. The shade line factors listed in the table are multiplied by the width of the overhang to find the vertical length of shading. For that part of the glass which is shaded the values from Table 6.18 for north facing glass are used.

No shade line factors for northwest and northeast are shown, because it is not feasible to shade those orientations with overhangs.

EXAMPLE 6.20

A south wall has a 6 ft high picture window with a roof overhang as shown in Figure 6.8. The house is in New Orleans, Louisiana. How much of the glass area is shaded?

Solution. New Orleans is at 30°N latitude. From Table 6.19, the shade line factor is 5.4. The vertical length of shade is 2 × 5.4 = 10.8 ft. The bottom of the glass is 8 ft below the overhang, so the glass is completely shaded! Note how effective the overhang is. Look what would happen if the wall faced southwest, however. The shade extends vertically 1.6 × 2 = 3.2 ft, and barely covers one foot of glass. The orientation and design of the building can have a major effect on energy use!

TABLE 6.18. DESIGN COOLING LOAD FACTORS THROUGH GLASS, BTU / HR-FT^2

Outdoor Design Temp	Regular Single Glass						Regular Double Glass						Heat Absorbing Double Glass						Clear Triple Glass		
	85	90	95	100	105	110	85	90	95	100	105	110	85	90	95	100	105	110	85	90	95
No Awnings or inside Shading																					
North	23	27	31	35	39	44	19	21	24	26	28	30	12	14	17	19	21	23	17	19	20
NE and NW	56	60	64	68	72	77	46	48	51	53	55	57	27	29	32	34	36	38	42	43	44
East and West	81	85	89	93	97	102	68	70	73	75	77	79	42	44	47	49	51	53	62	63	64
SE and SW	70	74	78	82	86	91	59	61	64	66	68	70	35	37	40	42	44	46	53	55	56
South	40	44	48	52	56	61	33	35	38	40	42	44	19	21	24	26	28	30	30	31	33
Horiz. Skylight	160	164	168	172	176	181	139	141	144	146	148	150	89	91	94	96	98	100	126	127	129
Draperies or Venetian Blinds																					
North	15	19	23	27	31	36	12	14	17	19	21	23	9	11	14	16	18	20	11	12	14
NE and NW	32	36	40	44	48	53	27	29	32	34	36	38	20	22	25	27	29	31	24	26	27
East and West	48	52	56	60	64	69	42	44	47	49	51	53	30	32	35	37	39	41	38	39	41
SE and SW	40	44	48	52	56	61	35	37	40	42	44	46	24	26	29	31	33	35	32	33	34
South	23	27	31	35	39	44	20	22	25	27	29	31	15	17	20	22	24	26	18	19	21
Roller Shades Half-Drawn																					
North	18	22	26	30	34	39	15	17	20	22	24	26	10	12	15	17	19	21	13	14	15
NE and NW	40	44	48	52	56	61	38	40	43	45	47	49	24	26	29	31	33	35	34	35	35
East and West	61	65	69	73	77	82	54	56	59	61	63	65	35	37	40	42	44	46	49	49	50
SE and SW	52	56	60	64	68	73	46	48	51	53	55	57	30	32	35	37	39	41	41	42	43
South	29	33	37	41	45	50	27	29	32	34	36	38	18	20	23	25	27	29	25	26	26
Awnings																					
North	20	24	28	32	36	41	13	15	18	20	22	24	10	12	15	17	19	21	11	12	13
NE and NW	21	25	29	33	37	42	14	16	19	21	23	25	11	13	16	18	20	22	12	13	14
East and West	22	26	30	34	38	43	14	16	19	21	23	25	12	14	17	19	21	23	12	13	14
SE and SW	21	25	29	33	37	42	14	16	19	21	23	25	11	13	16	18	20	22	12	13	14
South	21	24	28	32	36	41	13	15	18	20	22	24	11	13	16	18	20	22	11	12	13

Reprinted with permission from *1979 ASHRAE Load Calculation Manual*.

6.26. PEOPLE AND APPLIANCES

The sensible heat gain per person is assumed to be an average of 225 BTU/hr. If the number of occupants is not known in advance, it can be estimated as two times the number of bedrooms. Because the maximum load usually occurs in late afternoon, it is usual to assume that the occupants are in living and dining areas for purposes of load distribution.

A sensible heat gain of 1200 BTU/hr is considered a reasonable value for the kitchen area, allowing for intermittent use of appliances. Large special appliances should be evaluated separately.

TABLE 6.19. SHADE LINE FACTORS

Direction Window Faces	N Latitude, Deg						
	25	30	35	40	45	50	55
E/W	0.8	0.8	0.8	0.8	0.8	0.8	0.8
SE/SW	1.9	1.6	1.4	1.3	1.1	1.0	0.9
S	10.1	5.4	3.6	2.6	2.0	1.7	1.4

Note: Distance shadow line falls below the edge of the overhang equals shade line factor multiplied by width of overhang. Values are averages for 5 hr of greatest solar intensity on August 1.

Reprinted with permission from the *1979 ASHRAE Load Calculation Manual*.

Figure 6.8. Sketch for Example 6.20.

TABLE 6.20. SENSIBLE COOLING LOAD DUE TO INFILTRATION AND VENTILATION

Design Temperature, F	85	90	95	100	105	110
Infiltration, BTU/hr per ft² of gross exposed wall area	0.7	1.1	1.5	1.9	2.2	2.6
Mechanical ventilation, BTU/hr per CFM	11.0	16.0	22.0	27.0	32.0	38.0

Reprinted with permission from the *1985 Fundamentals, ASHRAE Handbook & Product Directory*.

6.27. INFILTRATION AND VENTILATION

The effect on sensible heat gains of infiltration or ventilation is shown in Table 6.20. The infiltration load is expressed in BTU/hr per ft² of exposed wall area. In the case where the air conditioning system is designed to bring in some outdoor ventilation air, this value is taken from the table, expressed in BTU/hr per CFM of ventilation air. Many residential air conditioning systems use 100% recirculated air, relying on infiltration for ventilation.

6.28. DUCT HEAT GAINS AND LEAKAGE

Heat gains to ducts are roughly estimated as 10% of RSHG for ducts in attics and 5% for ducts in unconditioned basements. Duct leakage can be estimated at 5% of RSHG. These figures are added to the building RSHG.

6.29. LATENT AND TOTAL COOLING LOAD

The latent heat gains are not calculated separately in residential applications. Unitary air conditioning equipment usually has a latent heat removal capacity of 0.3 times its sensible capacity. Therefore, to find the total cooling load simply multiply the sum of all the heat gains by 1.3. In very dry climates where the latent load is

less, a factor of 1.2 should be used, and suitable equipment should be installed.

6.30. SUMMARY OF RESIDENTIAL COOLING LOAD CALCULATION PROCEDURES

The steps in determining residential cooling loads can be summarized as follows.

1. Select inside and outdoor design temperatures from Tables 1.1 and A.9.
2. Use architectural plans to measure dimensions of all surfaces through which there will be external heat gain, for each room.
3. Calculate areas of all these surfaces.
4. Select heat transfer coefficient U each element from Tables A.7 or A.8, or calculate from individual R-values.
5. Calculate heat gains through walls, roof, and floors for each room using the ETD values from Table 6.17.
6. Calculate heat gains through glass, using Tables 6.18 and 6.19, for each room.
7. Determine occupancy and appliance load (Section 6.26).
8. Determine infiltration or ventilation load (Table 6.20).
9. Add individual gains to find RSHG for each room and building.
10. Add duct heat gains and leakage to RSHG of building.

11. Multiply RSHG by 1.3 (1.2 in dry climates) to find the RTHG for building. This is the refrigeration load.

A Residential Cooling Load Calculation form for recording data is shown in Figure 6.9 and in the Appendix.

EXAMPLE 6.21

Calculate the cooling load for the residence shown in Figure 6.10, located in Little Rock, Arkansas. Frame construction, naturally vented attic, dark roof.

Solution. The steps will be carried out as recommended in the summary. The results of each step are shown in Figure 6.9.

1. From Tables 1.1 and A.9 the indoor and outdoor design temperatures are 78 F and 96 F.

2 – 3. The dimensions are taken from the building plans and the gross and net areas of each element are calculated and recorded in Figure 6.9. Note that large closets in a room are included as part of the room. The hallway is included as part of the living room because there is no separating door.

4. Heat transfer coefficients for the materials listed are found from Tables A.7 and A.8 and recorded.

5. Select the ETD values from Table 6.17. Outdoor temperature range is 22 F, in the *M* class. The ETD listed for frame walls at 95 F outdoors and 75 F indoors is 23.6 F. This value is corrected for the actual design outdoor and indoor temperatures.

$$ETD = 23.6 + (96 - 95) - (78 - 75)$$
$$= 21.6 \text{ F}$$

The wall heat gains in each room are then calculated using Equation 6.10, and recorded

in Figure 6.9. Other elements are calculated in the same way.

6. From Table 6.18 for the windows on the south side the CLF for the type of glass and shading is 31 BTU/hr-ft^2 at 95 F outdoors and 75 F indoors. Correcting for actual design conditions

$$CLF = 31 + 1 - 3 = 29$$

The heat gains are calculated and recorded for all other windows.

7. For a two bedroom house, assume an occupancy of four, two in the living room, two in the dining area at peak load times, each at 225 BTU/hr. Assume a 1200 BTU/hr kitchen appliance load.

8. Assuming no mechanical introduction of outdoor air, the infiltration from Table 6.20 is 1.5 BTU/hr of gross exposed wall area. Multiplying by the areas for each room, the results are recorded.

9. The individual gains are added to find the RSHG for each room and the building.

10. The duct system is in the basement. Allow 5% for heat gain and 5% for leakage to add to the building RSHG.

11. Multiply the building RSHG by 1.3 to find the refrigeration load.

6.31. ENERGY CONSERVATION

Reducing the building cooling load provides a major opportunity for energy conservation. Some ways this can be achieved are:

1. Use high R-value insulation throughout the building.

2. Use the 2.5% summer outdoor design DB temperature and coincident WB temperature.

3. Use inside design DB temperatures of 78–80 F. These provide adequate comfort for most

RESIDENTIAL COOLING LOAD CALCULATIONS

Project *A. Jones Residence* Location *Little Rock, Ark.*
Outdoor 96 F. Inside 78 F. Outdoor Daily Range 22 F.
Engr. _____ Calc. *EP 2/3/87* Check *x.x. 2/4/87*

Room Name	Living Room				Dining Room				Bedroom No. 1				Bedroom No. 2			
Plan Size	21×12+12×3				9×14				14×12+3×4				10×11+3×4			
	U	A	ETD	BTU/hr	U	A	ETD	BTU/hr	U	A	ETD	BTU/hr	U	A	ETD	BTU/hr
Wall gross		264				184				208				168		
net	.20	202	21.6	870	.20	136	21.6	590	.20	180	21.6	780	.20	140	21.6	600
Roof/ceiling	.10	288	42.0	1210	.10	126	42.0	530	.10	168	42.0	710	.10	122	42.0	510
Floor																
Partition																
Door	.70	20	21.6	300												
	Dir.		CLF		Dir.		CLF		Dir.		CLF		Dir.		CLF	
Windows	S	42	29	1220	`E	28	54	1510	S	14	29	410	N	14	21	290
									W	14	54	760	W	14	54	760
					N	20	21	420								
Infiltration	1.5 × 264			400	1.5 × 184			280	1.5 × 208			310	1.5 × 168			250
People	2 × 225			450	2 × 225			450								
Appliances																
RSHG				4450				3780				2970				2410

Room Name	Kitchen				Bathroom											
Plan Size	9 × 10				6 × 7											
	U	A	ETD	BTU/hr	U	A	ETD	BTU/hr	U	A	ETD	BTU/hr	U	A	ETD	BTU/hr
Wall gross		72				48										
net	.20	58	21.6	250	.20	39	21.6	170								
Roof/ceiling	.10	90	42.0	380	.10	42	42.0	180								
Floor																
Partition																
Door																
	Dir.		CLF		Dir.		CLF		Dir.		CLF		Dir.		CLF	
Windows	N	14	21	290	N	9	21	190								
Infiltration	1.5 × 72			110	1.5 × 48			70								
People																
Appliances				1200												
RSHG				2230				610								

NOTES: *Ceiling ht. = 8'-0". Single clear glass, venetian blinds.*
U_w = 0.20, U_{roof} = 0.10, U_{door} = 0.70.
All windows 3.5'×4' except bath 3'×3'.

Building Total

Calc. RSHG	=	16,450	
Duct gain 5 %	=	820	
Duct leak 5 %	=	820	
Total RSHG	=	18,090	
Cool. Load = 1.3 × RSHG	=	23,520 BTU/hr	

Figure 6.9. Cooling load calculations form (Example 6.21).

Figure 6.10. Plan for Example 6.21.   Scale 1/16″ = 1′–0″

applications. Past practice of designing for 75 F or even lower is wasteful.

4. Consider use of heat absorbing glass.

5. Provide effective interior shading devices.

6. Minimize use of glass in building unless used on the south side for receiving solar heat in the winter.

7. Consider outside construction features that provide shading of glass.

8. Orient the building so that solar radiation in summer is minimum on sides with large glass areas.

9. Avoid unnecessarily excessive lighting levels.

10. Use types of lighting that more efficiently convert electrical energy into light.

11. Above all, use proper calculation procedures that account for heat storage and time lag.

A more detailed discussion of energy conservation in HVAC system design will be presented in Chapter 15.

PROBLEMS

6.1. A building with a 120 ft × 80 ft roof, located in Cincinnati, Ohio, has a roof constructed of 1 in. wood with R-5.5 insulation and a suspended ceiling. The inside

design condition is 80 F. Determine the net roof heat gain at

A. September 21 at noon.

B. Time of peak roof heat gain.

6.2. A southeast facing wall of a building located in Las Vegas, Nevada, is 90 ft × 24 ft. The wall is constructed of 8 in. concrete, R-5 insulation and $\frac{1}{2}$ in. gypsum wallboard. The inside design condition is 79 F. Determine the net heat gain through the wall at

A. June 15 at 11 AM.

B. Time of peak wall heat gain.

6.3. A building in Baltimore, Maryland, has 2300 ft^2 of exterior single glass with no interior shading. The inside design condition is 80 F. Determine the net conduction heat gain through the glass at 2 PM in summer.

6.4. A building in Dallas, Texas, has 490 ft^2 of windows facing west, made of $\frac{1}{4}$ in. single clear glass, with medium color interior venetian blinds. The building is of light construction. Find the maximum net solar heat gain through the windows. In what month and hour is this?

6.5. A room in a building in New York City has a 12 ft W by 6 ft H window facing south. The building is of medium weight construction. The window is $\frac{1}{4}$ in. single clear glass, with dark roller shades. There is a 3 ft outside projection at the top of the window. Find the solar heat gain through the window at 12 noon on July 1.

6.6. A room has four 40 W fluorescent lighting fixtures and two 200 W incandescent fixtures in use. The cooling system is shut down during unoccupied hours. What is the heat gain from the lighting?

6.7. Find the sensible and latent heat gain from 180 people dancing in the Get Down Disco. The temperature is 78 F.

6.8. The Squidgit factory, which is air conditioned 24 hours a day, operates from 8 AM to 5 PM. It has 76 (male and female) employees doing light bench work. What is the net heat gain at 1 PM? The temperature is 78 F.

6.9. The Greasy Spoon Cafe has a 20 ft² steam table, without a hood. What are the sensible and latent heat gains?

6.10. A room has the following exterior wall and window dimensions

WALL: *20 ft × 8 ft, U = 0.27 BTU / hr-ft² F.*
WINDOW: *8 ft W × 4 ft H, U = 0.69.*

Figure 6.11. Plan for Problem 6.11.

Determine the correction factor to be applied for heat transfer to the surroundings.

6.11. Find the peak heat gain for the general office shown in Figure 6.11, with the following conditions

LOCATION: *Sacramento, Cal. Inside conditions 80 F DB, 50% RH.*

Figure 6.12. Plan for Problem 6.12.

Office fixtures 200 W/each
Corridor fixtures 100 W/each
Ceiling ht. 10 ft
All windows 15 ft^2
Scale 1/8" = 1'-0"

Figure 6.13. Plan for Problem 6.16.

WALL: U = 0.28 BTU / hr-ft^2-F, Group E.
WINDOW: *20 ft W × 6 ft H single clear glass, dark interior blinds.*
OCCUPANCY: *10 people. Lights 4 W / ft^2 with ballast.*
FLOOR-TO-FLOOR HEIGHT: *10 ft.*

6.12. The building shown in Figure 6.12, located in Ottawa, Canada, has the following conditions:

DOORS: *7 ft H × 3 ft W, 1½ in. wood.*
WALLS: U = 0.18 BTU / hr-ft^2-F, Group C. *No windows. Light color.*
ROOF: *4 in. h.w. concrete, R-5.5 insulation, finished ceiling.*
INSIDE DESIGN CONDITIONS: *80 F, 55% RH.*
OCCUPANCY: *80 ft^2 / person. Lighting is 2.5 W / ft^2.*

Determine the peak heat gain and the heat gain at 11 AM June 30.

6.13. Repeat Problem 6.12 for the building turned 45° clockwise, and 90° clockwise, and located in or near your community.

6.14. A room in a building in Memphis, Tennessee, has one exposed wall, facing east, with the following conditions:

WALL: A = 68 ft^2, U = 0.21 BTU / hr-ft^2-F, Group E.
GLASS: A = 130 ft^2, *single heat absorbing glass, no shading.*

Lightweight construction. Room temperature = 78 F. Find the peak external heat gain.

6.15. The Beetle Concert Hall in London, England, seats 2300 people. Inside design conditions are 75 F and 50% RH. Calculate the ventilation loads in summer.

6.16. Perform a complete cooling load calculation for the one-story office building shown in Figure 6.13. Conditions are as follows

LOCATION: *Your town.*
WALLS: U = 0.20 BTU / hr-ft²-F. Group C.
GLASS: *Single heat absorbing glass, light interior blinds.*
ROOF: *2.5 in. wood, R-2.8 insulation.*

Orientation as assigned by instructor.
Light construction. People: 1 per 60 ft². Make assumptions based on recommendations in the text on all other data (ventilation, duct and fan heat gains, and so on).

6.17. Perform a complete cooling load calculation for the residence described in Problem 3.20, to be located in your town.

7

Psychrometrics

The atmospheric air that surrounds us is a mixture of dry air and water vapor, called moist air. Because this gas mixture is conditioned in environmental control systems, it is necessary to understand how it behaves. Psychrometrics is the name given to the study of air-water vapor mixtures. Hereafter, as conventionally done, we will use the word air to refer to the air-water vapor mixture that is the atmosphere.

In this chapter we will first learn how to determine the physical properties of air, and then examine how air is processed in air condi-

tioning systems. Because the water vapor content in air can change, these processes can be complex and their understanding may require a special effort by the student. Some comprehension of psychrometrics is an absolute necessity, however, in order to become a competent air conditioning practitioner. A psychrometric analysis is required in selecting the proper air conditioning equipment for a job, and in troubleshooting systems that are not performing properly. Examples of these uses will be demonstrated as we proceed.

OBJECTIVES

After studying this chapter you will be able to:

1. Read values of properties from the psychrometric chart.
2. Determine sensible and latent heat changes in air conditioning equipment.
3. Determine mixed air conditions.
4. Determine required supply air conditions.
5. Determine cooling coil performance specifications.
6. Determine reheat requirements.

7.1. PROPERTIES OF AIR

The physical properties of atmospheric air are defined as follows:

Dry Bulb Temperature (DB). The temperature of the air, as sensed by a thermometer. The words *temperature* and *dry bulb temperature* will be used to mean the same thing with regards to air.

Wet Bulb Temperature (WB). The temperature sensed by a thermometer whose bulb is wrapped with a water-soaked wick, in rapidly moving air.

Dew Point Temperature (DP). The temperature at which the water vapor in the air would begin to condense if the air were cooled, at constant pressure.

Humidity Ratio (W), also called the specific humidity. The weight of water vapor per pound of dry air, in lb/lb dry air, or grains/lb dry air.

Relative Humidity (RH). The ratio of the actual water vapor pressure in the air to the vapor pressure if the air were saturated at that dry bulb temperature. It is expressed in percent.

Specific Volume (v). The volume of air per unit weight of dry air, in ft^3/lb dry air.

Specific Enthalpy (h). The heat content of air per unit weight, in BTU/lb dry air.

The dry bulb temperature is the temperature in the conventional meaning that we are all accustomed to use. The meaning and use of wet bulb temperature will be explained when the process of evaporative cooling is described.

Note that the *specific* properties, those based on a unit weight, always refer to unit weight of dry air, although the air ordinarily is not dry. This is a convention that is generally used.

The weight unit of *grains* is often used in order to have more convenient numbers (7000 gr = 1 lb).

The specific enthalpy of air is the enthalpy of the dry air plus the water vapor, taken above an arbitrary reference temperature at which the enthalpy is given a value of zero. In most data used in air conditioning the arbitrary zero value point is 0 F for the dry air portion and 32 F for the water vapor portion of the air.

The term *saturated air* is used to describe the condition when air contains the maximum amount of water vapor that it can hold. When the amount of water vapor is less than the maximum possible, the condition is called *unsaturated*. The maximum quantity of water vapor that air can hold depends on the air temperature. This use of the word *saturated* does not have the same meaning as the *saturated state* of a pure substance, explained in Chapter 3.

7.2. DETERMINING AIR PROPERTIES

It is necessary to determine many of the physical properties of atmospheric air for air conditioning work. Some of the relationships between properties will now be explained.

Ideal Gas Laws   Both the dry air and water vapor in atmospheric air behave as ideal gases, and therefore Equation 2.15 applies to each

$$p_a V = m_a R_a T = 53.3\, m_a T \qquad (7.1)$$

$$p_w V = m_w R_w T = 85.7\, m_w T \qquad (7.2)$$

where

p_a = partial pressure of dry air in the mixture, lb/ft^2

p_w = partial pressure of water vapor in the mixture, lb/ft^2

m_a = weight of dry air, lb

m_w = weight of water vapor, lb

R_a, R_w = gas constants for air and water vapor, ft lb/lb R

V = volume of mixture, ft^3

T = temperature of mixture, R

The dry air and water vapor each exert only a

part of the total pressure, but each occupies the total volume. A useful principle that applies to the mixture, called *Dalton's Law*, is: *the total pressure equals the sum of the partial pressures*

$$p = p_a + p_w \qquad (7.3)$$

where

p = total (atmospheric) pressure
p_a = partial pressure of dry air
p_w = partial pressure of water vapor

Humidity Ratio   The ideal gas laws and law of partial pressures can be used to find a relationship for determining the humidity ratio. The definition of humidity ratio expressed as an equation is

$$W = \frac{m_w}{m_a} \qquad (7.4)$$

where

W = humidity ratio, lb water vapor/lb dry air
m_w = weight of water vapor, lb
m_a = weight of dry air, lb

Rearranging the ideal gas law as follows, for both air and water vapor

$$m_w = \frac{p_w V}{85.7T}$$

$$m_a = \frac{p_a V}{53.3T}$$

Dividing the first equation by the second results in a useful relationship for the humidity ratio

$$W = \frac{m_w}{m_a} = 0.622\frac{p_w}{p_a} \qquad (7.5)$$

In this equation p_a and p_w are in the same units.

─────────────────────────────

EXAMPLE 7.1

The partial pressure of the water vapor in the air is 0.20 psia on a day when the barometric (atmo-

spheric) pressure is 14.69 psi. Find the humidity ratio.

Solution.   From the law of partial pressures, Equation 7.3,

$$p_a = p - p_w = 14.69 - 0.20 = 14.49 \text{ psia}$$

Using Equation 7.5,

$$W = 0.622\frac{p_w}{p_a} = 0.622 \times \frac{0.20}{14.49}$$

$$= 0.0086 \text{ lb w./lb d.a.}$$

or

$$W' = \frac{7000 \text{ gr}}{1 \text{ lb}} \times 0.0086$$

$$= 60.2 \text{ gr w./lb d.a.}$$

─────────────────────────────

Relative Humidity and Dew Point   The relative humidity is defined by the equation

$$RH = \frac{p_w}{p_{ws}} \times 100 \qquad (7.6)$$

where

RH = relative humidity, %
p_w = partial pressure of water vapor at dry bulb temperature
p_{ws} = saturation pressure of water vapor at dry bulb temperature

The saturation pressure of the water vapor is found from the Steam Tables, at the dry bulb temperature.

The dew point temperature was defined as the temperature at which the water vapor in the air would condense if the air were cooled at constant pressure. Therefore, the water vapor is in a saturated condition at the dew point, and its partial pressure is the saturation pressure at the dew point. Example 7.2 illustrates this.

EXAMPLE 7.2

What is the relative humidity and humidity ratio of air at 80 F DB and 50 F DP? The barometric pressure is 14.7 psi.

Solution.   Using Table A.3 to find the saturation pressure and partial pressure of the water vapor

$$\text{at 80 F } p_{ws} = 0.507 \text{ psia}$$
$$\text{at 50 F } p_w = 0.178 \text{ psia}$$

Using Equation 7.6,

$$\text{RH} = \frac{p_w}{p_{ws}} \times 100 = \frac{0.178}{0.507} \times 100 = 35\%$$

Using Equation 7.5, where
$$p_a = p - p_w = 14.7 - 0.178 = 14.52 \text{ psia}$$

$$W = 0.622 \frac{p_w}{p_a}$$

$$= 0.622 \times \frac{0.178}{14.52}$$

$$= 0.0076 \text{ lb w./lb d.a.}$$

EXAMPLE 7.3

Find the specific volume (per lb dry air) of the air in Example 7.2.

Solution.   The ideal gas law will be used. Rearranging Equation 7.1, remembering that p_a must be in lb/ft^2,

$$V = \frac{53.3 \, m_a T}{p_a}$$

$$= \frac{53.3 \times 1 \times (80 + 460)}{14.52 \times 144}$$

$$= 13.8 \text{ ft}^3/\text{lb d.a.}$$

As the dry air and the water vapor both occupy the same volume, Equation 7.2 could also be

used. Note that the weight of water vapor per lb dry air is the humidity ratio, by definition.

$$V = \frac{85.7 \, m_w T}{p_w}$$

$$= \frac{85.7 \times 0.0076 \times (80 + 460)}{0.178 \times 144}$$

$$= 13.9 \text{ ft}^3/\text{lb d.a.}$$

Enthalpy (Heat Content)   The enthalpy of atmospheric air is the sum of the individual enthalpies of the dry air and water vapor. This includes the sensible heat of the dry air and the sensible and latent heat of the water vapor. Using specific heats of air and water vapor of 0.24 and 0.45 BTU/lb-F, and a latent heat value of 1061 BTU/lb for water, the equation for the specific enthalpy of the mixture, per pound of dry air, is

$$h = 0.24t + W(1061 + 0.45t) \qquad (7.7)$$

where

h = enthalpy of moist air, BTU/lb d.a.
t = dry bulb temperature of air, F.
W = humidity ratio, lb w./lb d.a.

EXAMPLE 7.4

Find the specific enthalpy of the air in Example 7.2.

Solution.   Using Equation 7.7,

$$h = 0.24t + W(1061 + 0.45t)$$
$$= 0.24 \times 80 + 0.0078(1061 + 0.45 \times 80)$$
$$= 27.8 \text{ BTU/lb d.a.}$$

In Example 7.2, the dry bulb temperature (DB) and dew point temperature (DP) were known. The value of p_w was found at the DP temperature, using Table A.3.

Generally, however, it is the wet bulb (WB) temperature and not the DP that is measured. This is partly because the apparatus for measuring the DP is cumbersome, whereas only a thermometer with a wetted wick is needed to measure the WB.

If the WB is known, the value of p_w can be calculated from the following equation, which was developed by Dr. Willis H. Carrier, a pioneer in the field of air conditioning:

$$p_w = p' - \frac{(p - p')(DB - WB)}{(2830 - 1.43 \times WB)} \quad (7.8)$$

where

p_w = partial pressure of water vapor at dry bulb temperature, psia

p = total air pressure, psia

p' = saturation water vapor pressure at the wet bulb temperature, psia

DB = dry bulb temperature, F

WB = wet bulb temperature, F

The following example illustrates the use of Equation 7.8.

EXAMPLE 7.5

A measurement of the air DB and WB gives readings of 90 F and 70 F, respectively. The atmospheric pressure is 14.70 psia. What is the RH?

Solution. Equation 7.8 will be used to find p_w, then Equation 7.6 to find the RH.

From Table A.3

$$\text{at 70 F, } p' = 0.363 \text{ psia}$$
$$\text{at 90 F, } p_{ws} = 0.698 \text{ psia}$$

From Equation 7.8

$$p_w = 0.363 - \frac{(14.70 - 0.36)(90 - 70)}{(2830 - 1.43 \times 70)}$$

$$= 0.258 \text{ psi}$$

From Equation 7.6

$$RH = p_w/p_{ws} \times 100$$

$$= 0.258/0.698 \times 100$$

$$= 37\%$$

As illustrated by the examples shown in this section, the above equations may be used to determine other unknown properties, after measuring the DB and WB or other conveniently measured properties.

To save repeated calculations, charts have been prepared from the most common values of the properties of air. Their use will be the subject of the next section.

7.3. THE PSYCHROMETRIC CHART

The properties of atmospheric air can be represented in tables or graphical form. The graphical form is called the *psychrometric chart*. It is universally used because it presents a great deal of information very simply and because it is helpful in studying air conditioning processes.

Construction of the Psychrometric Chart A psychrometric chart is shown in Figure 7.1. There are slight differences in the arrangement of charts furnished by different organizations. This should be studied before using one.

The location of the scales for each of the properties and the constant value lines for those properties are shown in Figure 7.2. Each figure is a sketch of the psychrometric charts, but not drawn to the actual scale.

Study these sketches until you are familiar with the scales and the lines of constant values for each property. When reading values or drawing lines, a sharp drafting-type pencil and straight edge must always be used. Values should be read to the best accuracy possible, interpolating between numbered values when necessary.

Figure 7.1. Psychrometric chart. (*Courtesy:* Carrier Corporation. Syracuse. N.Y.)

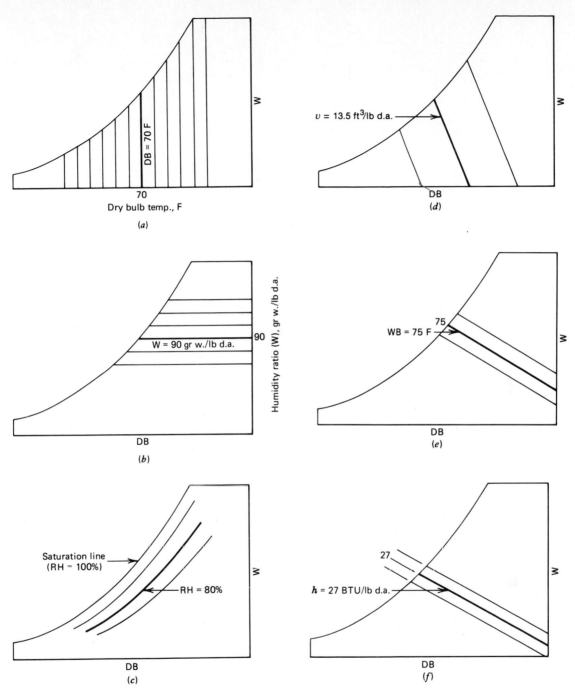

Figure 7.2. Construction of psychrometric chart, showing lines of constant property values. (*a*) Lines of constant dry bulb temperature (DB) on the psychrometric chart. (*b*) Lines of constant humidity ratio (W) on the psychrometric chart. (*c*) Lines of constant relative humidity (RH) on the psychrometric chart. (*d*) Lines of constant specific volume (*v*) on the psychrometric chart. (*e*) Lines of constant wet bulb temperature (WB) on the psychrometric chart. (*f*) Lines of constant enthalpy (*h*) on the psychrometric chart.

178

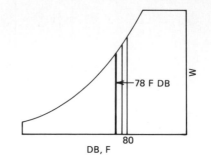

Figure 7.3. Solution to Example 7.6.

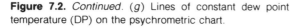

Figure 7.2. *Continued.* (*g*) Lines of constant dew point temperature (DP) on the psychrometric chart.

EXAMPLE 7.6

Draw a line of 78 F DB on the psychrometric chart.

Solution.   The solution is shown in Figure 7.3.

EXAMPLE 7.7

Draw a line of 76.5 F WB on the psychrometric chart.

Solution.   The solution is shown in Figure 7.4.

EXAMPLE 7.8

Draw a line of 45% RH on the psychrometric chart.

Solution.   The solution is shown in Figure 7.5.

Lines of constant enthalpy and constant wet bulb temperature, although actually not exactly parallel, can be considered parallel on the psychrometric chart used here. (See Section 7.10 for a related discussion.) The difference is because the enthalpy values shown on the chart are for saturated air instead of for the actual conditions.

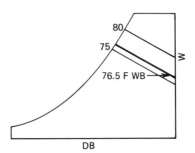

Figure 7.4. Solution to Example 7.7.

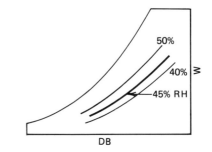

Figure 7.5. Solution to Example 7.8.

The error in using the listed values, however, is less than 2% when calculating enthalpy changes, and on the conservative side. The curved lines on the chart show the corrections to be made to the enthalpy for actual conditions when greater accuracy is necessary.

7.4. LOCATING THE AIR CONDITION ON THE CHART

Any condition of air is represented by a point on the chart. The condition can be located once two independent properties are known. Because each property is represented by a line, the intersection of the two lines establishes the point representing the condition of the air. Once the condition is located, any other properties can be read from the chart.

EXAMPLE 7.9

The weather report reads 90 F DB and 40% RH. What is the WB?

Solution. Using the psychrometric chart the condition of the air is at the point of intersection of the 90 F DB line and 40% RH line (Figure 7.6). Drawing a line of constant WB from this point, the WB temperature is read as 71.2 F.

EXAMPLE 7.10

The air leaving a cooling coil is at 60 F DB and 55 F WB. What is its humidity ratio and specific enthalpy?

Solution. On the chart the condition is found by the point of intersection of the 60 F DB and

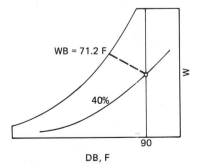

Figure 7.6. Solution to Example 7.9.

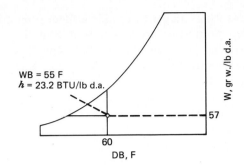

Figure 7.7. Solution to Example 7.10.

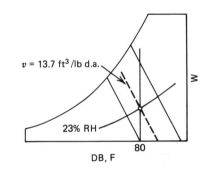

Figure 7.8. Solution to Example 7.11.

55 F WB lines (Figure 7.7). From this point, following a line of constant humidity ratio, read $W = 57$ gr w./lb d.a. Following a line of constant enthalpy from the point (parallel to WB lines) read $h = 23.2$ BTU/lb d.a.

EXAMPLE 7.11

Combustion air enters a furnace at 80 F DB and 23% RH. What is its specific volume?

Solution. The condition is located at the intersection of the 80 F DB and 23% RH lines (Figure 7.8). The specific volume is read as 13.7 ft^3/lb d.a. by interpolation.

The only circumstances under which two properties will not suffice to locate the air condition is when they are not independent properties. When properties are not independent they are measuring the same thing, even though it may not be apparent by their name or definition. Practically, this means that the property lines are parallel on the psychrometric chart. For example, if we know the DP and humidity ratio of an air sample, we could not establish a point, because this gives only one line. The reader should verify this by studying the chart.

The psychrometric chart of Figure 7.1 shows the properties of air at a pressure of 29.92 in. Hg, the standard atmospheric pressure at sea level. For pressures significantly different, some property readings from the chart will not be correct and it cannot be used. Two solutions are possible —either a chart for the actual pressure can be used, if available, or corrections can be made to the values. These corrections can also be made directly by applying the property equations. Geographical locations at high altitudes (e.g., Denver, Mexico City), which are at lower atmospheric pressures, will require these corrections.

7.5. CONDENSATION ON SURFACES

Who, as a child, does not remember drawing pictures on a fogged windowpane in winter? Moisture on the glass is condensed from the room air when the glass temperature is lower than the room air dew point. Air contacting the glass is cooled below its dew point. From the definition of dew point, the air is saturated with water vapor when cooled to that temperature. When cooled further, it can hold even less water vapor—some is condensed.

Understanding this concept enables us to determine the maximum humidity that can be maintained in a room in winter without condensation occurring on the windows. Condensation should be avoided because the water will stain or damage surfaces. For single-glazed windows the inside glass surface is only slightly higher than the outside temperature, because the thermal resistance of the glass is low. Example 7.12 illus-

Figure 7.9. Solution to Example 7.12.

trates the use of the psychrometric chart in relation to this problem.

EXAMPLE 7.12

A room with single-glazed windows is at 70 F DB. If the outside temperature is 30 F, what is the maximum RH that should be maintained in the room to avoid condensation on the windows?

Solution. The inside temperature of the glass can be assumed to be at the outside temperature. (The precise temperature can be calculated using the conduction heat transfer equation.)

Room air contacting the glass surface will be cooled to 30 F. Therefore, the dew point of the air must be less than 30 F to avoid condensation. Using the psychrometric chart (Figure 7.9), air with a 70 F DB and 30 F DP has an RH of 23%. This would be the maximum RH that should be maintained. Double glazing will of course increase the inside glass temperature and permissible RH.

The same problem occurs with bare cold water piping running through spaces. Chilled water lines are usually insulated so that the outside surface is well above the air dew point. A vapor barrier covering is necessary, however, to prevent the migration of any water vapor through the insulation to the cold pipe surface, where the water vapor would condense.

Air Conditioning Processes

7.6. PROCESS LINES ON THE PSYCHROMETRIC CHART

The purpose of air conditioning equipment is to change the condition of the entering air to a new condition. This change is called a *process*. Showing these processes on the psychrometric chart is very helpful in selecting equipment and in analyzing problems. Processes are shown by drawing a line from the initial air condition to its final condition. The air changes properties along this line. Most processes are represented by straight lines.

Sensible Heat Changes The *sensible heat changes* process is one where heat is added or removed from the air and the DB temperature changes as a result, but there is no change in water vapor content. The direction of the process must therefore be along a line of constant humidity ratio, as shown in Figure 7.10. Sensible heating (process 1-2) results in an increase in DB and enthalpy. Process 1-3, sensible cooling (heat removal), results in a decrease in DB and enthalpy.

Latent Heat Changes (Humidification and Dehumidification) The process of adding water vapor to the air is called *humidification*, and removal of water vapor from the air is called *dehumidification*, shown in Figure 7.11. Process 1-4, humidification, results in an increase in humidity ratio and enthalpy. In humidification the enthalpy of the air increases due to the enthalpy of the water vapor added. This is why it is called a latent heat change. In dehumidification, process 1-5, removal of water vapor results in a decrease in enthalpy.

These processes—pure humidification and dehumidification without a sensible heat change—do not occur often in practical air conditioning processes. However, the concept is important to understand in analyzing conditions.

Combination Sensible Heat and Latent Heat Change The following combined sensible and latent processes, shown in Figure 7.12, may occur in air conditioning:

1. Sensible heating and humidification (1-6)
2. Sensible heating and dehumidification (1-7)
3. Sensible cooling and humidification (1-8)
4. Sensible cooling and dehumidification (1-9)

Note that, generally, DB, W, and enthalpy all

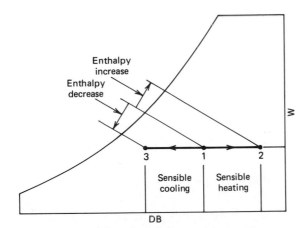

Figure 7.10. Sensible heating and sensible cooling processes.

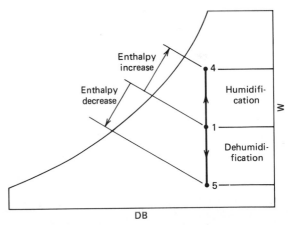

Figure 7.11. Humidification and dehumidification (latent heat change) processes.

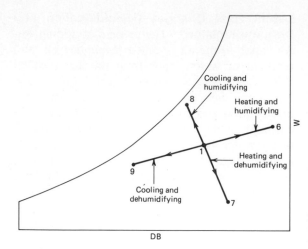

Figure 7.12. Combined sensible and latent heat change processes.

change. For example, in the cooling and dehumidification process 1–9 both the DB and W are decreased, and the enthalpy decreases due to both sensible and latent heat removal.

It is important to determine the amount of heat and water vapor to be added or removed in the conditioning equipment, and to determine the changes in properties. This can be done by using the sensible and latent heat equations (Chapter 2), with the aid of the psychrometric chart.

7.7. SENSIBLE HEAT CHANGE PROCESS CALCULATIONS (SENSIBLE HEATING AND COOLING)

The sensible heat equation applied to moist air is

$$Q_s = 0.24 m_a \times TC + 0.45 m_w \times TC \quad (7.9)$$

where

Q_s = sensible heat added to or removed from air, BTU/hr
m_a = weight of air, lb/hr
m_w = weight of water vapor, lb/hr

$TC = t_2 - t_1$ = temperature change, F
(Specific heats of 0.24 for air and 0.45 for water vapor.)

The first term in Equation 7.9 expresses the enthalpy change of the dry air and the second term expresses the enthalpy change of the water vapor. For approximate air conditioning calculations the second term is often small enough so that it can be neglected, and the sensible heat equation is written

$$Q_s = 0.24 m_a \times TC \quad (7.10)$$

EXAMPLE 7.13

An electric resistance heater is to be installed in a duct to heat 400 lb/hr of air from 60–90 F. What is the required capacity of the heater?

Solution. The electric energy in the resistance heater is converted into the required heat. Using Equation 7.10,

$$Q_s = 0.24 m_a \times TC = 0.24 \times 400(90 - 60)$$

$$= 2880 \text{ BTU/hr}$$

Because the capacity of electrical equipment is expressed in kilowatts or watts rather than BTU/hr

$$\text{Capacity} = 2880 \text{ BTU/hr} \times \frac{1 \text{ KW}}{3410 \text{ BTU/hr}}$$

$$= 0.84 \text{ KW}$$

EXAMPLE 7.14

The air entering the heater in Example 7.13 has an RH of 65%. How much error was there in neglecting the terms in the sensible heat equation that included the enthalpy change of the water vapor?

Solution. From the psychrometric chart the humidity ratio is 0.0072 lb w./lb d.a., and

$$m_w = 0.0072 \text{ lb w./lb d.a.} \times 400 \text{ lb air/hr}$$

$$= 2.9 \text{ lb w./hr}$$

The enthalpy change due to the water vapor is, from Equation 7.9,

$$Q_s = 0.45m_w \times TC = 0.45 \times 2.9 \times 30$$
$$= 39 \text{ BTU/hr}$$

Therefore, the correct amount of heat added is $2880 + 39 = 2919$ BTU/hr. The error from neglecting the enthalpy of the water vapor was about 1% of the total.

The sensible heating or cooling problem can also be solved using the psychrometric chart, as seen in Example 7.15.

EXAMPLE 7.15

Solve Example 7.14 using the psychrometric chart.

Solution. First plot the initial condition, process line, and final condition on the chart. The process and diagrammatic arrangement of the equipment are shown in Figure 7.13. The increase in enthalpy of each pound of air is

$$h_2 - h_1 = 29.6 - 22.3 = 7.3 \text{ BTU/lb d.a.}$$

and therefore the enthalpy increase of the total amount of air is

$$Q_s = m_a(h_2 - h_1) = 400 \text{ lb/hr} \times 7.3 \text{ BTU/lb}$$
$$= 2910 \text{ BTU/hr}$$

which is in close agreement with the previous result. Note how simple and convenient it is to use this method with the aid of the chart.

Always plot the process lines and sketch the equipment arrangement for every job. This will aid in understanding the system performance.

The sensible cooling process problem is handled in the same manner as sensible heating.

The flow rate of air is usually expressed in ft^3/min (CFM) rather than lb/hr in air conditioning work, because most instruments read CFM. Conversion between these units is therefore often necessary, as illustrated by Example 7.16.

EXAMPLE 7.16

What is the flow rate of air entering the duct heater in Example 7.15, expressed in CFM?

Solution. From the psychrometric chart, the specific volume of the entering air is 13.25 ft^3/lb d.a. Converting units to CFM,

$$\text{CFM}_{\text{in}} = 400 \frac{\text{lb}}{\text{hr}} \times \frac{1 \text{ hr}}{60 \text{ min}} \times 13.25 \frac{\text{ft}^3}{\text{lb}}$$
$$= 88.3 \text{ CFM}$$

The specific volume of a gas changes with temperature and pressure, as noted in the Gas

Figure 7.13. Sketch for Example 7.15.

Laws in Chapter 2. Therefore, the CFM leaving the duct heater in Example 7.16 will be different from that entering.

EXAMPLE 7.17

What is the CFM leaving the duct heater in Example 7.16?

Solution. The specific volume leaving, from the chart, is 14.0 ft^3/lb d.a. Therefore,

$$\text{CFM}_{\text{out}} = 400 \frac{\text{lb}}{\text{hr}} \times \frac{1\ \text{hr}}{60\ \text{min}} \times 14.0 \frac{\text{ft}^3}{\text{lb}}$$

$$= 93.3\ \text{CFM}$$

Of course the same quantity by weight of air is leaving the unit as is entering, but the CFM is greater. This variation in volume can lead to misunderstandings and error, such as selection of the wrong size equipment, unless clearly specified. Problems can be avoided by always indicating the temperature and pressure at which the CFM is specified.

However, when rating the capacity of equipment such as fans, coils, and air handling units, manufacturers do not know the conditions of temperature and pressure that each user will apply. In order to avoid misunderstandings, the CFM of the equipment is often expressed at standard air conditions. *Standard air* is defined as having a specific volume of 13.3 ft^3/lb d.a. (a density of 0.075 lb/ft^3 d.a.). This condition applies at 68 F and 29.92 in. Hg. In many air conditioning applications the range of temperatures is such that the specific volume is close enough to that of standard air so that no significant error occurs if the specific volume of standard air is used. For heating applications at high temperatures, or where pressures are considerably different from 14.7 psia, as at high altitudes, the actual specific volumes should be used.

The relation between air flow rate expressed in lb/hr and air flow rate expressed as ft^3/min,

using standard air conditions, is:

$$m_a \frac{\text{lb}}{\text{hr}} = \text{CFM} \frac{\text{cu ft}}{\text{min}} \times \frac{60\ \text{min}}{1\ \text{hr}} \times \frac{1\ \text{lb}}{13.3\ \text{ft}^3}$$

$$m_a = 4.5 \times \text{CFM} \qquad (7.11)$$

Substituting this in the sensible heat equation (7.9) and also assuming a typical average moisture content of air of 0.01 lb w./lb d.a. for air conditioning processes, the result is

$$Q_s = 1.1 \times \text{CFM} \times \text{TC}$$

$$= 1.1 \times \text{CFM} \times (t_2 - t_1) \quad (7.12)$$

This convenient form of the sensible heat equation is commonly used for moist air calculations in air conditioning.

EXAMPLE 7.18

Determine the capacity of the duct heater in Example 7.16, using Equation 7.12, based on entering air CFM.

Solution. Equation 7.12 is used.

$$Q_s = 1.1 \times 88.3(30) = 2914\ \text{BTU/hr}$$

The result agrees with that found by using the psychrometric chart.

Psychrometrics can be put to good use by the engineer or service technician in troubleshooting, as well as by the system designer. Example 7.19 will illustrate a case.

EXAMPLE 7.19

A cooling coil with a rated sensible cooling capacity of 50,000 BTU/hr while handling 2000 CFM of air entering at 80 F must be checked to see that it is performing properly. Thermometers at the air entrance and exit of the unit read 80 F and 62 F, and the air flow rate is measured and found to be 2000 CFM. Is the unit performing satisfactorily?

Solution. If the unit is cooling according to its rated capacity, the leaving air temperature will be at least as low as that predicted by the sensible heat equation. Solving for the temperature change in Equation 7.12,

$$TC = \frac{Q_s}{1.1 \times CFM} = \frac{50,000}{1.1 \times 2000} = 23 \text{ F}$$

$$t_2 = TC - t_1 = 80 - 23 = 57 \text{ F}$$

and we see that the unit is not performing as rated.

NOTE. When using equations with temperature changes, the reader should be careful not to make errors resulting from improper use of negative arithmetic signs. In Example 7.19, the temperature leaving, t_2, must be less than t_1, so 23 F is subtracted from 80 F.

7.8. LATENT HEAT CHANGE PROCESS CALCULATIONS (HUMIDIFYING AND DEHUMIDIFYING)

The amount of water vapor added to or removed from air in a humidifying or dehumidifying process is

$$m_w = m_a(W_2 - W_1)$$

where

m_w = water vapor added or removed, lb w./hr
m_a = air flow rate, lb/hr
$W_2 - W_1$ = change in humidity ratio, lb w./lb d.a.

As with the sensible heating process, it is usually acceptable to assume air at standard conditions. If the air flow rate is expressed in CFM, substituting in the above equation gives

$$m_w = 4.5 \times CFM(W_2 - W_1) \quad (7.13)$$

or, if the humidity ratio is given in gr w./lb d.a.,

dividing by 7000 gr/lb

$$m_w = \frac{CFM(W_2' - W_1')}{1556} \quad (7.14)$$

where W' = humidity ratio, gr w./lb d.a.

EXAMPLE 7.20

A water humidifier in a warm air heating duct handling 3000 CFM increases the moisture content of the air from 30 to 60 gr w./lb d.a. How much water must be supplied?

Solution. The flow diagram is shown in Figure 7.14. Using Equation 7.14,

$$m_w = \frac{CFM(W_2' + - W_1')}{1556} = \frac{3000(60 - 30)}{1556}$$

$$= 58 \text{ lb w./hr}$$

Latent Heat Change As discussed previously, the evaporation of water requires heat. The latent heat of vaporization of water at air conditioning temperatures is approximately 1055 BTU/lb. Using Equation 7.14,

$$Q_L = 1055 \times m_w = 1055 \times \frac{CFM(W_2' - W_1')}{1556}$$

$$Q_L = 0.68 \times CFM(W_2' - W_1') \quad (7.15)$$

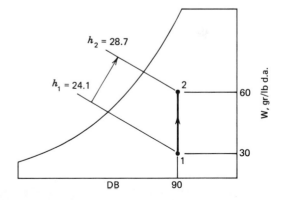

Figure 7.14. Sketch for Example 7.20 and 7.22.

where

$$Q_L = \text{latent heat change, BTU/hr}$$
$$W_2' - W_1' = \text{humidity ratio change, gr w./lb d.a.}$$

EXAMPLE 7.21

How much heat is required in the evaporation process in the process in the humidifier in Example 7.20?

Solution. Using Equation 7.15

$$Q_L = 0.68\, \text{CFM}(W_2' - W_1') = 0.68 \times 3000(30)$$
$$= 61{,}200 \text{ BTU/hr}$$

If water were simply evaporated in the air without providing the required heat, the air would cool down. (This process will be described shortly.) Therefore, the heating equipment must provide sufficient heat to prevent cooling of the air.

Another method of humidifying is to generate steam in a separate source and to inject it into the air to be humidified.

Humidification is a desirable process in winter air conditioning. An inspection of the psychrometric chart shows that natural air conditions in the winter have a very low humidity, a condition which is unacceptable for good comfort. A quality environment control system should include winter humidification.

Dehumidification is the reverse of the process described earlier. All of the equations for latent heat change hold true. In this case heat must be removed from the air to dehumidify it, that is, to condense water vapor from it. Pure dehumidification is not a very commonly used process, and the need arises only in specialized industrial air conditioning. It is usually combined with sensible cooling or heating.

The latent heat change problem can also be solved by using the psychrometric chart.

EXAMPLE 7.22

Find the amount of heat required for the humidifier in Example 7.21, if the air is at 80 F, by using the psychrometric chart.

Solution. From the psychrometric chart (see Figure 7.14)

$$h_2 - h_1 = 28.7 - 24.1 = 4.6 \text{ BTU/lb d.a.}$$

Using Equation 7.11

$$m_a = 4.5 \times \text{CFM} = 4.5 \times 3000 = 13{,}500 \text{ lb/hr}$$

The latent heat required is therefore

$$Q_L = m_a(h_2 - h_1) = 13{,}500(4.6)$$
$$= 62{,}100 \text{ BTU/hr}$$

The result agrees closely with that found previously.

7.9. COMBINED SENSIBLE AND LATENT PROCESS CALCULATIONS

In many air conditioning systems the air undergoes both sensible and latent heat changes. These changes may take place separately or may occur together. In either case the procedures for analysis use the sensible and latent heat equations.

The Cooling and Dehumidification Process Air conditioning for human comfort usually requires a process where both sensible and latent heat are removed from air—that is, the air is cooled and dehumidified. The sensible heat removed and latent heat removed are found from the Equations 7.12 and 7.15, respectively. The sum, $Q_T = Q_s + Q_L$, is the total heat removal for the process.

EXAMPLE 7.23

An air conditioning unit has a cooling coil that cools and dehumidifies 20,000 CFM of air from 82 F DB and 50% RH to 64 F DB and 61 F WB. Find the sensible, latent and total capacity of the cooling coil and the amount of moisture condensed.

Solution. The flow diagram is shown in Figure 7.15. The equations developed previously will provide the information. The sensible heat removed (Equation 7.12) is

$$Q_s = 1.1 \times CFM \times TC$$
$$= 1.1 \times 20,000 \ (18) = 396,000 \ BTU/hr$$

From the psychrometric chart $W_2' - W_1' = 82 - 75 = 7.0$ gr w./lb d.a.

$$Q_L = 0.68 \times CFM(W_2' - W_1')$$
$$= 0.68 \times 20,000 \ (7)$$
$$= 95,000 \ BTU/hr$$

The total heat removed is

$$Q_T = Q_s + Q_L = 396,000 + 95,000$$
$$= 491,000 \ BTU/hr$$

or

$$491,000 \ BTU/hr \times \frac{1 \ ton}{12,000 \ BTU/hr} = 41 \ tons$$

The total cooling capacity of the coil required for this job is 491,000 BTU/hr (41 tons) at the conditions specified. The amount of moisture

Figure 7.15. Sketch for Example 7.23.

condensed during the process is

$$m_w = \frac{CFM(W_2' - W_1')}{1556} = \frac{20,000 \ (7)}{1556}$$
$$= 90 \ lb/hr$$

Provision must be made for draining the water that is continually collecting in the air conditioning unit.

The problem in Example 7.23 can also be solved using the psychrometric chart. Although the sensible heat and latent heat are being removed simultaneously from the air in the conditioner, they can be shown separately on the chart.

EXAMPLE 7.24

Solve Example 7.23 using the psychrometric chart.

Solution. Referring to Figure 7.16, the process line is 1-2 (the actual line is curved, as explained later). However, the latent heat removal portion is shown by 1-a and the sensible heat removal is shown by a-2, even though these are not actual process lines.

The flow rate in lb/hr is

$$m_a = 4.5 \times CFM = 4.5 \times 20,000$$
$$= 90,000 \ lb/hr$$

The sensible heat removal is

$$Q_s = m_a(h_a - h_2) = 90,000 \ (31.6 - 27.2)$$
$$= 396,000 \ BTU/hr$$

The latent heat removal is

$$Q_L = m_a(h_1 - h_a) = 90,000 \ (32.7 - 31.6)$$
$$= 99,000 \ BTU/hr$$

and the total heat removal is

$$Q_T = Q_s + Q_L - 495,000 \ BTU/hr$$

Figure 7.16. Sketch for Example 7.24.

An equation for heat added or removed that is useful when solving problems on the psychrometric chart is

$$Q = 4.5 \times \mathrm{CFM}(h_2 - h_1) \qquad (7.16)$$

This applies to any process from any condition 1 to 2.

It is advisable to solve psychrometric problems both by the complete equations and from the psychrometric chart. This reduces the chances of errors, by comparing results.

The solution of any of the other combined sensible-latent processes is handled in the same manner as the cooling-dehumidification process. Heating and humidification are typical of winter air conditioning systems. The other processes are encountered less often. However, some industrial air conditioning applications may require them. It should be noted that some combinations of processes may have sensible and latent heat opposite in direction. For instance, the heating and dehumidification process has sensible heat added and latent heat removed.

7.10. THE EVAPORATIVE COOLING PROCESS AND THE WET BULB TEMPERATURE

One special cooling and humidification process called *evaporative cooling* requires a more detailed discussion. Referring to Figure 7.17, water is sprayed into the airstream. Some of the water evaporates, increasing the water vapor content of the air. The unevaporated water is recirculated continuously, and no external heat is added to the process.

If the (dry bulb) temperature of the air is measured entering and leaving the conditioning unit, it will be noted that the temperature leaving is lower than that entering, even though no external cooling source is used. This indicates that sensible heat was given up by the air. The important question here is: What caused this?

The evaporation of the water required heat. Because there is no external heat source, unlike the pure humidification process described earlier, this heat must be obtained from the air, lowering its temperature.

Figure 7.17. Evaporative cooling process.

The next important fact to note about the evaporative cooling process is that it is a *constant enthalpy* process. This must be so, because there is no heat added to or removed from the air-water vapor mixture. There is simply an exchange of heat. The sensible heat decreases and the latent heat increases by the same amount. A process in which there is no change in total heat content is called an *adiabatic* process. We can now determine the process line on the psychrometric chart for the evaporative cooling process; it is a line of constant enthalpy content, as seen in Figure 7.17.

Referring to the definition of wet bulb temperature—the temperature recorded by a thermometer whose stem is wrapped with a wetted wick, and placed in the airstream—it is seen that evaporative cooling is the process occurring at the thermometer stem. This results in a lower temperature reading of the wet bulb thermometer. The air passing through the wick becomes completely saturated.

Thus we can note that the evaporative cooling process is therefore a constant wet bulb temperature process. If wet bulb thermometers were placed in the airstream entering and leaving the evaporative cooling unit, they would have the same readings.

The process in an actual evaporative cooling unit does not usually reach saturation because it

would require too much spray coverage. Large evaporative cooling units, called air washers, are often used in textile mills, where high humidity and low DB temperature are desirable.

Air washers can reduce the dry bulb temperature of air considerably unless the air is already near saturation before entering the unit. This seems to offer a tempting bonus—conditioned air that could be used for summer cooling without any refrigeration machinery! This seldom works, however, because the incoming air usually has a high humidity.

In climates where the outside air has an extremely low humidity it may be possible that the air leaving the air washer still has a temperature and humidity low enough to be used for comfortable air conditioning. The southwest part of the United States often meets these conditions. Evaporative cooling was very popular at one time in that part of the country. However, on an occasional humid day the air conditioning is ineffective. For this reason, and because air washing systems are difficult to maintain, mechanical refrigeration is more commonly used today, even in dry climates.

The evaporative cooling process occurs in a cooling tower. *Cooling towers* are equipment used to cool water. The water is sprayed into an airstream, and a small portion of the water evaporates. The heat necessary to evaporate the water is taken both from the air and from the water that does not evaporate. The cooled water is then circulated to where it will be used. The use of cooling towers in refrigeration systems will be explained in Chapter 13.

7.11. THE AIR MIXING PROCESS

The air *mixing process* is one where two streams of air are mixed to form a third stream. This process occurs frequently in air conditioning, particularly in mixing outside air with return air from rooms. If the conditions of the two airstreams that are to be mixed are known, the conditions after mixing can be found. Referring to Figure 7.18, the procedures for finding the DB and W will be explained.

According to the Conservation of Energy Principle, the sensible heat content of the air before and after mixing is the same. That is

$$m_3 \times DB_3 = (m_1 \times DB_1) + (m_2 \times DB_2)$$

Solving for DB_3

$$DB_3 = \frac{(m_1 \times DB_1) + (m_2 \times DB_2)}{m_3} \quad (7.17)$$

If the specific volumes of the unmixed streams are not widely different, the equation can be written using flow rates in ft³/min, rather than lb/hr, without significant loss of accuracy:

$$DB_3 = \frac{(CFM_1 \times DB_1) + (CFM_2 \times DB_2)}{CFM_3} \quad (7.18)$$

The humidity ratio of the mixed air is found in a similar manner, applying the principle of conservation of mass—that is, the water vapor content before and after mixing is the same

$$m_3 \times W_3 = (m_1 \times W_1) + (m_2 \times W_2)$$

Solving for W_3

$$W_3 = \frac{(m_1 \times W_1) + (m_2 \times W_2)}{m_3} \quad (7.19)$$

As above, the following approximately correct equation can often be used

$$W_3 = \frac{(CFM_1 \times W_1) + (CFM_2 \times W_2)}{CFM_3} \quad (7.20)$$

For determining mixed air conditions, Equations 7.18 and 7.20 are accurate enough if the specific volumes of the unmixed airstreams are within 0.5 ft³/lb d.a. of each other. This occurs under most outside air (OA) and return air (RA) design conditions. For example, on the psychrometric chart, a RA condition of 78 F DB and 50% shows a specific volume of about 13.8 ft³/lb d.a. An OA condition of 94 F DB and 75 F WB (St. Louis, Missouri) shows a specific volume of about 14.3 ft³/lb d.a. Under these conditions, for most applications, Equations 7.18 and 7.20 would produce satisfactory answers for the mixed air condition.

EXAMPLE 7.25

Outside air and return air as shown in Figure 7.19 are mixed. Find the mixed air DB and WB.

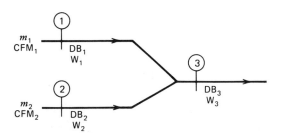

Figure 7.18. Air mixing process.

Figure 7.19. Sketch for Example 7.25.

Solution. The conditions are close enough so that the approximate equations can be used. Using Equation 7.18

$$DB_3 = \frac{(1000 \times 90) + (2000 \times 75)}{3000} = 80 \text{ F}$$

From the psychrometric chart $W_1' = 90$ and $W_2' = 65$ gr w./lb d.a. Using Equation 7.20

$$W_3' = \frac{(1000 \times 90) + (2000 \times 65)}{3000}$$

$$= 75 \text{ gr w./lb d.a.}$$

Locating the mixed air condition on the psychrometric chart, at 80 F DB and 73 gr w./lb d.a., the WB = 66.0 F.

The mixing process can also be solved graphically on the psychrometric chart, from the following two facts:

1. The mixed air condition will lie along a straight line connecting the two conditions of the unmixed airstreams.
2. The location of the mixed air condition on this mixing line will be inversely proportional to the quantities of the unmixed air streams to the total. (This is simply a graphical expression of Equations 7.18 and 7.20).

EXAMPLE 7.26

Solve Example 7.25 graphically, using the psychrometric chart.

Solution. The mixing process line is drawn between 1 and 2. The proportion of each airstream to the total is

$$\frac{CFM_1}{CFM_3} = \frac{1000}{3000} = \frac{1}{3}$$

$$\frac{CFM_2}{CFM_3} = \frac{2000}{3000} = \frac{2}{3}$$

Point 3, the mixed air condition, is therefore located one-third of the total distance, starting from point 2. It is usually convenient to use the DB scale to locate the mixed condition. Any proportional distance on the DB scale is the same as that on the mixing line. The total distance on the DB scale is $90 - 75 = 15$ spaces. The one-third distance from point 2 is $75 + 1/3 \times 15 = 80$ on the DB scale. This distance is projected vertically to the mixing line to locate

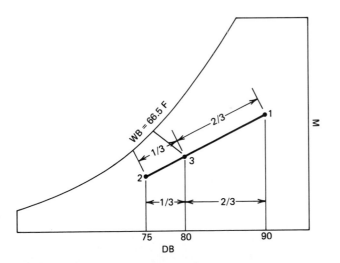

Figure 7.20. Sketch for Example 7.26.

point 3. The construction is shown in Figure 7.20.

Reading from point 3 on the psychrometric chart,

$$DB_3 = 80 \text{ F}, \qquad WB_3 = 66.5 \text{ F}$$

All of the air processes usually encountered in air conditioning have now been described. Our next task will be to learn how to put this information together in designing an air conditioning system or in analyzing performance problems. This requires determining the conditions of air to be supplied to the rooms. In studying this problem we will also learn some new concepts used in psychrometrics—the sensible heat ratio and the coil process line.

Psychrometric Analysis of the Air Conditioning System

In the following discussion we will use the basic psychrometric processes to analyze a complete air conditioning system, and will also briefly consider some more advanced psychrometric concepts. Some familiarity with types of air handling equipment and systems (Chapter 12) will aid in understanding the following material.

7.12. DETERMINING SUPPLY AIR CONDITIONS

The rooms in a building gain both sensible and latent heat in summer. (The procedures for cal-culating these heat gains have been discussed in Chapter 6.) In order to maintain the space temperature and humidity at planned comfortable values, the sensible and latent heat must be removed from each room.

When a space is conditioned by using air, this heat removal is accomplished by supplying air at a temperature and humidity lower than that of the room air. This *supply air* must have a sensible and latent heat content less than the room air by the same amount as the net room sensible and latent heat gains, according to the Energy Conservation Principle (Equation 2.11), in order to prevent the room conditions from changing. Referring to Figure 7.21, this is shown in the following analysis. Applying Equation 2.11,

Energy in − energy out = energy change

If the room temperature and humidity are to be prevented from changing, this means there must be no energy change in the room, and therefore,

Energy in = energy out

The energy flow into the room is made up of *room sensible* and *room latent heat gains* (RSHG and RLHG). The sum of these is the *room total heat gain* (RTHG). These gains must be offset by an equal energy flow out. This is accomplished by supplying air at a temperature and humidity low enough so that it can absorb these heat gains. The supply air takes care of removing both sensible and latent energy together. Noting that the energy content change of the supply air is expressed by the sensible and latent heat equa-

Figure 7.21. Energy flow for air conditioned room.

tions (7.12 and 7.15)

$$\text{Energy in} = \text{energy out}$$
$$\text{RSHG} = 1.1 \times \text{CFM}_{sa} \times \text{TC}$$
$$= 1.1 \times \text{CFM}_{sa}(t_2 - t_1)$$
$$\text{RLHG} = 0.68 \times \text{CFM}_{sa} \times (W_2' - W_1')$$

where W_1' and W_2' are in grains/lb.

From the above equations the required conditions of supply air to offset the known sensible and latent heat gains for a room or building can be determined.

It is usual practice to determine the supply air condition required for handling the sensible heat gain first, and then the condition for the latent gain.

In examining the sensible heat equation it is noted that we may choose different combinations of CFM and TC, all of which are satisfactory. One must be chosen first, from which the other is then calculated. Sometimes the CFM is chosen first, because the space requires a certain air quantity for proper ventilation. At other times the supply air temperature is chosen first, because this value will result in the best choice of air conditioning equipment. These choices will be examined in more detail when cooling coils are discussed. At present, Example 7.27 will illustrate the calculation of the supply air conditions.

EXAMPLE 7.27

The Unisex Hair Salon Shop has a sensible heat gain of 55,000 BTU/hr and latent heat gain of 22,000 BTU/hr. The room conditions are to be maintained at 78 F DB and 50% RH. If 2000 CFM of supply air is furnished, determine the required supply air DB and WB.

Solution. Applying Equation 7.12, solving for the supply air temperature change

$$t_2 - t_1 = \frac{\text{RSHG}}{1.1 \times \text{CFM}} = \frac{55,000}{1.1 \times 2000} = 25 \text{ F}$$

The supply air temperature is therefore

$$t_1 = 78 - 25 = 53 \text{ F}$$

The required humidity ratio of the supply air is then found from Equation 7.15

$$W_2' = W_1' = \frac{\text{RLHG}}{0.68 \times \text{CFM}}$$
$$= \frac{22,000}{0.68 \times 2000} = 16 \text{ gr w./lb d.a.}$$

From the psychrometric chart $W_1' = 72$ gr w./lb d.a., and therefore

$$W_1' = 72 - 16 = 56 \text{ gr w./lb d.a.}$$

Reading from the chart, $\text{WB}_2 = 52$ F.

7.13. SENSIBLE HEAT RATIO

If we recalculate the supply air conditions required in Example 7.27 for other CFM quantities, the conditions found would of course be different. Table 7.1 shows the results for two other assumed values of CFM, as well as the results already found.

If all three satisfactory supply air conditions are plotted on the psychrometric chart, as shown in Figure 7.22, a surprising fact is noted: All of the points lie on a straight line, and furthermore, this line also passes through the room air condition, R.

This line has considerable significance. If we were to assume still other air supply rates and

TABLE 7.1. SATISFACTORY SUPPLY AIR CONDITIONS FOR EXAMPLE 7.27

Supply Air Condition	CFM	DB, F	W' gr w./lb d.a.
1	2000	53	57
1A	2500	58	60
1B	3200	62.4	62.6

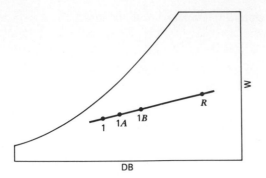

Figure 7.22. Satisfactory supply air conditions fall along a straight line.

then calculate the required supply air conditions, we would find that every one would lie on this same line. This is not a coincidence. Any supply air condition that will satisfactorily remove the proper proportion of room sensible and latent heat gains will be on this line, and in addition any supply air condition that is not on this line will not be satisfactory.

Figure 7.23, using the data from Example 7.27, will be used to explain this important fact.

We note from this figure that for air supply at either A or B, the ratio of sensible to total heat removal, h_s/h_t, is the same for both. It will also be true for any other point on line RA (this can be proven by the geometry of similar traingles).

The slope of line RA, which is defined as the ratio Ax/AR, can also be shown by geometry to

be equal to h_s/h_t. To sum up this idea:

$$\frac{Ax}{AR} = \frac{h_s}{h_t} = \frac{\text{RSHG}}{\text{RTHG}}$$

The ratio RSHG/RTHG is called the *room sensible heat ratio*, RSHR. (It is also called the *room sensible heat factor*, RSHF). That is

$$\text{RSHR} = \frac{\text{RSHG}}{\text{RTHG}}$$

7.14. THE RSHR OR CONDITION LINE

The RSHR line is defined as the line drawn through the room conditions with the room sensible heat ratio slope RSHG/RTHG.

A scale for sensible heat ratio slopes is shown on most psychrometric charts to make it easier to draw lines with this slope. The following example will illustrate how to plot the RSHR line. Following that, its importance will be explained.

EXAMPLE 7.28

The **Big Boy Hamburger Shop** has a sensible heat gain of 45,000 BTU/hr and a latent heat gain of 15,000 BTU/hr. The shop is maintained at 77 F DB and 45% RH. Draw the RSHR line.

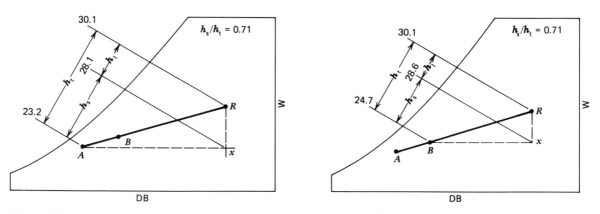

Figure 7.23. Sensible and latent heat removal for two different supply air conditions.

Figure 7.24. Plotting the RSHR line (Example 7.28).

Solution. The solution is shown in Figure 7.24. The following steps are carried out

1. Calculate the RSHR

$$RSHR = \frac{RSHG}{RTHG} = \frac{45,000}{45,000 + 15,000} = 0.75$$

2. On the SHR scale on the psychrometric chart, locate the 0.75 slope. There is also a guide point for the SHR scale, encircled on the chart (located at 80 F and 50% RH). Draw a guide line from SHR = 0.75 through the guide point.

3. Draw a line parallel to the guide line through the room condition point. This is the RSHR line, because it has the RSHR slope, and passes through the room condition. (Two drafting triangles will aid in drawing an exact parallel.)

The importance of the RSHR line is that it is *the line on which any satisfactory supply air condition must lie*. The reason for this is that it has the slope representing the correct proportion of sensible and latent heat removal. Therefore, if the supply air condition is on the line, it will remove the correct proportion of the RSHG and RTHG. The RSHR line is the same line that was developed from Table 7.1 by plotting a number of possible supply air conditions. The SHR scale on the chart enables us to plot the RSHR line in a much easier manner than was done there.

In selecting air conditioning equipment, the usual practice is to plot the RSHR line and then choose a supply air condition on this line. This procedure will be discussed later. First, let us look at an example of applying the RSHR line concept in troubleshooting a service problem.

EXAMPLE 7.29

Mrs. Van Astor, a regular patron of the swank Francais Restaurant, complains on one July day that it feels very "sticky." The manager turns down the thermostat. Mrs. Van Astor gets so cold she puts on her mink stole. Finally she stalks out. Soon all of the wealthy jet set customers leave. The manager calls I. Fixum, troubleshooter. Fixum looks up the air conditioning system design data, which are

RSHG = 150,000 BTU/hr
RLHG = 53,000 BTU/hr
Room Design Conditions = 78 F DB, 50% RH
Design Supply Air = 62 F DB

Solution. Fixum measures the actual supply air conditions and finds them to be 61 F DB and 60 F WB. Plotting the RSHR line (Figure 7.25) with a slope of

$$RSHR = \frac{150,000}{203,000} = 0.74$$

and then locating the room condition point on

Figure 7.25. Sketch for Example 7.29.

the chart, note that the supply air condition does not fall on the RSHR line, and therefore proper room design conditions will not be maintained, even though the supply air DB is adequate. Furthermore, noting that the supply air condition is *above* the line, Fixum knows that the air conditioning equipment is not removing enough latent heat, and the room humidity will be too high, even if the temperature is satisfactory. What happened in this case was that the manager in desperation lowered the supply air DB even lower than the design in order to lower the humidity, resulting in a room temperature that was too low. Yet leaving the supply air DB at proper design value results in uncomfortably high humidity.

There might be a number of causes of the problem in the preceding case. Perhaps the cooling coil was not the correct one, or the refrigerant temperature was not at the correct value. These matters will be discussed later in this chapter.

7.15. COIL PROCESS LINE

A line can be drawn on the psychrometric chart representing the changes in conditions of the air as it passes over the cooling and dehumidifying coil. This is called the coil process line. This line depends on the coil configuration, air velocity, and refrigerant temperature. It is a curved line, and is difficult to locate. However, it is possible to locate a straight line on the chart that, although it is not the true coil process line, will enable us to select a coil or check the performance of a coil. We will call this the coil process line anyway. The *coil process line* may then be defined as the straight line drawn between the air conditions entering and leaving the coil, as shown in Figure 7.26.

The capacity of a coil is defined as the sensible, latent, and total heat that it removes from the air it is conditioning. The required coil ca-

Figure 7.26. Coil process line.

pacity, called the cooling coil load or refrigeration load, can be determined from the coil process line, as illustrated in Example 7.30.

EXAMPLE 7.30

A cooling coil handles 24,000 CFM of air entering at 86 F DB and 73 F WB. The air leaves the coil at 59 F DB and 56 F WB. Determine the coil capacity.

Solution. The coil process line is drawn on the psychrometric chart (Figure 7.27) from entering condition 1 to leaving condition 2. The sensible, latent, and total heat content change per lb d.a.

Figure 7.27. Sketch for Example 7.30.

are as shown. The coil capacity is therefore

$$Q_s = 4.5 \times \text{CFM}(h_x - h_2)$$
$$= 4.5 \times 24,000 \, (30.6 - 23.8)$$
$$= 734,000 \text{ BTU/hr}$$
$$Q_L = 4.5 \times \text{CFM}(h_1 - h_x)$$
$$= 4.5 \times 24,000 \, (36.8 - 30.6)$$
$$= 670,000 \text{ BTU/hr}$$
$$Q_T = 734,000 + 670,000 = 1,404,000 \text{ BTU/hr}$$
$$= 117 \text{ tons}$$

The total capacity could also have been found directly

$$Q_T = 4.5 \times \text{CFM}(h_1 - h_2)$$
$$= 4.5 \times 24,000 \, (36.8 - 23.8)$$
$$= 1,404,000 \, \frac{\text{BTU}}{\text{hr}}$$

7.16. THE COMPLETE PSYCHROMETRIC ANALYSIS

We are now prepared to determine all of the required supply air conditions and the cooling coil capacity for proper conditioning of the space, based on the following known information (Chapter 6):

1. Room sensible and latent heat gains.
2. Outside and inside design conditions.
3. Ventilation (outside) air requirements.
4. Either CFM or DB temperature of the supply air. One of these is selected and the other is then determined from the sensible heat equation. However, both must be in a range that is considered satisfactory for "good practice."

Supply air temperature values are usually chosen so that the temperature difference between room and supply air is between 15–30 F. Factors such as the type and location of air supply

outlets will affect the temperature difference selected (see Chapter 10).

The supply air CFM must neither be too little nor too great, to prevent discomfort from staleness or drafts. Fortunately this is usually not a problem if the supply air temperature and ventilation air quantity are selected within acceptable values.

Example 7.31 will illustrate a complete psychrometric analysis.

EXAMPLE 7.31

The following design data has been established for the High Life Insurance Company office building

RSHG = 740,000 BTU/hr, RLHG = 150,000 BTU/hr

Outside design conditions—94 F DB, 75 F WB

Inside design conditions—78 F DB, 50% RH

Outside air required is 6730 CFM

Supply air temperature difference is 20 F

DETERMINE

a. Supply air CFM
b. Supply air conditions
c. Conditions entering cooling coil
d. Cooling coil sensible, latent, and total load

Solution. Each part of the problem will be solved in order. It is advisable to sketch a diagrammatic arrangement of the system and also each process on the psychrometric chart, as shown in Figure 7.28. Note that as some outside air (OA) is introduced, the same amount of air leaving the space must be exhausted (EA), and the remaining air is returned (RA) to the air conditioning unit.

a. Using Equation 7.12,

$$\text{CFM}_3 = \frac{Q_s}{1.1(t_4 - t_3)} = \frac{740,000}{1.1 \times 20}$$
$$= 33,640 \text{ CFM}$$

Figure 7.28. Sketch for Example 7.31.

b.
$$DB_3 = 78 - 20 = 58 \text{ F}$$

To find the remaining supply air conditions, plot the RSHR line. The slope is

$$RSHR = \frac{Q_s}{Q_T} = \frac{RSHG}{RTHG} = \frac{740,000}{890,000} = 0.83$$

The intersection of the RSHR line and 58 F DB line determines the remaining supply air condition $WB_3 = 56.3$ F.

c. The condition entering the cooling coil will be the mixed air condition of outside air and return air. Using Equation 7.18,

$$DB_2 = \frac{CFM_1 \times DB_1 + CFM_7 \times DB_7}{CFM_2}$$

$$= \frac{6730(94) + 26,910(78)}{33,640} = 81.2 \text{ F}$$

Draw the mixing line 1-7 on the chart. The intersection of this line with the 81.2 F DB line will locate point 2, the condition entering the coil. Read $WB_2 = 67.2$ F.

d. Draw the coil process line 2–3.

Using Equation 7.12 the coil sensible load is

$$Q_s = 1.1 \times CFM_2(DB_2 - DB_3)$$

$$= 1.1 \times 33,640(81.2 - 58)$$

$$= 858,500 \text{ BTU/hr}$$

Using Equation 7.15 the coil latent load is

$$Q_L = 0.68 \times CFM_2(W_2' - W_3')$$

$$= 0.68 \times 33,640(77.0 - 65.0)$$

$$= 274,500 \text{ BTU/hr}$$

The total cooling coil load (refrigeration load) is

$$Q_T = Q_s + Q_L = 1,133,000 \text{ BTU/hr}$$

$$= 94.4 \text{ tons}$$

The results should be checked by using the psychrometric chart

$$Q_s = 4.5 \times CFM(h_x - h_3)$$

$$= 4.5 \times 33,640(29.8 - 24.1)$$

$$= 862,900 \text{ BTU/hr}$$

$$Q_L = 4.5 \times CFM(h_2 - h_x)$$

$$= 4.5 \times 33,640(31.7 - 29.8)$$

$$= 287,600 \text{ BTU/hr}$$

$$Q_T = 4.5 \times CFM(h_2 - h_3)$$

$$= 4.5 \times 33,640(31.7 - 24.1)$$

$$= 1,150,000 \text{ BTU/hr}$$

which agrees quite well.

It is useful at this time for the student to study Figure 7.28 closely. We have drawn the RSHR line, the mixing line, and the coil process line. Identify these and relate each point to the equipment and duct arrangement.

Note that in the preceding example the coil loads are greater than the room loads. This is because the coil must remove the excess heat from the outside air, as well as remove the room heat gains. The heat removed from the outside air is called the *outside air load*.

EXAMPLE 7.32

Calculate the outside air load for Example 7.31. Determine the cooling coil load and compare the result with that found previously.

Solution. Using the psychrometric chart to find the total outside air load (we could also use the sensible and latent heat equations)

$$Q_T(OA) = 4.5 \times CFM_{OA}(h_1 - h_5)$$
$$= 4.5 \times 6730(38.6 - 30.1)$$
$$= 257,400 \frac{BTU}{hr}$$

The cooling coil load (i.e., the total load on the coil) must be the heat necessary to remove from the outside air plus the room loads, or

$$
\begin{aligned}
RSHG &= 740,000 \text{ BTU/hr} \\
RLGH &= 150,000 \\
Q_T(OA) &= \underline{257,400} \\
\text{Coil load} &= 1,147,400 \text{ BTU/hr,}
\end{aligned}
$$

which checks with the previous results.

Note in Example 7.32 the outside air load includes only removing the heat necessary to bring the outside air to room conditions—the excess heat in the outside air.

The psychrometric analysis explained previously provides information on the coil requirements—the coil sensible, latent, and total loads, entering and leaving conditions, and CFM. This data will enable the system designer to select the proper coil from the manufacturer's tables. However, a greater understanding of how a coil performs can be achieved from some further concepts that will now be explained. These ideas are

useful in troubleshooting as well as in selecting new equipment.

7.17. THE CONTACT FACTOR AND BYPASS FACTOR

When air passes across the outside surface of a coil, only part of the air actually contacts the surface and is cooled. This would be expected because there is a spacing between tubes.

The *Contact Factor* (CF) is defined as the proportion of air passing through the coil that touches the cooling surface and is thus cooled. The *Bypass Factor* (BF) is defined as the proportion of air that does not touch the surface, and is therefore not cooled. From this definition, it follows that

$$CF + BF = 1$$

It can be assumed that only the air that contacts the cooling surface is cooled and dehumidified. The bypass air is untreated—it leaves the coil at the same conditions as it entered.

7.18. THE EFFECTIVE SURFACE TEMPERATURE

The temperature of the outside surface of a cooling coil is not the same at all places along the coil tubing. It will vary due to a number of factors, which need not be discussed here. However, we can think of an average coil surface temperature that will be called the *Effective Surface Temperature* (EST). This can be considered as the temperature to which the air that contacts the surface is cooled. (It is also called the *Apparatus Dew Point*.)

From the definition it follows that if all the air passing over the coil contacted the surface (CF = 1), the air would leave at a temperature equal to the EST. This air would be saturated when the EST is below the air dew point, because moisture is being removed. Figure 7.29 shows this process.

Of course it is not possible for a coil to have a CF = 1, because some air through the unit must

Figure 7.29. Coil process line for a cooling coil with CF = 1.

Figure 7.31. Sketch for Example 7.33.

Figure 7.30. Determining CF for a cooling coil.

bypass the surface. Therefore, the air leaving the coil can never be saturated. The amount of air that bypasses the surface depends on tube size and spacing, air face velocity, and number and arrangement of rows. The CF and BF factors can be measured for a coil at each face velocity. Once this is known, we can predict the performance of the coil, based on the following fact:

The CF for a coil is the ratio of the length of the coil process line to the total length of that line extended to the effective surface temperature along the saturation line. Figure 7.30 illustrates this. The contact factor is CF = b/a.

EXAMPLE 7.33

Find the required contact factor and effective surface temperature for a cooling coil that is to cool air from 85 F DB and 69 F WB to 56 F DB and 54 F WB.

Solution. The cooling coil process line 1-2 is drawn on the psychrometric chart (Figure 7.31). The length of 1-2 and 1-EST are measured, using any scale. In this case the lengths were measured in tenths of an inch. The result is

$$CF = \frac{\text{length } 1\text{-}2}{\text{length } 1\text{-EST}} = \frac{2.9 \text{ in.}}{3.5 \text{ in.}} = 0.83$$

The intersection with the saturation line is the effective surface temperature, EST = 50 F.

We could have used the DB scale as a convenient scale, because only the ratio of lengths was needed. This gives the same result

$$CF = \frac{85 - 56}{85 - 50} = 0.83$$

Tables showing CF, BF, and EST values are available from some manufacturers. From these, and knowing the required coil performance, as found above, a proper coil can be selected. Manufacturers often present their data in a different form. In this case their recommended procedures should be followed. It is still useful to know the CF and EST for any apparatus. For instance, it can aid in troubleshooting, as seen in Example 7.34.

Figure 7.32. Sketch for Example 7.34.

EXAMPLE 7.34

What might be a solution to the problem that was shown to exist in Example 7.29, where the space humidity was too high?

Solution. Referring to the psychrometric chart in that example, we note that the supply air temperature was not on the RSHR line. The coil entering conditions could be measured and the coil process line drawn. Assume the coil process line 1-2-*A* is as shown in Figure 7.32. This indicates that the EST_A is too high. Lowering the coil refrigerant temperature results in a new coil process line 1-*B* and new EST_B that might result in a satisfactory supply air condition 2′, on the RSHR line. (The service engineer would have to check further to see if the refrigeration equipment would allow the lower refrigerant temperature and greater load, and if the coil CF was satisfactory.)

7.19. REHEAT

Many air conditioning systems have an apparatus arrangement with heating coils that are located after the cooling coil, which *reheat* the cooled air before delivery to the spaces. This seems both puzzling and an extravagant waste of energy. (Indeed, in many cases it is an unnecessary waste of energy.) However, the reheat process is sometimes required to provide supply air at satisfactory conditions.

Figure 7.33 illustrates the use of reheat. The required coil process line 1-*A* that intersects the RSHR line *R*-2 does not intersect the saturation curve. Therefore, there is no actual combination of coil and effective surface temperature that will meet the requirements. This problem can be resolved by providing a coil with a process line 1-3, and then reheating the air leaving the coil (process line 3-2) before supplying the air to the room. This results in a satisfactory supply air condition 2.

Reheat was required in this case because of the high latent heat gain in the space. This is reflected in a RSHR line with a steep slope, so that no coil process line drawn through it will intersect the saturation curve.

Figure 7.33 indicates the excess refrigeration needed because of reheat. The sensible heat removal 2-3 would not be required if the process could be accomplished in one step. Furthermore, the heating process 3-2 also requires additional energy consumption. In our discussion of energy

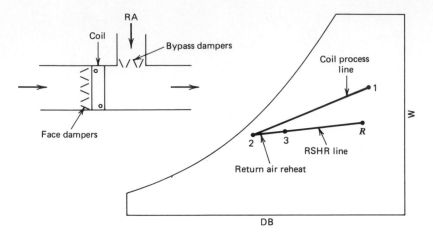

Figure 7.33. Reheat to provide satisfactory supply air condition.

conservation in Chapter 15 we shall determine whether this can be avoided.

7.20. PART LOAD CONTROL

Operation of the air conditioning system when the heat gains are less than the design maximum quantities requires that the equipment supply only part of its total capacity. Reheat is one method of accomplishing this. Usually the RSHG decreases at part load due to lower outside tem-

peratures, whereas the RLHG remains high. This increases the slope of the RSHR line. Figure 7.33 also shows how reheat is used to produce supply air at satisfactory conditions, even with a steep RSHR.

If the system described in Example 7.29 had suitable reheat control, perhaps Mrs. Van Astor would not have stalked out, leaving her $20 (unpaid for) portion of caviar on the plate!

Another part load control method is called *face-and-bypass* control. In this method some of the return air is bypassed around the cooling

Figure 7.34. Return air reheat with face-and-bypass control.

Figure 7.35. Effect of draw-through supply air fan heat gain.

coil. The psychrometric processes are shown in Figure 7.34.

Air through the cooling coil, 2, is mixed with return air that bypasses the coil, R, to form a mixed room supply air at the required DB temperature 3, at part load. The air conditioning unit face-and-bypass dampers vary the proportions of air, usually automatically, in response to a room thermostat. If the RSHR line becomes too steep, point 2 may not be on the RSHR line and the humidity will rise. For many applications, however, the RLHG is so low that the conditions remain satisfactory. Face-and-bypass control is a popular and inexpensive method of control.

7.21. FAN HEAT GAINS

The heat gains from the supply and return air fans have not been included in the psychrometric analysis we have explained. If these heat gains are a small proportion of the total, their effect can be neglected. There is no precise rule for determining when they should be considered. The greater the fan pressure, however, the greater the heat gain. Therefore, for small systems with short duct runs the effect can often be neglected. It is best to calculate the gain in each case and

then decide if it is significant. Heat gains that raise the air temperature one or more degrees F should usually be included in the analysis. Procedures for calculating this effect have been explained in Chapter 6.

The psychrometric processes with the supply air fan heat gains included are shown in Figure 7.35, for a draw-through type air handling unit (the fan is downstream from the coil). Note that the supply air condition, 1, is at a higher DB than the condition leaving the cooling coil, 7. Therefore, the cooling coil load is greater and the psychrometric analysis should include this.

If the air handling unit has a blow-through fan arrangement, the fan heat gain is imposed on the coil load but does not increase the supply air temperature. If supply or return duct heat gains are significant, these will also affect the process line locations on the psychrometric chart. The system design project in Chapter 17 will provide an opportunity for seeing some of these effects.

PROBLEMS

7.1. Using the psychrometric chart on the next page, for conditions (a) to (e), list the properties not shown.

	DB, F	WB, F	DP, F	RH, %	W', gr / lb d.a.	ft³ / lb d.a.	BTU / lb d.a.
(a)	80	60					
(b)	75			40			
(c)		65	50				
(d)			50	40			
(e)	70			70			

7.2. Air at 40 F DB and 60% RH is heated by an electric heater to 80 F. Find the DP, WB, and RH of the air leaving the heater. Draw the process line on the psychrometric chart.

7.3. A cold water pipe with a surface temperature of 52 F passes through a room that is at 75 F DB. At what RH in the room will moisture condense on the pipe?

7.4. Air initially at 90 F DB and 70 F WB is cooled and dehumidified to 56 F DB and 54 F WB. Draw the process line on the psychrometric chart and find the sensible, latent, and total heat removed per pound of dry air.

7.5. Using equations, solve Problem 7.4.

7.6. A cooling coil cools 5000 CFM of air from 80 F DB and 70% RH to 58 F DB and 56.5 F WB. Determine the sensible, latent, and total load on the coil, and the GPH of moisture removed.

7.7. An air handling unit mixes 1000 CFM of outside air at 92 F DB and 75 F WB with 4000 CFM of return air at 78 F DB and 45% RH. Determine the mixed air DB, WB, enthalpy, and humidity ratio.

7.8. A space has a SHG = 83,000 BTU/hr and a LHG = 31,000 BTU/hr. Determine the RSHR.

7.9. A space to be maintained at 75 F DB and 50% RH has a SHG = 112,000 BTU/hr and a LHG = 21,000 BTU/hr. An air supply of 5000 CFM is provided. Determine the supply air DB and WB.

7.10. A room has a SHG = 20,000 BTU/hr and a LHG = 9000 BTU/hr. The room design conditions are 77 F DB and 50% RH. Supply air is delivered at 58 F. Determine the required supply air flow rate in CFM. What is the supply air WB?

7.11. Air enters a cooling coil at 80 F DB and 66 F WB and leaves at 60 F DB and 57 F WB. Determine the coil CF, BF, and effective surface temperature.

7.12. Air at 82 F DB and 67 F WB passes through a coil with a CF of 0.91. The effective coil surface temperature is 55 F. Determine the DB and WB of the air leaving the coil.

7.13. An air conditioning unit is supplying 4000 CFM of air at 55 F DB and 53 F WB to a room maintained at 75 F DB and 55% RH. The outside air conditions are 95 F DB and 74 F WB, and 1000 CFM of ventilation air are furnished. Determine the following: RSHG, RLHG, cooling load, outside air load, and required coil CF. Sketch the apparatus arrangement and show conditions at all locations.

7.14. A room has design conditions of 78 F DB and 50% RH and a RSHG = 18,000 BTU/hr and RLHG = 8000 BTU/hr. An air conditioning unit supplies 900 CFM of air to the room at 58 F DB and 56 F WB. Will the unit maintain the room design DB and WB? What are the approximate conditions maintained in the room?

7.15. A space with a SHG = 172,000 BTU/hr and a LHG = 88,000 BTU/hr is to be maintained at 76 F DB and 50% RH. Conditioned air is supplied at 56 F DB and 54 F WB. Determine the wasted energy consumed if a reheat coil is used to maintain design conditions.

7.16. The following results have been found from a cooling load calculation for a building in Chicago, Illinois: SHG = 812,000 BTU/hr, LHG = 235,000 BTU/hr, ventilation air = 6000 CFM, supply air = 59 F DB, space design conditions = 77 F DB, 50% RH. Design the air system including:

A. Sketch apparatus arrangement

B. Determine supply air CFM and WB

C. Determine mixed air conditions

D. Determine the coil sensible, latent, and total load

E. Determine the outside air sensible, latent, and total load

F. Determine the required coil CF and BF

G. Sketch all psychrometric processes and label all points

7.17. On a day when the barometric pressure is 14.68 psi, the partial pressure of the water vapor in the air is 0.17 psia. Using equations, find the humidity ratio.

7.18. Using equations find the relative humidity, humidity ratio, and specific volume of air at 70 F DB and 60 F DP, when the barometric pressure is 14.7 psi. Compare the results with those found from the psychrometric chart.

7.19. Using equations, find the specific enthalpy of air in Problem 7.18. Compare the result with that found using the psychrometric chart.

7.20. There is 20,000 CFM at 80 F DB and 60% RH entering an air conditioning unit. Air leaving the unit is at 57 F DB and 90% RH. Determine the:

A. Cooling done by the unit in BTU/hr and tons

B. Rate that water is removed from the air in lb/hr

C. Sensible load on the unit in BTU/hr and tons

D. Latent load on the unit in BTU/hr and tons

E. Dew point of the air leaving the air conditioning unit

F. Effective surface temperature (apparatus dew point)

7.21. On a hot September day, a room has a sensible cooling load of 20,300 BTU/hr from occupants, lights, walls, windows, and so on. The latent cooling load for the room is 9000 BTU/hr. The room design conditions are 76 F DB and 50% RH. Supply air will enter the room at 58 F DB.

A. Sketch the equipment and duct arrangement, showing known information.

B. Determine the required air flow rate into the room in CFM.

C. For the above supply air, find the wet bulb temperature, enthalpy, relative humidity, and moisture content in gr/lb and lb/lb.

D. It is known that 260 CFM of outside air is required for ventilation in this room. The outside air is at 94 F DB and 76 F WB. It is mixed with return air from the room before it enters the air conditioning unit. For the mixed air, determine the dry bulb temperature, wet bulb temperature, enthalpy, and moisture content in gr/lb and lb/lb entering the unit.

E. Determine the required size of the refrigeration equipment required to condition this room, in BTU/hr and tons. (Include ventilation cooling load.)

F. Determine the savings in equipment capacity if the outside ventilation air requirement is reduced to 130 CFM. Give the answer in BTU/hr and percent.

7.22. An air conditioned space has a room sensible cooling load of 200,000 BTU/hr and a room latent cooling load of 50,000 BTU/hr. It is maintained at 76 F DB and 64 F WB. There is 1200 CFM of air vented through cracks and hoods in the space, or through a spill (exhaust) air opening. This means that the outdoor air flow rate is 1200 CFM. The outdoor air, which is at the design conditions of 95 F DB and 76 F WB, is mixed with return air before it enters the air conditioning unit.

A. Sketch the equipment and duct arrangement, showing known information.

B. Calculate the room sensible heat ratio (RSHR).

C. Find the required supply air flow rate in CFM for a supply air temperature of 60 F DB.

D. Determine the cooling load of the outside air in BTU/hr and tons.

E. Calculate the total cooling load in BTU/hr and tons.

F. What is the effective surface temperature (apparatus dew point)?

G. What is the coil CF and BF?

7.23. A refrigeration chiller supplies chilled water to an air conditioning unit. The unit takes in 3000 CFM of outdoor air at 95 F DB and 76 F WB. This outdoor air mixes with 20,000 CFM of return air at 78 F DB and 50% RH. Conditioned air leaves the cooling coil in the air conditioning unit at 52 F DB and 90% RH.

A. What is the load on the chiller due to the coil in the air conditioning unit? Give the answer in BTU/hr and tons.

B. Assume the conditioned air is reheated to 58.5 F DB with electric heaters. What is the operating cost per hour of these heaters if power costs 10 cents per kilowatt hour?

7.24. The supply air for Problem 6.12 is at 58 F DB. Determine:

A. The required air flow rate in CFM

B. Supply air WB, RH, enthalpy, and moisture content

C. Mixed air DB, WB, enthalpy, and moisture content

D. Coil sensible, latent, and total load

7.25. The supply air for Problem 6.16 is at 60 F. Determine:

A. The required air flow rate

B. Supply air WB, RH, enthalpy, and moisture content

C. Mixed air DB, WB, enthalpy, and moisture content

D. Coil sensible, latent, and total load

Fluid Flow in Piping and Ducts

In planning or servicing an HVAC system, it is often necessary to determine pump and fan pressure requirements and piping or duct pressure losses. These and related problems can be solved by an application of some principles of fluid flow which apply to the flow of water and air in air conditioning systems.

OBJECTIVES

After studying this chapter, you will be able to:

1. Use the continuity equation to find flow rate.
2. Use the energy equation to find pump and fan pressures.
3. Find velocity from total and static pressure.
4. Determine pipe and duct sizes.

8.1. THE CONTINUITY EQUATION

The flow of water through piping and air through ducts in HVAC systems is usually under conditions called *steady flow*. Steady flow means that the flow rate of fluid at any point in a section of pipe or duct is equal to that at any other point in the same pipe or duct, regardless of the pipe or duct's shape or cross section. That is, *the same quantity of fluid is passing through every section at a given moment.*

For example, in Figure 8.1, suppose the flow rate of water past section 1 were 10 GPM. If there is steady flow, there must also be 10 GPM flowing past section 2. To see this more clearly, if less than 10 GPM were flowing past section 2, say 4 GPM, ask what happened to the remaining 6 GPM that left section 1? It cannot disappear or be lost (unless there is a hole in the pipe!). Similarly, there cannot be more flow at section 2 than at section 1 because there was only 10 GPM available initially.

The student should not confuse the idea of steady flow *quantity* with the *speed* (velocity) at which fluid flows. At steady flow, fluid flows faster through sections of pipes or ducts that get smaller.

In steady flow, in HVAC systems, the density of the air or water flowing generally does not change significantly. When the density remains constant the flow is called *incompressible*.

Steady flow is a special case of a general principle called either the *conservation of mass principle* or the *continuity principle*. The continuity principle can be expressed as an equation, which is called the *continuity equation*. For incompressible steady flow, the continuity equation is

$$\text{VFR} = \text{constant} = A_1 \times V_1 = A_2 \times V_2 \quad (8.1a)$$

where

VFR = volume flow rate of fluid
A_1, A_2 = crosssectional area of pipe or duct at any points 1 and 2
V_1, V_2 = velocity of fluid at any points 1 and 2

(Figure 8.1 illustrates Equation 8.1a.)

The continuity equation can also be expressed using mass flow rate (MFR) rather than VFR. From Equation 2.1, mass equals density times volume. Then, with constant density (d),

$$\begin{aligned}
\text{MFR} &= \text{constant} \\
&= d \times \text{VFR} = d \times A_1 \times V_1 \\
&= d \times A_2 \times V_2 \quad (8.1b)
\end{aligned}$$

where

MFR = mass flow rate
d = density of fluid

Most flows in HVAC work are steady flow, so Equation 8.1a and 8.1b can be used. Occasionally, unsteady flow exists. (Analysis of this situation is beyond the scope of this book.)

Example 8.1 illustrates uses of the continuity equation.

EXAMPLE 8.1

A service engineer wishes to check if the proper flow rate is circulating in the chilled water piping on a job. This engineer measures a water velocity of 10 ft/sec. The cross-sectional area of pipe is 2 ft^2. What is the water flow rate through the pipe in GPM (gal/min)?

Solution.　Using Equation 8.1a

$$\text{VFR} = A_1 \times V_1 = 2 \text{ ft}^2 \times 10 \text{ ft/sec}$$

$$= 20 \text{ ft}^3/\text{sec}$$

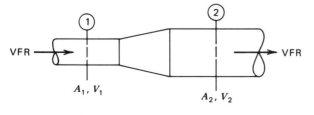

$$\text{VFR} = A_1 \times V_1 = A_2 \times V_2 = \text{constant}$$

Figure 8.1. The continuity equation for steady flow of air through a duct or water through a pipe.

Figure 8.2. Change of velocity with change in cross-sectional area of duct or pipe.

Converting from ft^3/sec to GPM

$$VFR = 20\frac{ft^3}{sec} \times \frac{60 \ sec}{1 \ min} \times \frac{7.48 \ gal}{1 \ ft^3}$$

$$= 8980 \ GPM$$

Area and Velocity Change The continuity equation may be used to demonstrate how velocity is affected by changes in the pipe or duct size. Since

$$VFR = constant = A_1V_1 = A_2V_2 \quad (8.1a)$$

Solving for V_1 (or V_2),

$$V_1 = \frac{A_2}{A_1} \times V_2; \qquad V_2 = \frac{A_1}{A_2} \times V_2 \quad (8.1c)$$

That is, the velocity changes inversely with the cross-sectional area. If the pipe or duct size increases the velocity decreases; if the size decreases, the velocity increases. This is illustrated in Figure 8.2 and Examples 8.2 and 8.3.

EXAMPLE 8.2

Air is flowing through a 1 ft × 2 ft (Figure 8.3) duct at a rate of 1200 CFM (ft^3/min). The duct changes size to 0.5 ft × 1 ft. What is the air velocity in the second section of duct?

Solution. Using Equation 8.1a to find V_1

$$V_1 = \frac{VFR}{A_1} = \frac{1200 \ ft^3/min}{2ft^2} = 600 \ ft/min$$

As the flow rate is constant, $A_1 \times V_1 = A_2 \times V_2$. Solving for V_2,

$$V_2 = \frac{A_1}{A_2} \times V_1 = \frac{2 \ ft^2}{0.5 \ ft^2} \times 600 \ ft/min$$

$$= 2400 \ ft/min$$

EXAMPLE 8.3

Air is flowing in a duct of 48 in.^{2} cross-sectional area at a velocity of 2400 ft/min. This high velocity results in a disturbing noise. The HVAC contractor wants to reduce the velocity to 1300 ft/min. What size duct should be substituted?

Solution. Using Equation 8.1, solving for A_2

$$A_2 = \frac{V_1}{V_2} \times A_1 = \frac{2400}{1300} \times 48 \ in.^2 = 88 \ in.^2$$

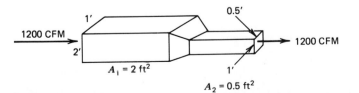

Figure 8.3. Sketch for Example 8.2.

$$H_{s1} + \frac{V_1^2}{2g} + H_{e1} + H_p = H_{s2} + \frac{V_2^2}{2g} + H_{e2} + H_f$$

Figure 8.4. The flow energy equation applied to flow in a duct or pipe.

A duct that has this cross-sectional area would be substituted (say 11 in. × 8 in.).

8.2. THE FLOW ENERGY EQUATION

When the energy balance principle (Section 2.11) is applied to flow in a pipe or duct, it may be stated as follows: between any two points 1 and 2 (Figure 8.2),

$$E_1 + E_{add} - E_{lost} = E_2$$

or

$$E_1 + E_{add} = E_2 + E_{lost}$$

where

E_1, E_2 = energy stored in fluid at points 1 and 2
E_{add} = energy added to fluid between points 1 and 2
E_{lost} = energy lost from fluid between points 1 and 2

The energy of the fluid at any point consists of pressure, velocity (kinetic energy), and elevation (potential energy). The energy added may be that of a pump or fan. The energy lost is due to friction. There may be other energy changes (e.g., a temperature change), but they are usually small and may be neglected. If the energy balance is expressed as an equation using units of head

pressure, as in Figure 8.4, it becomes

$$\underbrace{H_{s1} + \frac{V_1^2}{2g} + H_{e1}}_{E_1} + \underbrace{H_p}_{E_{add}} = \underbrace{H_{s2} + \frac{V_2^2}{2g} + H_{e2}}_{E_2} + \underbrace{H_f}_{E_{lost}}$$

$$(8.2)$$

where

H_s = static pressure of fluid (pressure at rest), ft
V = velocity, ft/sec
g = gravitational constant, 32.2 ft/sec^2
$\frac{V^2}{2g}$ = velocity pressure, ft
H_e = elevation, ft
H_p = pressure added by pump or fan, ft
H_f = pressure lost in piping or duct from friction, ft

Equation 8.2 is called the *flow energy equation* or *generalized Bernoulli equation*. It is used often to determine the pressure requirements of pumps and fans, and in testing and balancing systems.

EXAMPLE 8.4

The piping system shown in Figure 8.5 is to deliver water from the basement to the roof

Figure 8.5. Sketch for Example 8.4.

storage tank, 180 ft above, in the Abraham Lincoln Apartments. The friction loss in the piping, valves and fittings is 12 ft w. The water enters the pump at a pressure of 10 ft, and is delivered at atmospheric pressure (all values are gage pressure). The velocity at the pump suction is 2 ft/sec and at the piping exit is 10 ft/sec. What is the required pump pressure?

Solution. Writing Equation 8.2 to solve for the pump head

$$H_p = (H_{s2} - H_{s1}) + \frac{(V_2^2 - V_1^2)}{2g}$$

$$+ (H_{e2} - H_{e1}) + H_f$$

$$(H_{s2} - H_{s1}) = 0 - 10$$
$$= -10 \text{ ft (change in static pressure)}$$

$$\frac{(V_2^2 - V_1^2)}{2g} = \frac{(10)^2 - (2)^2}{64.4}$$
$$= 1.5 \text{ ft (change in}$$
$$\text{velocity pressure)}$$

$$(H_{e2} - H_{e1}) = 180 \text{ ft (change in elevation)}$$
$$H_f = 12 \text{ ft (friction pressure loss)}$$
$$H_p = -10 + 1.5 + 180 + 12$$
$$= 184 \text{ ft w.g.}$$

The additional pressure required because of the velocity pressure change in Example 8.4 was small. In some cases in piping systems it is small enough to be neglected.

8.3. PRESSURE IN CLOSED AND OPEN SYSTEMS

An *open* piping or duct system is one that is open to the atmosphere at some point. Example 8.4 is an open system. Note that any elevation change is included in determining the pump head. A condenser-cooling tower water system is also open. A *closed* system is one where the water is recirculated continuously and there is no gap in the piping. In a closed system there is no net change in elevation of the water around the whole circuit and therefore the change in H_e is 0 in the flow energy equation. A hydronic system is a closed system. Duct systems are generally open systems, but the change in elevation head is usually such a small value that it can be neglected (e.g., 1 in. w. = 70 ft air).

EXAMPLE 8.5

The pressure loss due to friction in the hydronic system shown in Figure 8.6 is 24 ft w. What is the pump head required?

Solution. This is a closed system. Starting at the pump discharge, point 1, and going around the complete loop back to 1,

$$H_{e1} - H_{e1} = 0 \text{ (no net change in elevation)}$$

$$V_1^2 - V_1^2 = 0$$

$$H_{s1} - H_{s1} = 0$$
$$H_p = H_f = 24 \text{ ft w.}$$

Figure 8.6. Sketch for Example 8.5.

Using Equation 8.2, with all the above terms = 0,

$$H_p = H_f = 24 \text{ ft w.}$$

Example 8.5 shows that the pump head in a closed hydronic system is simply equal to the pressure loss due to friction around the complete circuit.

Air duct systems are usually open systems. Although the effect of elevation change in air ducts can usually be neglected, the effect of velocity change is often quite large. Air pressures in ducts are usually measured in inches of water gage (in. w.g.). Example 8.6 illustrates these ideas.

EXAMPLE 8.6

The duct shown in Figure 8.7 has 8000 CFM flowing through it. The friction loss from point 1 to 2 is 0.43 in. w. If the static pressure at 1 is 1.10 in. w.g., what is the static pressure at 2?

Solution.   Writing Equation 8.2 to solve for H_{s2}

$$H_{s2} = H_{s1} + H_p - H_f + \left(\frac{V_1^2 - V_2^2}{2g} \right)$$

$$+ (H_{e1} - H_{e2})$$

$H_p = 0$ (because there is no fan between 1 and 2)

$H_{e1} - H_{e2} = 0$ (insignificant elevation change)

Finding V_1 and V_2 from Equation 8.1

$$V_1 = \frac{8000 \text{ ft}^3/\text{min}}{4 \text{ ft}^2}$$

$$= 2000 \text{ ft/min} \times \frac{1 \text{ min}}{60 \text{ sec}}$$

$$= 33.33 \text{ ft/sec}$$

$$V_2 = \frac{8000 \text{ ft}^3/\text{min}}{16 \text{ ft}^2}$$

$$= 500 \text{ ft/min} \times \frac{1 \text{ min}}{60 \text{ sec}}$$

$$= 8.33 \text{ ft/sec}$$

$$\frac{V_1^2 - V_2^2}{2g} = \frac{(33.33)^2 - (8.33)^2}{64.4} = 16.2 \text{ ft air}$$

Converting units to in. w.

$$16.2 \text{ ft air} \times \frac{1 \text{ in. w.}}{69.6 \text{ ft air}} = 0.23 \text{ in. w.}$$

$$H_{s2} = 1.10 + 0 - 0.43 + 0.23 + 0$$

$$= 0.90 \text{ in. w.g.}$$

In Example 8.6 it was found that the pressure decreased from point 1 to 2 because of the pressure loss due to friction, but there was a partial increase in pressure because of the velocity reduction (0.23 in. w.). This occurrence is of

Figure 8.7. Sketch for Example 8.6.

importance in air flow in ducts, as will be explained in the next sections.

8.4. TOTAL, STATIC, AND VELOCITY PRESSURE

The *total pressure* (H_t) of a flowing fluid is defined as

$$H_t = H_s + H_v \qquad (8.3)$$

where

H_t = total pressure
H_s = static pressure
H_v = velocity pressure

The *static pressure* is the pressure the fluid has at rest. The *velocity pressure* has been defined as

$$H_v = \frac{V^2}{2g} \qquad (8.4)$$

Thus the total pressure energy that a fluid has at any point can be considered to consist of two parts, its static pressure energy and its velocity pressure energy. This idea is useful in measuring velocities and flow rates in piping and ducts. If the velocity pressure can be measured, the velocity can be found by solving Equation 8.4 for V

$$V = \sqrt{2gH_v} \qquad (8.5)$$

where

V = velocity, ft/sec
g = gravitational constant, ft/sec^2
H_v = velocity head, ft of fluid

EXAMPLE 8.7

The total pressure and static pressure are measured as 66.5 ft w. and 64.8 ft w., respectively, in the condenser water pipe line from a refrigeration machine. What is the velocity in the line?

Solution.   Using Equations 8.3 and 8.5

$$H_v = H_t - H_s = 66.5 - 64.8 = 1.7 \text{ ft w.}$$

$$V = \sqrt{2gH_v} = \sqrt{2 \times 32.2 \text{ ft/sec}^2 \times 1.7 \text{ ft}}$$

$$= 10.5 \text{ ft/sec}$$

When measuring air flow, using in. w. as the unit of pressure and velocity in ft/min, if the conversion units are substituted in Equations 8.4 and 8.5, they become

$$H_v = \left(\frac{V}{4000} \right)^2 \qquad (8.6)$$

$$V = 4000\sqrt{H_v} \qquad (8.7)$$

where

V = air velocity, ft/min
H_v = velocity pressure, in. w.

Many testing and balancing instruments for measuring flow utilize the relationship between total, static, and velocity pressure. Figure 8.8 shows an example. In Figure 8.8*a* a manometer is connected to the duct to read static pressure. In Figure 8.8*b* a manometer reads total pressure because in addition to being exposed to the static

Figure 8.8. Manometer arrangement to read static, total, and velocity pressure.

Figure 8.9. Pitot tube used for measuring velocity pressure

pressure, the impact tube at the end of the manometer faces the oncoming airstream, and therefore receives the velocity pressure energy as well. By connecting the two manometers as shown in Figure 8.8c, the difference between total and static pressure—the velocity pressure —is read directly.

The *pitot tube* (Figure 8.9) is another air flow measuring device that works in the same manner. The probe that is inserted in the duct has two concentric tubes. The opening facing the airstream measures the total pressure and the concentric holes are exposed to static pressure, so the velocity pressure is read directly. A number of readings are usually taken across the duct to get an average velocity.

EXAMPLE 8.8

A contractor wishes to check the air flow rate in a 28 in. × 16 in. duct. The contractor takes a set of readings with a pitot tube, averaging 0.8 in. w. What is the air flow rate in the duct?

Solution. From Equation 8.7, the air velocity is

$$V = 4000\sqrt{0.8} = 3580 \text{ ft/min}$$

and the flow rate from Equation 8.1 is

$$A = 28 \text{ in.} \times 16 \text{ in.} \times \frac{1 \text{ ft}^2}{144 \text{ in.}^2} = 3.11 \text{ ft}^2$$

$$\text{VFR} = A \times V = 3.11 \text{ ft}^2 \times 3580 \text{ ft/min}$$
$$= 11,100 \text{ CFM}$$

8.5. CONVERSION OF VELOCITY PRESSURE TO STATIC PRESSURE (STATIC REGAIN)

One of the remarkable things that can occur in flow in a duct or pipe is that the static pressure can increase in the direction of flow if the velocity decreases. This is caused by a conversion of velocity energy to static energy, called *static regain*. It is a phenomenon that we have all experienced. If we hold a hand in front of the stream of water from a hose, we feel the pressure that is a result of reducing the velocity energy and converting it to pressure. Referring to the diverging transition of Figure 8.10, if Equation 8.6 is applied to points 1 and 2 and if we assume that the pressure loss due to friction H_f is negligible,

$$H_{s2} - H_{s1} = H_{v1} - H_{v2} = \left(\frac{V_1}{4000}\right)^2 - \left(\frac{V_2}{4000}\right)^2$$
$$(8.8)$$

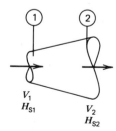

Figure 8.10. Sketch for Equation 8.8.

Because there is always some friction loss, the actual static pressure regain is never as high as that shown in Equation 8.8. The proportion of static regain that can be recovered, called the *recovery factor R*, depends on the shape of the transition that changes velocity. The actual static pressure regain (SPR) is therefore

$$SPR = R\left(\left[\frac{V_1}{4000}\right]^2 - \left[\frac{V_2}{4000}\right]^2\right) \quad (8.9)$$

Recovery factors of 0.7 to 0.9 are achieved with reasonably gradual transitions.

EXAMPLE 8.9

Find the increase in static pressure (regain) from point 1 to 2 in the duct system shown in Figure 8.11, if the recovery factor is 0.7.

Solution. Using Equation 8.9 with $R = 0.7$

$$H_{s2} - H_{s1} = 0.7\left[\left(\frac{1800}{4000}\right)^2 - \left(\frac{600}{4000}\right)^2\right]$$

$$= 0.13 \text{ in. w.}$$

The opposite event, a conversion of static pressure to velocity pressure, will occur in a

$V_1 = 1800$ ft/min $V_2 = 600$ ft/min

Figure 8.11. Sketch for Example 8.9.

converging transition, resulting in a decrease in static pressure (Figure 8.12).

8.6. PRESSURE LOSS FROM FRICTION IN PIPING AND DUCTS

We have seen from the flow energy equation that one of the effects the pump or fan must overcome is the pressure loss due to friction. Friction is a resistance to flow resulting from viscosity and from the walls of the pipe or duct. In previous examples we have assumed values of friction pressure loss. Actually we must be able to calculate it. For the type of flow usually existing in HVAC systems, called *turbulent flow*, the pressure loss or drop due to friction is given as

$$H_f = f\frac{L}{D}\frac{V^2}{2g} \quad (8.10)$$

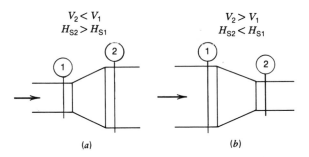

$V_2 < V_1$
$H_{S2} > H_{S1}$

$V_2 > V_1$
$H_{S2} < H_{S1}$

(a)

(b)

Figure 8.12. Conversion between velocity pressure and static pressure. (a) Diverging transition — velocity decreases, static pressure increases. (b) Converging transition — velocity increases, static pressure decreases.

where

H_f = pressure loss (drop) from friction in straight pipe or duct

f = a friction factor

L = length of pipe or duct

D = diameter of pipe or duct

V = velocity of fluid

The friction factor f depends on the roughness of the pipe or duct wall. Rougher surfaces will cause increased frictional resistance. This means that by using and maintaining smooth surfaces, friction decreases and less energy is used. The other terms in the equation also indicate useful information. Lower velocities and larger diameters reduce H_f and therefore result in lower energy consumption, although the pipe or duct cost increases.

Although H_f could be calculated each time from Equation 8.10, charts that are much easier to use and show the same information have been developed for water and air flow.

8.7. FRICTION LOSS FROM WATER FLOW IN PIPES

Pressure loss or drop from friction in straight pipe has been put in chart form for some commonly encountered materials and conditions. Three of these are presented in this book, Figures 8.13, 8.14, and 8.15. Figures 8.13 and 8.14 are suitable for water at 60 F flowing in steel Schedule 40 pipe (the Schedule number refers to the wall thickness). Figure 8.15 is suitable for water at 60 F flowing in copper tubing. Type K, L, M refer to different wall thickness (Chapter 9).

For water at 200 F the pressure loss is about 10% less than that shown, due to the change in viscosity and density.

Figure 8.13 is suitable for steel piping in a clean condition, which is usually true for a closed hydronic system that is well maintained. Figure

8.14 is suitable for open systems, such as condenser-cooling tower systems, where the steel pipe wall will become rougher with age, resulting in higher pressure drops. Figure 8.15 is suitable for open or closed systems. A few examples will illustrate some uses of the friction loss charts.

EXAMPLE 8.10

What is the pressure loss due to friction and the velocity in 500 ft of 2 in. Schedule 40 chilled water piping through which 40 GPM of water at 60 F is flowing?

Solution. The information can be found from Figure 8.13 (closed systems). The solution is indicated in Figure 8.16, at the point of intersection of a 40 GPM flow rate and $D = 2$ in. Note that the chart lists friction loss per 100 ft of pipe, which is then converted to the actual length of pipe. At 40 GPM and $D = 2$ in.,

$$H_f = \frac{3.2 \text{ ft w.}}{100 \text{ ft}} \times 500 \text{ ft} = 16.0 \text{ ft w.}$$

The velocity at the intersection point is $V = 3.9$ ft/sec.

EXAMPLE 8.11

A copper tubing system is to be used to circulate 30 GPM of water at 60 F. The system is to be designed to have a friction pressure drop no greater than 4 ft w. per 100 ft pipe. What is the smallest size tubing that can be used?

Solution. Figure 8.15 will be used. The solution is shown in Figure 8.17. The intersection point of 30 GPM and 4 ft w./100 ft pipe lies between a 2 in. and $1\frac{1}{2}$ in. diameter. If $1\frac{1}{2}$ in. is used, the pressure drop will be greater than 4 ft w./100 ft at 30 GPM, so this is unacceptable. If 2 in. is used, the pressure drop will be less than the maximum allowed; therefore, this is the correct

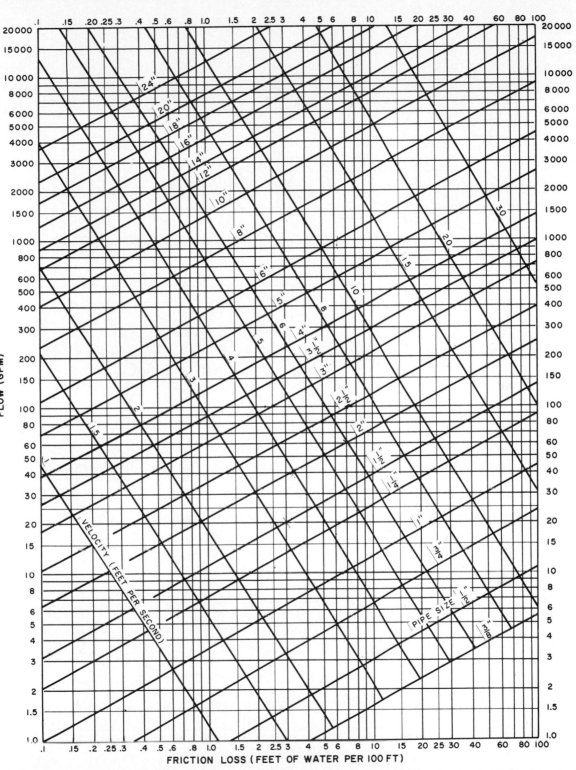

Figure 8.13. Friction loss for water in Schedule 40 steel pipe — closed system. (*Courtesy:* Carrier Corporation, Syracuse, N.Y.)

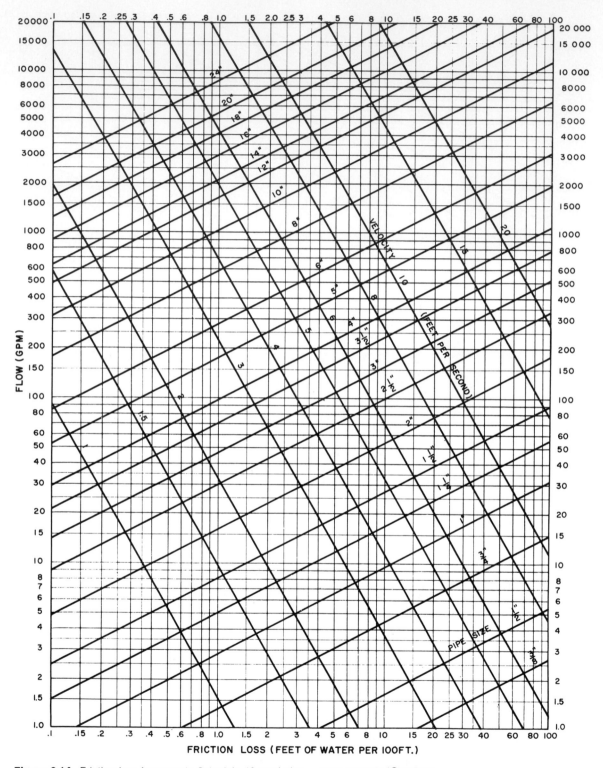

Figure 8.14. Friction loss for water in Schedule 40 steel pipe — open system. (*Courtesy:* Carrier Corporation, Syracuse, N.Y.)

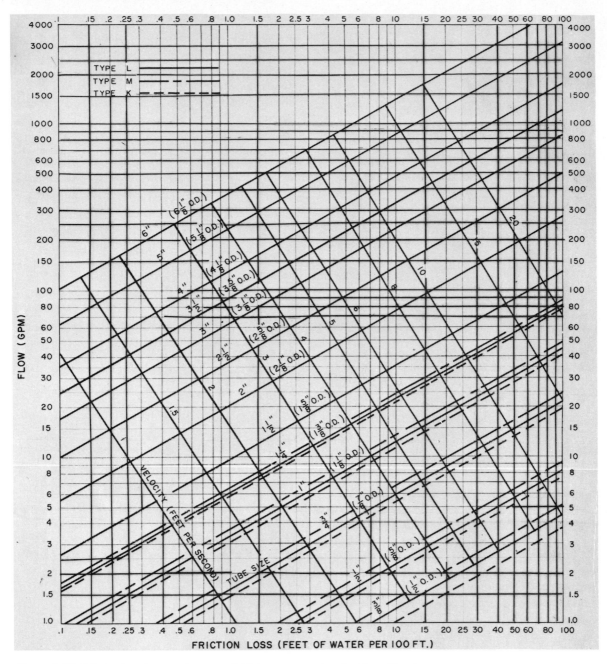

Figure 8.15. Friction loss for water in copper tubing — open or closed system. (*Courtesy:* Carrier Corporation, Syracuse, N.Y.)

Figure 8.16. Sketch for Example 8.10.

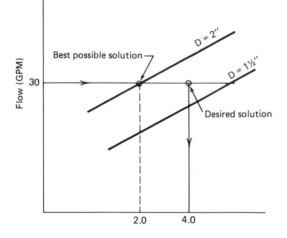

Figure 8.17. Sketch for Example 8.11.

solution. Note that the actual rather than the allowed pressure drop should be recorded. The solution is

$$D = 2 \text{ in.,} \quad H_f = 2.0 \text{ ft w.}/100 \text{ ft}$$

EXAMPLE 8.12

A 3 in. steel pipe, Schedule 40, is supposed to circulate 200 GPM in a chilled water system. I.

Fixit, a service troubleshooter, is asked to check if the flow rate is actually 200 GPM. Fixit puts two pressure gages on a horizontal run of the straight line, 200 ft apart. The first gage reads 40 ft w., the other 32 ft w. Is the system delivering the proper rate of water? How much is being circulated?

Solution. The actual pressure drop (equal to the friction loss) is

$$H_{s1} - H_{s2} = H_f = 40 - 32$$
$$= 8 \text{ ft w. per 200 ft pipe}$$

or

$$H_f/100 \text{ ft pipe} = \frac{8 \text{ ft w.}}{200 \text{ ft}} \times 100 \text{ ft} = 4 \text{ ft w.}$$

Using Figure 8.13 to compare the expected friction loss with that measured, at 200 GPM and 3 in. diameter

$$H_f/100 \text{ ft} = 9.0 \text{ ft w.}$$

Therefore, the flow rate is less than it should be. Reading at $H_f/100$ ft = 4.0 ft w. and D = 3 in., the actual flow rate is 130 GPM, seriously below normal.

For hot water systems the correction of 10% less pressure drop should be made. Otherwise oversized equipment or wasteful energy losses will result.

EXAMPLE 8.13

What would be the friction pressure drop in 800 ft of 2 in. copper tubing through which 50 GPM of hot water is flowing?

Solution. From Figure 8.15, the pressure drop for cold water is

$$H_f/100 \text{ ft} = 5 \text{ ft w.}$$

correcting this for hot water

$$H_f/100 \text{ ft} = 0.9 \times 5 = 4.5 \text{ ft w.}$$

For 800 ft the pressure drop is

$$H_f = \frac{4.5 \text{ ft w.}}{100 \text{ ft}} \times 800 \text{ ft} = 36.0 \text{ ft w.}$$

8.8. PRESSURE LOSS IN PIPE FITTINGS

In addition to the pressure loss in straight pipe, there will be pressure losses from turbulence and change of direction through fittings and valves. These are called dynamic losses. These pressure losses are shown in Table 8.1. The pressure losses are expressed in this table in a way that is called the *equivalent length*. The listings for a particular fitting of a given size show the equivalent length (E.L.) of straight pipe that would have the same pressure drop. After finding the E.L. from Table 8.1, the appropriate friction loss chart is used to find the actual pressure drop through the fitting.

EXAMPLE 8.14

Find the pressure drop through a 4 in. 90° cast iron (C.I.) standard elbow in a chilled water system through which 300 GPM of water is flowing.

Solution. From Table 8.1, find the equivalent length of the fitting

$$E.L. = 11.0 \text{ ft}$$

Using Figure 8.13, $H_f/100 \text{ ft} = 5.2 \text{ ft w.}$ The

TABLE 8.1. EQUIVALENT FEET OF PIPE FOR FITTINGS AND VALVES

	Nominal Pipe Size (inches)												
	$\frac{1}{2}$	$\frac{3}{4}$	1	$1\frac{1}{4}$	$1\frac{1}{2}$	2	$2\frac{1}{2}$	3	4	5	6	8	10
45° Elbow	0.8	0.9	1.3	1.7	2.2	2.8	3.3	4.0	5.5	6.6	8.0	11.0	13.2
90° Elbow standard	1.6	2.0	2.6	3.3	4.3	5.5	6.5	8.0	11.0	13.0	16.0	22.0	26.0
90° Elbow long	1.0	1.4	1.7	2.3	2.7	3.5	4.2	5.2	7.0	8.4	10.4	14.0	16.8
Gate valve open	0.7	0.9	1.0	1.5	1.8	2.3	2.8	3.2	4.5	6.0	7.0	9.0	12.0
Globe valve open	17	22	27	36	43	55	67	82	110	134	164	220	268
Angle valve	7	9	12	15	18	24							
Tee – side flow	3	4	5	7	9	12	14	17	22	28	34	44	56
Swing check valve	6	8	10	14	126	20	25	30	40	50	60	80	100
Tee – straight through flow	1.6	2.0	2.6	3.3	4.3	5.5	6.5	8.0	11.0	13.0	16.0	22.0	26.0
Radiator angle valve		5	5	5	5								
Diverting tee		20	14	11	12	14	14	14					
Flow check valve		27	42	60	63	83	104	125	126				
Air scoop		2	3	4	5	7	8	13	15				
Boiler (typical)	5	7	9	11	11								

pressure drop through the fitting is

$$H_f = \frac{5.2 \text{ ft w.}}{100 \text{ ft}} \times 11.0 \text{ ft} = 0.6 \text{ ft w.}$$

In addition to the equivalent length method of determining pressure drop through pipe fittings, there is another procedure called the loss coefficient method. A loss coefficient (called C, C_v, or K) for the fitting is determined from an appropriate table listing C-values. The loss coefficient method will not be used for pipe fittings here. It will be used for duct fittings (see Section 8.13).

8.9. PIPING SYSTEM PRESSURE DROP

A common problem is to determine the pressure loss from friction in a closed system, in order to determine the required pump head. The system pressure drop is simply the sum of the losses through each item in one of the paths or circuits from pump discharge to pump suction, including piping, fittings, valves, and equipment. Information on pressure drops through equipment is obtained from the manufacturer.

To determine the system pressure loss, the pressure losses through only one circuit are considered. This is because the pressure losses are the same through every circuit. This idea is quite similar to that in electric circuits, where the voltage loss through parallel electric circuits is the same; Figure 8.18 illustrates this. The pressure drop from A to D is indicated by the difference in readings on the two pressure gages located at A and D. Therefore, the pressure drop through the longer circuit ABD is the same as that through ACD.

It would seem from this explanation that it would not matter which circuit one chooses for actually calculating the system pressure drop. It usually does not work out that way, however. Most piping systems are designed to have equal friction loss per foot of length. It would seem, therefore, using our example above, that in this case the pressure drop ABD would be greater than that through ACD. This certainly is not possible, because each pressure gage has one fixed reading. What actually happens is that in balancing the system the contractor will throttle valves in ABD to increase the pressure drop and produce the proper flow in that circuit.

For this reason the longest circuit is usually chosen to calculate system pressure drop. A shorter circuit usually has valves partially closed, through which there is no way of predicting pressure drops.

To find the system pressure drop, therefore, first inspect the piping layout to see which circuit is the longest. Occasionally a circuit that is not the longest is chosen. This occurs when a circuit has an unusual number of items in it with a high pressure drop, and that circuit is then the one with the greatest equivalent length.

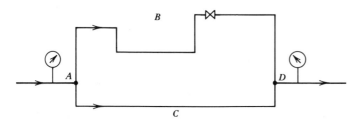

Figure 8.18. Pressure drop between any two points is always the same.

Figure 8.19. Sketch for Example 8.15.

EXAMPLE 8.15

For the steel piping system shown in Figure 8.19 determine the required pump head.

Solution. From the energy equation, the required pump head rise *F–A* is equal to the pressure drop due to friction from *A–F* through the system. But only the circuit with the greatest pressure drop is chosen. This is probably *ABCDEF*, the longest circuit. Circuits *ABEF* and *ACDF* are ignored. (It is possible in unusual cases that one of these may have a greater pressure drop if it has items with great resistance.

TABLE 8.2. PIPING PRESSURE DROP CALCULATIONS FOR EXAMPLE 8.15

Section	Item	D, in.	GPM	V FPS	E.L., ft	No. of Items	Total Length, ft	Friction Loss H_f ft w. 100 ft	ft w.
EFAB	Pipe	4	300	7.8			800	5.2	
EFAB	Gate valve				4.5	1	5		
EFAB	Globe valve				110	1	110		
EFAB	90° std ell				11	2	22		
B	Tee				11	1	11		
Subtotal							948 ×	5.2 =	49.3
BC	Pipe	$3\frac{1}{2}$	200	7.0			100	4.8	
C	Tee				9	1	9		
Subtotal							109 ×	4.8 =	5.2
CD	Pipe	$2\frac{1}{2}$	70	4.8			600	3.7	
CD	Globe valve				67	1	67		
CD	90° std ell				6.5	2	13		
D	Tee				6.5	1	7		
Subtotal							687 ×	3.7 =	25.4
DE		$3\frac{1}{2}$	200	7.0			100	4.8	
E	Tee				9	1	9		
Subtotal							109 ×	4.8 =	5.2
								Total H_f =	85.1

Such cases must be checked.) Using Figure 8.13 and Table 8.1, the friction pressure loss for each item is found and summed up. This information is presented in an organized form in Table 8.2. Review each entry carefully, comparing it with the piping diagram and the pressure drop charts, to see if you arrive at the same results.

A quick estimate method sometimes used for determining the system pressure drop is to multiply the straight pipe friction loss (in the longest circuit) by 1.5, to allow for fittings and valves. The author does not recommend this rule of thumb, except for preliminary studies.

8.10. SYSTEM PIPE SIZING

One of the tasks in designing a hydronic system is to determine the size of each section of pipe. This is usually done by a procedure called the *equal friction method*. The steps in this procedure are as follows.

1. Prepare a sketch of the piping system.

2. Find the flow rate in each section of pipe by adding up the required flow rates for each terminal unit, working back from the last unit to the main from the pump.

3. Choose a value of friction loss rate for the main from the pump based on the following limits.

   (a) The friction loss rate should be between 1-5 ft w./100 ft pipe.

   (b) The velocity in the main should not exceed 4-6 FPS in small systems, or 8-10 FPS in larger systems. The velocity in occupied areas should not exceed 4 FPS, to prevent noise.

4. Select a pipe size for the main from the pump from the friction loss charts based on the above limits. If there is more than one choice, select the smaller pipe size if initial costs are

more critical or the larger size if operating costs are more critical.

5. Continuing along the circuit chosen, select succeeding pipe sizes. This should be done according to the following guides.

   (a) When the flow rate decreases significantly, select the next smaller pipe size from the chart so that the friction loss rate is approximately equal to the value chosen for the initial section (equal friction).

   (b) Do not let the velocity decrease below 1.5-2 FPS, to prevent the settling of dirt and air in the lines.

6. Select the pipe sizes for the branches in other circuits. If a circuit is not extremely short, use the same friction loss rate. For a very short circuit, this procedure would result in a total pressure drop much less than that for the longest circuit, and cause problems of balancing flow. Therefore, select smaller pipe sizes in these branches to create increased pressure drop. This problem does not arise in a reverse return type system.

7. Determine by inspection which circuit will be the longest, or have the greatest equivalent length. Calculate the pressure drop in the longest circuit.

Example 8.16 will illustrate pipe sizing procedures.

EXAMPLE 8.16

Select the pipe sizes for the chilled water piping system shown in Figure 8.20. Type L copper tubing is used.

Each terminal unit takes 10 GPM. The branches to each terminal unit are 10 ft long. Lengths of mains are shown on the sketch.

Solution. The stepwise procedure explained previously will be used to size the pipe.

1. The piping sketch is first drawn as in Figure 8.20.

Figure 8.20. Sketch for Example 8.16.

2. The flow rate in each section is found by adding the flow rates from each unit, starting with the last as shown.

3. *ABCDEFGHIJ* is clearly the longest circuit.

4. Main *AB* has 40 GPM. Using Figure 8.15, either a $2\frac{1}{2}$ in. or 2 in. size will result in the friction loss rate between 1-5 ft w./100 ft.

5. It is decided that the 2 in. pipe size will be selected to minimize initial costs. The friction loss is 3.3 ft w./100 ft.

6. The tabulation of pipe sizes selected for each remaining section in the longest circuit is shown in Table 8.3. Note that the pipe size is decreased gradually as flow rate decreases, to maintain a friction loss reasonably close to the initial friction loss, within limits of available pipe sizes.

TABLE 8.3. PIPE SIZING PROCEDURES FOR EXAMPLE 8.16

Section	GPM	D, in.	L, ft	Friction, ft w./100 ft	V, FPS
AB	40	2	80	3.3	4.3
BC	30	2	60	2.0	3.2
CD	20	$1\frac{1}{2}$	100	3.7	3.7
DE	10	$1\frac{1}{4}$	60	2.5	2.7
FG	10	$1\frac{1}{4}$	80	2.5	2.7
GH	20	$1\frac{1}{2}$	100	3.7	3.7
HI	30	2	40	2.0	3.2
IJ	40	2	80	3.3	4.3
BRANCHES					
BI	10	1	20	7.0	4.0
CH	10	1	20	7.0	4.0
DG	10	1	20	7.0	4.0

7. The piping system shown is a direct return arrangement, which in this case will have some circuits much shorter than others. The water flow will tend to short-circuit through *ABIJ* and other short loops, starving the last units. By increasing the pressure drop in the branches, this unbalance can be improved somewhat. A check should be made that the velocities are not excessive. The pipe size selected for branches is 1 in. rather than $1\frac{1}{4}$ in. in order to accomplish this. This is still probably not enough to solve the problem completely, and balancing valves would be required. In reality, a reverse return system would be a better choice of piping arrangements than the one shown.

8. The pressure drop in the longest circuit can now be calculated by the same procedures as used in Example 8.15. This is left as a problem for the student.

8.11. FRICTION LOSS FROM AIR FLOW IN DUCTS

Pressure loss from friction in straight round ducts is shown in charts in a manner similar to water flow in piping. Figure 8.21 shows this information. This chart is suitable for clean galvanized steel ducts with about 40 joints per 100 ft, and with air at standard conditions. It can be used for the general range of HVAC temperatures and for altitudes up to 2000 ft.

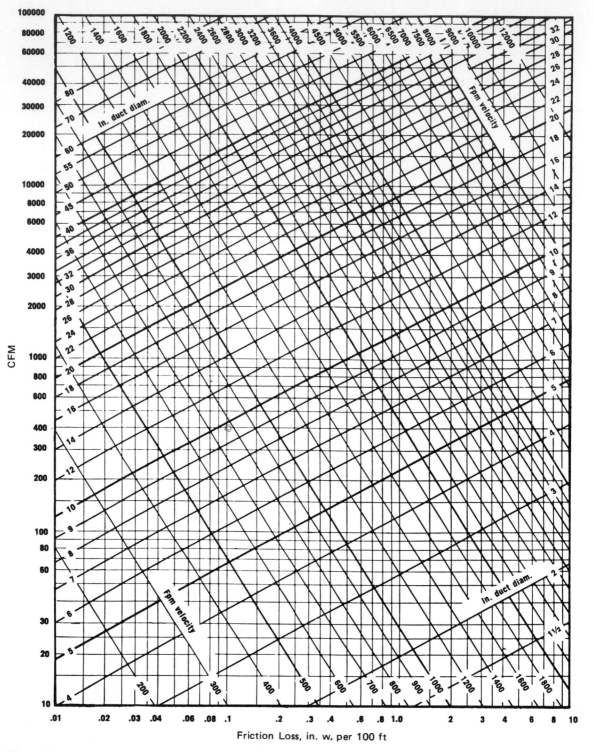

Figure 8.21. Friction loss for air flow in galvanized steel round ducts.

Figure 8.22. Sketch for Example 8.17.

EXAMPLE 8.17

A 12 in. diameter round galvanized duct 250 ft long has 1000 CFM of air flowing through it. What is the pressure loss due to friction and the velocity in the duct?

Solution. The solution is found from Figure 8.21, as seen in the sketch in Figure 8.22

$$H_f/100 \text{ ft} = 0.2 \text{ in. w.}$$

$$H_f = \frac{0.20 \text{ in. w.}}{100 \text{ ft}} \times 250 \text{ ft} = 0.50 \text{ in. w.}$$

$$V = 1300 \text{ FPM}$$

To find the friction loss in rectangular section ducts, Figure 8.23 must first be used. This chart shows *equivalent round duct* sizes. The equivalent round duct is the round duct that would have the same friction loss as a rectangular duct found in the chart.

EXAMPLE 8.18

A 30 in. by 19 in. rectangular duct is delivering 7000 CFM of air. What is the friction loss per 100 ft?

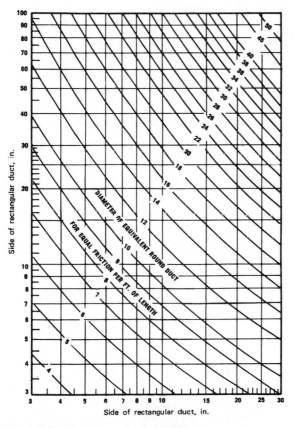

Figure 8.23. Equivalent round duct sizes.

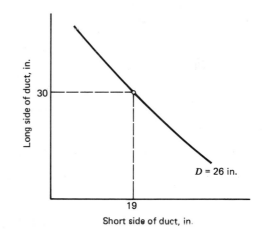

Figure 8.24. Sketch for Example 8.18.

Solution.   Referring first to Figure 8.23, as seen in the sketch in Figure 8.24, the equivalent round diameter to a 30 in. by 19 in. duct is

$$D = 26 \text{ in.}$$

Figure 8.21 can now be used to find the friction loss in the rectangular duct, as shown previously

$$H_f/100 \text{ ft} = 0.17 \text{ in. w.}$$

8.12. ASPECT RATIO

At first consideration it might seem that the equivalent round duct would have the same cross-sectional area as a rectangular duct for the same friction loss. This is not quite true. A rectangular duct with the same friction loss will have a greater area than a round duct. This is because the rectangular shape, with a greater ratio of surface to cross section, causes more friction. This problem becomes worse as the aspect ratio increases. The *aspect ratio* is the ratio of the dimensions of the two adjacent sides of a rectangular duct. As a general rule, the aspect ratios of rectangular ducts should be as low as possible to keep friction losses reasonably low and thereby avoid excess energy consumption. An increasing aspect ratio will also mean more sheet metal and therefore a more expensive system. Unfortunately, the clear space available for horizontal ducts is often limited by the depth of hung ceilings, resulting in high aspect ratios.

EXAMPLE 8.19

Ace Sheet Metal, a contractor, wants to install a duct handling 3000 CFM in a hung ceiling that has 12 in. of vertical clear space available for the duct. The velocity in the duct is not to exceed 1600 FPM to avoid excessive noise. What size duct should Ace install?

Solution.   Ace wants to keep the aspect ratio as low as possible, to reduce friction loss and also to save money on sheet metal cost, so they will try to use as much of the 12 in. as possible. Let us say that Ace is going to put 1 in. of insulation on the duct; therefore, the maximum duct depth can be 10 in.

From Figure 8.21, with 3000 CFM at 1600 FPM, a 19 in. round duct is found. From Figure 8.23, for a round duct of ʻ19 in., the equivalent rectangular duct with one side 10 in. is 33 in. by 10 in. This is a reasonably good solution, because the aspect ratio is $33/10 = 3.3$.

The friction loss charts can be used for testing and troubleshooting as well as design and installation, as illustrated in Example 8.20.

EXAMPLE 8.20

A 20 in. by 11 in. duct is supposed to be handling 3000 CFM. The engineer from Top Testing Co. is assigned to check the performance. The engineer takes readings on manometers 50 ft. apart in the duct and reads 1.75 in. w. and 1.63 in. w. Is the system handling the proper air flow? If not, what is the flow?

Solution.   From Figure 8.23 the equivalent round duct diameter to a 20 in. by 11 in. rectangular duct is 16 in. Using Figure 8.21, the friction loss for this duct at 3000 CFM is

$$H_f/100 \text{ ft} = 0.37 \text{ in. w.}$$

and for 50 ft is

$$H_f = \frac{0.37}{100} \times 50 = 0.19 \text{ in. w.}$$

The friction loss is actually

$$H_f = 1.75 - 1.63 = 0.12 \text{ in. w.}$$

and therefore the duct is supplying less than 3000 CFM. The actual conditions are

$$H_f/100 \text{ ft} = \frac{0.12 \text{ in. w.}}{50 \text{ ft}} \times 1200 \text{ ft} = 0.24 \text{ in. w.}$$

From Figure 8.21, at this friction loss

$$\text{Flow rate} = 2400 \text{ CFM}$$

Of course this check is accurate only if the installation is similar to the one on which the friction charts are based, as described previously.

8.13. PRESSURE LOSS IN DUCT FITTINGS

The pressure losses in duct fittings resulting from direction change can be expressed in two ways. One is the equivalent length method, as used previously for pipe fittings. Another procedure is called the *loss coefficient* method, which will be used here.

TABLE 8.4. LOSS COEFFICIENTS, ELBOWS
Use the velocity pressure (H_v) of the upstream section. Fitting loss (H_f) = $C \times H_v$

A. Elbow, Smooth Radius (Die Stamped), Round

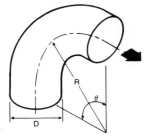

Coefficients for 90° Elbows: (See Note)

R/D	0.5	0.75	1.0	1.5	2.0	2.5
C	0.71	0.33	0.22	0.15	0.13	0.12

Note: For angles other than 90° multiply by the following factors:

θ	0°	20°	30°	45°	60°	75°	90°	110°	130°	150°	180°
K	0	0.31	0.45	0.60	0.78	0.90	1.00	1.13	1.20	1.28	1.40

B. Elbow, Round, 3 to 5 pc — 90°

Coefficient C

No. of Pieces	R/D				
	0.5	0.75	1.0	1.5	2.0
5	—	0.46	0.33	0.24	0.19
4	—	0.50	0.37	0.27	0 24
3	0.98	0.54	0.42	0.34	0.33

C. Elbow, Round, Mitered

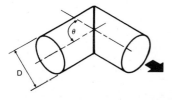

Coefficient C

θ	20°	30°	45°	60°	75°	90°
C	0.08	0.16	0.34	0.55	0.81	1.2

TABLE 8.4. *(Continued).*

D. Elbow, Rectangular, Mitered

	Coefficient C										
θ	H/W										
	0.25	0.5	0.75	1.0	1.5	2.0	3.0	4.0	5.0	6.0	8.0
20°	0.08	0.08	0.08	0.07	0.07	0.07	0.06	0.06	0.05	0.05	0.05
30°	0.18	0.17	0.17	0.16	0.15	0.15	0.13	0.13	0.12	0.12	0.11
45°	0.38	0.37	0.36	0.34	0.33	0.31	0.28	0.27	0.26	0.25	0.24
60°	0.60	0.59	0.57	0.55	0.52	0.49	0.46	0.43	0.41	0.39	0.38
75°	0.89	0.87	0.84	0.81	0.77	0.73	0.67	0.63	0.61	0.58	0.57
90°	1.3	1.3	1.2	1.2	1.1	1.1	0.98	0.92	0.89	0.85	0.83

E. Elbow, Rectangular, Smooth Radius without Vanes

Coefficients for 90° elbows: (See Note)

	H/W										
R/W	0.25	0.5	0.75	1.0	1.5	2.0	3.0	4.0	5.0	6.0	8.0
0.5	1.5	1.4	1.3	1.2	1.1	1.0	1.0	1.1	1.1	1.2	1.2
0.75	0.57	0.52	0.48	0.44	0.40	0.39	0.39	0.40	0.42	0.43	0.44
1.0	0.27	0.25	0.23	0.21	0.19	0.18	0.18	0.19	0.20	0.27	0.21
1.5	0.22	0.20	0.19	0.17	0.15	0.14	0.14	0.15	0.16	0.17	0.17
2.0	0.20	0.18	0.16	0.15	0.14	0.13	0.13	0.14	0.14	0.15	0.15

F. Elbow, Rectangular, Mitered with Turning Vanes

Push-on Vane Runner

Embossed Vane Runner

AIR FLOW

TRAILING EDGE

SINGLE THICKNESS VANES

*No.	Dimensions, inches			Coeff.
	R	S	L	C
1*	2.0	1.5	0.75	0.12
2	4.5	2.25	0	0.15
3	4.5	3.25	1.60	0.18

*Numbers are for reference only.

*When extension of trailing edge is not provided for this vane, losses are approximately unchanged for single elbows, but increase considerably for elbows in series.

DOUBLE THICKNESS VANES
Coefficient C

*No.	Dimensions, in.		Velocity (V), fpm				Remarks
	R	S	1000	2000	3000	4000	
1	2.0	1.5	0.27	0.22	0.19	0.17	Embossed Vane Runner
2	2.0	1.5	0.33	0.29	0.26	0.23	Push-On Vane Runner
3	2.0	2.13	0.38	0.31	0.27	0.24	Embossed Vane Runner
4	4.5	3.25	0.26	0.21	0.18	0.16	Embossed Vane Runner

*Numbers are for reference only.

Reprinted with permission from the SMACNA "HVAC Systems — Duct Design" manual. Second Edition, 1981.

TABLE 8.5. LOSS COEFFICIENTS, TRANSITIONS (DIVERGING FLOW)
Use the velocity pressure (H_v) of the upstream section. Fitting loss (H_f) = $C \times H_v$

A. Transition, Round, Conical

When: θ = 180°

R_e = 8.56 DV

where:

D = Upstream Diameter (inches)
V = Upstream Velocity (fpm)

		Coefficient C (See Note)							
R_e	A_1/A	θ							
		16°	20°	30°	45°	60°	90°	120°	180°
0.5×10^5	2	0.14	0.19	0.32	0.33	0.33	0.32	0.31	0.30
	4	0.23	0.30	0.46	0.61	0.68	0.64	0.63	0.62
	6	0.27	0.33	0.48	0.66	0.77	0.74	0.73	0.72
	10	0.29	0.38	0.59	0.76	0.80	0.83	0.84	0.83
	≥16	0.31	0.38	0.60	0.84	0.88	0.88	0.88	0.88
2×10^5	2	0.07	0.12	0.23	0.28	0.27	0.27	0.27	0.26
	4	0.15	0.18	0.36	0.55	0.59	0.59	0.58	0.57
	6	0.19	0.28	0.44	0.90	0.70	0.71	0.71	0.69
	10	0.20	0.24	0.43	0.76	0.80	0.81	0.81	0.81
	≥16	0.21	0.28	0.52	0.76	0.87	0.87	0.87	0.87
≥6×10^5	2	0.05	0.07	0.12	0.27	0.27	0.27	0.27	0.27
	4	0.17	0.24	0.38	0.51	0.56	0.58	0.58	0.57
	6	0.16	0.29	0.46	0.60	0.69	0.71	0.70	0.70
	10	0.21	0.33	0.52	0.60	0.76	0.83	0.84	0.83
	≥16	0.21	0.34	0.56	0.72	0.79	0.85	0.87	0.89

B. Transition, Rectangular, Pyramidal

When θ = 180°

	Coefficient C (See Note 1)							
A_1/A	θ							
	16°	20°	30°	45°	60°	90°	120°	180°
2	0.18	0.22	0.25	0.29	0.31	0.32	0.33	0.30
4	0.36	0.43	0.50	0.56	0.61	0.63	0.63	0.63
6	0.42	0.47	0.58	0.68	0.72	0.76	0.76	0.75
≥10	0.42	0.49	0.59	0.70	0.80	0.87	0.85	0.86

Note: A = Area (Entering airstream), A_1 = Area (Leaving airstream)

Reprinted with permission from the SMACNA ''HVAC Systems — Duct Design'' manual. Second Edition, 1981.

By using this method the pressure loss through a fitting can be calculated as follows

$$H_f = C \times H_v = C \times \left(\frac{V}{4000}\right)^2 \qquad (8.11)$$

where

H_f = total pressure loss through fitting, in. w.
C = a loss coefficient
H_v = velocity pressure at fitting, in. w.
V = velocity, ft/min

Some values of C for various fitting are shown in Tables 8.4–8.8. The losses through fittings depend on the shape, of which there are many variations. A more extensive list can be found in the ASHRAE Fundamentals.

EXAMPLE 8.21

A 12 in. by 8 in. 90° smooth radius elbow without vanes has a center-line radius R = 16 in. It has 1500 CFM flowing through it. Find the pressure loss through the fitting.

Solution. The loss coefficient is found in Table 8.4. Referring to Figure 8.25 on page 238, H = 12 in., W = 8 in., and R = 16 in. This gives

$$\frac{H}{W} = \frac{12}{8} = 1.5 \qquad \frac{R}{W} = \frac{16}{8} = 2.0 \text{ and } C = 0.14$$

TABLE 8.6. LOSS COEFFICIENTS, TRANSITIONS (CONVERGING FLOW)
Use the velocity pressure (H_v) of the downstream section. Fitting loss (H_f) = $C \times H_v$

A. Contraction, Round and Rectangular, Gradual to Abrupt

Coefficient C (See Note)

A_1/A	θ						
	10°	15°-40°	50°-60°	90°	120°	150°	180°
2	0.05	0.05	0.06	0.12	0.18	0.24	0.26
4	0.05	0.04	0.07	0.17	0.27	0.35	0.41
6	0.05	0.04	0.07	0.18	0.28	0.36	0.42
10	0.05	0.05	0.08	0.19	0.29	0.37	0.43

When θ = 180°

Note: A_1 = Area (Entering airstream), A = Area (Leaving airstream)

Reprinted with permission from the SMACNA "HVAC Systems—Duct Design" manual. Second Edition, 1981.

TABLE 8.7. LOSS COEFFICIENTS, CONVERGING JUNCTIONS
Use the velocity pressure (H_v) of the downstream section. Fitting loss (H_f) = $C \times H_v$

A. Converging Tee, Round Branch to Rectangular Main

	Branch, Coefficient C (See Note)									
V_c	Q_b/Q_c									
	0.1	0.2	0.3	0.4	0.5	0.6	0.7	0.8	0.9	1.0
< 1200 fpm	-.63	-.55	0.13	0.23	0.78	1.30	1.93	3.10	4.88	5.60
> 1200 fpm	-.49	-.21	0.23	0.60	1.27	2.06	2.75	3.70	4.93	5.95

When:

A_b/A_s	A_s/A_c	A_b/A_c
0.5	1.0	0.5

B. Converging Tee, Rectangular Main and Branch

	Branch, Coefficient (See Note)									
V_c	Q_b/Q_c									
	0.1	0.2	0.3	0.4	0.5	0.6	0.7	0.8	0.9	1.0
< 1200 fpm	-.75	-.53	-.03	0.33	1.03	1.10	2.15	2.93	4.18	4.78
> 1200 fpm	-.69	-.21	0.23	0.67	1.17	1.66	2.67	3.36	3.93	5.13

When:

A_b/A_s	A_s/A_c	A_b/A_c
0.5	1.0	0.5

Note: A = Area (sq. in.), Q = Airflow (cfm), V = Velocity (fpm)

C. Converging Tee, 45° Entry Branch to Rectangular Main

When:

A_b/A_s	A_s/A_c	A_b/A_c
0.5	1.0	0.5

	Branch, Coefficient C (See Note)									
V_c	Q_b/Q_c									
	0.1	0.2	0.3	0.4	0.5	0.6	0.7	0.8	0.9	1.0
< 1200 fpm	-.83	-.68	-.30	0.28	0.55	1.03	1.50	1.93	2.50	3.03
> 1200 fpm	-.72	-.52	.23	0.34	0.76	1.14	1.83	2.01	2.90	3.63

TABLE 8.7. *(Continued).*

D. Converging Wye, Rectangular

$\frac{R}{W} = 1.0$

Branch, Coefficient (See Note)

A_b/A_s	A_b/A_c	Q_b/Q_c								
		0.1	0.2	0.3	0.4	0.5	0.6	0.7	0.8	0.9
0.25	0.25	-.50	0	0.50	1.2	2.2	3.7	5.8	8.4	11
0.33	0.25	-1.2	-.40	0.40	1.6	3.0	4.8	6.8	8.9	11
0.5	0.5	-.50	-.20	0	0.25	0.45	0.70	1.0	1.5	2.0
0.67	0.5	-1.0	-.60	-.20	0.10	0.30	0.60	1.0	1.5	2.0
1.0	0.5	-2.2	-1.5	-.95	-.50	0	0.40	0.80	1.3	1.9
1.0	1.0	-.60	-.30	-.10	-.04	0.13	0.21	0.29	0.36	0.42
1.33	1.0	-1.2	-.80	-.40	-.20	0	0.16	0.24	0.32	0.38
2.0	1.0	-2.1	-1.4	-.90	-.50	-.20	0	0.20	0.25	0.30

Main, Coefficient C (See Note)

A_s/A_c	A_b/A_c	Q_b/Q_c								
		0.1	0.2	.0.3	0.4	0.5	0.6	0.7	0.8	0.9
0.75	0.25	0.30	0.30	0.20	-.10	- .45	-.92	-1.5	-.2.0	-2.6
1.0	0.5	0.17	0.16	0.10	0	-0.08	-.18	-.27	-.37	-.46
0.75	0.5	0.27	0.35	0.32	0.25	0.12	-.03	-.23	-.42	- .58
0.5	0.5	1.2	1.1	0.90	0.65	0.35	0	-.40	-.80	-1.3
1.0	1.0	0.18	0.24	0.27	0.26	0.23	0.18	0.10	0	-.12
0.75	1.0	0.75	0.36	0.38	0.35	0.27	0.18	0.05	-.08	-.22
0.5	1.0	0.80	0.87	0.80	0.68	0.55	0.40	0.25	0.08	-.10

Reprinted with permission from the SMACNA "HVAC Systems — Duct Design" manual. Second Edition, 1981.

The duct cross-sectional area and velocity are

$$A = 12 \text{ in.} \times 8 \text{ in.} \times \frac{1 \text{ ft}^2}{144 \text{ in.}^2} = 0.667 \text{ ft}^2$$

$$V = 1500 \frac{\text{ft}^3}{\text{min}} \times \frac{1}{0.667 \text{ ft}^2} = 2250 \text{ ft/min}$$

Using Equation 8.11, the pressure loss is

$$H_f = 0.14 \left(\frac{2250}{4000} \right)^2 = 0.04 \text{ in. w.}$$

The pressure loss in transition pieces is calculated in the same manner. With converging transitions the downstream velocity is used and with diverging transitions the upstream velocity is used.

EXAMPLE 8.22

The transition piece in Figure 8.26 on page 238 is handling 12,000 CFM. Find the pressure loss through the fitting.

Solution. From Table 8.4, with $A_2/A_1 = 2.0$, read $C = 0.20$. Using Equation 8.11

$$V_1 = 12,000 \frac{\text{ft}^3}{\text{min}} \times \frac{1}{8 \text{ ft}^2} = 1500 \text{ ft/min}$$

$$H_f = 0.20 \left(\frac{1500}{4000} \right)^2 = 0.03 \text{ in. w.}$$

The conversion between velocity pressure and static pressure that takes place in a transition is separate from the effect of pressure loss. The effect of a pressure loss is a decrease in total and static pressure. The conversion between velocity and static pressure will result in an increase in static pressure in a diverging transition and a decrease in static pressure in a converging transition. The separate effects are added algebraically to find the net effect on the static pressure.

EXAMPLE 8.23

The total pressure at point 1 for the fitting in Example 8.22 is 2.35 in. w.g. Find the static pressure at 1 and the total and static pressure at point 2.

Solution. Refer to Figure 8.27 (page 238), which shows the results of the calculations following, as well as a profile of the pressure changes in the fitting. The static pressure at 1 is

$$H_{s1} = H_{t1} - H_{v1} = 2.35 - \left(\frac{1500}{4000}\right)^2 = 2.21 \text{ in. w.g.}$$

The total pressure loss was found in Example 8.22 to be $H_f = 0.03$ in. w. The total pressure at

2 is therefore

$$H_{t2} = H_{t1} - H_f = 2.35 - 0.03 = 2.32 \text{ in. w.g.}$$

The static pressure at 2 is

$$H_{s2} = H_{t2} - H_{v2} = 2.32 - \left(\frac{750}{4000}\right)^2$$
$$= 2.28 \text{ in. w.g.}$$

Example 8.23 illustrates a case of static pressure regain (SPR). The actual regain was

$$\text{SPR} = 2.28 - 2.21 = 0.07 \text{ in. w.}$$

TABLE 8.8. LOSS COEFFICIENTS, DIVERGING JUNCTIONS
Use the velocity pressure (H_v) of the upstream section. Fitting loss (H_f) = $C \times H_v$

A. Tee, 45° Entry, Rectangular Main and Branch

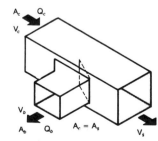

	Branch, Coefficient C (See Note)								
					Q_b/Q_c				
V_b/V_c	0.1	0.2	0.3	0.4	0.5	0.6	0.7	0.8	0.9
0.2	0.91								
0.4	0.81	0.79							
0.6	0.77	0.72	0.70						
0.8	0.78	0.73	0.69	0.66					
1.0	0.78	0.98	0.85	0.79	0.74				
1.2	0.90	1.11	1.16	1 23	1.03	0.86			
1.4	1.19	1.22	1.26	1.29	1.54	1.25	0.92		
1 6	1.35	1.42	1.55	1.59	1.63	1.50	1.31	1.09	
1.8	1.44	1.50	1.75	1.74	1.72	2.24	1.63	1.40	1.17

B. Tee, 45° Entry, Rectangular Main and Branch with Damper

	Branch, Coefficient C (See Note)								
					Q_b/Q_c				
V_b/V_c	0.1	0.2	0.3	0.4	0.5	0.6	0.7	0.8	0.9
0.2	0.61								
0.4	0.46	0.61							
0.6	0.43	0.50	0.54						
0.8	0.39	0.43	0.62	0.53					
1.0	0.34	0.57	0.77	0.73	0.68				
1.2	0.37	0.64	0.85	0.98	1.07	0.83			
1.4	0.57	0.71	1.04	1.16	1.54	1.36	1.18		
1.6	0.89	1.08	1.28	1.30	1.69	2.09	1.81	1.47	
1.8	1.33	1.34	2.04	1.78	1.90	2.40	2.77	2.23	1.92

Note: A = Area (sq. in.), Q = Airflow (cfm), V = Velocity (fpm)

Figure 8.25. Sketch for Example 8.21.

Figure 8.26. Sketch for Example 8.22.

If there has been no loss in total pressure, the theoretical SPR would be

SPR = 0.07 + 0.03 = 0.10 in. w.

The proportion of actual to theoretical SPR is

$$R = \frac{0.07}{0.10} = .70$$

R is the recovery factor, as defined in Section 8.5. To put this in another way, 70% of the pressure loss in the fitting is recovered.

Where there is a combined transition and branch in a duct system, the pressure loss in the straight main run and in the branch are separate, and the value of each depends on the shape. If it is important to minimize losses, the shape should be as shown in Figure 8.28. In this case the pressure loss through the straight run can usually be neglected, and the branch pressure loss can be calculated as an elbow.

In order to reduce the fabrication cost of fittings, they are often made as shown in Figure

8.29. In this case the pressure loss through the branch is usually significant, particularly at high velocities.

A further discussion of recommended duct fitting practice can be found in Chapter 9.

8.14. PRESSURE LOSS AT FAN INLET AND OUTLET

There will be a pressure loss at the fan inlet and outlet, the value of which depends on the shape of the fan-duct connection. This is called the *system effect*. Some values of the resulting loss coefficient C are shown in Table 8.9. An inspection of the types of connections in Table 8.9 will show the importance of considering the system effect and of installing the fan with good connections. A list of system effects can be found in the *Air Moving and Conditioning Association* (AMCA) Manuals.

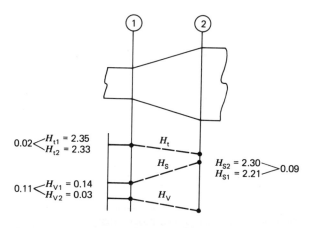

Figure 8.27. Sketch for Example 8.23.

Figure 8.29. Branch with high pressure loss.

Figure 8.28. Branch with low pressure loss.

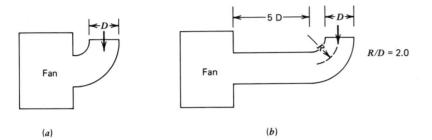

(a)   (b)

Figure 8.30. Sketch for Example 8.24. (a) No straight inlet duct. (b) Long straight inlet duct.

EXAMPLE 8.24

A contractor installs the inlet connection to a fan as shown in Figure 8.30(a) instead of as shown in Figure 8.30(b). The fan inlet velocity is 2000 ft/min. What is the pressure loss inlet to the fan in each case?

Solution. From Table 8.9, we read the values of $C = 1.2$ and $C = 0.25$ for the poor and good connection. The pressure loss for each, using Equation 8.11, is

$$\text{Poor } H_f = 1.2\left(\frac{2000}{4000}\right)^2 = 0.30 \text{ in. w.}$$

$$\text{Good } H_f = 0.25\left(\frac{2000}{4000}\right)^2 = 0.06 \text{ in. w.}$$

Note the greatly increased pressure loss with the poor connection, resulting in wasted energy.

TABLE 8.9. LOSS COEFFICIENTS (C) FOR STRAIGHT ROUND DUCT FAN INLET CONNECTIONS

	Loss coefficient C		
	Length of Inlet in Diameters		
R/D	0	2D	5D
0.75	1.6	0.8	0.4
1.0	1.3	0.7	0.3
2.0	1.2	0.5	0.25
3.0	0.7	0.4	0.20

Reprinted with permission from the SMACNA "HVAC System — Duct Design" manual. Second Edition, 1981.

8.15. DUCT SYSTEM PRESSURE LOSS

The duct pressure losses must be found in order to determine fan capacity, check equipment performance, and balance air quantities. The system total pressure loss is defined as the total pressure loss through the duct path that has the largest pressure losses. This path is often the longest one, but it may be a shorter path that contains an unusual number of fittings with large pressure losses.

It is simpler to work with total pressure loss rather than static pressure loss when analyzing duct pressure losses. This gives a better understanding of the total pressure available at any point in case problems exist. To find the system total pressure loss, the losses are summed up for each section of straight duct and each fitting in the path chosen. Pressure losses through any equipment (coils, filters, diffusers) must be included. The manufacturer will furnish this data.

EXAMPLE 8.25

For the duct system shown in Figure 8.31, determine the system total pressure loss and fan requirements. The fan inlet and outlet connections are not shown, but it has been found that the system effect inlet loss is 0.20 in. w. and the outlet loss is 0.08 in. w. The total pressure re-

quired at each air outlet for proper distribution is 0.1 in. w.g.

Solution. From inspection of the duct layout, path *XABCDEF* is the longest. It is also the path with the greatest pressure loss, because none of the shorter paths have unusual pressure losses.

The pressure losses for the straight ducts and fittings are read from the appropriate tables. The results are listed in Table 8.10. Each item should be checked by the student. The diverging transitions have a negligible pressure loss due to the gradual transformation and relatively low velocities.

The fan selected for the system would be specified for a total pressure of 0.79 in. w.g.

The method used in Example 8.25 is also valuable in solving balancing problems, because the total pressure can be found anywhere in the duct, as shown in Example 8.26.

EXAMPLE 8.26

The system in Example 8.25 is installed in Governor Jawbone's offices. There is a complaint about drafts near outlet *G*. What is the trouble?

Solution. We will find the total pressure at outlet *G* and check it against the value (0.1 in. w.g.) required at the air outlet. The pressure loss is

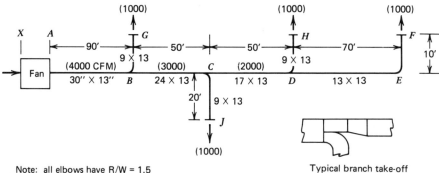

Note: all elbows have R/W = 1.5

Figure 8.31. Sketch for Example 8.25.

TABLE 8.10. SUMMARY OF RESULTS FOR EXAMPLE 8.25

Section	Item	Flow Rate, CFM	Duct Size, in.	Equivalent Round, in.	V, FPM	Friction Loss / 100 ft, in. w.	Length, ft	Loss Coeff, C	Velocity Press., in. w.	Pressure Loss, in. w.
X	Fan inlet	4000								0.20
A	Fan outlet				1477					.08
AB	Duct		30 × 13	21	1477	0.17	90			.15
B	Diverging transitions				1385					0
BC	Duct	3000	24 × 13	19	1385	0.17	50			.09
C	Diverging transitions				1303					0
CD	Duct	2000	17 × 13	16	1303	0.18	50			.09
D	Diverging transitions									0
DEF	Duct	1000	13 × 13	14	852	0.09	80			.07
E	Elbow							.17		.01
F	Outlet									.10
									Total system pressure loss =	0.79

calculated through path *XABG*. The pressure loss *XAB* has already been found. From Table 8.10 it is

$$XAB \ H_f = 0.20 + 0.08 + 0.15 = 0.43 \text{ in. w.}$$

The pressure loss from *B* to *G* is now found. For the transition elbow at *B*, $H/W = 0.7$. If $R/W = 1.5$, $C = 0.19$

$$\text{Elbow } H_f = 0.19 \left(\frac{1700}{4000} \right)^2 = 0.03 \text{ in. w.}$$

$$\text{For duct } BG \ H_f/100 \text{ ft} = 0.2 \text{ in. w.}$$

$$\text{Duct } H_f = 0.2 \times \left(\frac{10}{100} \right) = 0.02 \text{ in. w.}$$

The total pressure loss in this path is therefore

$$XABG \ H_f = 0.43 + 0.03 + 0.02 = 0.48 \text{ in. w.}$$

The fan total pressure is 0.79 in. w.g. Therefore, the total pressure in the duct at outlet *G* is

$$H_{tG} = 0.79 - 0.48 = 0.31 \text{ in. w.g.}$$

This pressure is much greater than the pressure required (0.1 in. w.g.) and will result in excess air at uncomfortable velocities being delivered through outlet *G*.

8.16. DUCT DESIGN METHODS

The solution to the excess pressure at the outlet in Example 8.26 might be handled by partially closing a damper in the branch duct, if one had been installed. This might create noise problems, however. A better solution would be to design the duct system so that excess pressures are dissipated in losses. Procedures for doing this will be explained shortly.

In the previous section we showed how to find pressure losses in ducts once the sizes were known. In planning a duct system, however, the sizes must first be determined. Two methods of designing duct systems will be explained here, the *equal friction* method and the *static regain* method.

8.17. EQUAL FRICTION METHOD

The basis for this method of sizing ducts is that a value for the friction loss rate per length of duct is selected and then kept constant for all duct sections in the system. The value selected is usually based on the maximum velocity permissible in the main trunk duct from the fan to prevent excessive air noise. Table 8.11 lists some suggested velocities.

EXAMPLE 8.27

Find the size of each duct section for the system shown in Figure 8.32, using the equal friction design method. Use rectangular ducts. The system serves an industrial building.

Solution.

1. Sum up the CFMs backward from the last outlet, to find the CFM in each section. The results are shown in Table 8.12.

2. Select a design velocity for the main from the fan, using Table 8.11. A velocity of 1400 ft/min. will be chosen, which should be reasonably quiet for the application.

3. From Figure 8.21, the friction loss rate for the main section *AB* is read as 0.13 in. w./100 ft. The equivalent round duct diameter is read as 20.5 in.

4. The equivalent round duct diameter for each duct section is read from Figure 8.21 at the intersection of the design friction loss rate (0.13 in. w./100 ft) and the CFM for the section.

5. The rectangular duct sizes are read from Figure 8.23. In the actual installation, the duct proportions chosen would depend on space available.

6. The pressure loss in the system can be calculated as shown previously.

When packaged equipment is used, the fan may have only a limited pressure capacity. In this case, instead of starting with a selected velocity, it may be necessary to find the total equivalent length of the system, then divide the fan pressure by this to find the maximum design friction loss rate. This limit usually does not occur because systems with packaged equipment are small, but it should be checked.

TABLE 8.11. SUGGESTED VELOCITIES IN LOW VELOCITY AIR CONDITIONING SYSTEMS

	Recommended Velocities, FPM			Maximum Velocities, FPM		
Designation	**Residences**	**Schools Theaters, Public Buildings**	**Industrial Buildings**	**Residences**	**Schools, Theaters, Public Buildings**	**Industrial Buildings**
Outside air intakes[a]	500	500	500	800	900	1200
Filters[a]	250	300	350	300	350	350
Heating coils[a]	450	500	600	500	600	700
Air washers	500	500	500	500	500	500
Suction connections	700	800	1000	900	1000	1400
Fan outlets	1000 – 1600	1300 – 2000	1600 – 2400	1700	1500 – 2200	1700 – 2800
Main ducts	700 – 900	1000 – 1300	1200 – 1800	800 – 1200	1100 – 1600	1300 – 2200
Branch ducts	600	600 – 900	800 – 1000	700 – 1000	800 – 1300	1000 – 1800
Branch risers	500	600 – 700	800	650 – 800	800 – 1200	1000 – 1600

[a]These velocities are for total face area, not the net free area; other velocities are for net free area.
Reprinted with permission from the 1967 Systems and Equipment, ASHRAE Handbook & Product Directory.

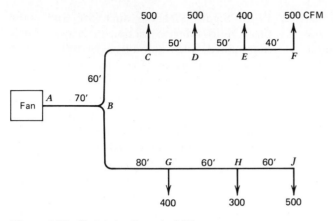

Figure 8.32. Sketch for Example 8.27.

The equal friction method of designing ducts is quite easy, and is probably the most popular one used. For systems that do not have great distances between the first and the last outlets it works quite satisfactorily. If there are long distances between the outlets at the beginning and the end of the system, however, those near the fan will be over pressured. This was demonstrated in Example 8.26. The result may cause difficulties in balancing the flow rates, and possibly excess noise.

If the outlets closest to the fan are on long separate branches, this problem may be overcome by modifying the equal friction design method. The longest run is sized by the design friction loss rate, but the branches are chosen at a higher friction loss rate, thus using up the excess pressure. To reduce extreme pressure differences throughout the system, the static regain method of duct design may also be used.

8.18. STATIC REGAIN METHOD

This method is used to design high velocity systems, above 2500–3000 FPM. The basis of the static regain method of determining duct sizes is to reduce velocities at each section of duct so that the resulting increase in static pressure is

TABLE 8.12. SUMMARY OF RESULTS FOR EXAMPLE 8.27

Section	CFM	V, ft / min	Friction Loss, in. w. per 100 ft	Eq. D, in.	Rect. Duct Size, in.
AB	3100	1240	0.13	20.5	24 × 15
BC	1900	1140	0.13	17	20 × 12
CD	1400	1050	0.13	15	16 × 12
DE	900	900	0.13	12.5	16 × 9
EF	500	889	0.13	10	9 × 9
BG	1200	1029	0.13	14	14 × 12
GH	800	914	0.13	12	14 × 9
HJ	500	889	0.13	10	9 × 9

enough to provide for the frictional losses in the next section. The static pressure is therefore the same at each junction in the main run. Example 8.28 will illustrate how to size the ducts by this method.

EXAMPLE 8.28

Determine the duct sizes for the system shown in Figure 8.33, using the static regain method. Round duct will be used.

Solution.   The results of the work are summarized in Table 8.13. The steps are as follows.

1. A velocity in the initial section is selected. (This system is a high velocity system, so the noise level will not determine the maximum velocity. Sound attenuating devices must be used.) A velocity of 3200 ft/min will be chosen.

2. From Figure 8.21 the duct size and static pressure loss due to friction in section AB is determined. The friction loss per 100 ft is 0.56 in. w. and therefore the friction loss in the section is $0.56 \times 50/100 = 0.28$ in. w.

3. The velocity must be reduced in Section BC so that the static pressure gain will be equal to the friction loss in BC. There will not be a complete regain, due to dynamic losses in the transition at B. We will assume a 75% regain factor for the fittings. A trial and error procedure is necessary to balance the regain against the friction loss. Let us try a velocity of 2400 ft/min. in section BC. The friction loss is

$$\text{Loss in } BC = \frac{0.32 \text{ in. w.}}{100 \text{ ft}} \times 40 \text{ ft}$$

$$= 0.13 \text{ in. w.}$$

The static pressure regain available to over-

Figure 8.33. Sketch for Example 8.28.

TABLE 8.13. SUMMARY OF RESULTS FOR EXAMPLE 8.28

Section	CFM	V, ft / min	Eq. D, in.	Velocity Pressure, in. w.	Friction Loss, in. w./ 100 ft	Length, ft	Friction Loss, in. w.	Static Pressure Regain, in. w
AB	8000	3200	22	0.64	0.56	50	0.28	
B								0.16
BC	6000	2600	21	0.43	0.40	40	0.16	
C								0.09
CD	4000	2200	18	0.30	0.33	30	0.10	
D								0.09
DE	2000	1700	15	0.18	0.26	35	0.09	

come this loss, using Equation 8.9, is

$$\text{Regain at } B = 0.75\left[\left(\frac{3200}{4000}\right)^2 - \left(\frac{2400}{4000}\right)^2\right]$$

$$= 0.21 \text{ in. w.}$$

This is too large a regain. Try a velocity of 2600 ft/min.

$$\text{Loss in } BC = \frac{0.40 \text{ in. w.}}{100 \text{ ft}} \times 40 \text{ ft}$$

$$= 0.16 \text{ in. w.}$$

$$\text{Regain at } B = 0.75\left[\left(\frac{3200}{4000}\right)^2 - \left(\frac{2600}{4000}\right)^2\right]$$

$$= 0.16 \text{ in. w.}$$

This trial is satisfactory. The regain at B is precisely enough to overcome the loss in section BC. The duct size of BC is 21 in.

4. Continue the same procedure at transition C. Let us try a velocity of 2200 ft/min. in CD. The results are

$$\text{Loss in } CD = \frac{0.33 \text{ in. w.}}{100 \text{ ft}} \times 30 \text{ ft}$$

$$= 0.10 \text{ in. w.}$$

$$\text{Regain at } C = 0.75\left[\left(\frac{2600}{4000}\right)^2 - \left(\frac{2200}{4000}\right)^2\right]$$

$$= 0.09 \text{ in. w.}$$

This first guess is satisfactory. No further trial is needed. The duct size is 18 in.

5. The trial-and-error process at D results in a duct size of 15 in. for section DE. The reader should check this.

The result of this method is that the static pressure in the duct at outlets B, C, D, and E will be the same. Assuming that these outlets all required the same static pressure for proper air distribution, the static regain procedure provided duct sizes that will reduce air balancing difficulties. On the other hand, if the equal friction method had been used, the static pressure at B would be considerably higher than at E.

The static pressure loss in the system is simply the loss in section AB. This would be the required fan static pressure.

One disadvantage of the static regain method of duct design is that it usually results in a system with some of the duct sections larger than those found by the equal friction method. For systems at high velocities, however, this method is recommended.

For return air duct systems the equal friction duct sizing method is generally used.

PROBLEMS

8.1. The average velocity of air flowing in a 24 × 18 in. duct is 1300 ft/min. What is the volume flow rate of air in the duct in CFM?

8.2. A pipe with a cross-sectional area of 8.4 in.^{2} has 12 GPM of water flowing through it. What is the water velocity in ft/sec?

8.3. A duct is to be installed that will carry 3600 CFM of air. To avoid excess noise the maximum velocity of air allowed in the duct is 1750 ft/min. The depth of the duct is to be 10 in. What is the minimum width of the duct?

8.4. A 2 in. pipe has water flowing through it at 4 FPM. The diameter is increased to 3 in. What is the velocity in the 3 in. section?

8.5. A 42 in. wide by 20 in. deep duct has a flow rate of 18,000 CFM of air. It is desired to reduce the velocity to 1800 ft/min. The depth of the duct is fixed. What should be the width of the new section?

8.6. Cooling water is pumped from a river to a refrigeration machine condenser 80 ft above the pump intake. The condenser requires 920 GPM. The friction loss through all of the piping, valves, fittings, and condenser is 31 ft w. The water leaves the condenser at atmospheric pressure and flows back to the river by gravity. The velocity of the water entering and leaving the system is the same. What is the required pump head?

8.7. For the piping system shown in Figure 8.34, determine the required pump head. The friction loss is 27 ft w.

8.8. For the piping system shown in Figure 8.35, determine the pressure drop due to friction between points, 1 and 2, if the pressure gages read as shown.

8.9. For the piping system shown in Figure 8.36, the friction loss between points 1 and

Figure 8.34. Sketch for Problem 8.7.

2 is 18 ft w. The pressure gage at 1 reads 23 psig. What would be the reading of the pressure gage at 2?

8.10. A hydronic cooling system has a pressure drop due to friction of 41 ft w. The pump

Figure 8.35. Sketch for Problem 8.8.

Figure 8.36. Sketch for Problem 8.9.

Figure 8.37. Sketch for Problem 8.11.

$V_1 = 2000$ ft/min $V_2 = 1200$ ft/min

Figure 8.38. Sketch for Problem 8.13.

discharge pressure is 83 ft w.g. What would be the reading on a pressure gage at the pump suction, in psig?

8.11. In the duct system shown in Figure 8.37, what is the pressure change from point 1 to 2, if the friction loss is 2.1 in. w.g. between the points?

8.12. The average velocity pressure in a 48 by 18 in. duct is 0.5 in. w. Determine the flow rate in CFM.

8.13. The duct transition shown in Figure 8.38 has a recovery factor of 0.8. Determine the static pressure regain.

8.14. Find the pressure loss due to friction and the velocity in a 250 ft straight section of 4 in. Schedule 40 steel pipe in a hydronic cooling system through which 200 GPM of water is flowing.

8.15. What would be the pressure loss in the piping in Problem 8.14 if it were in a hydronic heating system?

8.16. Determine the friction loss per 100 ft of pipe for 10 GPM flowing through $1\frac{1}{4}$ in. Type L copper tubing.

8.17. Water flows through a clean 3 in. Schedule 40 steel pipe at a velocity of 4 ft/sec. What is the flow rate and friction loss per 100 ft?

8.18. A Schedule 40 steel pipe is to be used to deliver 150 GPM to a cooling tower. It shall have a friction pressure loss no greater than 4 ft w. per 100 ft. What is the minimum pipe size that can be used? What is the actual friction loss?

8.19. Determine the pressure drop through a $1\frac{1}{2}$ in. cast iron globe valve through which 40 GPM is flowing.

8.20. Find the pressure drop through the cold water piping system shown in Figure 8.39, composed of Type L copper tubing.

8.21. Find the pressure drop through the condenser-cooling tower water system shown in Figure 8.40, composed of 8 in. Schedule 40 steel pipe, through which 1100 GPM is flowing.

Figure 8.39. Sketch for Problem 8.20.

8.22. For the piping arrangement shown in Example 8.16, size the piping by the constant friction loss method, using a value of about 1.5 ft w. per 100 ft.

8.23. For the piping arrangement shown in Example 8.16, with each unit circulating 6 GPM, size the piping system, using Type L copper tubing, at 4 ft w. per 100 ft.

6.24. Determine the pressure drop in the piping system in Problem 8.23. Assume a typical set of valves and auxiliaries. The boiler has a pressure drop of 3 ft w. The length of each branch is 12 ft.

8.25. Find the equivalent round diameter of a 36 in. by 12 in. rectangular duct.

8.26. A 28 by 14 in. galvanized steel duct has a flow rate of 5000 CFM of air. Find the friction loss per 100 ft and the velocity.

8.27. A straight length of duct 420 ft long has a flow rate of 2000 CFM of air. The friction loss in the duct must be limited to 1.6 in. w. What is the smallest round duct size that could be used?

8.28. Find the loss coefficient of a 30 in. by 12 in. 90° duct elbow with a mean radius of 12 in., without vanes. If the elbow is carrying 5000 CFM, what is the pressure drop?

8.29. Find the static pressure change from A to B in the duct shown in Figure 8.41, assuming a recovery factor of 0.75.

8.30. A 24 by 12 in. duct has a flow rate of 4000 CFM. The static pressure is 3 in. w.g. at a certain location. What is the total pressure at this location?

8.31. A 24 in. by 14 in. galvanized duct is supposed to be delivering 5000 CFM. Test readings on static pressure manometers 80 ft apart in a straight section of the duct read 2.10 in. w.g. and 1.90 in. w.g. Determine if the duct is suplying the proper air quantity.

8.32. Find the total pressure loss in the duct system shown in Figure 8.42 from the fan outlet at A to F and also to G. The pressure loss through each air outlet is 0.20 in. w.g.

Figure 8.40. Sketch for Problem 8.21.

Figure 8.41. Sketch for Problem 8.29.

Figure 8.42. Sketch for Problem 8.32.

Figure 8.43. Sketch for Problem 8.33.

8.33. Size the rectangular ducts in the system shown in Figure 8.43 by both the equal friction method and by the static region method. All ducts shall be 10 in. deep.

8.34. Find the pressure drop in the duct system in Example 8.27. Outlets have a pressure loss of 0.10 in. w. Assume reasonable shapes of transitions and elbows.

8.35. Size the ducts in Example 8.27 by the static regain method, starting with the same initial velocity. Compare the duct sizes and pressure drop found by the two methods.

8.36. Determine the pipe sizes and system pressure drop for the house in Problems 3.20 and 5.14.

8.37. Determine the pipe sizes and system pressure drop for the building in Problems 3.21 and 5.15.

9

Piping, Valves, Ducts, and Insulation

*T*he selection of the correct types of piping, valves, ducts, and insulation for a particular HVAC system is an important task. In addition, the proper method of installation should be un- derstood, considering such problems as expansion, hanging, supporting, anchoring, and vibrations.

OBJECTIVES

After studying this chapter, you will be able to:
1. Specify the appropriate materials, fittings, and values for an HVAC piping installation.
2. Determine the need for pipe expansion and vibration treatment.
3. Use correct practices for pipe installations.
4. Specify the appropriate materials for HVAC duct installations.
5. Use correct practices for duct installations.

9.1. PIPING MATERIALS AND SPECIFICATIONS

Piping is made of many possible materials, and the proper choice depends on the *service* for which the piping is intended. The service includes:

1. Properties of the fluid being carried
2. Temperature
3. Pressure
4. Exposure to oxidation or corrosion

In addition, cost and availability also affect choice of materials. Finally, codes and regulations usually limit the choice of materials for a given use. It is the responsibility of the HVAC specialist to be aware of the codes that apply to each installation before selecting materials. The piping material most commonly used in hydronic systems is either low carbon ("black") steel pipe or copper tube. For severe problems of oxidation or corrosion other materials may be necessary.

Physical specifications of steel pipe and copper tubing are standardized by the *American Society of Testing Materials* (ASTM). The term *black steel* pipe is used in the trade, but this actually refers to ASTM A-120 or ASTM A-53 low carbon steel. The engineer should always specify the pipe intended by the ASTM number. For the interested reader, the chemical composition of those materials can be found in the ASTM publications.

Some specifications for black steel pipe are shown in Table 9.1. The wall thickness is re-

ferred to by a *Schedule number*, such as 20, 30, 40, or 80. These numbers supersede a former description called standard, extra strong, and double extra strong.

The choice of the correct Schedule number of piping depends on the pressure and temperature service. The allowable pressures can be calculated from formulas established in the *American Standard Code for Pressure Piping*. In hydronic systems at pressures commonly encountered, Schedule 40 pipe is usually specified, except in very large diameters, where Schedule 30 or 20 is sometimes used.

The allowable pressure is only part of the story, however. The engineer should recognize that corrosion and erosion may reduce the pipe wall thickness over a period of years. Therefore, selecting piping with a substantial wall thickness may mean a longer life system.

The wall thickness of copper tubing is specified by a lettering system, Type K, L, or M. Type K has the thickest wall, and is used with high

TABLE 9.1. SPECIFICATIONS OF STEEL PIPE

Pipe Size (inches)	Schedule	Outside Diameter (inches)	Inside Diameter (inches)	Wall Thickness (inches)	Weight, lbs / ft	Volume, gal / ft
$\frac{1}{2}$	40	0.840	0.622	0.109	0.850	0.0158
$\frac{3}{4}$	40	1.050	0.824	0.113	1.130	0.0276
1	40	1.315	1.049	0.133	1.678	0.0449
$1\frac{1}{4}$	40	1.660	1.380	0.140	2.272	0.0774
$1\frac{1}{2}$	40	1.900	1.610	0.145	2.717	0.106
2	40	2.375	2.067	0.154	3.652	0.174
$2\frac{1}{2}$	40	2.875	2.469	0.203	5.790	0.248
3	40	3.500	3.068	0.216	7.570	0.383
$3\frac{1}{2}$	40	4.000	3.548	0.226	9.110	0.513
4	40	4.500	4.026	0.237	10.790	0.660
5	40	5.563	5.047	0.258	14.620	1.039
6	40	6.625	6.065	0.280	18.970	1.501
8	40	8.625	7.981	0.322	28.550	2.597
10	40	10.750	10.020	0.365	40.480	4.098
12	30	12.750	12.090	0.330	43.800	5.974
14	30	14.000	13.250	0.375	54.600	7.168
16	30	16.000	15.250	0.375	62.400	9.506

pressures and refrigerants. Type L has an intermediate thickness wall. It is usually adequate for hydronic system piping. Type M is used for low pressure plumbing. The outside diameter (OD) is the same for any size for all three types; the inside diameter (ID) changes. The pressure drop will therefore be greatest for Type K. Figure 8.15 confirms this. The specifications for Type L tubing are shown in Table 9.2.

Hard temper as opposed to *soft temper* tubing has greater rigidity and will not sag as much as soft tubing when hung horizontally.

The decision to choose between steel piping or copper tubing for an installation is based primarily on cost. Copper is more expensive than steel, but in smaller sizes the labor cost for copper is often less. It is common to see larger installations in steel and smaller ones in copper.

Copper tubing has two advantages that should be noted. First, the frictional resistance is less than for steel, resulting in the possibility of smaller pumps and less power consumption. Second, it is not subject to oxidation and scaling to the same extent as steel. On the other hand, steel is a stronger material and therefore does not damage as easily. Sometimes the larger piping is made of steel and smaller branches to units are copper. When this is done, a plastic bushing should be used to separate the copper and steel, because otherwise corrosion may occur at the joint due to electrolytic action.

In open piping systems, such as piping to a cooling tower, oxidation may occur if black steel is used. *Galvanized* steel pipe is sometimes used in these applications. This is black steel pipe that has a coating of a tin alloy which resists oxidation. In very severe corrosion applications galvanized piping is not adequate, and *wrought iron* or *cast iron* pipe is used. These are quite costly, however.

TABLE 9.2. SPECIFICATIONS OF COPPER TUBING (TYPE L)

Nominal Size (inches)	Outside Diameter (inches)	Inside Diameter (inches)	Wall Thickness	Weight, lbs / ft	Volume, gal / ft
$\frac{3}{8}$	$\frac{1}{2}$	0.430	0.035	0.198	0.00753
$\frac{1}{2}$	$\frac{5}{8}$	0.545	0.040	0.285	0.0121
$\frac{5}{8}$	$\frac{3}{4}$	0.660	0.042	0.362	0.0181
$\frac{3}{4}$	$\frac{7}{8}$	0.785	0.045	0.455	0.0250
1	$1\frac{1}{8}$	1.025	0.050	0.655	0.0442
$1\frac{1}{4}$	$1\frac{3}{8}$	1.265	0.055	0.884	0.0655
$1\frac{1}{2}$	$1\frac{5}{8}$	1.505	0.060	1.140	0.0925
2	$2\frac{1}{8}$	1.985	0.070	1.750	0.1610
$2\frac{1}{2}$	$2\frac{5}{8}$	2.465	0.080	2.480	0.2470
3	$3\frac{1}{8}$	2.945	0.090	3.330	0.3540
$3\frac{1}{2}$	$3\frac{5}{8}$	3.425	0.100	4.290	0.4780
4	$4\frac{1}{8}$	3.905	0.110	5.380	0.6230
5	$5\frac{1}{8}$	4.875	0.125	7.610	0.9710
6	$6\frac{1}{8}$	5.845	0.140	10.200	1.3900
8	$8\frac{1}{8}$	7.725	0.200	19.300	2.4300
10	$10\frac{1}{8}$	9.625	0.250	30.100	3.7900
12	$12\frac{1}{8}$	11.565	0.280	40.400	5.4500

Note that pipe and tubing diameters are specified by a *normal* size, which does not always correspond exactly to either the inside diameter or outside diameter.

The specification tables contain much useful information, as shown in Example 9.1.

EXAMPLE 9.1

A 60 ft. long 5 in. Schedule 40 chilled water steel pipe is to be hung horizontally from a floor slab above. The structural engineer asks the HVAC contractor to determine how much extra weight the floor will have to carry.

Solution.   The weight includes the pipe and the water it carries. Using Table 9.1,

$$\text{Pipe weight} = 14.6 \text{ lb/ft}$$
$$\text{Water weight} = 1.04 \text{ gal/ft} \times 8.3 \text{ lb/gal}$$
$$= 8.6 \text{ lb/ft}$$
$$\text{Total weight} = (14.6 + 8.6) \text{ lb/ft} \times 60 \text{ ft}$$
$$= 1392 \text{ lb}$$

9.2. FITTINGS AND JOINING METHODS FOR STEEL PIPE

In hydronic systems, joining of steel pipe is usually done with either *screwed*, *welded*, or *flanged* fittings. Specifications for fittings are established by the *American National Standards Institute (ANSI)* for both steel pipe and copper tube.

Screwed fittings for steel piping are generally made of cast or malleable iron. For typical hydronic systems, fittings with a 125 psi pressure rating are usually adequate. If in doubt, the system pressure should be checked. Typical pipe fittings are shown in Figure 9.1.

Elbows (ells), used for changing direction, are available in 30°, 45°, and 90° turns. *Long radius* ells have a more gradual turn than *standard* ells,

and thus have a lower pressure drop. Sometimes, however, tight spaces require standard ells. *Tees* are used for branching, and *couplings* are used to join straight lengths of threaded pipe. When joining to equipment, however, *unions* should be used so that the connection may be disconnected for service. With welded fittings, mating welding flanges on the pipe and equipment serve the same purpose as unions. *Bushings* are used when connecting from a pipe of one size to a piece of equipment that has a different size opening.

Welding is a process where the two metal ends to be joined are melted and then fused together with a metal welding rod that also liquifies and fuses. The heat may come from either a gas torch or an electric arc. Welding makes a very strong joint. Welding fittings similar to screwed fittings are available. Straight lengths of pipe may be butt welded directly together without couplings.

In hydronic installations, screwed steel pipe joints are commonly used up to about 2–3 in., and welded joints are used for larger sizes. This generally results in the lowest cost of labor plus materials in most locations. Furthermore, it is more difficult to make pressure-tight threaded joints in very large sizes.

9.3. FITTINGS AND JOINING METHODS FOR COPPER TUBING

Copper tubing joints in hydronic systems are made either by soldering (also called *sweating*) or by *flaring*. Typical solder fittings are shown in Figure 9.2.

Soldering is a process where a metal alloy called solder is melted (between 400 and 1000 F) and when it solidifies, it forms a pressure-tight joint between the two parts to be joined. Solder fittings are made to slip over the tubing with enough clearance for the solder to flow in the annular space between the fitting and the tube. Surfaces must be clean of all oxidation. A chemical coating called *flux* is then used to prevent further oxidation. When the joints must with-

Figure 9.1. Steel pipe fittings. (*a*) A 90° elbow, threaded. (*b*) Tee, threaded. (*c*) Coupling, threaded. (*d*) Bushing, threaded. (*e*) Nipple. (*f*) Union. (*g*) A 90° elbow, flanged. (*h*) Tee, welding. (*i*) 90° elbow, welding. (*Courtesy:* Grinnell Corporation, Providence, R.I.)

Figure 9.2. Copper tube solder fittings. (*a*) A 90° copper elbow. (*b*) Copper tee.

stand high temperatures and pressures or severe vibrations, a soldering process called *brazing* is used. Basically it is no different from lower temperature soldering except that a different soldering material is used which melts at a higher temperature (above 1000 F) and makes a stronger joint.

Flared joints are made by flaring out the end of the copper tubing and using a flare fitting union that will make a pressure-tight seal when

Figure 9.3. Copper tube flaring fittings. (*a*) Tee. (*b*) Nut. (*c*) Tee with nuts assembled.

Figure 9.4. Y-type strainer. (*Courtesy:* Grinnell Corporation, Providence, R.I.)

9.4. VALVES

There are many types and uses of valves. We will discuss mainly *general service* valves, types that are used widely in piping systems. Automatic control valves will be discussed in Chapter 14. Valves for controlling flow may be grouped into three classes according to their function.

Stopping Flow   Valves in this group are used only to shut off flow. This procedure is useful in isolating equipment for service or in isolating sections of a system so that it may be serviced, yet allowing the rest of the system to operate. *Gate* valves (Figure 9.5) are used for this purpose. Note that a gate valve has a straight through flow passage, resulting in a low pressure loss.

tightened against the flare (Figure 9.3). Flared fittings are expensive but are removable, and therefore should be used when access to equipment is required for service or maintenance.

Strainers are used to remove solid particles from the circulating system. The water passes through a perforated plate or wire mesh in which the particles are trapped. The strainer is cleaned at regular intervals. In small pipe sizes the Y-type is generally used and in large sizes a basket type is used (Figure 9.4). Strainers are usually installed at the suction side of pumps and before large automatic control valves.

Regulating Flow Rate   Valves in this group are used to adjust flow rate manually. This is desirable in setting proper flow rates through equipment and different circuits in a system. *Globe* valves (Figure 9.6), *angle* valves, *plug* valves, *needle* valves, and *butterfly* valves can be used to regulate flow.

Gate valves should not be used to regulate flow, and they must be closed or left completely open. Their internal construction is not suitable for throttling flow. If partially closed, *wiredrawing* (erosion of the valve seat) may occur. How-

WHEEL NUT
LOCK WASHER
IDENTIFICATION PLATE
HAND WHEEL
STEM
STUFFING NUT
GLAND
PACKING
STUFFING BOX
BONNET

ONE PIECE WEDGE

BODY

(a)

WHEEL NUT
LOCK WASHER
IDENTIFICATION PLATE
HAND WHEEL
STEM
STUFFING NUT
GLAND
PACKING

BONNET

UNION BONNET RING

ONE PIECE WEDGE
SEAT RING

BODY

(b)

Figure 9.5. Gate valves. (*a*) Screwed bonnet, nonrising stem. (*b*) Union bonnet, rising stem.

WHEEL NUT
LOCK WASHER
IDENTIFICATION PLATE
HANDWHEEL
STEM
STUFFING NUT
GLAND
PACKING
BONNET

DISC LOCKNUT

DISC
BODY

Figure 9.6. Globe valve.

ever, most flow regulation valves can be used to stop flow. This should only be done in an emergency, as the system must then be balanced again.

Limiting Flow Direction Valves that allow flow in only one direction are called *check* valves. In water circulating systems reverse flow could occur when the system is not operating, particularly if there is a static head of water. Reverse flow may damage equipment or empty out a line or equipment unintentionally. Figure 9.7 shows some types of check valves. The *swing check* can be installed only in horizontal lines. A *vertical lift check* or *spring loaded check* can be used in vertical lines. Check valves are usually installed at the discharge of a pump.

9.5. PRESSURE REGULATING VALVES

Where water pressure may exceed safe limits for equipment, a *pressure regulating valve* (PRV) is used. This valve limits the discharge pressure to a preset value. These valves are often used in the make-up water supply line to a system where the make-up is from a city water supply at high pressures.

Pressure relief valves are not the same as a PRV. A relief valve opens when the valve inlet pressure exceeds a preset value. Relief valves are used as safety devices to relieve excess pressure in boilers and other equipment (Chapter 4).

A *flow check valve* is used in hydronic heating systems (see Chapter 4). This is a lift check type valve that closes when the pump stops, thus preventing natural convection flow of hot water through the system, which could result in over-heating.

9.6. VALVE CONSTRUCTION

The construction of gate and globe valves will be discussed in more detail here. Refer to Figures 9.5 and 9.6 to see what each part looks like and how it is assembled. Knowledge of valve construction will enable you to select the correct valve for each application and to understand how to service it.

Disc or Wedge   The part that closes off flow is called the disc or wedge. A *solid* wedge is simplest and is often used in gate valves. A two-piece *split* wedge is less subject to sticking, and is often used at high temperatures and pressures. Globe valve discs made in bevel or plug shapes are best for throttling service. Flat *composition discs* are not recommended for close throttling, but they have the advantage of being replaceable without removing the valve. Bevel and plug discs can usually be reground in place when they wear.

Stem   The stem lifts and lowers the disc or wedge. The valve may be constructed with a *rising* or *nonrising* stem. In the rising stem type (Figure 9.6) the disc and stem rise together. The length of exposed stem therefore provides a quick visual indication of whether the valve is open or closed. This is advantageous to the operating engineer, particularly if the service is one where an incorrect position of the valve might result in immediate harm to some process or even danger to people. In the nonrising stem type the disc travels on the stem, and the stem stays in one position. There is therefore no visual indication of whether the valve is open or closed. This arrangement takes up less space, however, because the valve stem does not travel.

The valve may be constructed so that the stem has an inside screw or *outside screw and yoke*

CAP	CAP
HINGE PIN	UNION CAP RING
HINGE	DISC HOLDER
DISC HINGE NUT	DISC
DISC NUT SPLIT PIN	BODY
DISC	DISC RETAINING NUT
BODY	

(a)   *(b)*

Figure 9.7.   Check valves. (*a*) Swing check. (*b*) Horizontal lift check.

HANDWHEEL —
YOKE NUT —
BONNET —
GASKET —
SEAT RING —
— LOCK NUT
— STEM
— GLAND
— PACKING
— BODY
— DISC

Figure 9.8. OS & Y globe valve.

(OS & Y). In the inside screw arrangement the stem threads are inside the valve body. In the OS & Y type (Figure 9.8) the threads are outside the valve body and are held by a yoke. With the inside arrangement the threads are exposed to the system fluid. The OS & Y arrangement might be used when corrosive fluids or extreme temperatures and pressures exist, to prevent damage to the threads.

Packing Nut and Stuffing The valve must have a means of sealing around the stem to prevent the fluid from leaking out under high pressure. This is accomplished with the *stuffing* or *packing*, usually made of a soft material impregnated with graphite. The packing is held in place and compressed against the stem by tightening the packing nut.

Bonnet The bonnet connects the nut to the body of the valve. *Screwed bonnets*, which screw directly to the valve body (Figure 9.6), are common in small valves. A screwed *union* type bonnet is used when frequent disassembly is expected. Bolted bonnets are used on larger valves.

Valve Materials For hydronic service, either *all bronze* valves or *iron body* with bronze parts are generally used. For many applications valves with a 125 psi pressure rating are suitable, but

this should be checked before selection. At extremely high pressures steel valves may be required. Valves are available with screwed, flanged, or welded ends.

9.7. VALVE SELECTION

Although the types of valves were described previously, specific recommendations on choosing the proper valve for an application will be discussed here.

Gate valves should be used only for stopping flow, and not for throttling. Use them for isolating equipment and sections of a system.

Regulating or throttling flow can be accomplished with globe, angle, plug, needle, or butterfly valves. Use these valves in any section, branch, or unit where flow needs to be manually adjusted or balanced. For large diameters butterfly valves have become popular in hydronic systems because they cost considerably less than globe valves. However, they should not be used where extremely tight shut-off or very close regulation is required.

9.8. PIPE EXPANSION AND ANCHORING

Most materials (unless constrained) will change their length when their temperature is changed. Pipe lines in hydronic heating systems will therefore tend to expand from their initial lengths when brought up to system temperature. This fact must be accounted for in the piping design and installation. Failure to do so may result in broken piping and damaged equipment.

The engineer provides for *expansion* where desirable, and for proper *anchoring* of piping where expansion is undesirable. Each system layout is unique and must be analyzed to determine the correct solution. Some guidelines will be suggested here. If pipe expansion is completely prevented, considerable forces and stresses may re-

Figure 9.9. Expansion loop and offset.

sult in the piping, which could rupture it. Therefore, it is usual to provide for some expansion. This may be done by the following methods:

1. Using expansion *loops* or *offsets* (Figure 9.9) allows the pipe to bend at the loop or offset, thus accommodating the expansion. Of course, the bending itself results in stresses on the pipe, so the size of the loop or offset must be adequate, and depends on the length, size, material, and temperature change. Often a run of piping will have enough natural offsets to accommodate expansion, particularly in small installations. Cast iron fittings should not be used in expansion loops, because cast iron is brittle and may crack.

2. Expansion joints, which are manufactured items, may be used. There is a *slip* type where the pipe slides inside the joint, and there is a *bellows* type where the joint is a movable bellows. Expansion joints are subject to wear and leakage and must be periodically inspected and maintained. Therefore, they should not be installed in inaccessible locations. Access doors must be provided if they are located in shafts or otherwise closed in.

The effect that expansion of a long run has on branch piping must be provided for. Branch connections must be provided with sufficient flexibility so as not to break. This is done by offsets—changing directions at a branch connection. Movement at the offset then prevents a break. This type of offset is called a *swing joint* (Figure 9.10).

The provisions for expansion described above will result in reduced forces at points where the

Figure 9.10. Swing joint.

piping is anchored, but does not exclude the need to provide proper anchoring methods. Equipment should not be used as anchoring points. Do not allow rigid connections to equipment where expansion occurs; use offsets or flexible connections. Anchoring connections must be made so that any force is transmitted to a part of the building structure adequate to take the force. It is sometimes necessary to solve this problem with the aid of the structural engineer. Location of anchoring points can be determined only by studying the particular installation. It may be best to anchor at both ends, or only in the middle, or at a number of points, depending on the length of the run, where expansion loops are provided, and where equipment connections are located.

Pipe *supports* are necessary to carry the weight of the piping and water. Vertical piping may be supported at the bottom, or at one or more points along the length. The supports also may or may not serve as anchors. Horizontal piping is supported by *hangers*. Supports must be provided at frequent enough intervals to not only

(a)

(b)

Figure 9.11. Pipe supports. (*a*) Clevis support and hanger. (*b*) Roller support and hanger.

carry the weight, but to prevent sag. The hanger usually consists of a *rod* and a *cradle*.

There are various types of cradles, as shown in Figure 9.11. Where considerable movement occurs, *roll-type* hangers should be used. If the floor slab is concrete, hanger inserts are installed before the concrete is poured. This requires careful planning and coordination between the HVAC and structural engineers. If extra hangers are needed after the construction is completed,

inserts can be driven into the concrete slab with a gunlike tool.

9.9. VIBRATION

Consideration must be given to possible vibrations occurring in the piping system. Pumps and compressors usually are the source of vibrations. Reciprocating machinery generally creates more

(a)

(b)

Figure 9.12. Vibration isolation mountings. (*a*) Rubber pad, (*b*) Spring. (*Courtesy: Vibration Mountings and Controls, Inc.*)

vibrations than rotating machinery. Vibrations may be transmitted to the building structure or to piping. Both of these problems must be examined.

In many cases the intensity of vibration produced by machinery may not be great enough to result in significant transmission to the piping or structure and no further consideration is necessary. When vibration transmitted to the structure requires treatment, it may be reduced by use of heavy concrete foundations and by suitable machinery locations. There are cases, however, where prevention of any transmission of vibration to the building structure is critical. An example is where machinery is located on a lightweight penthouse floor above office spaces. In this case the equipment is mounted on *vibration isolators*. Isolating supports may be made of cork, rubber, or steel springs (Figure 9.12).

One or more of the following procedures may be used when vibrations are transmitted to the piping:

1. Use of isolation hangers. Using an isolation material (cork, felt) between the cradle and pipe may sometimes be adequate. For a more serious problem, spring hangers can be used.
2. Use of flexible pipe connections to the troublesome machinery. However, flexible connections tend to become inflexible at high pressures.

Difficult vibration problems may require the aid of a specialist in these fields, and the HVAC engineer should not hesitate to call on such help when necessary.

9.10. PIPE INSULATION

Thermal insulation should be used on all cold or hot hydronic system piping. On both hot and chilled water systems thermal insulation serves two purposes:

1. To reduce energy waste and possible increased size of heating or refrigerating equipment.

2. To reduce incorrect distribution of heat. Uninsulated piping may result in the water being at an unsatisfactory temperature when it reaches the conditioned spaces.

In chilled water systems, it is also necessary to prevent condensation of moisture from the air on the outside of the cold piping, which could both damage the insulation and drip onto surrounding surfaces. The prevention of condensation is achieved by covering the insulation with a material that serves as a *vapor barrier*. It is impervious to the flow of the water vapor in the air.

There are a great many materials from which pipe insulation may be made. A good insulation should have the following characteristics:

1. Low thermal conductivity
2. Noncombustible
3. Not subject to deterioration or rot
4. Adequate strength

Pipe insulation may be made from natural materials such as wool, felt, rock, or glass fibers, cork, and rubber. In recent years synthetic materials such as polyurethane have been developed, which have an extremely low thermal conductivity and other excellent properties. Pipe insulation may be furnished in blanket form or premolded to the size of the pipe to be covered. The latter is preferable because it requires less labor and will have a neater appearance.

Vapor barriers are made from treated paper or aluminum foil. Usually the manufacturer furnishes it already wrapped on the insulation (Figure 9.13).

Molded sponge rubber insulation is very popular on small diameter chilled water lines. The rubber serves as both a thermal insulation and vapor barrier. It is very easy to cut and install, thus resulting in low labor costs.

When exposed, the surface of insulation is often provided with a canvas cover. The surface is then *sized*, that is, painted with a material which makes the surface smooth and stiff.

Fittings and valves are also insulated. This is usually done by trowelling on an insulating ce-

Figure 9.13. Pipe insulation with vapor barrier covering. (*Courtesy:* Owens Corning Fiberglas Corporation).

ment mixture. Premolded, shaped insulation is also available for some typical fittings. The installer must take special care not to cover operating parts of valves or removable flanges when applying the insulation.

The question of what thickness of insulation to use is an important one. The greater the thickness, the less the energy losses and thus operating costs are reduced. However, the insulation costs increase with thickness. Therefore, the correct insulation thickness is generally that which provides the minimum owning and operating cost.

Exposed piping and equipment, whether insulated or not, is usually painted when the installation is completed, both for appearance and protection. When lines carrying different fluids exist, the piping is often *color-coded*, that is, each system is painted a different color. In any case, lines should be stenciled with their proper names and direction of flow at reasonable intervals, to make operation and maintenance easier.

Brass name tags should always be furnished and attached to valves. These are usually numbered and a key list is made identifying them by cross-referencing. This list should be mounted in a highly visible place, such as the operating engineer's office.

9.11. THE PIPING INSTALLATION

Some good general practices for installing the piping system will be listed here.

1. Piping should generally be parallel to building walls.
2. Direction changes should be minimized to reduce the number of fittings.
3. The installation should provide simple access to and maintenance of equipment. For example, do not run piping in front of a control panel.
4. Piping should avoid penetration of beams or other structural members. Where this is unavoidable, the structural engineer must be consulted.
5. The piping must not interfere with installations of other trades. This must be checked with the plans of ducts, lighting, and so on.
6. The piping location must not affect the building function. An obvious example is running piping across a door opening.
7. Install horizontal piping with a slight pitch and take all branch connections from the top so that any entrapped air will flow to high points.
8. Provide air vent devices at all high points (see Chapter 11).
9. Provide a short pipe connection and gate valve at all low points in order to drain the system.

This list does not include special features peculiar to each project, nor safety and code requirements.

9.12. DUCT CONSTRUCTION

The most commonly used material for general HVAC ducts is galvanized steel sheet metal. In recent years molded glass fiber ducts have also come into use. When the air being carried is corrosive, more corrosion resistant materials are used, such as stainless steel, copper, or aluminum. Exhausts from kitchens and chemical laboratories are examples where special materials would be required.

SMACNA (*Sheet Metal and Air Conditioning Contractors National Association*) has established standards for construction of ductwork. These standards specify the sheet metal thickness (gage), methods of bracing and reinforcing the duct to prevent collapse or sagging, and methods of joining sections. The standards depend on the air pressure in the duct. High pressure systems require stronger construction. The details of recommended duct construction can be found in *SMACNA* publications. Glass fiber ducts are recommended only for low pressure systems.

Rectangular-shaped sheet metal duct is most commonly used in lower pressure HVAC applications (up to 3 in. w.g. static pressure). Machine-made round sheet metal duct is popular in high velocity, high pressure systems, although heavy gage rectangular duct is also used. Round flexible duct is often used to make final connections to air diffusers, because this permits the contractor to make small adjustments in the location of the diffuser.

Rectangular duct is usually made to order for each job. Round duct is fabricated by machinery in standard diameters. Both for this reason and because round duct is lighter in gage, it is less expensive than rectangular duct for high pressure systems.

Rectangular duct fittings are very expensive because of the labor cost involved, and should be as simple as possible, unless minimum pressure loss is important. Figure 9.14 shows a transition and branch fitting for minimum pressure loss. Figure 9.15 shows a simpler lower cost fitting, but it will have a greater branch pressure loss.

When changing duct shapes, the transition should preferably have a slope of 7 : 1, and should not be less than 4 : 1, to minimize pressure loss (Figure 9.16).

When changing direction, round elbows with a wide sweep radius should be used to keep

Figure 9.14. Branch connection with low pressure loss.

Figure 9.15. Branch connection with high pressure loss.

Slope *b* : *a* of 7 : 1 preferred
Slope *b* : *a* of 4 : 1 minimum recommended

Figure 9.16. Recommended slope for duct transition.

Figure 9.17. Square elbow with turning vane.

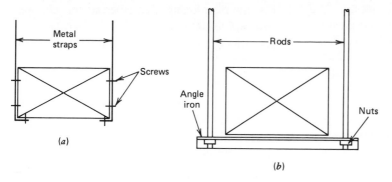

Figure 9.18. Duct hangers. (*a*) Strap hanger. (*b*) Trapeze hanger.

pressure loss low. If short radius or square elbows are necessary to save space, turning vanes should be installed in the elbow (Figure 9.17).

Duct joints should be made as tight as possible to reduce air leakage, using a sealant if necessary. It is not unusual to find installations losing 10% or more of the design air flow due to poor installation.

Small horizontal ductwork is supported by sheet metal straps. Heavier ducts require angle iron support cradles suspended from rods (Figure 9.18).

The ductwork standards described here apply to commercial applications. Ducts for residential

use are simpler in construction and will not be described here. For further information see the *ASHRAE Systems Volume*. Duct connections to fans and air distribution devices will be discussed in Chapter 10.

9.13. DUCT INSULATION

Ducts carrying hot or cold air are covered with thermal insulation to reduce heat loss. In addition, the insulation is covered with a vapor barrier to prevent condensation of water on cold ducts. Glass fiber or similar material with a high

(*a*)   (*b*)

Figure 9.19. Duct insulation. (*a*) Blanket. (*b*) Rigid board. (*Courtesy:* Owens Corning Fiberglas Corporation).

thermal resistance is used for insulation. The vapor barrier is usually aluminum foil. Insulation comes in either *rigid board* or *blanket* form (Figure 9.19). The rigid board costs considerably more, and is used only when the duct is exposed and appearance is important or abuse is likely.

Ducts are frequently lined internally with acoustical insulation to absorb sound. In this case the acoustical lining often also serves as thermal insulation.

REVIEW QUESTIONS

1. Describe the systems for specifying wall thicknesses of both steel pipe and copper tubing.

2. List the pipe fittings described in this chapter and explain their uses.

3. Describe the joining methods used for copper tubing and steel pipe.

4. List the major types of valves and their uses.

5. List the types of materials used for HVAC ducts and their applications.

6. Discuss the features and uses of rectangular and round ducts.

7. List the good practices for duct design and installation.

8. What are the two forms of duct insulation? What is a vapor barrier?

9. List recommended good practices for pipe installation.

Fans and Air Distribution Devices

*F*ans are necessary to distribute air through equipment and through ductwork to spaces that are to be air conditioned. In the first part of this chapter, we will study types of fans, their performance, selection, and application, construction, and some installation and energy conservation recommendations. After that, we will discuss air distribution devices and their selection, and sound control in air distribution systems.

OBJECTIVES

After studying this chapter, you will be able to:

1. Distinguish the types of fans and their characteristics.

2. Select a fan.

3. Use the fan laws to determine the effect of changed conditions.

4. Distinguish the types of air distribution devices and their applications.

5. Select an air distribution device.

6. Analyze the sound conditions in an air distribution system.

10.1. FAN TYPES

Fans may be classified into two main types, *centrifugal* fans and *axial flow* fans, which differ in the direction of air flow through the fan. In a

Figure 10.1. Centrifugal fan, airfoil blade type. (*Courtesy:* Buffalo Forge Company.)

Figure 10.2. Vaneaxial fan. (*Courtesy:* Buffalo Forge Company.)

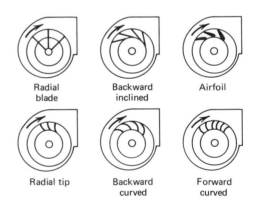

Figure 10.3. Types of centrifugal fan impeller blades.

centrifugal fan, air is pulled along the fan shaft and then blown radially away from the shaft. The air is usually collected by a scroll and concentrated in one direction (Figure 10.1). In an axial flow fan, air is pulled along the fan shaft and then blown along in the same direction (Figure 10.2).

Centrifugal Fans Centrifugal fans may be subclassified into *forward curved*, *radial*, *backward curved*, and *backward inclined* types, which differ in the shape of their impeller blades (Figure 10.3). In addition, backward curved blades with a double thickness blade are called *airfoil* blades.

Axial Fans Axial fans may be subclassified into *propeller*, *tubeaxial*, and *vaneaxial* types. The propeller fan shown in Figure 10.4(*a*) is like a window fan, consisting of a propeller-type wheel mounted on a ring or plate. The tubeaxial fan shown in Figure 10.4(*b*) has a vaned wheel mounted in a cylinder. The vaneaxial fan in

Figure 10.4. Types of axial flow fans. (*a*) Propeller. (*b*) Tubeaxial. (*c*) Vaneaxial.

Figures 10.4(*c*) and 10.2 is similar to the tubeaxial type, except that it also has guide vanes behind the fan blades which improve the direction of air flow through the fan.

10.2. FAN PERFORMANCE CHARACTERISTICS

In the general discussion on fluid flow, we noted that there is a resistance, caused by friction, to the flow of air through ducts. To overcome this resistance, energy in the form of pressure must be supplied to the air. This is accomplished by the rotating fan impeller, which exerts a force on the air, resulting in both flow of the air and an increase in its pressure.

The volume flow rate of air delivered and the pressure created by the fan are called *performance characteristics*. Other performance characteristics of importance are *efficiency* and *brake horsepower* (BHP).

Knowledge of the fan performance is useful for correct fan selection and proper operating and troubleshooting procedures.

The following symbols and definitions will be used in discussing fan performance.

$$CFM = \text{volume flow rate, ft}^3/\text{min}$$

H_s = static pressure, inches of water gage (in. w.g.)

H_v = velocity pressure, in. w.g.

H_t = total pressure, in. w.g.

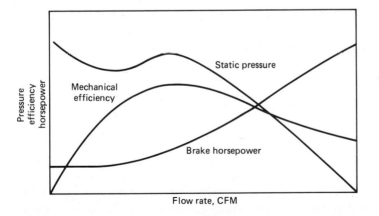

Figure 10.5. Typical performance characteristics of a forward curved blade centrifugal fan.

Figure 10.6. Typical performance characteristics of a backward curved blade centrifugal fan.

BHP = brake horsepower input

N = speed, revolutions per min (RPM)

d = air density, lb/ft^3

ME = mechanical efficiency

= air horsepower output/BHP input

Fan performance is best understood when presented in the form of curves. Figures 10.5 and 10.6 are typical performance curves for forward and backward curved bladed centrifugal fans. Some important features seen are:

1. For both forward curved and backward curved blade centrifugal fans, the pressure developed has a slight peak in the middle range of flow, then the pressure drops off as flow increases.

2. The BHP required for the forward curved blade fan increases sharply with flow, but with the backward curved blade fan the BHP increases only gradually, peaks at a maximum, and then falls off.

3. Efficiency is highest in the middle ranges of flow.

4. A higher maximum efficiency can often be achieved with a backward curved blade fan.

10.3. FAN SELECTION

The choice of the best type of fan to be used for a given application depends on the fan performance characteristics and other features that will be discussed.

Propeller fans cannot create a high pressure, and are thus used where there is little or no ductwork. They are low in cost, and typical applications are as wall- or window-installed exhaust fans. Centrifugal fans are the most commonly used type in ducted air conditioning systems.

Forward curved blade fans are usually lower in initial cost than backward blade types for the same performance. The operating cost will often be higher, however, due to lower efficiency. The rising BHP characteristic curve could result in overloading the motor if operated at a condition beyond the selected CFM. These fans are often used in packaged air conditioning units because of low cost.

Backward (curved or inclined) blade centrifugal fans are generally more expensive than forward curved types, but usually have lower operating costs due to higher efficiency. The limiting horsepower characteristic reduces the possibility of overloading the motor or electrical distribution system if the fan is delivering more air than it was designed for. Airfoil bladed fans have the highest efficiency of any type.

Tubeaxial and vaneaxial fans can be used in ducted systems. The air distribution from tubeaxial fans is uneven, thus making them undesirable for air conditioning systems. Vaneaxial fans are suitable for ducted air conditioning systems. They usually produce a higher noise level than centrifugal fans and therefore may require greater sound reduction treatment. Their compact physical construction is useful when space is limited.

10.4. FAN RATINGS

After the best type of fan is selected for an application, the next task is to determine the proper size to be used.

Manufacturer's fan ratings are presented as either performance curves (Figure 10.7) or tables (Table 10.1) each fan size. Curves and tables each have their good and bad points.

Performance curves enable the engineer to visualize changes in static pressure, BHP, and efficiency easily. Note that each fan curve represents the performance at a specific fan speed and air density. Fans are usually rated with air at *standard conditions*: a density of 0.075 lb/ft^3 at

70 F and 29.92 in. Hg. Performance curves at different air conditions may be available from the manufacturer, but if not, they may be predicted from the *fan laws* to be described in a later section.

Tables list fan performance at different speeds, and therefore replace a large number of curves. For this reason tables are used more often than curves for fan selection. However, the operating condition of maximum efficiency is not apparent when using tables. Some manufacturers resolve this by noting the point of maximum efficiency on their tables (usually in boldface).

Operating near maximum efficiency generally results in the lowest noise output by a fan.

To select a fan, the duct system static pressure resistance (duct H_s) is first calculated, using the procedures explained in Chapter 8. Manufacturer's data are then used to select a fan that will produce the required CFM against the system static pressure resistance.

In effect, the fan must develop a static pressure (fan H_s) and CFM equal to the system requirements. (We will discuss the fan system interaction in more detail shortly, but for now we will focus on the fan selection.)

The fan may also be selected on the basis of total pressure rather than static pressure. Either basis is satisfactory for low velocity systems. For

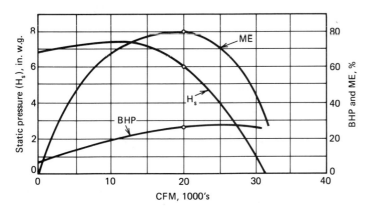

Figure 10.7. Performance curves of a 33 in. diameter backward inclined blade centrifugal fan at 1440 RPM.

TABLE 10.1. PERFORMANCE CHARACTERISTICS OF TYPICAL AIRFOIL BLADE CENTRIFUGAL FANS (27 in. WHEEL DIAMETER)

CFM	OV	¼" SP		⅜" SP		½" SP		⅝" SP		¾" SP		1" SP		1¼" SP		1½" SP		1¾" SP		2" SP	
		RPM	BHP	RPM	BHP	RPM	BHP	RPM	BHP	RPM	BHP	RPM	BHP	RPM	BHP	RPM	BHP	RPM	BHP	RPM	BHP
2085	500	325	.10	376	.15	438	.24	481	.30												
2502	600	351	.13	395	.18	458	.28	495	.35	532	.42										
2919	700	382	.16	421	.22	484	.33	516	.40	548	.47										
3336	800	414	.19	450	.26	484	.33	516	.40	548	.47	613	.63								
3753	900	447	.23	481	.31	513	.38	542	.46	572	.54	629	.70	686	.89	744	1.09				
4170	1000	482	.28	514	.36	543	.45	571	.53	598	.62	650	.79	702	.98	753	1.18	805	1.40		
4587	1100	518	.34	547	.42	575	.52	602	.61	627	.70	676	.89	723	1.08	770	1.29	816	1.51	863	1.74
5004	1200	555	.40	582	.50	609	.59	634	.69	658	.79	703	1.00	748	1.20	791	1.42	834	1.64	876	1.87
5421	1300	592	.47	618	.57	642	.68	667	.79	689	.89	733	1.11	775	1.33	815	1.56	855	1.79	895	2.03
5838	1400	629	.56	654	.66	677	.78	700	.89	722	1.00	765	1.23	804	1.47	842	1.71	880	1.96	917	2.21
6255	1500	668	.65	691	.76	714	.88	734	1.00	755	1.12	796	1.37	834	1.62	870	1.88	906	2.13	941	2.40
6672	1600	707	.76	728	.87	749	1.00	770	1.13	789	1.25	828	1.51	866	1.78	900	2.05	934	2.32	967	2.60
7089	1700	745	.87	765	1.00	786	1.13	806	1.26	824	1.40	861	1.67	897	1.95	932	2.24	964	2.53	996	2.81
7506	1800	785	.99	804	1.14	823	1.27	842	1.41	860	1.55	895	1.84	930	2.13	963	2.43	995	2.73	1025	3.05
7923	1900	823	1.13	842	1.29	860	1.43	878	1.57	896	1.72	930	2.02	963	2.34	995	2.64	1027	2.96	1056	3.28
8340	2000	863	1.28	881	1.45	898	1.61	915	1.75	933	1.91	965	2.23	997	2.55	1028	2.87	1058	3.19	1088	3.53
8757	2100	903	1.45	920	1.62	936	1.80	952	1.94	969	2.11	1000	2.44	1031	2.77	1061	3.12	1090	3.45	1119	3.79
9174	2200	942	1.65	960	1.81	975	2.00	990	2.16	1006	2.31	1036	2.66	1066	3.01	1095	3.37	1123	3.72	1151	4.08
9591	2300	982	1.86	998	2.01	1014	2.21	1028	2.40	1043	2.55	1073	2.91	1102	3.28	1130	3.63	1157	4.02	1184	4.38
10008	2400	1023	2.09	1037	2.23	1053	2.43	1068	2.64	1080	2.81	1110	3.17	1137	3.54	1165	3.93	1191	4.32	1217	4.71
10425	2500	1062	2.33	1077	2.47	1092	2.68	1106	2.89	1119	3.09	1147	3.45	1173	3.84	1200	4.25	1226	4.63	1251	5.05
10842	2600	1102	2.59	1117	2.74	1131	2.94	1145	3.16	1158	3.38	1184	3.74	1210	4.16	1235	4.56	1261	4.98	1285	5.40
11259	2700	1142	2.85	1156	3.04	1170	3.22	1184	3.45	1196	3.68	1222	4.07	1247	4.49	1271	4.90	1296	5.35	1320	5.76
11676	2800	1183	3.13	1196	3.36	1209	3.51	1223	3.76	1236	4.00	1259	4.40	1284	4.83	1308	5.28	1331	5.72	1355	6.16
12093	2900	1223	3.44	1237	3.69	1249	3.84	1262	4.09	1274	4.33	1297	4.75	1321	5.20	1344	5.67	1367	6.11	1390	6.59
12510	3000	1264	3.78	1277	4.04	1289	4.21	1301	4.43	1313	4.69	1335	5.14	1359	5.60	1381	6.07	1403	6.54	1425	7.01
12927	3100	1304	4.14	1316	4.42	1328	4.60	1340	4.79	1353	5.07	1373	5.55	1397	6.02	1419	6.49	1440	6.99	1461	7.46
13344	3200	1345	4.52	1356	4.81	1368	5.02	1379	5.19	1392	5.47	1412	5.98	1434	6.44	1456	6.95	1477	7.45	1498	7.95
13761	3300	1385	4.92	1396	5.21	1408	5.46	1419	5.63	1430	5.88	1450	6.45	1472	6.92	1494	7.43	1514	7.94	1535	8.47
14178	3400	1426	5.35	1437	5.62	1449	5.92	1459	6.10	1469	6.32	1490	6.91	1510	7.41	1531	7.94	1552	8.45	1571	8.99
14595	3500	1466	5.80	1477	6.07	1489	6.40	1499	6.61	1509	6.79	1529	7.40	1549	7.94	1569	8.44	1590	8.99	1608	9.54

CFM	OV	¼" SP		⅜" SP		½" SP		⅝" SP		¾" SP		1" SP		1¼" SP		1½" SP		1¾" SP		2" SP	
		RPM	BHP	RPM	BHP	RPM	BHP	RPM	BHP	RPM	BHP	RPM	BHP	RPM	BHP	RPM	BHP	RPM	BHP	RPM	BHP
2575	500	292	.13	339	.19	394	.29	433	.37												
3090	600	316	.16	356	.22	412	.27	445	.43	478	.51										
3605	700	344	.19	379	.27	435	.41	465	.49	494	.58	551	.78								
4120	800	373	.24	405	.32	435	.41	465	.49	494	.58	551	.78								
4635	900	403	.29	433	.38	461	.47	488	.57	514	.67	566	.87	617	1.09	669	1.35				
5150	1000	434	.35	463	.45	489	.55	514	.66	538	.76	585	.98	631	1.20	678	1.45	725	1.73		
5665	1100	466	.42	492	.52	518	.64	542	.75	565	.86	608	1.10	651	1.34	693	1.59	735	1.86	777	2.15
6180	1200	499	.49	524	.61	548	.73	570	.85	592	.97	633	1.23	673	1.49	712	1.75	750	2.02	789	2.31
6695	1300	533	.58	556	.71	578	.84	600	.97	620	1.10	660	1.37	697	1.64	734	1.92	770	2.21	805	2.51
7210	1400	566	.69	589	.82	609	.96	630	1.09	650	1.24	688	1.52	723	1.82	758	2.11	792	2.42	825	2.72
7725	1500	601	.81	622	.94	642	1.09	661	1.23	680	1.39	716	1.69	751	2.00	783	2.32	815	2.63	847	2.96
8240	1600	636	.93	655	1.08	674	1.23	693	1.39	710	1.54	745	1.87	779	2.20	810	2.53	841	2.87	871	3.21
8755	1700	671	1.07	689	1.24	707	1.39	725	1.56	742	1.73	775	2.07	807	2.40	839	2.76	868	3.12	897	3.47
9270	1800	707	1.23	723	1.41	741	1.56	758	1.74	774	1.92	806	2.28	837	2.63	867	3.00	895	3.37	923	3.76
9785	1900	741	1.39	758	1.59	774	1.76	791	1.94	807	2.13	837	2.50	867	2.88	896	3.26	924	3.65	951	4.05
10300	2000	777	1.58	793	1.79	808	1.99	824	2.16	839	2.36	869	2.75	897	3.15	925	3.54	952	3.94	979	4.36
10815	2100	812	1.79	828	2.00	843	2.22	857	2.40	872	2.60	900	3.01	928	3.42	955	3.85	981	4.26	1007	4.69
11330	2200	848	2.03	864	2.24	878	2.47	891	2.67	906	2.86	933	3.29	960	3.72	986	4.16	1011	4.60	1036	5.04
11845	2300	884	2.30	899	2.48	913	2.73	926	2.96	939	3.15	966	3.59	992	4.05	1017	4.49	1041	4.96	1065	5.41
12360	2400	921	2.58	934	2.75	948	3.01	961	3.26	973	3.48	999	3.92	1023	4.38	1049	4.85	1072	5.33	1095	5.82
12875	2500	956	2.88	969	3.05	983	3.31	996	3.57	1007	3.82	1032	4.26	1056	4.74	1080	5.24	1103	5.72	1126	6.24
13390	2600	992	3.20	1005	3.38	1018	3.64	1030	3.91	1043	4.18	1066	4.62	1089	5.14	1112	5.63	1135	6.15	1157	6.66
13905	2700	1028	3.53	1041	3.75	1053	3.98	1066	4.26	1077	4.54	1100	5.02	1122	5.54	1144	6.06	1167	6.61	1188	7.12
14420	2800	1065	3.87	1077	4.15	1088	4.34	1101	4.65	1112	4.94	1136	5.44	1156	5.97	1177	6.52	1198	7.06	1220	7.61
14935	2900	1101	4.25	1113	4.56	1124	4.75	1136	5.05	1147	5.35	1168	5.87	1190	6.43	1210	7.00	1231	7.55	1251	8.14
15450	3000	1138	4.67	1149	5.00	1160	5.20	1171	5.47	1182	5.79	1202	6.35	1224	6.92	1243	7.50	1263	8.08	1283	8.66
15965	3100	1174	5.12	1185	5.46	1196	5.69	1206	5.92	1218	6.26	1236	6.86	1257	7.44	1277	8.02	1297	8.64	1315	9.22
16480	3200	1211	5.58	1221	5.95	1232	6.20	1242	6.41	1253	6.75	1271	7.39	1291	7.96	1311	8.58	1330	9.21	1348	9.83
16995	3300	1247	6.08	1257	6.44	1268	6.74	1278	6.95	1288	7.26	1306	7.96	1325	8.55	1345	9.18	1363	9.80	1382	10.46
17510	3400	1284	6.60	1294	6.95	1304	7.31	1313	7.54	1323	7.80	1341	8.53	1360	9.16	1379	9.81	1397	10.43	1415	11.11
18025	3500	1320	7.16	1330	7.50	1340	7.91	1349	8.16	1358	8.39	1377	9.14	1394	9.81	1413	10.42	1431	11.11	1448	11.78

TABLE 10.1. *(Continued)* (33 in. WHEEL DIAMETER)

CFM	OV	¼" SP		⅜" SP		½" SP		⅝" SP		¾" SP		1" SP		1¼" SP		1½" SP		1¾" SP		2" SP	
		RPM	BHP	RPM	BHP	RPM	BHP	RPM	BHP	RPM	BHP	RPM	BHP	RPM	BHP	RPM	BHP	RPM	BHP	RPM	BHP
3130	500	**261**	**.15**	302	.22	....	....	....	....	....	....	....	....	....	....	....	....	....	....	....	....
3756	600	282	.19	**317**	**.26**	351	.35	385	.44	....	....	....	....	....	....	....	....	....	....	....	....
4382	700	306	.23	337	.32	**367**	**.41**	397	.50	426	.61	....	....	....	....	....	....	....	....	....	....
5008	800	332	.28	360	.38	388	.48	414	.58	440	.69	491	.93	....	....	....	....	....	....	....	....
5634	900	359	.35	386	.45	410	.56	435	.68	458	.79	**504**	**1.03**	550	1.30	595	1.60	....	....	....	....
6260	1000	388	.42	412	.54	436	.65	457	.77	479	.90	521	1.16	**563**	**1.43**	604	1.73	645	2.06	....	....
6886	1100	416	.50	440	.63	461	.76	482	.89	502	1.02	541	1.30	580	1.59	**617**	**1.89**	655	2.21	692	2.57
7512	1200	446	.59	468	.74	488	.87	508	1.02	527	1.16	564	1.46	599	1.76	634	2.08	**669**	**2.40**	703	2.75
8138	1300	476	.70	496	.86	516	1.01	535	1.16	553	1.31	587	1.62	621	1.95	653	2.28	686	2.63	**718**	**2.98**
8764	1400	506	.82	526	.99	544	1.15	562	1.31	579	1.48	612	1.81	644	2.15	675	2.51	705	2.87	735	3.23
9390	1500	536	.95	555	1.13	573	1.32	590	1.49	607	1.66	638	2.01	668	2.37	698	2.74	726	3.13	754	3.51
10016	1600	567	1.10	585	1.30	602	1.49	618	1.68	634	1.86	665	2.24	693	2.61	721	2.99	749	3.40	776	3.81
10642	1700	598	1.27	615	1.47	631	1.68	647	1.89	663	2.08	692	2.48	720	2.87	746	3.28	772	3.68	798	4.12
11268	1800	630	1.45	645	1.67	661	1.89	676	2.11	691	2.32	719	2.73	746	3.16	772	3.57	797	4.00	822	4.44
11894	1900	661	1.66	676	1.88	691	2.11	706	2.34	720	2.58	747	3.01	773	3.46	799	3.90	822	4.35	846	4.80
12520	2000	693	1.90	707	2.12	721	2.36	735	2.61	749	2.84	776	3.33	801	3.76	825	4.25	849	4.71	871	5.19
13146	2100	724	2.13	738	2.38	752	2.63	766	2.88	779	3.13	804	3.65	829	4.12	852	4.60	875	5.11	897	5.59
13772	2200	756	2.39	770	2.64	783	2.91	796	3.17	808	3.46	833	3.99	857	4.50	880	4.98	902	5.52	924	6.04
14398	2300	787	2.69	801	2.95	814	3.23	826	3.51	838	3.79	862	4.34	886	4.90	908	5.41	929	5.93	950	6.50
15024	2400	819	3.02	833	3.30	845	3.58	857	3.85	868	4.13	892	4.73	914	5.32	936	5.87	957	6.40	977	6.97
15650	2500	851	3.36	864	3.66	876	3.92	888	4.21	899	4.53	922	5.15	943	5.74	965	6.35	985	6.90	1005	7.45
16276	2600	884	3.73	896	4.01	908	4.29	919	4.64	930	4.94	951	5.59	973	6.20	993	6.69	1013	7.44	1033	8.01
16902	2700	917	4.13	928	4.40	939	4.74	950	5.07	960	5.36	980	6.01	1002	6.69	1022	7.35	1042	8.00	1061	8.60
17528	2800	950	4.57	959	4.85	971	5.22	981	5.50	992	5.84	1012	6.51	1032	7.22	1051	7.88	1071	8.58	1089	9.22
18154	2900	981	5.03	991	5.34	1002	5.69	1013	5.97	1023	6.37	1044	7.04	1061	7.77	1081	8.45	1099	9.16	1118	9.86
18780	3000	1013	5.51	1023	5.86	1034	6.16	1044	6.51	1054	6.89	1074	7.58	1090	8.30	1111	9.05	1129	9.77	1146	10.52
19406	3100	1046	6.02	1055	6.37	1066	6.68	1076	7.10	1085	7.40	1104	8.14	1122	8.89	1140	9.55	1158	10.43	1175	11.19
20032	3200	1078	6.55	1088	6.95	1097	7.26	1107	7.70	1117	7.98	1136	8.78	1154	9.55	1169	10.36	1188	11.11	1204	11.89
20658	3300	1111	7.13	1122	7.54	1129	7.89	1139	8.28	1148	8.65	1169	9.50	1184	10.23	1198	11.00	1217	11.86	1234	12.63
21284	3400	1144	7.74	1154	8.19	1161	8.57	1171	8.89	1180	9.37	1201	10.24	1214	10.90	1230	11.72	1247	12.64	1263	13.40
21910	3500	1177	8.42	1186	8.87	1193	9.27	1203	9.57	1211	10.08	1231	10.97	1245	11.63	1262	12.51	1275	13.37	1293	14.23

CFM	OV	¼" SP		⅜" SP		½" SP		⅝" SP		¾" SP		1" SP		1¼" SP		1½" SP		1¾" SP		2" SP	
		RPM	BHP	RPM	BHP	RPM	BHP	RPM	BHP	RPM	BHP	RPM	BHP	RPM	BHP	RPM	BHP	RPM	BHP	RPM	BHP
3830	500	**235**	**.18**	273	.27	....	....	....	....	....	....	....	....	....	....	....	....	....	....	....	....
4596	600	255	.23	**287**	**.32**	317	.43	348	.54	....	....	....	....	....	....	....	....	....	....	....	....
5362	700	277	.28	305	.39	**332**	**.50**	359	.67	385	.75	....	....	....	....	....	....	....	....	....	....
6128	800	300	.35	326	.46	350	.59	374	.71	**398**	**.84**	444	1.13	....	....	....	....	....	....	....	....
6894	900	325	.42	349	.55	371	.68	393	.83	414	.96	**456**	**1.26**	497	1.59	538	1.96	....	....	....	....
7660	1000	350	.51	373	.65	394	.80	414	.94	433	1.10	471	1.42	**509**	**1.75**	546	2.12	583	2.52	....	....
8426	1100	377	.61	397	.77	417	.93	436	1.09	454	1.25	489	1.59	524	1.94	**558**	**2.31**	592	2.71	625	3.14
9192	1200	403	.73	423	.90	442	1.07	460	1.24	477	1.42	510	1.78	541	2.16	573	2.54	**605**	**2.94**	636	3.37
9958	1300	430	.86	449	1.05	467	1.23	484	1.42	500	1.60	531	1.98	562	2.39	590	2.79	620	3.21	**649**	**3.64**
10724	1400	457	1.00	475	1.21	492	1.41	508	1.60	524	1.81	553	2.21	582	2.63	610	3.07	637	3.51	665	3.96
11490	1500	485	1.17	502	1.39	518	1.61	534	1.82	548	2.03	577	2.46	604	2.90	631	3.36	657	3.82	682	4.30
12256	1600	513	1.34	529	1.59	544	1.82	559	2.05	574	2.29	601	2.74	627	3.19	652	3.66	677	4.16	701	4.66
13022	1700	541	1.56	556	1.80	571	2.05	585	2.31	599	2.55	625	3.03	651	3.51	674	4.01	698	4.51	722	5.04
13788	1800	569	1.77	584	2.04	598	2.31	612	2.58	625	2.84	651	3.33	675	3.87	698	4.37	720	4.90	743	5.43
14554	1900	598	2.03	611	2.30	625	2.59	638	2.87	651	3.16	676	3.69	699	4.23	722	4.78	743	5.32	765	5.87
15320	2000	626	2.32	640	2.59	652	2.88	665	3.19	678	3.48	701	4.07	724	4.61	746	5.21	767	5.76	787	6.34
16086	2100	655	2.61	668	2.92	680	3.22	692	3.53	704	3.83	727	4.47	749	5.04	771	5.63	791	6.26	811	6.84
16852	2200	684	2.92	696	3.23	708	3.56	719	3.88	731	4.23	753	4.88	775	5.51	796	6.10	815	6.76	835	7.39
17618	2300	712	3.29	725	3.61	736	3.95	747	4.29	758	4.64	780	5.31	801	6.00	821	6.62	840	7.25	859	7.96
18384	2400	741	3.69	753	4.04	764	4.30	775	4.71	785	5.06	807	5.78	827	6.51	846	7.18	865	7.83	884	8.53
19150	2500	770	4.12	782	4.47	792	4.80	803	5.15	813	5.55	834	6.30	853	7.03	872	7.77	890	8.44	909	9.11
19916	2600	799	4.57	810	4.91	821	5.26	831	5.67	841	6.05	860	6.84	880	7.59	898	8.38	916	9.10	934	9.80
20682	2700	830	5.06	839	5.39	849	5.80	859	6.21	868	6.56	886	7.36	906	8.18	924	8.99	942	9.79	959	10.52
21448	2800	859	5.59	867	5.94	878	6.38	887	6.73	897	7.15	915	7.96	933	8.84	951	9.64	968	10.50	985	11.28
22214	2900	887	6.16	896	6.53	906	6.96	916	7.30	925	7.79	944	8.62	959	9.55	977	10.34	994	11.21	1011	12.07
22980	3000	916	6.75	925	7.17	935	7.54	944	7.97	953	8.43	971	9.28	986	10.15	1004	11.07	1020	11.96	1037	12.88
23746	3100	946	7.37	954	7.80	964	8.17	973	8.69	981	9.06	998	9.96	1014	10.89	1031	11.89	1047	12.76	1063	13.70
24512	3200	975	8.02	984	8.50	992	8.89	1001	9.42	1010	9.77	1027	10.74	1043	11.69	1057	12.68	1074	13.60	1089	14.55
25278	3300	1005	8.73	1014	9.23	1021	9.66	1030	10.14	1038	10.59	1055	11.63	1071	12.52	1084	13.46	1101	14.51	1116	15.46
26044	3400	1034	9.48	1043	10.02	1050	10.49	1059	10.88	1067	11.46	1086	12.53	1098	13.35	1112	14.35	1127	15.47	1142	16.40
26810	3500	1064	10.30	1072	10.86	1078	11.34	1087	11.72	1095	12.34	1114	13.43	1126	14.24	1141	15.32	1153	16.36	1169	17.42

TABLE 10.1. *(Continued)* (40¼ in. WHEEL DIAMETER)

CFM	OV	¼" SP		⅜" SP		½" SP		⅝" SP		¾" SP		1" SP		1¼" SP		1½" SP		1¾" SP		2" SP	
		RPM	BHP	RPM	BHP	RPM	BHP	RPM	BHP	RPM	BHP	RPM	BHP	RPM	BHP	RPM	BHP	RPM	BHP	RPM	BHP
4655	500	213	.22	248	.33	....	....	....	....	....	....	....	....	....	....	....	....	....	....	....	....
5586	600	230	.28	259	.39	288	.52	318	.66	....	....	....	....	....	....	....	....	....	....	....	....
6517	700	248	.34	276	.47	301	.61	325	.75	350	.91	....	....	....	....	....	....	....	....	....	....
7448	800	268	.42	293	.56	317	.72	339	.87	360	1.03	404	1.38	....	....	....	....	....	....	....	....
8379	900	289	.50	313	.67	335	.84	355	1.01	375	1.17	413	1.54	452	1.94	492	2.37	....	....	....	....
9310	1000	312	.61	333	.78	354	.97	373	1.16	392	1.35	427	1.72	461	2.13	496	2.58	532	3.05	....	....
10241	1100	335	.73	354	.92	373	1.12	392	1.32	409	1.53	443	1.94	475	2.36	505	2.81	537	3.30	570	3.81
11172	1200	358	.87	376	1.07	394	1.28	412	1.50	428	1.72	460	2.18	490	2.62	519	3.09	547	3.58	577	4.11
12103	1300	382	1.03	399	1.25	416	1.47	432	1.70	448	1.94	479	2.43	507	2.92	535	3.40	561	3.91	587	4.43
13034	1400	406	1.20	422	1.45	438	1.68	453	1.92	468	2.17	497	2.70	525	3.22	552	3.75	577	4.26	602	4.81
13965	1500	431	1.40	446	1.67	460	1.91	475	2.17	489	2.43	517	2.99	544	3.54	569	4.10	594	4.67	617	5.22
14896	1600	455	1.63	470	1.91	484	2.18	497	2.44	511	2.72	537	3.29	563	3.89	588	4.48	611	5.08	634	5.69
15827	1700	480	1.87	494	2.17	508	2.47	520	2.74	533	3.04	558	3.62	583	4.26	606	4.89	630	5.52	652	6.16
16758	1800	504	2.13	519	2.46	532	2.78	544	3.08	555	3.37	580	3.99	603	4.63	626	5.32	648	5.99	670	6.65
17689	1900	529	2.43	543	2.78	555	3.11	568	3.44	579	3.75	602	4.40	624	5.05	646	5.76	667	6.48	688	7.18
18620	2000	554	2.76	567	3.13	580	3.47	591	3.83	602	4.17	624	4.83	645	5.52	666	6.22	687	6.99	707	7.74
19551	2100	579	3.13	592	3.51	604	3.87	615	4.24	626	4.61	647	5.28	667	6.02	687	6.74	707	7.51	727	8.31
20482	2200	605	3.52	616	3.90	629	4.31	639	4.68	650	5.08	670	5.81	689	6.56	709	7.30	728	8.07	747	8.90
21413	2300	630	3.95	641	4.34	653	4.77	664	5.16	674	5.57	693	6.36	712	7.09	731	7.91	749	8.69	767	9.52
22344	2400	656	4.43	666	4.81	677	5.27	688	5.69	698	6.09	717	6.94	735	7.71	753	8.56	771	9.37	788	10.19
23275	2500	682	4.94	691	5.34	702	5.78	713	6.25	722	6.66	741	7.57	758	8.40	776	9.19	793	10.09	810	10.92
24206	2600	707	5.50	716	5.91	727	6.34	737	6.85	747	7.29	765	8.21	782	9.09	798	9.91	815	10.84	832	11.72
25137	2700	733	6.09	742	6.50	751	6.95	761	7.47	771	7.95	789	8.88	806	9.83	821	10.72	838	11.58	854	12.56
26068	2800	758	6.70	768	7.13	776	7.60	786	8.11	796	8.65	813	9.61	830	10.62	845	11.54	860	12.41	876	13.41
26999	2900	784	7.35	793	7.83	802	8.31	811	8.80	820	9.38	838	10.37	854	11.42	869	12.39	883	13.35	899	14.27
27930	3000	809	8.06	819	8.58	827	9.07	836	9.56	845	10.15	861	11.19	878	12.25	893	13.32	907	14.31	921	15.23
28861	3100	835	8.85	845	9.39	852	9.86	861	10.36	869	10.92	886	12.08	902	13.14	917	14.27	931	15.28	944	16.30
29792	3200	861	9.69	870	10.23	878	10.68	886	11.23	894	11.77	911	12.99	926	14.09	941	15.24	955	16.34	968	17.40
30723	3300	887	10.59	896	11.12	903	11.59	911	12.16	919	12.69	935	13.94	950	15.07	965	16.24	979	17.44	992	18.51
31654	3400	913	11.54	921	12.04	929	12.57	936	13.13	944	13.66	960	14.93	974	16.14	989	17.31	1003	18.56	1016	19.70
32585	3500	939	12.51	947	12.98	955	13.59	962	14.12	969	14.70	985	15.96	999	17.26	1013	18.46	1027	19.71	1040	20.96

high velocity systems, it is sometimes more accurate to use total pressure (see Chapter 8). In our examples, static pressure will be used.

The system static pressure resistance is often called the external static pressure, and is frequently abbreviated SP or ESP in manufacturers' literature (as in Table 10.1).

Examples 10.1 and 10.2 illustrate the use of fan manufacturers' curves and tables.

EXAMPLE 10.1

What static pressure (H_s) will the fan whose performance curves are shown in Figure 10.7 develop at a delivery of 20,000 CFM? What will be the brake horsepower (BHP) and mechanical efficiency (ME) at this condition?

Solution. Using Figure 10.7, at the intersection of 20,000 CFM and the H_s curve,

$$H_s = 6 \text{ in. w.g.}$$

At this CFM, we also note

$$\text{BHP} = 27 \text{ HP}$$

$$\text{ME} = 80\%$$

This example does not indicate if there are better choices of fans to deliver 20,000 CFM at 6 in. w.g. static pressure. Perhaps a more efficient choice exists, or if the emphasis is on initial cost, a smaller fan might be found. Other fan performance curves could be studied to determine these possibilities. However, it is simpler to use fan tables for selection, as Example 10.2 illustrates.

EXAMPLE 10.2

Select an airfoil blade centrifugal fan to supply 8400 CFM at 1½ in. w.g. static pressure.

Solution. The selection will be made from Table 10.1. Assume that energy conservation is important. The following possible selections are noted from the data in the table. (Interpolation between listed values is carried out where necessary.)

Wheel Size, in.	CFM @1½ in.	BHP
27	8400	3.0
30	8400	2.6
33	8400	2.4
$36\frac{1}{2}$	8400	2.3
$40\frac{1}{4}$	8400	2.4

The best selection is probably a 33 in. fan. The saving on energy use is negligible with the $36\frac{1}{2}$ in. fan. The 30 in. fan uses 10% more energy, but if initial cost were the most important consideration, it might be selected. It will be noisier, however, because it is less efficient.

10.5. SYSTEM CHARACTERISTICS

In a manner similar to considering fan performance characteristics of CFM versus pressure developed, we can examine the duct *system characteristics* of CFM versus pressure loss (H_f). The pressure loss due to frictional resistance in a given duct system varies as the CFM changes as follows:

$$H_{f2} = H_{f1}\left(\frac{CFM_2}{CFM_1}\right)^2 \qquad (10.1)$$

Equation 10.1 can be used to find the changed pressure loss in a duct system for a changed CFM flow, if the pressure loss is known at some other flow rate.

EXAMPLE 10.3

The ductwork in a certain ventilating system has a pressure loss of 2 in. w.g. with 5000 CFM of air flowing. What would be the pressure loss if the air flow were 7000 CFM?

Solution. Using Equation 10.1,

$$H_{f2} = 2\left(\frac{7000}{5000}\right)^2 = 3.9 \text{ in. w.g.}$$

By plotting a few of such H_f versus CFM points, a system characteristic curve can be determined. Note that the pressure loss rises sharply with CFM for any duct system, as shown in Example 10.4.

EXAMPLE 10.4

Plot the system CFM versus H_f curve for the duct system of Example 10.3.

Solution. Using Equation 10.1 to plot a few points, the results are shown in Table 10.2 and plotted in Figure 10.8.

TABLE 10.2. RESULTS FOR EXAMPLE 10.4

CFM	0	2500	4000	5000	6000	7000
H_f, in. w.g.	0	0.5	1.3	2.0	2.9	3.9

Figure 10.8. Sketch for Example 10.4 (system characteristic curve).

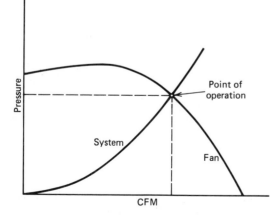

Figure 10.9. Fan and system curves plotted together — intersection is point of operation.

10.6. FAN-SYSTEM INTERACTION

By plotting both the fan and system characteristic pressure versus flow curves together, we can find the condition of operation of the fan and system (Figure 10.9). Because the fan can only perform at conditions on the fan curve, and the system can only perform at conditions on the system curve, the following important principle is always true: *The point of intersection of the system and fan curves is the operating condition of the system.*

EXAMPLE 10.5

For the fan whose performance characteristics are shown in Figure 10.10, what will be the

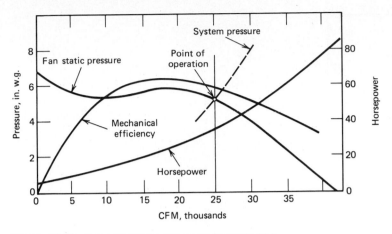

Figure 10.10. Fan and system curves for Example 10.5.

operating conditions when used with the duct system whose characteristic curve is also shown in Figure 10.10?

Solution. The intersection point of the fan and system pressure characteristic curves is the operating condition, 25,000 CFM and a static pressure of 5.3 in. w.g. The fan BHP = 35, ME = 60% at this condition.

Examining the fan and system curves is not only useful for selecting the operating condition, but aids in analyzing changed conditions and in finding causes of operating difficulties. A common occurrence in air conditioning systems is that the actual system resistance for a design CFM is different from that calculated by the designer. Some reasons this may happen are:

1. An error in calculating pressure loss.

2. Allowance of an extra resistance as a "safety factor" by the designer.

3. The contractor installs the ductwork in a manner different from that planned.

4. Filters may have a greater than expected resistance due to excess dirt.

5. An occupant may readjust damper positions.

The result of this type of condition is that the duct system has a different characteristic than planned. An examination of the fan and system curves will aid in analyzing these situations.

EXAMPLE 10.6

For the fan whose performance is shown in Figure 10.11, the system required CFM is 5000 and calculated pressure loss is 4.4 in. w.g. The design operating condition is therefore point 1. The system design performance curve A, calculated from Equation 10.1, is also shown.

The system designer has allowed an extra 1.0 in. w.g. as a "safety factor" in calculating the pressure loss. Assuming that there is no real extra pressure loss, what will the actual operating conditions be?

Solution. The actual system pressure loss is 4.4 − 1.0 = 3.4 in. w.g. at 5000 CFM (point 2). The real system pressure curve can be plotted (curve

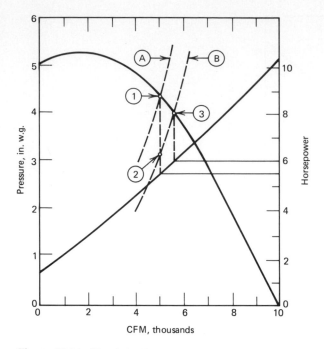

Figure 10.11. Sketch for Example 10.6.

B). The actual operating condition is the intersection of the fan and real system pressure curves (point 3).

At this point, the flow rate is 5500 CFM. The system will be delivering too much air to the spaces. Also note that the BHP = 6.0 instead of 5.5, as would have occurred with the design condition. This is a considerable waste of energy and money. Of course, the actual flow rate could be throttled by using dampers. Another solution might be to change the fan speed.

10.7. SYSTEM EFFECT

The fan that is chosen for a given application must develop, at the design CFM, a pressure at least equal to the duct system pressure loss. The manufacturer's fan rating curves or tables are used to find a suitable fan.

However, the fan ratings are based on testing the fans in a manner prescribed by the *Air Movement and Control Association* (*AMCA*). Because every fan installation is unique, each arrangement of inlet and outlet connections to the fan will probably differ from that of the laboratory arrangement in which the fan was tested. The result will be that there are additional pressure losses at the fan inlet and fan outlet, which must be added to the system pressure loss before selecting the required fan pressure.

The precise loss at fan inlet and outlet depends on the shape, size, and direction of the connections to the fan. These are called *system effects*, which we have shown how to determine in Chapter 8. In any case, connections should be made to minimize losses, as described later.

For fans that are part of a packaged unit, the manufacturer usually has allowed for the system effect pressure loss for the unit.

10.8. SELECTION OF OPTIMUM FAN CONDITIONS

Often a number of fans of different sizes, or operating at different speeds, will satisfy the pressure and CFM requirements; therefore the next step in selecting fans is to decide what further criteria should be used in selecting the "best" choice. Some of these factors will now be examined.

1. Fans should be chosen for close to maximum efficiency, particularly with high energy costs. This will fall in the middle ranges of the pressure-CFM curve. Avoid the temptation of selecting a fan far out on the CFM curve, near maximum delivery. This temptation is great, because it may mean a smaller fan and therefore lower initial cost, but the operating cost will be high.

2. Fans should not be selected to the left of the peak pressure on the fan curve (Figure 10.12). At these conditions the system operation may be unstable—there may be pressure fluctuations and excess noise generated.

Figure 10.13. Steep and flat fan pressure curves.

3. When using forward curved blade centrifugal fans, check the system to see if it might operate at significantly greater than design CFM. If so, the motor horsepower required will increase, and a larger motor may be necessary.

4. Allow for system effect according to the duct inlet and outlet connections as explained previously.

5. Fans may have pressure curves of varying steepness (Figure 10.13). If it is expected that there will be considerable changes in system resistance, but constant CFM is required, a fan with a steep curve is desirable. For variable air volume (VAV) systems (Chapter 12), a flat curve type is desirable.

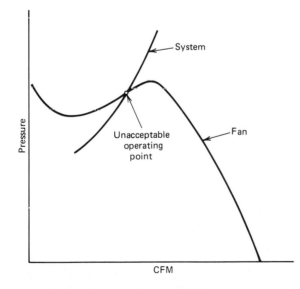

Figure 10.12. Unstable operating condition.

10.9. FAN LAWS

There are a number of relationships between fan performance characteristics for a given fan operating at changed conditions, or for different size fans of similar construction; these are called *fan laws*. These relations are useful for predicting performance if conditions are changed. We will present some of these relations and their possible uses.

$$CFM_2 = CFM_1 \times \frac{N_2}{N_1} \quad (10.2)$$

$$H_{t2} = H_{t1}\left(\frac{N_2}{N_1}\right)^2 \text{ and } H_{s2} = H_{s1}\left(\frac{N_2}{N_1}\right)^2 \quad (10.3)$$

$$BHP_2 = BHP_1\left(\frac{N_2}{N_1}\right)^3 \quad (10.4)$$

$$H_{t2} = H_{t1} \times \frac{d_1}{d_2} \quad \text{and} \quad H_{s2} = H_{s1} \times \frac{d_1}{d_2}$$
$$(10.5)$$

EXAMPLE 10.7

A ventilating fan is delivering 8000 CFM while running at a speed of 900 RPM and requiring 6.5 BHP. The operating engineer wants to increase the air supply to 9000 CFM. At what speed should the fan be operated? What must be checked first before making such a change?

Solution. Using Equation 10.2 to find the new speed

$$N_2 = \frac{CFM_2}{CFM_1} \times N_1$$

$$= \left(\frac{9000}{8000}\right) \times 900 = 1010 \text{ RPM}$$

Equation 10.4 shows that the horsepower will increase, so this must be checked.

$$BHP_2 = \left(\frac{1010}{900}\right)^3 \times 6.5 = 9.2 \text{ HP}$$

This is a considerable increase in power. If the fan had a 7.5 HP motor, which would be likely because it originally required only 6.5 HP, the motor would now be overloaded. The motor would have to be changed, as might the wiring,

and the fan capability itself should be checked with the manufacturer.

10.10. CONSTRUCTION AND ARRANGEMENT

The *AMCA* has established standards of centrifugal fan construction and arrangement that are generally followed in the United States.

Pressure Rating   In order to construct fans of sufficient strength to withstand air pressures to which they will be subjected, and yet not to overdesign the fan, resulting in unnecessary cost, centrifugal fans are classified into groups of different allowable maximum pressures (Figure 10.14). The engineer should select and specify fans in the pressure class required for the job. A different pressure classification has been established for fans mounted in cabinets (see *AMCA* standards).

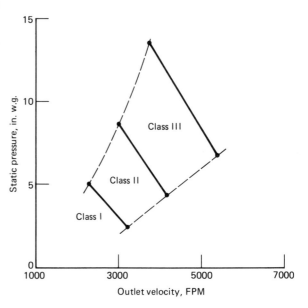

Figure 10.14. Fan construction classes for allowable pressures, backward inclined type. (*Courtesy:* Air Movement & Control Association, Inc. [AMCA].)

Figure 10.15. Single Width Single Inlet and Double Width Double Inlet fans.

Arrangement Centrifugal fans are available in nine different arrangements of bearings and air inlets. Arrangements 1, 2, and 3 (Figure 10.16) are often used in HVAC applications because of lower cost and convenience, particularly Arrangement 3.

Inlet Single Width Single Inlet (SWSI) fans have the air inlet on one side; Double Width Double Inlet (DWDI) fans have air inlets on both sides (Figure 10.15). DWDI fans would thus be suitably installed in a plenum type cabinet.

Rotation, Discharge, Motor Position Direction of rotation is described when viewed from the opposite side of an inlet, and is referred to as clockwise (CW) or counterclockwise (CCW). Several discharge arrangements are available, as shown in Figure 10.17. The standard motor positions available are shown in Figure 10.18.

Centrifugal fans are usually belt driven. Different size pulleys make it possible to change

Figure 10.16. Examples of bearing arrangements commonly used in HVAC system fans.

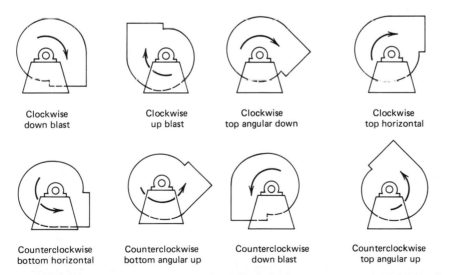

Figure 10.17. Examples of discharge arrangements. (*Note:* Rotation direction determined from drive side of fan).

Figure 10.18. Motor positions. (*Note:* Specify motor position by letter, from drive side.)

speeds. Adjustable pulleys are available so that a limited speed adjustment can be made on the job.

10.11. INSTALLATION

The above specifications of fan construction, rotation, discharge, and motor position must be decided upon when planning the HVAC system installation. In addition, there are other general installation procedures that should be followed:

1. Inlet and discharge connections to the fan should be made to create air flow with minimum pressure loss and equal velocity across the duct section. Some good and poor examples are shown in Figure 10.19.

2. Connection between the inlet and discharge duct and fan should be made with canvas, to reduce vibration transmission. Fans should be mounted or hung on vibration isolators. Spring and rubber type isolators are available. The manufacturer should be consulted about correct choice of isolator.

3. Access openings should be provided if needed, for inspection and service of bearings.

4. A belt guard with a mesh front should be provided so that the belts can be seen without removing the guard.

5. Adequate clearance for inspection and removal should be provided on all sides of the fan. Allow space for the motor installation.

10.12. ENERGY CONSERVATION

1. Airfoil blade centrifugal fans have the highest efficiency and therefore use the least power.

Figure 10.19. Examples of good and poor inlet and discharge connections.

Figure 10.20. Inlet guide vanes for volume control. (*Courtesy:* Buffalo Forge Company.)

2. Do not allow extra pressure loss as a "safety factor" in the duct system.

3. Select fans in the mid-range of total flow, where efficiency is highest.

4. If volume control of the fan is to be used, inlet guide vane dampers (Figure 10.20) are preferable to outlet dampers; less power will be used when the volume flow rate is reduced.

5. Reducing fan speed to reduce flow is the most efficient method for reduced power consump-

tion. However, multispeed fan drives are expensive.

6. Inlet and discharge connections should be arranged to provide minimum pressure loss.

Air Distribution Devices

The conditioned air that is being supplied to each room must be distributed throughout the space in a certain manner; if it is not, uncomfortable conditions will result. This is an aspect of the environmental control system that is often neglected, because it seems simple. Air at the proper flow rate and conditions may be supplied, and yet often the occupants are quite uncomfortable. This is because the air is not distributed properly in the room. We will investigate some of the principles of air distribution, and then will look at some of the devices (terminal units) that are used to provide proper air distribution. We will consider the types available, their features, and selection.

10.13. ROOM AIR DISTRIBUTION

Good room air distribution requires the following characteristics for comfort:

1. Temperatures throughout the *occupied zone* of the room within ± 2 F (1 C) of the design temperature. Temperature fluctuations greater than this will usually result in discomfort. The occupied zone of most spaces is considered to be from the floor to an elevation of 6 ft. Above this height greater temperature fluctuations are permissible.

2. Air velocities throughout the occupied zone (called *residual velocities*) between 25–35 FPM for applications where people are seated. Higher velocities (drafts) cause discomfort. Lower velocities also result in discomfort, usually a feeling of stuffiness, or staleness. In applications where people are moving around and occupancy is for a short period, as in

Figure 10.21. Description of terms used in air distribution.

department stores, higher air velocities are acceptable (50–70 FPM).

10.14. AIR PATTERNS

There are a few facts about how an air supply to a room will behave which are important to understand in order to select and locate air supply devices properly and to balance and adjust the devices.

1. When air lower in temperature than room air is supplied (as in summer), it will drop.
2. When air higher in temperature than room air is supplied (as in winter), it will rise.
3. When air is supplied parallel to and near a ceiling, it will tend to "hug" the ceiling for some distance. This is called the *ceiling effect*.
4. The supply air to the room (called the *primary air*) when distributed from an air supply device, will *induce* room air (called *secondary*

air) into the airstream, thus rapidly mixing the supply air and the room air.

There are certain other terms used in studying air distribution that need to be defined. Figure 10.21 illustrates these terms. The *throw* from a supply air device is the distance that the supply air travels before reaching a relatively low velocity, called the *terminal velocity*. Terminal velocities of 75–200 FPM are recommended, resulting in residual velocities of 20–70 FPM. The *drop* is the vertical distance the (cold) supply air drops by the end of its throw. The *temperature differential* is the temperature difference between the supply air and the room air. The *spread* is the horizontal divergence of the airstream.

10.15. LOCATION

The location of air distribution devices in the room is an important consideration in achieving good air distribution.

Cooling—good air distribution

Heating—poor air distribution

Figure 10.22. High wall outlet location.

Figure 10.23. Ceiling outlet location for cooling provides good distribution.

Figure 10.24. Floor or sill location under window for heating provides good air distribution.

1. *High wall* (Figure 10.22). This is a good location for cooling, because the cold air will drop naturally and adequate air circulation throughout the occupied zone will occur. It is not a good location for heating, because the warm air will rise, leaving a stagnant zone in the occupied area. Separate heating (under the window) should be used in this case.

2. *Ceiling* (Figure 10.23). This is an excellent location for cooling, because the cold air will drop naturally. It is not a very good location for heating, because the warm air will rise, unless forced down at a high velocity.

3. *Low Wall.* This is a good location for heating, because the warm air will rise naturally, but is not desirable for cooling, because the cold air will tend to remain near the floor.

4. *Floor or Sill* (Figure 10.24). This is an excellent location for heating if located under windows, because it counteracts the cold air downdraft that would otherwise result near the glass. It can also be used for cooling if an adequate outlet velocity is achieved, forcing the cold air to rise and circulate.

Beams and ceiling mounted lighting fixtures create a problem for ceiling or high wall air outlets. The primary air hugs the ceiling due to the ceiling effect, and then bounces off the obstruction, sending a cold draft down to the occupied zone (Figure 10.25). In this case a ceiling outlet should be mounted below the obstruction.

Figure 10.25. Effect of obstruction at ceiling.

For a high wall outlet the air should be directed to clear the obstruction.

For very high ceilings it is usually better to install ceiling and high wall outlets below or at the level of lighting. In this way cooling of unused space near the ceiling is reduced, saving energy.

When locating floor or sill outlets, care must be taken not to let drapes or furniture block the air flow.

10.16. TYPES OF AIR SUPPLY DEVICES

There are four types of air supply devices used for creating proper air distribution in the conditioned space:

1. *Grilles and registers*
2. *Ceiling diffusers*

Figure 10.26. Supply register, double deflection type. (*Courtesy:* Tuttle & Bailey®. Division of Interpace Corporation.)

3. *Slot diffusers*
4. *Plenum ceilings*

Grilles and Registers (Figure 10.26) These devices consist of a frame and parallel bars, which may be either fixed or adjustable. The bars serve to deflect the supply air in the direction the bars are set, and if the bars are adjustable, to adjust the throw and spread of the air. Grilles with two sets of bars at right angles to each other are available and are called *double deflection* grilles. They enable control of the air distribution in both directions, if needed. Grilles with volume control dampers mounted behind the grille are called *registers*.

Ceiling Diffusers (Figure 10.27) These devices usually consist of a series of separated concentric rings or louvers with a collar or neck to connect to the duct. They may be round, square, or rectangular in shape. In addition to those that distribute air equally in all directions, they can be designed to distribute air in any desired direction. Ceiling diffusers are also available in the form of perforated panels. This type is sometimes used because it blends architec-

turally with the appearance of a hung panel ceiling.

Slot Diffusers (Figure 10.28) This is a long strip-shaped outlet with one or more narrow openings, depending on the number of bars or vanes.

Plenum Ceilings Hung ceilings are available with slots or perforations throughout most or all of the ceiling, which can serve as supply air outlets. The space above the ceiling is used as a large plenum through which the supply air is delivered. In this way air can be distributed evenly throughout the whole space to be conditioned. The design and balancing of plenum ceiling systems is a specialized procedure and will not be discussed further here. Manufacturers of these ceilings will aid the interested designer or contractor.

10.17. APPLICATIONS

Grilles When used for cooling, a high sidewall location is one of the preferred locations. The air can be directly slightly arched upward, and then will follow the ceiling due to the ceiling effect, mixing well with induced secondary air. In this way the mixed air temperature will not be unacceptably lower than room temperature before it drops to the occupied zone. When used for warm air heating, the high sidewall outlet may result in stratification of the warm air. However, a careful selection of outlet throw, vertical spread, and return air locations could make the installation satisfactory in mild climates. In any case, caution is urged.

Grilles and registers can also be used at ceilings with results similar to high sidewall locations. They are not installed in ceilings as often, however, because their appearance is not considered pleasing in a ceiling. Adjustable deflection vanes are used to set proper air direction.

For warm air heating, a perimeter location under windows discharging vertically upward from the floor is ideal in cold climates. This not only provides good mixing of primary and secondary air, but blankets the glass with warm air, offsetting cold down drafts. This location is popular in residential installations. It usually results in low installation costs in this application because the ductwork in the basement below is relatively simple. When used for cooling, the discharge air velocity must be adequate to overcome the gravity effect of the denser air.

Ceiling Diffusers   These are usually located at the ceiling. The air is discharged horizontally when used for cooling, and follows the ceiling for some distance due to the ceiling effect. They are also often installed in the bottom of horizontal ductwork below the ceiling when a hung ceiling is not used.

Round and square diffusers that have equal openings all around are used to cover a square floor area. A part of the outlet can be blanked off

Figure 10.27. Ceiling diffusers. (*Courtesy:* Tuttle & Bailey®. Division of Interpace Corporation.)

Figure 10.28. Slot diffuser. (*Courtesy:* Tuttle & Bailey®. Division of Interpace Corporation.)

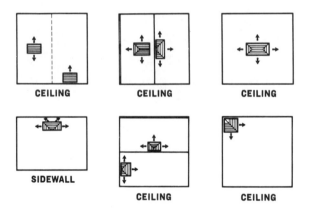

Figure 10.29. Use of 1-, 2-, or 3-way blow diffusers.

with a piece of sheet metal to get directional air patterns. However, diffusers are available with 1-, 2-, or 3-way blow to cover rectangular shaped room areas (Figure 10.29).

Some types of ceiling diffusers can be used for heating by adjusting the air pattern to discharge vertically downward at a high velocity. This is more common in industrial applications.

Slot Diffusers   These are available in arrangements enabling them to be used either at ceilings or sidewalls. In addition, they are popular in perimeter applications discharging vertically up from the floor under sills. This popularity is due to the use of low sills and long expanses of glass in many modern buildings.

10.18. SELECTION

The air outlets chosen for a project depend on the following:

1. *Architectural requirements*. The architect usually wishes the space to have a certain appearance, which may guide the type used and their location.
2. *Structural requirements*. The building structure behind walls, ceilings, and floors may restrict the location of ductwork and thus the air outlets. The structural engineer must be consulted on this.
3. *Temperature differentials*. For large cooling temperature differentials (TD) between supply and room air, the danger of unacceptable air temperatures entering the occupied zone increases. Ceiling diffusers generally have a high induction ratio of room air, therefore lessening this problem when large temperature differential is needed. Some types can be used up to 35 F TD. Grilles are generally limited to 25 F TD.
4. *Location*. When distributing cold air, supply outlets may be located at the ceiling or high on sidewalls. If there are exposed beams they may deflect the air down to the occupied zone. In this case the diffuser may have to be located below the beam line. An alternate location for cold air supply is a perimeter location discharging vertically upward from below the window sill. In this case the return air location becomes important—it should be located in the interior of the room, preferably at a low elevation, to avoid short circuiting. Warm air supply outlets are preferably lo-

cated at the perimeter discharging vertically upward.

5. *Quantity*. Often more than one supply outlet is located in a room. This choice depends on a number of factors—air quantity, cost, and architectural requirements. Generally, in a large space, better air distribution is achieved by using a number of diffusers. However, this will increase the cost of the installation. The architect will also set certain requirements, and will want the ceiling to have a certain

TABLE 10.3. RANGES OF INDOOR DESIGN GOALS FOR AIR CONDITIONING SYSTEM SOUND CONTROL

Type of Area	Range of A-Sound Levels, Decibels	Range of NC Criteria Curves	Type of Area	Range of A-Sound Levels, Decibels	Range of NC Criteria Curves
RESIDENCES			**CHURCHES AND SCHOOLS** (Cont'd)		
Private homes (rural and suburban)	25–35	20–30	Laboratories	40–50	35–45
Private homes (urban)	30–40	25–35	Recreation halls	40–55	35–50
Apartment houses, 2- and 3-family units	35–45	30–40	Corridors and halls	40–55	35–50
			Kitchens	45–55	40–50
HOTELS					
Individual rooms or suites	35–45	30–40	**PUBLIC BUILDINGS**		
Ballrooms, banquet rooms	35–45	30–40	Public libraries, museums, court rooms	35–45	30–40
Halls and corridors, lobbies	40–50	35–45	Post offices, general banking areas,		
Garages	45–55	40–50	lobbies	40–50	35–45
Kitchens and laundries	45–55	40–50	Washrooms and toilets	45–55	40–50
HOSPITALS AND CLINICS			**RESTAURANTS, CAFETERIAS, LOUNGES**		
Private rooms	30–40	25–35	Restaurants	40–50	35–45
Operating rooms, wards	35–45	30–40	Cocktail lounges	40–55	35–50
Laboratories, halls and corridors			Night clubs	40–50	35–45
Lobbies and waiting rooms	40–50	35–45	Cafeterias	45–55	40–50
Washrooms and toilets	45–55	40–50			
OFFICES			**STORES, RETAIL**		
Board room	25–35	20–30	Clothing stores	40–50	35–45
Conference rooms	30–40	25–35	Department stores (upper floors)		
Executive office	35–45	30–40	Department stores (main floor)	45–55	40–50
Supervisor office, reception room	35–50	30–45	Small retail stores		
General open offices, drafting rooms	40–50	35–45	Supermarkets	45–55	40–50
Halls and corridors	40–55	35–50			
Tabulation and computation	45–65	40–60	**SPORTS ACTIVITIES, INDOOR**		
AUDITORIUMS AND MUSIC HALLS			Coliseums	35–45	30–40
Concert and opera halls			Bowling alleys, gymnasiums	40–50	35–45
Studios for sound reproduction	20–30	15–25	Swimming pools	45–60	40–55
Legitimate theaters, multi-purpose halls	30–35	25–30	**TRANSPORTATION (RAIL, BUS, PLANE)**		
Movie theaters, TV audience studios			Ticket sales offices	35–45	30–40
Semi-outdoor amphitheaters	35–45	30–35	Lounges and waiting rooms	40–55	35–50
Lecture halls, planetarium			**EQUIPMENT ROOMS**		
Lobbies	40–50	35–45	8 hr/day exposure	<90	
CHURCHES AND SCHOOLS			3 hr/day exposure	<97	
Sanctuaries	25–35	20–30	(or per OSHA requirement)		
Libraries	35–45	30–40			
Schools and classrooms	35–45	30–40			

NOTE. These are for unoccupied spaces with all systems operating.
Reprinted with permission from the *1976 ASHRAE Handbook & Product Directory*.

TABLE 10.4. PERFORMANCE DATA FOR TYPICAL ROUND CEILING DIFFUSERS

Size	Neck Velocity, fpm	400	500	600	700	800	900	1000	1200	1400	1600
	NC					20		30		40	
	Vel. Press. in W.G.	.010	.016	.023	.031	.040	.051	.063	.090	.122	.160
	Total Press — Horizontal	.021	.034	.048	.065	.084	.107	.132	.189	.256	.346
	Total Press — Vertical	.027	.044	.063	.085	.109	.139	.172	.246	.333	.437
6″	Flow Rate, cfm	80	100	120	140	160	180	200	235	275	315
	Radius of Diff., ft.	1-2-3	2-3-4	2-3-5	2-4-6	3-4-7	3-5-7	4-5-8	4-6-10	5-7-11	6-8-11
	NC	--	—	15	20	24	27	31	36	41	45
8″	Flow Rate, cfm	140	175	210	245	280	315	350	420	490	560
	Radius of Diff., ft.	2-3-4	2-3-5	3-4-7	3-5-8	4-5-9	4-6-10	5-7-11	5-8-13	6-9-15	7-11-17
	NC	—	—	16	21	26	29	32	38	43	47
10″	Flow Rate, cfm	220	270	330	380	435	490	545	655	765	870
	Radius of Diff., ft.	2-3-5	3-4-7	3-5-8	4-6-9	4-7-11	5-8-12	6-8-14	7-10-16	8-12-19	9-13-22
	NC	—	—	17	22	27	30	33	39	44	48
12″	Flow Rate, cfm	315	390	470	550	630	705	785	940	1100	1255
	Radius of Diff., ft.	3-4-7	3-5-8	4-6-10	5-7-11	5-8-13	6-9-15	7-10-16	8-12-19	9-14-23	11-16-26
	NC	—	—	18	23	27	31	34	40	45	50
14″	Flow Rate, cfm	425	530	635	745	850	955	1060	1270	1490	1695
	Radius of Diff., ft.	3-5-8	4-6-9	5-7-11	5-8-13	6-9-15	7-11-17	8-12-19	9-14-22	11-16-26	13-19-30
	NC	—	13	19	24	28	32	35	41	46	60
16″	Flow Rate, cfm	560	700	840	980	1120	1260	1400	1680	1960	2240
	Radius of Diff., ft.	4-5-9	5-7-11	5-8-13	6-9-15	7-11-17	8-12-20	9-14-22	11-16-26	13-19-30	14-22-35
	NC	—	14	19	25	29	32	36	41	46	51
18″	Flow Rate, cfm	710	885	1060	1240	1420	1590	1770	2120	2480	2830
	Radius of Diff., ft.	4-6-10	5-8-12	6-9-15	7-11-17	8-12-20	9-14-22	10-15-24	12-18-29	14-21-34	16-24-39
	NC	—	15	20	26	30	33	36	42	47	52
20″	Flow Rate, cfm	875	1100	1310	1530	1750	1970	2190	2610	3060	3500
	Radius of Diff., ft.	4-7-11	6-9-14	7-10-16	8-12-19	9-14-22	10-15-24	11-17-27	13-19-32	16-24-38	18-27-43
	NC	—	15	21	26	30	34	37	43	48	52
24″	Flow Rate, cfm	1260	1570	1880	2200	2510	2820	3140	3770	4400	5020
	Radius of Diff., ft.	5-8-13	7-10-16	8-12-19	9-14-23	11-16-26	12-18-29	14-20-32	16-24-39	19-28-45	22-32-52
	NC	—	16	22	27	31	35	38	44	49	53
30″	Flow Rate, cfm	1960	2450	2940	3430	3920	4410	4900	5880	6860	7840
	Radius of Diff., ft.	7-10-16	8-13-20	10-15-24	12-18-28	13-20-32	15-23-36	17-25-41	20-30-49	24-35-57	27-40-65
	NC	—	17	23	27	32	36	39	45	50	54
36″ **	Flow Rate, cfm	2820	3520	4230	4930	5630	6340	7040	8450	9850	11,260
	Radius of Diff., ft.	8-12-20	10-15-24	12-18-29	14-21-34	16-24-39	18-27-44	20-30-49	24-36-58	28-42-68	32-48-78
	NC	—	18	24	28	33	37	40	46	51	55

NOTE. 1. Minimum radii of diffusion are to a terminal velocity of 150 F pm, middle to 100 F pm, and maximum* to 50 F pm. If diffuser is mounted on exposed duct, multiply radii of diffusion shown by 0.70.
2. The NC values are based on a room absorption of 18 dB, Re 10^{-13} watts or 8 dB, Re 10^{-12} watts. Values shown are for a horizontal pattern, add 1 dB for a vertical pattern.
Reprinted with permission from Environmental Elements Corporation, Baltimore, Maryland.

appearance in regard to location of diffusers. (This is called the *reflected ceiling plan.*) If there is a likelihood of rearranging partitions in the future, diffuser location and quantity may be chosen to allow changes without having to move ducts and diffusers.

6. *Size.* When the above decisions regarding selection have been made, the proper sizes of outlets can be chosen. This is done with the air of manufacturers' rating tables. The variable performance characteristics of ceiling diffusers that are of major importance are CFM, throw, mounting height, and sound level. The CFM is the quantity previously determined as required to condition the space. The throw or radius of diffusion is the horizontal distance that the diffuser projects the air. The maximum throw allowable is the distance to a wall or to the edge of the zone of the next diffuser. The minimum throw for adequate circulation is recommended by the manufacturer (usually $\frac{3}{4}$ of maximum). A selection should never be made with a throw greater than the maximum for this will result in drafts bounding off walls.

Most manufacturers give ratings acceptable for a mounting height between 8 and 10 ft, and show corrections for other heights. The sound level produced by a diffuser depends on the air velocity. Therefore, for a given required CFM, a smaller diffuser will mean a higher sound level, but a lower cost. The engineer must balance these needs according to the applications. The diffuser manufacturer usually lists sound ratings of the diffusers by NC (*noise criteria*) levels. This is a weighted perceived sound level. Recommended NC levels are shown in Table 10.3. Ratings for one type of round ceiling diffuser are shown in Table 10.4.

EXAMPLE 10.8

Select a single round ceiling diffuser for Betina's Boutique (Figure 10.30). The RSHG is 18,000

Figure 10.30. Sketch for Example 10.8.

BTU/hr. The supply air temperature differential is 25 F.

Solution. The required CFM must first be determined. Using the sensible heat equation (3.7)

$$\text{CFM} = \frac{18,000}{1.1 \times 25} = 650 \text{ CFM}$$

The maximum radius of diffusion permitted is 10 ft (from the center of the room to the wall). The NC level suggested for a small store is from 40 to 50. From Table 10.4, a Size 10 diffuser has the following listed rating:

> 650 CFM, 7 − 16 ft radius of diffusion,
> NC-39 sound level

This is a satisfactory selection. A Size 8 would have too high a sound level and throw.

Note that the pressure requirements are also given. This information is needed when balancing the air flow.

EXAMPLE 10.9

Determine suitable diffuser locations and throws for a room with a 60 ft by 30 ft floor plan if four diffusers are to be used.

Figure 10.31. Sketch for Example 10.9.

(a)

(b)

Figure 10.32. Use of equalizing grid in duct collar. (*a*) Poor air distribution in duct collar. (*b*) Equalizing grid evens flow.

Solution. The diffusers will have equal throws in all directions. Figure 10.31 shows the floor plan divided into squares. The diffusers will be located in the center of each square. The maximum throw therefore is 15 ft. Acceptable throws are from 12–15 ft when the diffuser size is selected.

Of course the space could also have been divided into eight areas each 15 ft square, and eight diffusers could be used. This would greatly increase the cost. Another possible solution might be to use one rectangular diffuser that had different throw distances on each side.

Selection procedures for supply grilles and slot diffusers are similar. The interested student can refer to manufacturers' catalogs.

10.19. ACCESSORIES AND DUCT CONNECTIONS

There are a number of accessories that are used with air supply devices to control or improve air distribution.

Equalizing Grids When an air outlet is connected to a duct as shown in Figure 10.32(*a*), the air may flow unevenly from the outlet, resulting in poor room air distribution. An equalizing grid installed in the duct collar as shown in Figure 10.32(*b*), can be used to equalize the air flow pattern to the outlet.

Splitter Dampers (Figure 10.33) These are sometimes used to direct air into the outlet and to control volume, but they can cause both uneven flow and additional noise.

Control Damper (Figure 10.34) These are used to adjust the volume rates of flow to the desired quantity. Opposed blade dampers are

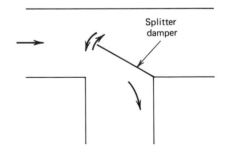

Figure 10.33. Splitter damper to control flow-poor air distribution.

Figure 10.34. Control damper.

TABLE 10.5. RECOMMENDED RETURN AIR INLET FACE VELOCITIES

Location	Velocity, FPM
Above occupied zone	800
Within occupied zone, not near seats	600 – 800
Within occupied zone, near seats	400 – 600
Door or wall louvers	200 – 300
Door undercuts	200 – 300

preferable to those that rotate in the same direction, because they will not result in uneven flow. The dampers can usually be adjusted from the face of the outlet with a special key.

Anti-Smudge Rings A strip of dirt on the ceiling surrounding a ceiling diffuser is a common sight. A ring that surrounds the diffuser is available which will reduce this problem.

Cushion Head For a diffuser that is at the end of a duct, it is advisable to extend the duct beyond the outlet neck (about one-half the take-off diameter, with a 6-in. minimum); this results in more even air distribution to the outlet.

10.20. RETURN AIR DEVICES

All of the devices used for air supply are suitable for return air. Grilles are the most commonly used, however, because of their lower cost. The location of return air devices is not as critical as supply devices because the air will not short-circuit in most cases. It is nevertheless a wise rule to locate return air inlets far from outlets, and, if this is not possible, to check the possibility of short-circuiting.

When air is being returned to corridors or adjacent spaces, transfer grilles or louvers may be located in partitions or doors, or the door may be undercut.

The selection of return air devices is usually quite simple. Recommended face inlet velocities

that generally provide acceptable noise levels are shown in Table 10.5. The face velocity refers to the velocity calculated by using the overall dimensions of the grille. The actual velocity is higher, because of the grille bars.

EXAMPLE 10.10

A return air grille is to be located high on a sidewall in a room, exhausting 1500 CFM. What should be the grille face area?

Solution. Referring to Table 10.5, a velocity of 800 FPM is acceptable. The required face area is therefore

$$\text{Area} = \frac{1500 \text{ ft}^3/\text{min}}{800 \text{ ft}/\text{min}} = 1.9 \text{ ft}^2 \text{ (use 2 ft}^2\text{)}$$

10.21. SOUND

Air conditioning systems generate sound, which may be objectionable in some cases. It is therefore the responsibility of the designer and contractor to provide adequate sound control when necessary. We will examine only a small part of this complex subject, but will present enough information so that the student will have an understanding of the problems that occur and how they may be resolved.

The magnitude of sound is measured in a unit called the *decibel* (dB). *Sound power* is the sound level generated by a noise source. We are interested in the sound transmitted and received, however, particularly by humans. This is called *sound pressure*. For our purposes we do not need to be overly concerned with distinctions between these terms.

In addition to *magnitude*, sound also has *frequency*. Most sound generated has a range of frequencies. The audible range is from about 20–20,000 Hz (cycles per second). We are particularly interested in frequency because the human ear has less sensitivity to lower frequencies (low pitch) than to higher pitch sounds. That is, a higher pitch sound of the same dB level as a lower one seems louder to the individual.

In studying an actual sound problem, the sound levels at each frequency sometimes need to be analyzed. However, an average of the levels at each frequency is often sufficient to work with. A *weighted average* is used to account for the change in sensitivity of the ear to different frequencies.

The weighted average that corresponds well to human response to sound is called the *A-scale network* (dB-A). Sound level measuring meters are available that read dB-A levels. This provides a simple means of measuring effective surrounding sound levels. Table 10.3 lists recommended dB-A levels as well as NC-levels. Both are used in setting standards.

EXAMPLE 10.11

What are the average recommended dB-A and NC sound levels for a hotel room?

Solution. From Table 10.3, the recommended range of dB-A is 35–45 and of NC is 30-40. The average dB-A is 40 and NC is 35. This would be a suitable sound level in a hotel room.

When two sources produce sound the combined level is found from Table 10.6.

TABLE 10.6. EFFECT OF COMBINING TWO SOUND LEVELS

Difference Between Levels, dB	0–1	2–4	5–9	10 or more
dB addition to highest level to obtain combined level	3	2	1	0

EXAMPLE 10.12

The sound power level in a duct approaching an air diffuser is 52 dB. The sound power level of the diffuser is 49 dB. What is the sound power exiting into the room?

Solution. Using Table 10.6,

$$\text{Difference} = 52 - 49 = 3 \text{ dB}$$
$$\text{dB to be added to higher level} = 2 \text{ dB}$$
$$\text{Combined sound power level} = 52 + 2 = 54 \text{ dB}$$

10.22. SOUND CONTROL

The main sources of sound generation in an air conditioning system are the fan and the noise generated by air in the ductwork. Often the resultant sound levels in the rooms are satisfactory and no special treatment is necessary. In any case the system design and installation should be carried out to minimize sound problems. Some general recommendations are:

1. Select fans near their most efficient operating point. Some of the wasted energy is otherwise converted into noise.
2. Isolate fans from their supports by using vibration isolators, and from the ductwork by using flexible connections.
3. Make duct connection transitions as gradual as possible.
4. Use duct velocities recommended for quietness.

TABLE 10.7. NATURAL DUCT ATTENUATION

Ducts		Radius Elbow	
Size, in.	dB	Size, in.	dB
(small) 6 × 6	0.10	10 × 10	1
(med.) 24 × 24	0.05	20 × 20	2
(large) 72 × 72	0.01	over 20	3

Branches					
Ratio of branch to main CFM, %	5	10	20	30	50
dB attenuation	13	10	7	5	3

Figure 10.35. Sketch for Example 10.13.

Solution. From Table 10.7, the attenuation is

$$\text{Duct } 0.05 \text{ dB/ft} \times 500 \text{ ft} = 2.5 \text{ dB}$$

$$\text{Elbow} = 3$$

$$\text{Natural attenuation} = \overline{5.5 \text{ dB}}$$

5. Avoid abrupt changes in direction in ducts. Use wide radius elbows or turning vanes.

6. Avoid obstructions in the ductwork. Install dampers only when required.

7. Balance the system so that throttling of dampers is minimized.

8. Select air outlets at sound levels as recommended by the manufacturer.

In many applications special sound treatment must be carried out, particularly in high velocity systems where considerable noise is generated. In a thorough sound analysis a series of calculations are made. This involves first determining the sound level generated at each frequency and then the amount of sound attenuation (reduction) required to meet the sound level required in the room.

Ductwork, branches, and elbows provide some natural sound attenuation, and the attenuation varies with the sound frequency. Tables listing this information can be found in the *ASHRAE Systems Volume*. Tables that list average attenuation for all frequencies are also available (Table 10.7). They are not precise, but are suitable for noncritical applications.

EXAMPLE 10.13

Determine the natural attenuation in the duct system shown in Figure 10.35.

Rooms also have sound absorbing characteristics, which depend on the size and sound absorbing qualities of the surfaces and furnishings. Values range from 0 up to 20 or 25 dB. Tables for determining this effect are also available.

The following procedure can be used for predicting sound levels and required sound treatment:

1. Determine the sound level generated by the fan. This information is usually available from the manufacturer.

2. Determine the amount of natural sound attenuation from ducts and fittings, using tables.

3. Subtract item 2 (natural sound attenuation) from item 1 (sound generated). This will be the sound level in the ducts to the air outlet.

4. Determine the sound level of the air outlet from the manufacturer.

5. Combine the sound power level in the duct (item 3) to that of the outlet (Table 10.6). This will be the sound generated at the outlet exit to the room.

6. Determine the room effect from tables. This is the sound absorption from room materials.

7. Subtract item 6 from item 5. This will be the room sound level if no special sound treatment is used.

8. Determine the recommended dB and NC level for the type of room, using tables.

9. If item 7 is less than item 8 (the desired room sound level), no treatment is required. If item 7 is greater, then the difference between them is the amount of additional sound attenuation that must be added.

As an exercise in understanding the procedure we will assume some figures in the following example, which would be taken from the tables recommended.

EXAMPLE 10.14

Find the additional average sound attenuation required (if any) for an air conditioning system for a private office, with the following data:

1. Fan sound power level   = 72 dB
2. Attenuation in ducts   = 17 dB
3. Sound power to diffuser (item 1 less 2)   = 55 dB
4. Diffuser sound power level   = 51 dB
5. Sound power level from diffuser, using Table 10.6, is 55 + 2 = 57 dB
6. Room effect attenuation   = 10 dB
7. Room sound level (item 5 less 6)   = 47 dB
8. Recommended room sound level (executive office) = 40 dB
9. Required additional sound attenuation   = 7 dB

Solution. For more accurate results the above analysis would be carried out at each sound frequency in order to determine how much additional sound attenuation is needed.

There are a few methods for achieving sound attenuation. It can be accomplished by lining ducts internally with a sound absorbing material. It is also very effective to internally line the air handling unit casing. Manufactured sound traps can also be used. These devices have special internal configurations of sound absorbing materials and perforated plates. They are quite effec-

tive and are used frequently in high velocity systems. Often a combination of these sound attenuation methods is used on a system.

Vibrations from fans, pumps, and compressors can also transmit sound if not isolated. This subject is discussed in Chapter 9.

REVIEW QUESTIONS

1. List the types of centrifugal fans and sketch their blade positions.

2. What is the difference between a vaneaxial and tubeaxial fan?

3. Sketch the three performance curve shapes for a backward and a forward curved blade centrifugal fan.

4. How is the point of operation of a duct system determined?

5. What criteria should be used in selecting a fan?

6. Describe the effect of changing fan speed on the CFM, BHP, and total pressure.

7. What are the main features of centrifugal fan construction?

8. What good practices should be followed when installing fans?

9. What energy conservation practices should be considered with fans?

10. Describe what is meant by the terms throw, drop, spread, terminal velocity, residual velocity, ceiling effect, and primary and secondary air.

11. List the considerations in choosing the location of air supply devices.

12. List the types of air supply devices and their applications.

13. Describe the use of equalizing grids, splitter dampers, control dampers, anti-smudge rings, and cushion heads.

14. List the recommended procedures to minimize sound generated by a duct system.

Figure 10.36. Sketch for Problem 10.8.

PROBLEMS

10.1. Find the static pressure developed, BHP, and ME of the fan whose performance curves are shown in Figure 10.7, at a flow of 25,000 CFM.

10.2. A 30 in. centrifugal fan of the type shown in Table 10.1 is specified to deliver 10,000 CFM at 1 in. SP (static pressure). At what speed should the TAB technician set it? What would be the expected motor BHP?

10.3. A fan is delivering 8100 CFM at $1\frac{1}{2}$ in. SP while running at a speed of 650 RPM and using 2.3 BHP. The fan speed is changed to 700 RPM. Find the new CFM, SP, and BHP.

10.4. Select a ceiling diffuser to deliver 2000 CFM in a 30 ft by 30 ft classroom.

10.5. A 100 ft by 50 ft pharmacy requires 8000 CFM of air. Locate and select three ceiling diffusers.

10.6. A louver is to be installed in a door to return 400 CFM of air to a corridor. What is the recommended louver area?

10.7. What are the recommended NC-level and dB-A level for a classroom?

10.8. The sound lower level of the fan shown in Figure 10.36 is 44 dB, and 40 dB for the diffuser. Determine the duct attenuation and sound level exiting from the diffuser.

10.9. If the room attenuation in Problem 10.8 is 8 dB, what additional attenuation would be required for a conference room?

10.10. Select air diffusers for the warehouse shown in Figure 6.12.
A. Use 1.1 CFM/ft^2. First select two, then four diffusers.
B. Use the results of your solution to Problem 7.24. Select two, then four diffusers.

10.11. Select air diffusers for the office building shown in Figure 6.13.
A. Use 1.5 CFM/ft^2.
B. Use the results of your solution to Problem 7.25.

11

Centrifugal Pumps, Expansion Tanks, and Venting

A pump is a device that circulates liquids through piping systems. The centrifugal pump is the type most widely used in circulating water in HVAC systems. In this chapter we will discuss the principles of operation, selection, construction, installation, and maintenance of cen- *trifugal pumps. The subject of controlling and venting air from the circulating water system will also be discussed, including use of the expansion tank, because this subject is closely related to how the pump is used.*

OBJECTIVES

After studying this chapter, you will be able to:

1. Identify the basic parts and construction of a centrifugal pump.
2. Use pump characteristic curves for rating and selecting a pump.
3. Use pump similarity laws to find the effect of changing speed.

4. Determine how to locate and size expansion tanks.

11.1. TYPES OF PUMPS

A pump provides the pressure necessary to overcome the resistance to flow of a liquid in a piping system. Pumps can be classified in two groups according to the way they develop this pressure —either by positive displacement or centrifugal force. In the first group are included reciprocating, gear, vane, screw, and rotary pumps. They are used only in specialized cases in HVAC work, and will not be discussed further here. The centrifugal pump is generally used in both hydronic and cooling tower water systems. It is very reliable, rugged, and efficient.

11.2. PRINCIPLES OF OPERATION

The centrifugal pump increases the pressure of the water by first increasing its velocity, and then converting that velocity energy to pressure energy. Figure 11.1 shows the operating elements of a centrifugal pump.

The *impeller* is the part that transmits energy to the water. Flowing from the pump suction line, water enters the opening in the center of the impeller called the *eye*. The impeller rotates, driven by a motor or other prime mover. The water is forced in a centrifugal direction (radially outward) by the motion of the impeller vanes. The velocity of the water is increased considerably by this action. The pump casing contains and guides the water toward the discharge opening.

The action of the impeller has increased the velocity of the water, but not its pressure. The velocity energy is converted into pressure energy by decreasing its velocity. This is accomplished by increasing the flow area, called the *volute* section of the pump casing. Refer to Chapter 8 where this principle is explained.

11.3. PUMP CHARACTERISTICS

The items of major importance in the performance of a pump are the pressure (*head*) it will

Figure 11.1. Operating elements of a centrifugal pump.

Figure 11.2. Typical performance characteristic curves for a centrifugal pump.

develop, the *flow rate* it will deliver, the *horse-power* required to drive the pump, and its *efficiency*. These are called the *pump characteristics*. The characteristics are usually presented in the form of curves for each pump, and can be used to select the correct pump for an application. The general shape of these curves is similar for all centrifugal pumps. Analyzing these curves is often quite useful in troubleshooting operating problems. Three curves are usually presented:

1. Flow rate versus head
2. Flow rate versus brake horsepower
3. Flow rate versus efficiency

The curves for a typical centrifugal pump are shown in Figure 11.2.

Notice that the head-flow curve indicates that a centrifugal pump develops less head at greater flow. The condition of no flow is called *shut-off*, at which the head is at or close to a maximum.

The power required to drive the pump is called the *brake horsepower* (BHP). The flow rate-BHP curve indicates that the BHP increases with flow rate for any centrifugal pump.

The BHP is the power input to a pump. The power output is the power transmitted to the water, given by the following equation

$$\text{WHP} = \frac{\text{GPM} \times H \times s.g.}{3960} \quad (11.1)$$

where

WHP = water horsepower (output), HP
GPM = flow rate, GPM

H = total pump head, ft of liquid
s.g. = specific gravity of liquid,
     (s.g. = 1 for water)

The power input to a pump is always greater than the power output because of friction and other unavoidable losses. The efficiency (E) of a pump is defined as

$$E = \frac{\text{power output}}{\text{power input}} \times 100 = \frac{\text{WHP}}{\text{BHP}} \times 100$$
$$(11.2)$$

Figure 11.2 also shows the flow rate-efficiency curve for the same centrifugal pump. Note that at shutoff efficiency is zero because there is no flow, then it rises to a maximum, and then decreases again at the pump's maximum flow rate.

EXAMPLE 11.1

A chilled water pump for the air conditioning system in the Five Aces Casino is delivering 200 GPM at a total head of 36 ft of water. The manufacturer lists the pump efficiency as 60% at this condition. What is the minimum size motor that should be used to drive the pump?

Solution. We must find the required power input (BHP). Using Equation 11.1 to find the power output

$$\text{WHP} = \frac{200 \times 36}{3960} \times 1 = 1.82 \text{ HP}$$

Using Equation 11.2

$$\text{BHP} = \frac{\text{WHP}}{E} \times 100 = \frac{1.82}{60} \times 100 = 3.0 \text{ HP}$$

This would be the minimum power needed for a motor. In reality a larger nominal size motor might be used to prevent possible overloading of the motor, especially for reasons that will be discussed shortly.

The pump performance characteristics, also called ratings, are shown for a particular pump in Figure 11.3. They are determined by the manufacturer by testing the pump. The performance depends on the speed at which the pump is operated. The ratings of the pump in Figure 11.3 are for a speed of 1750 RPM. (Its performance would be different at other speeds.) This speed and 3500 RPM are the most commonly used in the United States because they are the natural speeds resulting from direct connection to a 60 Hz motor. (In countries that use 50 Hz current,

1450 RPM and 2900 RPM would be the usual speeds.)

The performance of a given pump is found from its curves, as Example 11.2 illustrates.

EXAMPLE 11.2

If the pump whose ratings are shown in Figure 11.3, operating at 1750 RPM, is delivering 120 GPM, find the head it is developing, the BHP it uses, and its efficiency.

Solution.   Using Figure 11.3, at 120 GPM, proceed vertically up to the intersection with the head, BHP, and efficiency curves. Reading horizontally across,

$$H = 35.5 \text{ ft w.} \qquad \text{BHP} = 1.6 \text{ HP} \qquad E = 64\%$$

To conserve data space a manufacturer may show the performance curves for a number of different size pumps together. Figure 11.4 is a set of flow-head curves for a number of small pumps.

Figure 11.3. Performance curves for a $6\frac{1}{2}$ in. pump at 1750 RPM.

Figure 11.4. Performance curves for a group of small inline pumps.

BHP and efficiency are not indicated. Each pump is furnished with a motor large enough to handle the maximum BHP.

Another form of presenting pump curves is shown in Figure 11.5. In this case the flow-head curves are shown for a few pumps with impeller sizes ranging from 5–7 in. in diameter, all using the same casing. Instead of BHP and efficiency curves, lines of constant BHP and constant efficiency are shown.

Figure 11.5. Performance curves for a group of pumps with impeller sizes ranging from 5–7 in. at 1750 RPM. (*Courtesy:* ITT Fluid Handling Division.)

EXAMPLE 11.3

A 6 in. pump of the type shown in Figure 11.5 is operating at 1750 RPM. Suction and discharge gages at the pump read 30 psig and 45 psig, respectively. How much water is the pump circulating, what BHP is it using, and what is its efficiency?

Solution. The pump head is the difference between the suction and discharge pressures. Converting this to ft of water

$$H = (45 - 30) \text{ psi} \times \frac{2.3 \text{ ft water}}{1 \text{ psi}} = 34.5 \text{ ft w.}$$

Using the head-flow curve for the 6 in. pump from Figure 11.5, at a pump head of 34.5 ft of water the flow rate is 78 GPM. At this flow we read the BHP and efficiency, interpolating if necessary, as

$$\text{BHP} = 1.0 \text{ HP} \qquad E = 67\%$$

11.4. PUMP SELECTION

In the previous section we learned how to determine the performance of a pump from its curves. This is useful for the operator or TAB engineer in testing an existing pump. Another situation is the need to select a pump for a system. The pump must have a capacity equal to the system flow rate and a head equal to the system pressure loss. These two system characteristics are the primary ones in selecting a pump.

EXAMPLE 11.4

For the group of pumps shown in Figure 11.4, which one would be the best choice, based on size, for a system with a required flow rate of 15 GPM and 10 ft water pressure loss?

Solution. Locating the system point *A* in the figure, at the required flow rate we see that the smallest pump that will provide the required head is a size 103, and the head it develops is 11.5 ft water, which is more than enough. The size 102 pump does not develop adequate head at the required flow rate, and a size 104 develops much more than required. That pump might be used by throttling flow rate, but would be unnecessarily expensive.

In Example 11.4 there was only one suitable pump for the application. Usually, however, there are a number of factors that should be considered in selecting the most appropriate pump.

1. A pump that is operating near the point of maximum efficiency should be selected. This generally falls in the mid-range of pump flow capacity.

2. For hydronic systems it is preferable to select a pump operating at 1750 RPM rather than 3500 RPM. At 3500 RPM a smaller pump can be used, but the higher speed results in higher noise levels that may be disturbing in occupied areas.

3. It is not advisable to select a pump operating near its maximum capacity, even though a smaller pump results from this choice. If the system flow rate actually required is greater than designed for, the pump will not have the extra needed capacity. Select a pump in the vicinity of 50–75% of maximum flow.

4. The steepness of the head-flow curves varies among centrifugal pumps, depending on their design. Figure 11.6 shows examples of a *flat* head curve and *steep* head curve pump. It is recommended that pumps with flat head characteristic curves be used for hydronic systems. If there is a large change in flow rate, there will be a corresponding small change in pump head. This makes balancing and controlling flow rates easier. A steep head curve pump might be used in a system where the system pressure resistance is expected to gradually

Figure 11.6. Flat versus steep pump head characteristics.

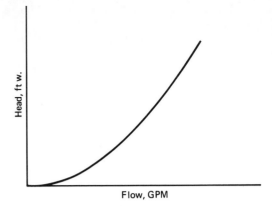

Figure 11.7. Typical system characteristic curve.

increase with time, yet where it is desired to maintain reasonably constant flow rate. A cooling tower circuit might be an example, where the pipe will roughen with age, increasing frictional resistance, and therefore also increasing the required pump head.

11.5. SYSTEM CHARACTERISTICS

The pressure loss in a piping system changes with the flow rate through the system. The pressure loss-flow rate relation is called the *system characteristic*, which can be determined from the following equation

$$\frac{H_{f2}}{H_{f1}} = \left(\frac{GPM_2}{GPM_1} \right)^2 \qquad (11.3)$$

where

H_{f2}, H_{f1} = pressure loss in piping due to friction at conditions 2 and 1.

GPM_2, GPM_1 = flow rates at conditions 2 and 1.

If the pressure loss is calculated at one flow rate, it can be found at any other flow rate, using Equation 11.3.

EXAMPLE 11.5

A piping system has a pressure loss due to friction of 30 ft water when the flow rate is 60 GPM.

If the flow rate were 80 GPM, what would be the pressure loss due to friction?

Solution. Using the system characteristic Equation 11.3

$$H_{f2} = H_{f1} \left(\frac{GPM_2}{GPM_1} \right)^2 = 30 \left(\frac{80}{60} \right)^2 = 53 \text{ ft w.}$$

A system charcteristic curve can be plotted for any piping system by calculating the pressure loss at a few different conditions (Figure 11.7). Note that the system frictional resistance rises very sharply with increased flow rate.

This characteristic curve includes frictional pressure loss only, not any static head, if there is any. Therefore, it applies to a closed circuit only (see Chapter 8). If the circuit is open, then to find the total system resistance, the static head would be added for the net height the water is lifted.

11.6. SYSTEM CHARACTERISTICS AND PUMP CHARACTERISTICS

The system and pump characteristic curves can both be plotted together (Figure 11.8). This is very useful in analyzing operating problems.

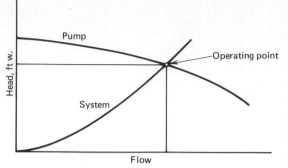

Figure 11.8. System and pump head characteristic curves indicating point of operation.

The head developed by the pump must be exactly equal to the system pressure loss. The only point where this is true is where the system and pump head curves intersect. Therefore, the following important statement always holds true:

The point of intersection of the system characteristic and pump characteristic head-flow curves is always the actual operating condition for the system-pump combination.

It is not usually necessary to plot both curves to make a pump selection, but many types of problems encountered in balancing and operating systems can be understood by studying the curves together. For instance, consider the situation where the actual system pressure loss is less than the design pressure loss (Figure 11.9).

The system design pressure loss and flow rate is given by point 1. A pump is selected to develop this head. Point 1 therefore represents the expected point of operation. Curve A, the expected system characteristic curve, could also be plotted. Suppose, however, that the actual system pressure loss at the design GPM was only that indicated by point 2. This might occur because the system designer allowed for a "safety factor" in calculating the piping friction loss. Point 2 of course cannot be the system operating point because it is not a point of intersection with the pump curve. Then what is the operating condition? To find it, we plot a new system characteristic curve, the real curve B, through point 2, using Equation 11.3. The real operating condition must be where this curve intersects the pump curve, point 3. But notice what has happened. The pump is actually delivering more flow than is desired. This may overcool or overheat the building. Furthermore, we know that the real operating point is farther out on the pump curve than expected, and the pump will thus use more power than expected. This is a waste of energy, and if the motor has not been oversized, we may have a burned out motor.

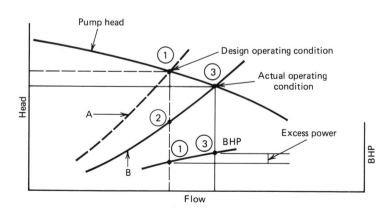

Figure 11.9. Illustration of excess power use and incorrect operation condition by use of "safety factor."

Instead of the safety factor, the condition may be less safe!

The problem could be resolved after installation by adding resistance in the circuit, say by throttling a balancing valve. This would bring us back to point 1. However, if the excess and unnecessary pressure loss had not been allowed for originally, a smaller pump might have been chosen (at point 2), using less power.

Even though it is proper to select a pump with head close to the actual system pressure loss, it is often advisable to select a motor for nonoverloading conditions, with a capacity greater than the BHP at maximum flow. The extra motor cost is a nominal part of the total cost, and there are many variables that might cause operation at higher than design flow.

11.7. PUMP SIMILARITY LAWS

There are a number of relations concerning flow rate, speed, power, and head for any given centrifugal pump that are sometimes useful to the HVAC engineer. A few of these are:

$$\frac{GPM_2}{GPM_1} = \frac{N_2}{N_1} \quad (11.4)$$

$$\frac{H_2}{H_1} = \left(\frac{N_2}{N_1}\right)^2 \quad (11.5)$$

$$\frac{BHP_2}{BHP_1} = \left(\frac{N_2}{N_1}\right)^3 \quad (11.6)$$

where

H = pump head, ft w.
N = pump speed, RPM
BHP = brake horsepower
1, 2 = any two operating conditions

EXAMPLE 11.6

The Pumpernickel Pump Co. ships a centrifugal pump to Argentina, where 50 Hz electric current

is used. The pump is rated at 380 GPM and 40 ft head, using 6 BHP, at a speed of 1750 RPM. What will be the pump's rating in Argentina?

Solution. The pump will operate at 1450 RPM on the 50 Hz current. Using the similarity laws,

$$GPM_2 = 380 \text{ GPM} \times \frac{1450}{1750} = 315 \text{ GPM}$$

$$H_2 = 40 \text{ ft} \times \left(\frac{1450}{1750}\right)^2 = 27.5 \text{ ft w.}$$

$$BHP_2 = 6 \text{ BHP} \times \left(\frac{1450}{1750}\right)^3 = 3.4 \text{ HP}$$

There are also pump similarity laws for determining the effect of a change in impeller diameter. It is advisable, however, to consult the manufacturer if a change in impellers is being considered.

Note that the centrifugal pump similarity laws are identical to those for a centrifugal fan (see Chapter 10).

11.8. PUMP CONSTRUCTION

Centrifugal pumps are available in varied arrangements and features of construction, each having different applications.

The *in-line pump* (Figure 11.10) is used for small, low head applications. The pump and motor are mounted integrally, and the pump suction and discharge connections are in a straight line (in-line). Because of this arrangement and the relatively light weight the pump can be supported directly by the piping, and is inexpensive and simple to install. In-line pumps, sometimes called booster pumps or *circulators*, are popular for small hydronic heating systems.

The *close coupled* pump (Figure 11.11) has the impeller mounted on and supported by the mo-

Figure 11.10. Inline type pump. (*Courtesy:* ITT Fluid Handling Division.)

Figure 11.11. Close coupled pump. (*Courtesy:* ITT Fluid Handling Division.)

Figure 11.12. Pump and motor connected by flexible coupling and mounted on common baseplate. (*Courtesy:* ITT Fluid Handling Division.)

tor shaft. The motor has a mounting flange for supporting the motor-pump combination from a suitable base. The pump has an end suction connection. The close coupled pump is relatively inexpensive and is available from small to medium capacities and heads.

Aside from the close coupled and in-line arrangements, the pump and motor shafts are connected by a flexible coupling and are individually supported. Some medium sizes are available preassembled by the manufacturer on a common baseplate (Figure 11.12) for convenience of installation. For larger pumps the contractor mounts the pump and motor separately and couples them. The coupling is flexible to allow adjustments in alignment of the two shafts and to reduce vibrations.

Centrifugal pumps can have either *single* or *double suction* construction, in which water enters either through one or both sides of the

pump. Larger pumps are constructed with double suction inlets.

The pump casing can be one cast piece or can be split—manufactured in two halves that are bolted together. The split casing makes repairs more convenient. The pump can be opened on the job for access to bearings or other internal parts. Both *horizontal split case* and *vertical split case* construction are used. The horizontal split case is used on very large pumps so that the very heavy upper part of the casing can be lifted vertically by a mechanical hoist.

If the impeller has walls (called *shrouds*) on both sides, it is called a *closed* impeller; if it has shrouds on one side, it is called *semi-open*; if it has no shrouds, it is called an *open* impeller. Open-type impellers are not generally used in HVAC applications because their purpose is to permit handling of liquids containing solids, such as sewage.

The *bronze fitted* pump is generally the combination of materials used for hydronic systems. The casing is cast iron and the impeller is bronze or brass.

Sleeve- and *ball*-type bearings are both used. Sleeve bearings are lubricated with oil. One ar-

Figure 11.13. Sleeve and ball bearings. (*a*) Sleeve bearings. (*b*) Ball bearings. (*Courtesy:* ITT Fluid Handling Division.)

Figure 11.14. Packed and mechanical seals. (*a*) Packed seal. (*b*) Mechanical seal. (*Courtesy:* ITT Fluid Handling Division.)

rangement has a reservoir of oil and an oil ring that flings the oil around as it rotates. Another arrangement uses cotton waste packing that is impregnated with oil. Large pumps may have an oil pump for forcing the oil to the bearings.

Ball bearings are lubricated with grease. Some motor ball bearings are sealed and cannot be lubricated in the field (Figure 11.13).

Ball bearings are generally used on smaller pumps. Larger pumps may have either sleeve or ball bearings. Sleeve bearings are quieter and are therefore recommended for hydronic service with larger pumps.

Seals are required to prevent leakage of water under pressure. Either *packing* or *mechanical* seals are used (Figure 11.14). Packed seals use a soft material that presses against the shaft, which has a tightness that is adjustable. The packing will wear and must be inspected and replaced at intervals. A small leakage is expected and normal, and a drip pan and drain lines should be provided to handle this. Mechanical seals have two hard, very smooth mating surfaces, one that is stationary and one that rotates. Properly applied, they will prevent any significant leakage. They cannot be used where there are any solid particles in the system, because the surfaces will become scored and the seal will be lost. They are very popular in hydronic systems because no maintenance is required and they can last many years.

11.9. NET POSITIVE SUCTION HEAD

Under certain conditions in circulating water systems a phenomenon called *cavitation* may occur in the pump suction, causing operating problems and possible damage to the pump. It results when the water pressure at the pump suction is too low. If this happens, the vapor pressure of the water in the pump may fall below its corresponding saturation temperature (Chapter 2) and the water will flash into steam.

The steam bubbles formed may collapse in the pump, momentarily leaving pockets or cavities. The water will rush into these cavities at great force, causing erratic operation, noise, and may damage the pump.

To avoid this possible occurrence a minimum pressure must be maintained at the pump suction, called *net positive suction head* (NPSH). The *required* NPSH for a pump can be obtained from the manufacturer. The *available* NPSH is calculated from examining the suction system arrangement of pressure loss, lift, and temperature. The available NPSH must be greater than the pump requires. If it is too low, the piping arrangement must be changed.

The possibility of cavitation is usually of concern in an open system where there is suction lift to the pump and where the temperature is high. It can be a problem in cooling tower systems if the pump is elevated to a location requiring a high suction lift. If this occurs, the pump or tower may have to be relocated. The problem is not usually encountered in hydronic systems because the static head in a closed system acts on the suction. The placement of the compression tank is an important factor relating to this, however, and will be explained later.

Boiler condensate return systems are also subject to cavitation if they are not designed and installed in accordance with the NPSH requirements.

The available NPSH can be determined from the following equation

$$H_n = H_a \pm H_z - H_f - H_v \qquad (11.7)$$

where

H_n = available NPSH
H_a = absolute pressure at surface of liquid where pump takes suction (atmospheric pressure, if open)
H_z = elevation of the liquid suction above (+) or below (−) the centerline of the impeller

H_f = friction and velocity head loss in the suction piping

H_v = absolute vapor pressure of water corresponding to the temperature

The units in Equation 11.7 are in feet of water.

EXAMPLE 11.7

A pump takes water at 180 F from an open tank that is 8 ft below the pump centerline. Friction and velocity head loss in the piping is 2.5 ft w. Atmospheric pressure is 14.7 psi. Determine the available NPSH.

Solution.   The vapor pressure of water at 180 F is 7.5 psia (Table A.3). Changing all units to ft w.

$$H_v = 7.5 \text{ psi} \times \frac{2.31 \text{ ft w.}}{1 \text{ psi}} = 17.3 \text{ ft w.}$$

$$H_a = 14.7 \times 2.31 = 34 \text{ ft w.}$$

Substituting in Equation 11.7

$$H_n = H_a \pm H_z - H_f - H_v$$
$$= 34 - 8 - 2.5 - 17.3 = 6.2 \text{ ft w.}$$

The pump used must have a required NPSH less than 6.2 ft w.

11.10. THE EXPANSION TANK

Water expands when its temperature increases, unless it is restricted. In a hydronic system an allowance must be made for this. If the piping system is completely filled and there is no space for the water to expand, the piping or equipment might break. An *open expansion* tank can be provided at the highest point in the system to solve this problem. Figure 11.15(*a*) shows that as the water temperature increases, the total volume of water in the system increases, the effect being a rise in the water level in the tank. Because the tank is open to the atmosphere, however, the system has some of the defects of open hydronic

Figure 11.15. Expansion and compression type tanks.

systems. Particularly undesirable is the continual exposure to air and its possible corrosive effects. A much better solution is to use a closed expansion tank containing a gas (air or nitrogen). When the water expands, it partially fills the tank, compressing the gas. For this reason the closed expansion tank such as the one shown in Figure 11.15(*b*) is usually called a *compression tank*.

The compression tank serves an additional purpose beyond that of providing for the water expansion—it aids in controlling system pressure. For these reasons compression tanks have largely replaced open expansion tanks in hydronic systems.

11.11. SYSTEM PRESSURE CONTROL

The pressure in a hydronic system must be controlled within certain maximum and minimum limits. This is a subject that is often not understood correctly, leading to operating difficulties and possible equipment damage.

The maximum allowable pressures are usually based on the permissible equipment pressures. In a low temperature hydronic heating system, for example, the boiler relief valve is usually set at 30 psig, which would therefore be the maximum allowable pressure at the boiler.

The minimum pressure requirement is based on two factors:

1. The pressure at any location must not be lower than the saturation pressure of the water. If this happens, the water will boil and

the steam will cause operating problems. As mentioned previously, this may occur particularly at the pump suction.

2. The pressure at any location should not be lower than atmospheric pressure. If this happens, air may enter the system.

Control of maximum and minimum pressures to ensure that none of these problems occur is achieved by proper sizing and location of the compression tank and by correctly pressurizing the system when filling. To know how to accomplish this we must understand how the compression tank functions.

The compression tank acts similarly to a spring or an air cushion. The water in the tank will be at the same pressure as the gas in the tank. The value of this pressure will depend on how much the gas in the tank is compressed. Once the system is filled with water and the water is heated to operating temperature, the total volume of water in the system remains constant. The volume of gas in the tank therefore also remains constant, and its pressure does not change. This holds true regardless of where the tank is located and whether or not the pump is operating. The following statement summarizes this fact:

The point at which the compression tank is connected to the system is the point of no pressure change.

By assuming two different tank locations and utilizing the above principle, we can see what effect the tank's location has on controlling system pressure. Consider first the compression tank

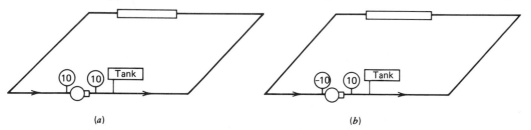

(a) (b)

Figure 11.16. Effect of compression tank location at pump discharge. (*a*) Tank at pump discharge, pump not operating. (*b*) Tank at pump discharge, pump operating.

located at the discharge side of the pump for the system shown in Figure 11.16. All of the piping is on the same level. Assume that the pressure throughout the system initially is 10 psig (25 psia) without the pump running as seen in Figure 11.16(*a*). Figure 11.16(*b*) shows what happens to the pressures when the pump runs. Assume the pump has a head of 20 psi. The pressure at the tank must be the point of no pressure change, so when the pump runs the pressure at this point is still 25 psia. The pressure at the pump suction must therefore be 20 psi less than this value, or 5 psia, because the pump adds 20 psi. But 5 psia is —10 psig, which is far below atmospheric pressure. Air would undoubtedly leak into the system at the pump suction. Cavitation in the pump might also occur in a heating system, because the boiling point of water at 5 psia is only 160 F.

Let us see what happens if the tank is located at the suction side of the pump. In Figure 11.17(*a*) the initial pressure is at 25 psia throughout the system, as before. When the pump runs, the pressure at the tank location remains at 25 psia. As the pump adds 20 psi, the pressure at the pump discharge must be 45 psia (30 psig), as seen in Figure 11.17(*b*). The pressure throughout the system is well above atmospheric.

This example shows that the compression tank should be connected to the system at the pump suction, not at the pump discharge. If the pump head is very low, the pressure at the pump suction might not fall below atmospheric, even if the tank is located at the pump discharge, but this arrangement is not advisable.

To keep the pressure exerted on the boiler as low as possible, the pump should be located to discharge away from the boiler, so that the boiler is not subject to the pump discharge pressure. On a high-rise building the static head at the boiler, if it is in the basement, might be above the maximum pressure. In this case the boiler could be located in a penthouse, or a steam boiler and hot water heat exchanger might be used (Chapter 5).

(The recommended arrangement of pump, boiler, and compression tank is shown in Figure 11.18. General service valves are not shown.)

11.12. COMPRESSION TANK SIZE

The size of the compression tank for a system must be adequate to receive the increased volume of water from expansion and also to keep the pressures within minimum and maximum limits. The size depends on the following sources of pressure:

1. *Static Pressure.* This is the pressure due to the height of water above any point. Usually the critical point is the boiler, which is often at the bottom of the system.

2. *Initial Fill Pressure.* If the system were initially filled without pressure, the pressure at the highest point in the system would be atmospheric. In order to provide a safety margin to prevent the pressure from going below atmospheric and thus leaking air in,

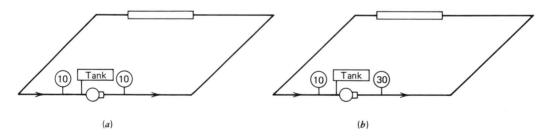

(*a*)   (*b*)

Figure 11.17. Effect of compression tank location at pump suction. (*a*) Tank at pump suction, pump not operating. (*b*) Tank at pump suction, pump operating.

the contractor should fill the system under pressure. A pressure of 4–5 psig at the top of the system is adequate for hydronic systems.

3. *Pressure-Temperature Increase.* After the system is filled with cold water and pressurized, when the temperature is raised in a hydronic heating system the pressure will increase further due to expansion of the water compressing air in the tank.

4. *Pump Pressure.* When the pump is operated, the pressures change in the system by the value of the pump head. As explained earlier, this depends on where the compression tank is located. If the tank is connected at the pump suction the head is added at every point. If the tank is connected at the pump discharge the pump head is subtracted from the pressure at every point.

These factors have been combined in the following formula developed by the ASME for determining tank size

$$V_t = \frac{(0.00041t - 0.0466)V_s}{H_a/H_t - H_a/H_0} \qquad (11.8)$$

where

V_t = required volume of compression tank, gallons
V_s = volume of system, gallons
t = design average water temperature, F
H_a = atmospheric pressure, ft water absolute
H_t = minimum pressure at tank, equal to the fill pressure plus static pressure at tank, ft water absolute
H_0 = maximum pressure at tank, ft water absolute

The term in the parentheses in the equation represents the expansion of water. The system volume is determined from the pipe sizes and from equipment volumes, information that can be obtained from manufacturers. In using the formula, pressure loss from friction is usually

Figure 11.18. Sketch for Example 11.8.

neglected. Example 11.8 will illustrate how the compression tank is sized.

EXAMPLE 11.8

Determine the required size of the closed expansion tank for the hydronic heating system shown in Figure 11.18. The system volume is 600 gal. The high point is 25 ft above the boiler. The pump head is 20 ft of water. The pressure relief valve on the boiler is set at 30 psig. The design average water temperature is 200 F. Fill pressure is 5 psig. The tank is located at the boiler elevation.

Solution. Equation 11.8 will be used. From the previous information, the terms in the equation are:

$t = 200$ F
$V_s = 600$ gallons
$H_a = 14.7$ psi $\times$ 2.3 ft w./1 psi $= 34$ ft w.
$H_t = 5 \times 2.3 + 25 = 38$ ft w.g. $+ 34 = 72$ ft w. absolute
$H_0 = 30$ psig $= 45$ psia $\times$ 2.3
$= 104$ ft w. absolute

Substituting in the equation

$$V_t = \frac{[0.00041(200) - 0.0466]600}{34/72 - 34/104} = 146 \text{ gal}$$

A compression type expansion tank of approximately 150 gal capacity would be used on this system, if connected at the suction side of the pump. If the tank is connected to the discharge side of the pump, the minimum pressure at the tank must be increased by the amount of pump head. That is,

$$H_t = 72 + 20 = 92 \text{ ft w. absolute}$$

The required tank size in this case becomes

$$V_t = \frac{21.2}{34/92 - 34/104} = 493 \text{ gal}$$

The tank size must be increased greatly because of its location.

For residential and other small hydronic heating systems, it is usually not necessary to calculate the size of the compression tank. Tables are available from manufacturers that list the appropriate tank size according to the building heating load.

11.13. AIR CONTROL AND VENTING

When the system is initially filled, air unavoidably enters the system. Dissolved air is in the fill water and the compressed air is in the tank. Eventually, further air will enter even a carefully designed and installed system, through make-up water and when the system is drained for maintenance and repair. Control of air in the system is necessary for two reasons: The presence of air will block the flow of water, and air and water together may result in corrosion.

When the water in the system is first heated to operating temperature, the dissolved air is released from the water as air bubbles. Much of

Figure 11.19. Automatic air venting valve for terminal units. (*Courtesy:* Taco, Inc.)

this air will find its way to the compression tank. Sometimes an *air separator* device (Figure 11.15) is located at the tank connection to divert air to the tank. Because air in a piping system will rise, some air will not be collected in the tank, but will find its way to other high points in the piping system. This air must be *vented* from the system or it will block flow through terminal units. *Air vents*, which are small valves, must be provided at all high points in the system. Vent valves may be manual or automatic (Figure 11.19).

Ideally, a system should have only one high point, but often there are many rises and drops in piping, each resulting in a high point. It is helpful to pitch horizontal piping slightly up toward high points when installing it. It is advisable to install a vent at each terminal unit. After the system is filled and put into operation the contractor opens each vent and bleeds air until none are left. Automatic air vents are convenient, but if they stick in an open position, they could cause considerable water damage to a building.

An eccentric type reducing fitting (Chapter 9) is recommended when changing pipe size, which

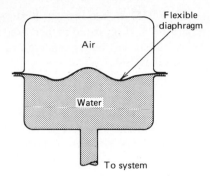

Figure 11.20. Diaphragm type compression tank.

often occurs at the pump connections. This prevents the trapping of air at the top of the fitting.

For small systems a convenient flexible *diaphragm type* compression tank (Figure 11.20) is available.

11.14. ENERGY CONSERVATION

1. Select pumps in their range of greatest efficiency, which is usually in the range of 50–70% of their maximum capacity.
2. Do not allow an extra pressure loss in the piping as a "safety factor."
3. Provide for venting air from the system in design, installation, and maintenance. Air will block flow and prevent proper operation, therefore it will indirectly reduce efficiency.

REVIEW QUESTIONS

1. List the basic parts of a centrifugal pump and their functions.
2. On one sketch, show the typical centrifugal pump characteristics curves.
3. What factors should be considered in selecting a pump for a hydronic system?
4. What is the relationship between the pump and system characteristics?

5. Explain the following terms: in-line, close coupled, single or double suction, and closed or open impeller.
6. Name the two types of pump bearings for hydronic systems and their features.
7. List the methods of bearing lubrication and the types of bearings to which they apply.
8. List the two types of pump seals and their features.
9. What is the importance of NPSH?
10. What are the functions of a compression tank?
11. What considerations are important for minimum and maximum pressure control?
12. Where should air vents be installed?
13. How should the piping be installed to improve air venting?

PROBLEMS

11.1. A pump for a hydronic solar heating system is circulating 60 GPM of an antifreeze solution that has a specific gravity of 1.14, against a total head of 23 ft w. The pump efficiency is 55%. What is the minimum size motor required to drive the pump?

11.2. The pump whose characteristic curves are shown in Figure 11.3 is circulating 100 GPM. What is the head, BHP, and efficiency?

11.3. Pump $6\frac{1}{2}$ (Figure 11.5) is operating with a total head of 38 ft w. Find the GPM it is circulating, the power it is using, and its efficiency.

11.4. A hydronic system requires 20 GPM of water. The system pressure loss is 6 ft w. at this flow rate. Select a pump from Figure 11.4 for this application.

11.5. The pressure loss due to friction in a hydronic system is 10 ft w. when the flow rate is 15 GPM. Pump 103 (Figure 11.4) is used. It is decided to increase the flow rate to 18 GPM. Will pump 103 be satisfactory?

11.6. For a system requiring 110 GPM, the engineer calculates the pressure loss at 37 ft w., selecting the pump in Figure 11.3. The actual pressure drop is only 25 ft w. Determine the actual operating conditions of GPM and BHP if no changes are made in the system.

11.7. For Problem 11.6 how could the proper flow rate be achieved? Determine the BHP and head at this condition.

11.8. A centrifugal pump has a rating of 120 GPM and 36 ft w. total head at a speed of 1150 RPM. It requires 2 BHP at these conditions. If its speed is increased to 1450 RPM, what will be its expected rating and required BHP?

11.9. A hydronic heating system has a design average water temperature of 220 F. The high point is 20 ft above the boiler. The pump head is 30 ft w. The system volume is 290 gal. The pressure relief valve on the boiler is set at 30 psig. The system is filled under a pressure of 4 psig. Find the required compression tank volume if it is located at the (a) pump suction, (b) pump discharge, and (c) high point of the system.

Air Conditioning Systems and Equipment

*T*here are a large number of variations in the types of air conditioning systems and the ways they can be used to control the environment in a building. In every application the planner must consider the features of each type of system and decide which is the best choice. Load changes, zoning requirements, space available, and costs are some of the variables that determine which type of system is to be used.

OBJECTIVES

After studying this chapter you will be able to:

1. Identify the components of single zone central system air conditioning equipment and their functions.
2. Identify the types of zoned air conditioning systems and their features.
3. Describe the features of all-water and air-water systems.
4. Describe the features of the different types of unitary equipment.
5. Select a cooling coil.
6. Identify the types and performance characteristics of air cleaners.

12.1. SYSTEM CLASSIFICATIONS

Two ways that systems can be classified will be discussed here. One way is by the cooling/heating fluid distributed, whether it is air or water.

319

The following groups are possible:

1. All-air systems
2. All-water (hydronic) systems
3. Air-water combination systems

Another way of classifying systems is according to whether the equipment is packaged together (unitary system) or separately (central system). Recall from Chapter 1 that any air conditioning system consists basically of a heating and/or cooling source, a fluid moving device, fluid distribution system, and terminal units. In unitary systems most or all of this equipment is located in one package. Unitary systems are usually of the all-air type. Central systems may be of any of the three types listed. Heating and cooling source equipment is discussed in Chapters 4 and 13. Hydronic terminal units and systems are discussed in Chapter 5, although further material will be explained here. Fans and air terminal units (distribution devices) are discussed in Chapter 10, and pumps are covered in Chapter 11. Most of the material following will therefore deal with types of distribution equipment and systems and their energy use.

The need for *zoning* is the main reason why different types of air conditioning systems exist. The heating or cooling load varies in each room due to changes in outside temperatures, solar radiation, occupancy, and other causes. Often these changes occur differently in each room. If this is the case, a system that changes its output in response to a load change in one room will satisfy the needs in that room, but not others. Systems can be designed to vary their output separately for each room or group of rooms (zone). This point should be kept in mind as we discuss each type of system. A further discussion of zoning will be included in Chapter 14.

12.2. ALL-AIR SYSTEMS

The essence of these types of systems is that they distribute only air to the spaces. Some of the variations of all-air central systems will be explained, including their applications.

12.3. SINGLE ZONE SYSTEM

This is the simplest all-air system. A unit conditions and then distributes a constant volume of air through one duct to a group of rooms (Figure 12.1). The conditions in each room therefore cannot be controlled separately.

The equipment shown in Figure 12.1 provides a complete year-round air conditioning system to control both temperature and humidity. Not all of the components are used in all circumstances.

The *supply air fan* is necessary to distribute air through the unit, ductwork, and air distribution devices to the rooms.

The *cooling coil* cools and dehumidifies the air in summer. It receives chilled water or refrigerant from the cooling source.

The *reheat coil* partially reheats the cooled air when the room heat gain is less than maximum, thus providing humidity control in summer. If no reheat coil is used, temperature, but not humidity can be controlled in summer (see Chapter 7).

The ductwork is generally arranged so that the system takes in some outside ventilation air, the rest being return air recirculated from the rooms. The equivalent amount of outside air must then be exhausted from the building. Provisions are often made in the arrangement of dampers (Figure 12.2) so that 100% outside air can be drawn in and exhausted. This would be done in intermediate season cool weather to obtain cooling without operating refrigeration machinery. Some systems use 100% outside air and no return air at all times, even though this increases the refrigeration load considerably. Examples might be operating rooms or laboratories where contaminated exhaust air cannot be recirculated.

The *return air fan* takes the air from the rooms and distributes it through return air ducts

Figure 12.1. Arrangement of single zone central system air conditioning equipment.

back to the air conditioning unit or to the out-doors. In small systems with little or no return air ducts, the return air fan is not required, because the supply fan can be used to draw in the return air.

The *preheat coil* is required in cold climates (below freezing) to temper air so that chilled water cooling coils cannot freeze up. It is op-tional in milder climates and when DX (dry expansion) cooling coils are used. The preheat coil may be located to receive either outside air or the mixed air.

When the system is used for winter heating, both the preheat and reheat coils can be utilized.

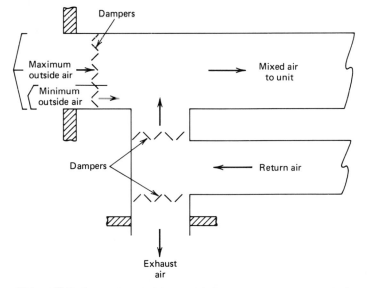

Figure 12.2. Arrangement of ducts and dampers to vary proportion of outside and return duct air.

The *filters* are required to clean the air.

Bypassing air around the cooling coil (Figure 12.3) provides another method of controlling humidity (see Chapter 7), but does not give as good a humidity control in the space as with a reheat coil.

To achieve separate temperature or humidity control in different zones, individual single zone units can be used. This of course multiplies equipment costs and therefore other schemes are often used.

Figure 12.3. Arrangement of face and bypass dampers to provide reheat for humidity control.

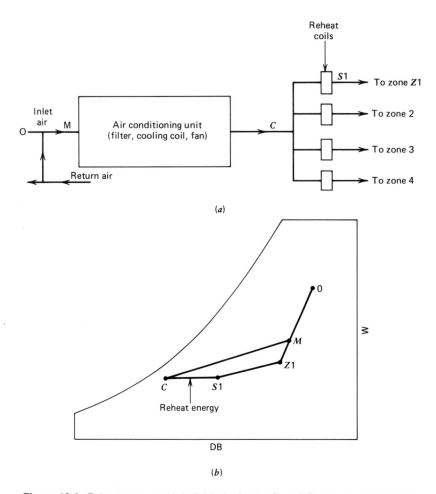

Figure 12.4. Reheat system with individual reheat coils. (*a*) Equipment arrangement. (*b*) Psychrometric processes for reheat system.

12.4. REHEAT SYSTEM

In this arrangement separate single ducts from the air handling unit are distributed to each zone or room that is to be controlled separately (Figure 12.4). A reheat coil is used in each of these ducts. In this way separate control of both temperature and humidity can be achieved in each zone. The basic air conditioning unit is the same as with a single zone system, except perhaps the main reheat coil can be eliminated.

The reheat system provides good control of each zone. As seen in Figure 12.4, however, it is very wasteful of energy because the air must always be completely cooled to C and then often reheated (to $S1$ as shown for zone 1)—a double waste of energy.

12.5. MULTIZONE SYSTEM

The multizone system uses an air handling unit that has a heating coil (*hot deck*) and cooling coil (*cold deck*) in parallel (Figure 12.5). Zone dampers are provided in the unit across the hot and cold deck at the outlet of the unit.

Separate ducts are run from each set of dampers to each zone (Figure 12.6). Cold and hot air are mixed in varying proportions by the dampers according to zone requirements.

The psychrometric processes for the multizone system are the same as that for the dual duct system, described in the next section.

The multizone system can provide good zone temperature control, but because mixed air is bypassed around the dehumidifying coil, humid-

Figure 12.5. Multizone unit.

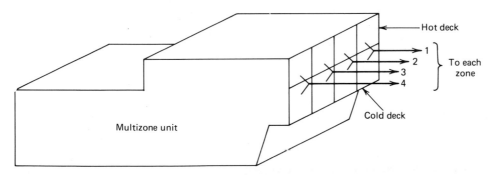

Figure 12.6. Duct arrangement for multizone system.

ity control may not be satisfactory in applications with high proportions of outside air. Because of the limit on the size of units available, each air handling unit is limited to about 12–14 zones. It is a relatively inexpensive system for small and medium size applications where a few separate zones are desired and humidity conditions are not critical.

12.6. DUAL DUCT SYSTEM

In this arrangement separate hot and cold main ducts are run from heating and cooling coils in the air handling unit (Figure 12.7).

Mixing boxes (Figure 12.8) are provided in each zone, tapping air from the hot and cold ducts. Dampers in the mixing box respond to a

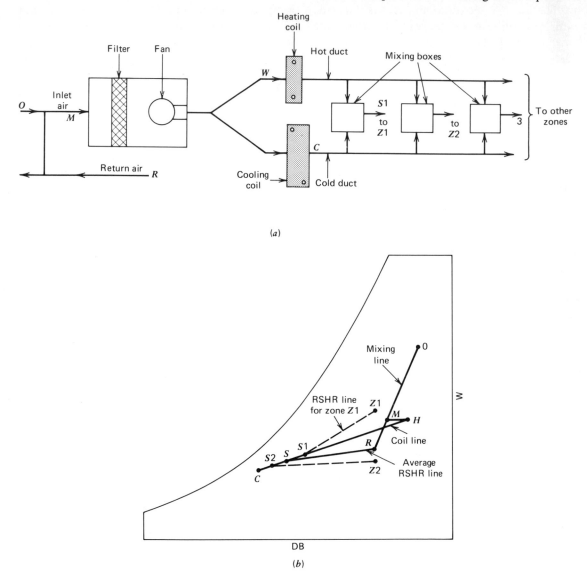

Figure 12.7. Dual duct system arrangement. (*a*) Equipment arrangement. (*b*) Psychrometric processes.

Figure 12.8. Mixing box for dual duct system. (*Courtesy:* Environmental Elements Corporation, Baltimore, Maryland.)

room thermostat to mix the proper proportion of hot and cold air delivered to the zone.

The psychrometric processes for summer cooling zone control are shown in Figure 12.7. Mixed air at *M* is heated to *H* from the fan heat. Chilled air leaves the cooling coil at *C*. This air mixes with air from the hot duct to produce supply air at *S*. RSHR line *RS* is an average condition for all zones, not an actual room line. Line *Z1-S1* is an actual room line for a zone *Z1*, with a less-than-peak sensible cooling load and high latent load. Warm air and cold air are supplied in the correct proportion from the zone mixing box to provide zone supply air *S1*. Note that the room humidity is higher than the average. In most applications the humidity increase is not great enough to be uncomfortable. *Z2* is an example of a room condition with a higher sensible and lower latent load.

As the outside air temperature falls it may be necessary for the reheat coil to operate to maintain an adequate hot duct temperature, so that humidity does not rise too high. This is one of the instances where extra energy may be used. In any case the hot duct temperature control should be set at the minimum required to provide comfort. Many installations have been designed or operated with hot duct temperatures that result in considerable excess energy use.

In order to simplify the explanation, duct and return air fan heat gains are not shown in the psychrometric processes that have been described.

Dual duct systems are usually designed as high-velocity air systems in order to reduce duct sizes. The mixing boxes therefore have a sound attenuating section built into them. The air downstream from the mixing box is run at conventional low velocities.

The availability of cold and warm air at all times in any proportion gives the dual duct system great flexibility in handling zones with widely varying loads. The installed cost of the dual duct system is usually quite high, and fan horsepower requirements are high because large volumes of air are moved at high pressure.

12.7. VARIABLE AIR VOLUME (VAV) SYSTEM

In this system a single main duct is run from the air handling unit, the same as with a single zone system. However, the quantity of air supplied to each branch and zone is varied (Figure 12.9).

A room thermostat operates a damper or other throttling device that controls the air flow rate to the zone in response to the load.

Early VAV systems were limited to systems where the cooling load did not decrease below approximately 75% of design, because the reduction in air supply volume resulted in inadequate air circulation and distribution. The development of improved terminal air units, capable of throttling air volume considerably while maintaining adequate room air circulation, has extended the range of VAV systems to applications with large variations in loads.

Because the volume of supply air is reduced in response to the room sensible cooling load, the room humidity will usually rise at part load, if

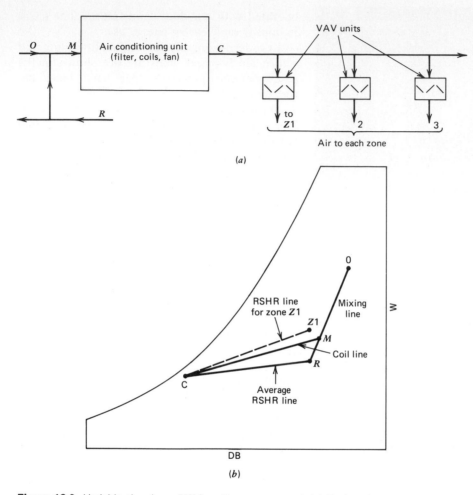

Figure 12.9. Variable air volume (VAV) system arrangement. (*a*) Equipment arrangement.
(*b*) Psychrometric processes.

the latent load remains constant (zone Z1). For applications with a high sensible heat ratio the effect is negligible.

A combination of VAV with reheat results in an energy efficient system, yet also provides good humidity control even at very low loads. In this arrangement supply air volume is reduced to a minimum of perhaps 50–60%. If the sensible load falls further, a reheat coil in the VAV terminal unit is activated. In this way better humidity control is obtained.

Duct and fan heat gains are not shown in the psychrometric chart, to simplify the explanation.

Later in the chapter system types will be discussed from the viewpoint of energy consumption.

12.8. ALL-WATER SYSTEMS

The basic concept of all-water systems, that is, hydronic systems, was introduced in Chapter 5.

However, further material will be discussed here. Hydronic systems distribute hot or chilled water from the central plant to each space. No air is distributed from the central plant. Hydronic terminal units such as fan-coil units heat or cool the room air. Ventilation air can be brought through the outside wall and the terminal unit.

All-water systems for commercial use can be considerably less expensive and take up much less space than all-air systems (this is not necessarily true for residential use). Water has a much higher specific heat and density than air. This means that considerably less volume of water needs to be circulated for the same amount of heat transfer. The result is that the cross-sectional area of piping is much smaller than the ductwork would be for the same job.

A hydronic cooling system is therefore useful when space is extremely limited, particularly space in shafts and ceilings. An important example is installation of air conditioning systems in existing large buildings that were not originally designed to include air conditioning.

The lack of need for ductwork and central air handling equipment, and the saving on using much valuable building space, all result in the fact that hydronic systems are often less expensive initially than all-air systems for large jobs, particularly in high-rise buildings. On the other hand, all-water systems have certain disadvantages. The multiplicity of fan-coil units means a great deal of maintenance work and costs. Control of ventilation air quantities is not precise with the small fans in the units. Control of humidity is limited. All-water systems are popular as low cost central systems in multiroom high-rise applications.

12.9. AIR-WATER SYSTEMS

Combination air-water systems distribute both chilled and/or hot water and conditioned air from a central system to the individual rooms. Terminal units in each room cool or heat the room.

Air-water systems utilize the best features of all-air systems and all-water systems. Most of the energy is carried in the water. Usually the air quantities distributed are only enough for ventilation. Therefore, the total shaft and ceiling space required is small. In addition, the air is usually carried at high velocities.

Fan-coil units can be used as the room terminal units, arranged to receive the centrally distributed air, or the air can be supplied directly to the room. However, the most common air-water system uses terminal units called *induction units*. This unit has been described in detail in Chapter 5. The central air delivered to each unit is called *primary air*. As it flows through the unit at high velocity, it induces room air (secondary air) through the unit and across the water coil. Therefore, no fans or motors are required in this type of unit, reducing maintenance greatly.

The induction unit air-water system is very popular in high-rise office buildings and similar applications. Its initial costs are relatively high.

The primary air quantity in the induction system may be only about 25% or less than the total of the air volume rate of a conventional all-air system. Because of this it is often not adequate for outside air cooling in mild or even cold weather. At these times chilled water must be supplied to the room unit coils. This is particularly true on southern exposures. There are buildings with air-water induction systems requiring refrigeration at outdoor temperatures as low as 30 F. In some cases, this energy inefficient situation may be improved by utilizing another source of chilled water, such as an outside air heat exchanger (see Chapter 15).

Any of the 2, 3, and 4 pipe hydronic system arrangements described in Chapter 5 can be applied to air-water systems. The factors in choosing a particular arrangement are discussed in that chapter.

12.10. UNITARY VERSUS CENTRAL SYSTEMS

As stated previously, air conditioning systems can also be classified into either *unitary* or *central* systems. This classification is not according to how the system functions, but how the equipment is arranged. A unitary system is one where the refrigeration and air conditioning components are factory selected and assembled in a package. This includes refrigeration equipment, fan, coils, filters, dampers, and controls. A central or remote system is one where the components are all separate. Each is selected by the designer and installed and connected by the contractor. Unitary equipment is usually located in or close to the space to be conditioned. Central equipment is usually remote from the space, and each of the components may or may not be remote from each other, depending on the desirability.

Unitary or central systems can both in theory be all-air, all-water, or air-water systems, but practically, unitary systems are generally all-air systems, and limited largely to the more simple types such as single zone with or without reheat or multizone. This is because they are factory assembled on a volume basis. Unitary systems and equipment can be divided into the following groups:

1. *Room units*
2. *Unitary conditioners*
3. *Rooftop units*

These names are not standardized in the industry. For example, *unitary* conditioners are also called *self-contained* units or *packaged* units. Sometimes packaged equipment is used as the name for all unitary equipment.

12.11. ROOM UNITS

Room units (Figure 12.10) are available in two types; *window* units and *through-the-wall* units. The window unit fits in the sash opening of an existing window, resting on the sill. The through-the-wall unit fits in an outside wall opening, usually under the window sill.

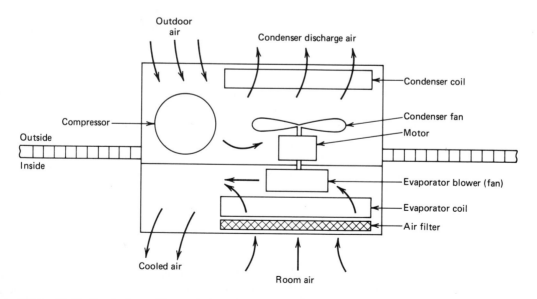

Figure 12.10. Room air conditioner equipment arrangement.

Compressor, evaporator cooling coil, condenser, filter, motors, fan and controls are assembled in the unit casing. Dampers can be adjusted so that only room air is used, or so that some outside ventilation air can be brought through the conditioner. Room units are available up to about 3 tons of refrigeration capacity. Their advantages are low cost and simplicity of installation and operation. Window units are particularly applicable to existing buildings. Through-the-wall units are often used in new apartment houses where low cost is primary. In existing buildings, of course, electrical services may have to be increased to take the added electrical load.

Room units have no flexibility in handling high latent heat gains or changed sensible heat ratios, and therefore do not give good humidity control. Sound levels are higher than with remote equipment. Air cleaning quality is minimal because the filters remove only large particles, in order that the resistance to air flow be low. When used in multiroom buildings, maintenance of the large number of units can be very burdensome and expensive. These units are inherently energy wasteful in multiple use, because they cannot modulate capacity.

Figure 12.11. Unitary air conditioner.

12.12. UNITARY AIR CONDITIONERS

This type of unit is designed to be installed in or near the conditioned space (Figure 12.11). The components are contained in the unit. Heating components are sometimes included.

Unitary conditioners are available in vertical or horizontal arrangements, according to the space available for the equipment. Although they often discharge air directly into the space, a limited amount of ductwork can be connected if air distribution with outlets is desired. These units are popular in small commercial applications.

Units are available that have all components packaged except the condenser. This is a popular arrangement in private residential applications. The condenser is located outdoors and compressor, coil, and fan package in an attic or basement. Another common arrangement is a *split system*. The condenser and compressor are in one package, located outdoors, and the fan and cooling coil are another package located indoors. This is especially popular with small residential heat pumps.

Unitary conditions have the same advantages and disadvantages as room units. In larger units multiple compressors are used. Units are available in sizes up to about 50 tons.

12.13. ROOFTOP UNITS

Ths type of unitary equipment (Figure 12.12) is designed to be located outdoors, and is generally installed on roofs. Usually, all of the refrigeration, cooling, and air handling equipment is assembled together, although the compressor and condenser may be remote. Heating equipment may be incorporated in the unit.

Rooftop units may be used with ductwork and air outlets. They must have weatherproofing features not required with equipment located indoors. All electrical parts must be moistureproof, and the casing and any other exposed parts must be corrosion protected.

The advantages of rooftop units are that they do not use valuable building space and they are relatively low in cost. Units are available with multizone arrangement, thereby offering zone controls, but humidity control is limited. Rooftop systems are extremely popular in low cost, one-story building applications, such as supermarkets and suburban commercial buildings.

12.14. AIR HANDLING UNITS

The air handling unit refers to the combination of coils, fan, filters, dampers, and casing. It is also sometimes called the central air conditioning apparatus. There are basically two arrangements: single zone units and multizone units. The distinction between these has been discussed in Sections 12.3 and 12.5. In small- and medium-capacities air handling units are factory made in sections—fan section, coil sections, mixing box, filter section—in numerous sizes. Those parts that are required are selected by the user. For large systems, separate coils, filters, and fans are selected by the engineer and casings are fabricated by the contractor to suit the equipment.

Casings are usually made of galvanized sheet metal. The casing should be insulated to prevent energy losses. When cooling and dehumidifying, drain pans must be included under the coil to collect condensed moisture, and a piping drain connection must be provided which is run to a

Figure 12.12. Rooftop unit. (*Courtesy:* McQuay Group, McQuay-Perfex, Inc.)

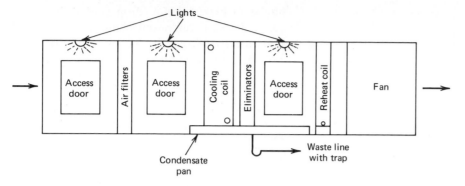

Figure 12.13. Maintenance accessories in large air handling unit.

waste drain. The pipe should have a deep seal trap so that a water seal always exists (Figure 12.13).

Access doors should be provided to permit maintenance. They should be located on both sides of coils and filters. In large equipment, lights should be provided inside each section.

When the fan is located downstream of the cooling coil, the unit is called a *draw-through* type. When the fan is upstream of the coils, it is called a *blow-through* type. Draw-through is preferable because the air will flow more uniformly across the coil section when drawn through by the fan. Multizone units are blow-through types. To aid in distributing the air more evenly across the heating and cooling coil in blow-through units, a perforated plate is sometimes located between the fan and coils.

12.15. COOLING AND HEATING COILS

Cooling coils may use either chilled water or evaporating refrigerant. The latter are called *dry expansion* (DX) coils.

Cooling coils are usually made of copper tubing with aluminum fins, but copper fins are sometimes used. The coils are arranged in a serpentine shape, in a number of rows, depending on the need (Figure 12.14).

The fins increase the effective surface area of tubing, thus increasing the heat transfer for a given length of tube. The coil may be constructed either with tubes in series or in parallel to reduce water pressure drop.

When cooling coils have a number of rows, they are usually connected so that the flow of water and air are opposite to each other, called

Figure 12.14. Cooling coil (chilled water type).

Figure 12.15. Counterflow arrangement of air and water flow for cooling coil.

counterflow (Figure 12.15). In this way the coldest water is cooling the coldest air and fewer rows may be needed to bring the air to a chosen temperature than if parallel flow were used, and the chilled water temperature can be higher.

The water inlet connection should be made at the bottom of the coil and the outlet at top, so that any entrapped air is carried through more easily. In addition, an air vent should be located at the outlet on top.

12.16. COIL SELECTION

Coil selections are made from manufacturers' tables or charts based on the required performance. The performance of a cooling coil depends on the following factors:

1. The amount of sensible and latent heat that must be transferred from the air.
2. Conditions of air entering and leaving, DB and WB.
3. Coil construction—number and size of fins size and spacing of tubing, number of rows.
4. Water (or refrigerant) velocity.
5. Air face velocity. The face velocity is the air flow rate in CFM divided by the projected (face) area of the coil.

Water velocities from 1-8 FPS are used. High water velocity increases heat transfer but also

results in high pressure drop, and therefore requires a larger pump and increased energy consumption. Velocities in the midrange of about 3–4 FPS are recommended.

High air velocities also result in better heat transfer and also more CFM handled. However, if the coil is dehumidifying, the condensed water will be carried off the coil into the airstream above 500–550 FPM face velocity, and *eliminator baffles* must be used to catch the water droplets.

The form in which manufacturers present their coil rating data varies greatly one from another. Using these ratings does not give much insight into how a coil performs. For this reason we will not present any rating data. However, the procedures described in Chapter 7 give all the basic data necessary to select a coil. Indeed, that procedure has the advantage of being suitable for any manufacturer's coils, as seen in Example 12.1.

Table 12.1 lists typical contact factors (CF) for finned cooling coils. With this type of table the number of rows of coil needed for given entering and leaving air conditions can be directly determined.

EXAMPLE 12.1

A cooling coil has 3200 CFM of air flowing across it at a face velocity of 400 FPM. Air

TABLE 12.1. TYPICAL CONTACT FACTORS FOR HELICAL FINNED COOLING COILS

| | Face Velocity, FPM | | | | | |
| | 8 fins / in. | | | 14 fins / in. | | |
No. of Rows	400	500	600	400	500	600
2	0.60	0.58	0.57	0.73	0.69	0.65
3	0.75	0.73	0.71	0.86	0.82	0.80
4	0.84	0.82	0.81	0.93	0.90	0.88
6	0.94	0.93	0.92	0.98	0.97	0.96
8	0.98	0.97	0.96			

enters the coil at 85 F DB and 69 F WB and leaves at 56 F DB and 54 F WB. Determine the required number of rows and face area of an 8 fin/in. coil.

Solution. The required CF is 0.83, as worked out in Example 7.32. From Table 12.1, for that type of coil, a 4 row coil will do the job. The face area needed is

$$3200 \text{ CFM}/400 \text{ FPM} = 8 \text{ ft}^2$$

12.17. AIR CLEANING DEVICES (FILTERS)

Air conditioning systems that circulate air generally have provisions for removing some of the objectionable air contaminants. Most systems have devices that remove particles commonly called dust or dirt, which result largely from industrial pollution. Occasionally gases that have objectionable odors are also removed from the air.

Air cleaning is often treated casually when designing and operating an air conditioning system. The incorrect type of filter may be chosen, or the filters may not be maintained properly. This is a serious neglect, because we are dealing with a question of air pollution and human health. Proper air cleaning is necessary for the following reasons:

1. *Protection of human health and comfort.* Dust particles are related to serious respiratory ailments (emphysema and asthma).

2. *Maintaining cleanliness of room surfaces and furnishings.*

3. *Protection of equipment.* Some equipment will not operate properly or will wear out faster without adequate clean air. Some manufacturing processes are particularly sensitive.

4. *Protection of the air conditioning machinery.* Dust collecting on coils will increase their resistance to heat transfer.

12.18. METHODS OF DUST REMOVAL

Air cleaners can remove dust in three principal ways:

1. *Impingement.* The dust particles in the airstream strike the filter media and are therefore stopped.

2. *Straining.* The dust particles are larger than the space between adjacent fibers and therefore do not continue with the airstream.

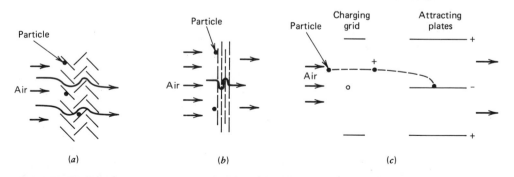

Figure 12.16. Methods of removing particles from air. (*a*) Impingement. (*b*) Straining. (*c*) Electrostatic precipitation.

3. *Electrostatic precipitation*. The dust particles are given an electric charge. The filter media is given the opposite charge, and therefore the particles are attracted to the media.

A filter may remove particles by one or more of the above methods. This will be discussed when specific types of filters are described. Figure 12.16 shows each of the methods.

12.19. METHODS OF TESTING FILTERS

Understanding how air filter performance is evaluated is important because only in this way can a proper filter be selected. Only in recent years have standard test methods developed. Without standard procedures filters cannot be compared with each other. The problem is complicated because filter performance depends on the concentration and sizes of dust particles in the air. This varies greatly from one location to another and at different times.

The following tests are generally accepted and recommended in the industry:

1. *Weight*. The weight of dust captured by the air filter is measured. A standard dust of fixed concentration and particle sizes is used. This test is useful in comparing ability to remove larger particles. It does not indicate ability to remove small particles because the small particles comprise such a small proportion of the total weight of atmospheric dust.

2. *Dust spot discoloration*. In this test air is first passed through the air cleaning device and then a white filter paper. The degree to which the filter paper is discolored is an indicator of the amount of smaller dust particles not removed by the air cleaner. This test is important because these particles cause soiling of room surfaces.

3. *DOP penetration*. This test is used to measure the ability of air cleaners to remove extremely small particles. A cloud of particles of a substance called DOP is chemically generated. The size of these particles is 0.3 microns in diameter (one micron is about 1/25,000th of one inch). A cloud of DOP particles in an airstream is passed through the air cleaner. The concentration of particles not removed is measured downstream of the cleaner by using a light-scattering technique. In this way the filter's effectiveness in removing very small particles is tested. As an example, bacteria range from about 0.3–30 microns in diameter, and cigarette smoke particles from 0.01–1 micron. The DOP test is used only on air cleaners that are designed to have a high efficiency in removing very small particles.

4. *Dust holding capacity*. The above three tests all measure efficiency of an air cleaner in removing particles. What they do not measure is how much the filter air resistance will increase with dust accumulation. A filter that will hold a considerable amount of dust before resistance increases considerably is preferable to one that has a lesser capacity before buildup up to a given resistance. The dust holding capacity test compares weight of dust collected with increase in air resistance through the filter.

12.20. TYPES OF AIR CLEANERS

Air cleaners can be classified in a number of ways.

Type of Media The viscous impingement air filter has a media of coarse fibers that are coated with a viscous adhesive. Glass fibers and metal screens are two commonly used media materials. Air velocities range from 300–600 FPM. The pressure drop when clean is low, around 0.1 in. w.g.; the filter should be serviced when the resistance reaches 0.5 in. w.g. This type of filter will

Figure 12.17. Viscous impingement disposable filter. (*Courtesy:* American Air Filter Co., Louisville, Kentucky.)

Figure 12.18. High efficiency dry type (HEPA) filter. (*Courtesy:* American Air Filter Co., Louisville, Kentucky.)

remove larger dust particles satisfactorily but not small particles. It is low in cost (Figure 12.17).

The dry-type air filter uses uncoated fiber mats. Glass fibers and paper are two commonly used materials. The media can be constructed of either coarse fibers loosely packed or fine fibers densely packed. By varying density, dry-type air filters are available that have good efficiency only on larger particles, as with the viscous impingement type, or are also available with medium or high efficiency for removing very small particles.

The HEPA filter (High Efficiency Particulate Air) is a very high efficiency dry-type filter for removing extremely small particles (Figure 12.18). For example, it is the only type filter that will effectively remove viruses as small as 0.05 micron (1/500,000th of an inch!). Air face velocities through HEPA filters are very low, about 50 FPM, and resistance rises to about 2.0 in. w.g. before servicing. They are quite expensive.

The media in air filters can be arranged in the form of random fiber mats, screens, or corrugated sinuous strips.

Permanent or Disposable Air filters may be designed so that they are discarded (disposable) when filled with dust, or are cleaned and reused. Permanent types have metal media that will withstand repeated washings, but they cost more than disposable types.

Stationary or Renewable Stationary air filters are manufactured in rectangular panels that are placed alongside each other and stacked, according to the size needed. The panels are removed and either replaced or cleaned when dirty. Renewable air filters consist of a roll mounted on a spool that moves across the airstream (Figure 12.19). The media is wound on a take-up spool, driven by a motor. The movement of the media is often controlled by a pressure switch which senses the pressure drop across the media. When the resistance increases to a set value because of the dirt collected, the motor moves the curtain,

Figure 12.19. Automatic renewable filter. (*Courtesy:* American Air Filter Co., Louisville, Kentucky.)

Figure 12.20. Electronic air cleaner. (*Courtesy:* American Air Filter Co., Louisville, Kentucky.)

exposing clean media. Renewable air filters are considerably more expensive than the stationary types, but maintenance costs are greatly decreased. Either fibrous materials or metal screens are used as media.

Electronic Air Cleaners (Figure 12.20). In this type there is no fibrous media to entrap dust. Dust particles are given a high voltage charge by an electric grid. A series of parallel plates are given the opposite electric charge. As the dust laden airstream passes between the plates the particles are attracted to the plates. The plates may be coated with a viscous material to hold the dust. After an interval of time the air cleaner must be removed from service in order to clean the plates and remove the dirt. Electronic air cleaners are expensive, but are very efficient for removing both large and very small particles.

12.21. SELECTION OF AIR CLEANERS

The selection of the proper air cleaner depends on the degree of contamination of the air to be cleaned and the cleanliness requirements.

For applications that require minimum cleanliness and low cost, inexpensive viscous impingement type disposable air filters would be used. A private residence or apartment house might be an example. For applications that require a greater degree of cleanliness, and where contamination is greater, perhaps intermediate efficiency dry-type filters would be used. Another choice might be to use electronic air cleaners, particularly where smoking is heavy. Often electronic cleaners are used in conjunction with a prefilter, a coarse viscous impingement cleanable filter that removes the large particles first, so that they do not cause quick build-up of dirt on the electronic air cleaner. This arrangement is popular in large commercial buildings.

Where removal of extremely small particles is critical, such as viruses, bacteria, or radioactive particles, HEPA filters are used. These are also usually backed up with a coarse prefilter to remove large particles.

For removing gases with objectionable odors from the air *activated carbon* (charcoal) filters are used. The carbon absorbs the gas molecules. These filters are sometimes used in restaurants to remove odorous gases resulting from cooking.

12.22. ENERGY REQUIREMENTS OF DIFFERENT TYPES OF AIR CONDITIONING SYSTEMS

A comparative energy use analysis of some of the major types of air conditioning systems will

be made in this section. This will illustrate the opportunities for energy conservation by the proper choice of a system for a given application.

A comparison will be made of energy requirements for constant volume reheat, dual duct, and variable air volume systems. A further comparison will be made with an air-water system such as the induction or fan-coil type. A typical office building will be specified. Some simplifications will be made to avoid unnecessary details that would detract from following the analysis. Of course, in an actual energy study every factor would be included.

The following are the design specifications:

Office building of 200,000 ft² area.

N, E, S, W zones are 25,000 ft² each. Interior zone is 100,000 ft².

Outside condition is 97 F DB, 74 F WB. Inside 78 F DB, 45% RH.

Ventilation air is 45,000 CFM.

Lights and power are 12 BTU ft².

Occupants 3300.

Building peak load is in July at 4 PM.

Air off cooling coils is at 54 F DB.

PEAK SENSIBLE HEAT GAINS FOR EACH ZONE, BTU/hr.

	N	E	S	W	I
Solar + trans.	150,000	800,000	450,000	1,000,000	
Lights	300,000	300,000	300,000	300,000	1,200,000
People	90,000	90,000	90,000	90,000	350,000
Totals	540,000	1,190,000	840,000	1,390,000	1,550,000

Sum = 5,510,000 BTU / hr

BUILDING PEAK SENSIBLE HEAT GAINS

	N	E	S	W	I
Solar + trans.	150,000	180,000	200,000	1,000,000	
Lights	300,000	300,000	300,000	300,000	1,200,000
People	90,000	90,000	90,000	90,000	350,000
Totals	540,000	570,000	590,000	1,390,000	1,550,000

Sum = 4,460,000 BTU / hr

Supply fan temperature rise is 4 F. Pressure is 6 in. w.g. Duct and return fan heat gains are neglected; RSHR is 0.88.

1. *Constant volume reheat system.* The air handling unit psychrometric processes are shown in Figure 12.4. Supply air temperature is 58 F if the air off cooling coil is 54 F DB. The air supply rate must satisfy the sum of each zone peak.

$$\text{CFM} = \frac{5,510,000}{1.1 \times (78 - 58)} = 250,000$$

The refrigeration capacity of the reheat system must satisfy the sum of the zone peaks

Sum of zone sensible peaks, BTU/hr	=	5,510,000
People latent	=	660,000
Outside air 45,000 × 4 5(37 7 − 29.0)	=	1,760,000
Fan heat 250,000 × 1.1 × 4	=	1,100,000
Refrigeration load, BTU/hr	=	9,030,000
	=	752 tons

The reheat system must furnish heat from the heating coil for all zones except those at peak loads. The heating required at design conditions is

$$5,510,000 - 4,460,000 = 870,000 \text{ BTU/hr}$$

2. *Dual duct system.* The air handling unit psychrometric processes are shown in Figure 12.7. The total air supply rate must satisfy the sum of each zone peak load. Furthermore, additional air is required because of leakage through closed dampers in the zone mixing box. The equation for finding total CFM is

$$\text{CFM}_t = \frac{\text{sum of zone peaks}}{1.1(t_r - t_c)}$$
$$\times \frac{1}{1 - x - x\left(\dfrac{t_w - t_r}{t_r - t_c}\right)}$$

where

CFM_t = total air supply rate, hot and cold ducts
t_r = average room temperature
t_c = cold air supply temperature at mixing dampers
t_w = warm air supply temperature at mixing dampers
x = fraction of air leakage through closed mixing damper

Using a leakage rate of 5%, and noting that $t_w = 85.3$ F DB (Figure 12.7)

$$\text{CFM}_t = \frac{5,510,000}{1.1(78 - 54)}$$
$$\times \frac{1}{1 - 0.05 - 0.05\left(\dfrac{85.3 - 78}{78 - 54}\right)}$$
$$= 223,000$$

Although the dual duct system must supply an air rate to satisfy each zone peak, the refrigeration capacity must only satisfy the building peak

Building sensible peak	=	4,640,000 BTU/hr
People latent	=	660,000
Outside air	=	1,760,000
Fan heat 223,000 × 1.1 × 4	=	980,000
Refrigeration load	=	8,040,000 BTU/hr
	=	670 tons

3. *Variable air volume system.* The air handling unit psychrometric processes are shown in Figure 12.9. The maximum air supply rate must satisfy only the building peaks

$$\text{CFM} = \frac{4,640,000}{1.1(78 - 58)} = 211,000$$

The refrigeration capacity must satisfy only the building peak, 670 tons. (We will neglect the slight difference in fan heat.)

The difference in the energy requirements of the three systems at full load is summarized as follows:

	Reheat	Dual duct	VAV
Tons of refrigeration	752	670	670
Refrigeration KW	677	603	603
CFM	250,000	223,000	211,000
Fan KW	200	178	169
Heating BTU / hr	870,000		
Total KW	877	781	772
Extra cost $ / hr	16.23	0.72	0

In the above estimates, figures of 0.9 KW/ton, $0.08/KWH, $9.00/10^6 BTU, and 0.8 KW/1000 CFM at 6 in. w.g. were used.

Note the huge extra expense and energy waste from using the reheat system. If we were to assume this energy difference for a full load equivalent of 1200 hours a year, the extra cost would be

Reheat $19,480

Dual duct   860

This is not the actual situation, however. The energy consumption differences are much greater at part load, both because the VAV system throttles air flow with load reduction and the reheat system must add even more heat. We will make a rough analysis of this situation. Assume a mild day with no solar, outside air, or transmission loads.

1. *Reheat system.* The cooling coil still cools all the air to set temperature, 54 F. The outside air load is the only one not required. The refrigeration load is therefore

Full load for reheat system	9,030,000 BTU/hr
Outside air design load	− 1,760,000
Reheat system part load	7,270,000 BTU/hr
	= 606 tons

The actual building peak load is

Peak design load	8,040,000 BTU/hr
Outside air design load	− 1,760,000
Solar + trans. design load	− 1,530,000
Building part load	4,750,000 BTU/hr
	= 396 tons

The reheat system must provide external heat source for all of the difference between the building peak and sum of zone peaks

$$7,270,000 - 4,750,000 = 2,520,000 \text{ BTU/hr}$$

2. *Dual duct system.* The refrigeration capacity required is the building peak part load, 396 tons, since the cold air is throttled as needed in each zone.

3. *VAV system.* The refrigeration capacity required is 396 tons, because the air supply rate is throttled as required. The total air suply rate is

$$CFM = \frac{\text{room sensible heat}}{1.1 \times 20} = \frac{3,110,000}{1.1 \times 20}$$
$$= 141,000$$

The differences in energy requirements at the part load condition are summarized as follows:

	Reheat	Dual duct	VAV
Tons of refrigeration	606	396	396
Refrigeration KW	545	356	356
CFM	250,000	223,000	141,000
Fan KW	200	178	113
Total KW	745	534	469
Heating BTU / hr	2,270,000		
Extra cost $ / hr	42.50	5.20	0

Assuming an operating period of 2000 hours a season with this part load condition as an average, the extra energy costs would be

Reheat $85,000
Dual duct   10,400

The part load condition selected of course does not necessarily represent the average of an actual installation. To carry out an accurate yearly energy use, hourly weather and equipment performance data are needed. However, the general conclusions from our analysis hold true.

We have not attempted to show variations in energy use due to types of controls selected. Other factors such as diversity or operation of reheat coils in the dual duct and VAV systems have not been included.

If the systems analyzed were used for winter heating, further sharp differences in energy consumption would appear for similar reasons. Many buildings have a separate radiation heating system for perimeter zones, however, which reduces the penalty of excessive use of air heating. This is particularly true during unoccupied hours, where otherwise fans would have to be operated continually.

We have not made a comparative analysis of the energy use of an air-water system. At full load this system has about the same energy requirements as the VAV system, because the room units handle only the load for their zones. There may be a small advantage in auxiliary energy use for the air-water system because only the primary (ventilation) air is moved. The additional pump energy is usually less than that of the greater air quantity. At lower outdoor temperatures, however, the situation may change significantly. All-air systems of course have as much air for cooling as needed in cold weather. For the air-water system, however, the primary air quantity is not adequate to cool zones with large heat gains, and chilled water must be used.

Consider the south side of the building on a sunny November day with an outdoor temperature of 40 F. With fan heat, the primary air supply temperature is 44 F. The peak refrigeration load is

Solar gain	= 1,200,000 BTU/hr
People sensible	= 90,000
Lighting	= 300,000
Primary air 1.1 × 5000 × 30	= −165,000
Transmission losses	= −255,000
Refrigeration load	= 1,170,000 BTU/hr
	= 98 tons

This is a heavy penalty to pay for using this type of system without heat recovery, because there will be many hours in the heating season when refrigeration is needed. Chilled water and condenser water pumps will also have to be operated.

12.23. ENERGY CONSERVATION

1. Systems that mix air (dual duct or multizone) or mix water (three pipe) may result in energy waste, although they can be designed to minimize the loss.

2. When using air-water systems, care should be taken that they are not producing opposite effects and therefore wasting energy. For example, in an induction system the primary air may be warm while at the same time chilled water is being distributed to the induction units. This should be avoided where possible by proper design and operation.

3. Reheat is unavoidably wasteful and should be avoided except for special applications, unless the reheat would come from otherwise wasted energy (see Chapter 15).

4. Systems should be designed and operated to use all outside air for cooling when it is adequate (see Chapter 15).

5. Replace or clean filters on a regular schedule to limit pressure losses to those recommended, thereby avoiding excessive fan power.

6. Clean coils regularly, thereby maintaining maximum heat transfer.

REVIEW QUESTIONS

1. Sketch and label all elements of a single zone air handling unit and a typical duct arrangement.

2. Sketch and label all elements of a reheat system arrangement and air handling unit.

3. Sketch and label all elements of a multizone system arrangement and air handling unit.

4. Sketch and label all elements of a dual duct system arrangement and air handling unit.

5. Sketch and label all elements of a VAV duct system arrangement and air handling unit.

6. Prepare a list of advantages and disadvantages of reheat, multizone, dual duct, VAV, and unitary systems.

7. Explain the following terms:
   A. Split system
   B Draw-through unit
   C Blow-through unit

8. List four purposes of air cleaning devices.

9. Explain the following terms related to air cleaners:
   A. Impingement
   B. Straining
   C. Electrostatic precipitation

10. List and explain the four methods of testing and rating air cleaners.

11. What is a HEPA air filter? What are its applications?

12. What type of air cleaner is used to remove undesired gases?

PROBLEMS

12.1. Select a 14 fin/in. cooling coil to cool 12,000 CFM of air from 82 F DB and 70 F WB to 55 F DB and 54 F WB. The coil face velocity is 600 FPM.

12.2. Select an 8 fin/in. coil for the same requirements as described in Problem 12.1.

12.3. A 4 row, 8 fin/in. cooling coil is handling air at a face velocity of 600 FPM. Air enters at 87 F DB, and 72 F WB and leaves at 59 F DB. What is the leaving air WB?

12.4. If the air velocity for the coil in Problem 12.3 is reduced to 400 FPM, assuming the same leaving DB, what is the leaving air WB?

13

Refrigeration Systems and Equipment

*A*n environmental control system that includes cooling and dehumidification will require a means of removing heat from the conditioned spaces. Because heat flows only from a higher to a lower temperature, a fluid with a temperature lower than the room design temperature must be made available, to which the excess room heat can be transferred. Refrigeration produces this low temperature fluid.

Occasionally a natural low temperature fluid is available. The ancient Roman rulers had slaves transport snow from the high mountains to cool their food and beverages. Cold well water has often been used in modern air conditioning systems. Many communities now restrict the use of well water for air conditioning, however, because of the depleted supply. Furthermore, well water temperatures are often 50 – 60 F, which is too high to accomplish adequate dehumidification. Another natural heat sink that

is used occasionally for cooling water is atmospheric air. In climates where the humidity is extremely low, evaporative cooling of air may reduce both the water and air temperature low enough so that either can be used for cooling (Chapter 7). Well water or evaporative cooling should be considered for refrigeration when available.

Usually there is no natural heat sink at a temperature lower than the desired space temperature when cooling is required. In this case refrigeration systems that require machinery are used to provide a cold fluid for cooling or dehumidification.

Vapor compression and absorption refrigeration systems are both used widely for producing refrigeration required for air conditioning. In this chapter we will explain how each system functions, the types of equipment used, and some equipment selection procedures. We will not *discuss the calculations related to the thermo-dynamic cycle or to compressor performance. Our purpose here will be to relate the refrigeration equipment to the complete air conditioning system. However, a further understanding of refrigeration theory is necessary for the well-trained air conditioning practitioner. Text-*

books on refrigeration should be consulted, such as the author's text Refrigeration Principles and Systems: An Energy Approach.

Other refrigeration methods, such as the air cycle, thermoelectric cooling, and steam jet refrigeration are not widely used in commercial air conditioning, and will not be discussed.

OBJECTIVES

After studying this chapter you will be able to:
1. Describe and sketch the vapor compression refrigeration system.
2. Identify the types of compressors, condensers, evaporators, and flow control devices.
3. Select packaged refrigeration equipment.
4. Describe and sketch the absorption refrigeration system.
5. Describe and sketch the heat pump system.

Vapor Compression Refrigeration System

13.1. PRINCIPLES

A schematic flow diagram showing the basic components of the *vapor compression refrigeration* system is shown in Figure 13.1. To aid in understanding it, some typical temperatures for air conditioning applications are indicated. Refrigerant fluid circulates through the piping and equipment in the direction shown. There are four processes (changes in the condition of the fluid) that occur as it flows through the system:

PROCESS 1-2. *At point (1), the refrigerant is in the liquid state at a relatively high pressure and*

high temperature. It flows to (2) through a restriction, called the flow control device *or* expansion device. *The refrigerant loses pressure going through the restriction. The pressure at (2) is so low that a small portion of the refrigerant* flashes *(vaporizes) into a gas. But in order to vaporize it must gain heat (which it takes from that portion of the refrigerant that did not vaporize), thus cooling the mixture and resulting in a low temperature at (2).*

PROCESS 2-3. *The refrigerant flows through a heat exchanger called the* evaporator. *This heat exchanger has two circuits. The refrigerant circulates in one, and in the other the fluid to be cooled (usually air or water) flows. The fluid to be cooled is at a slightly higher temperature than the refrigerant, therefore heat is transferred from it to the refrigerant, producing the cooling effect desired. The refrigerant boils because of the heat it receives in the evaporator. By the time it leaves the evaporator (4) it is completely vaporized.*

PROCESS 3-4. *Leaving the evaporator, the refrigerant is a gas at a low temperature and low pressure. In order to be able to use it again to achieve the refrigerating effect continuously, it must be brought back to the conditions at (1)—a liquid at a high pressure. The first step in this process is to increase the pressure of the refrigerant gas by using a* compressor. *Compressing the gas also results in increasing its temperature.*

Figure 13.1. The vapor compression refrigeration system.

PROCESS 4-1. The refrigerant leaves the compressor as a gas at high temperature and pressure. In order to change it to a liquid, heat must be removed from it. This is accomplished in a heat exchanger called the condenser. *The refrigerant flows through one circuit in the condenser. In the other circuit a cooling fluid flows (air or water), at a temperature lower than the refrigerant. Heat therefore transfers from the refrigerant to the cooling fluid, and as a result, the refrigerant condenses to a liquid (1).*

The refrigerant has returned to its initial state and is now ready to repeat the cycle. Of course the processes are actually continuous as the refrigerant circulates through the system.

13.2. EQUIPMENT

As noted in the explanation of how the vapor compression refrigeration system functions, the major equipment components are the *compressor, evaporator, condenser,* and *flow control de-* vice, some types of which will now be described. The complete refrigeration plant has many other additional components (e.g., valves, controls, piping) that will not be discussed in detail here.

13.3. EVAPORATORS

These may be classified into two types for air conditioning service—*dry expansion* (DX) evaporators or *flooded* evaporators. In the dry expansion type, refrigerant flows through tubing and there is no liquid storage of refrigerant in the evaporator. In the flooded type of evaporator, a liquid pool of refrigerant is maintained.

Dry expansion (DX) evaporators exist in two types—DX *cooling coils* or DX *chillers.* Cooling coils are used for cooling air and chillers for cooling water or other liquids. Flooded evaporators are called flooded chillers.

When cooling air, dry expansion (DX) cooling coils are used. The tubing is arranged in a serpentine coil form, and is finned to produce more heat transfer from a given length. The air flows

Figure 13.2. Flooded chiller. (Reprinted with permission from the *1979 Equipment ASHRAE Handbook & Products Directory*.)

Figure 13.3. Dry expansion chiller. (Reprinted with permission from the *1979 Equipment ASHRAE Handbook & Products Directory*.)

across the coils. Cooling coils are discussed in more detail in Chapter 12.

Evaporators for cooling water or other liquids are called *chillers*. In the *shell and tube* type a bundle of straight tubes is enclosed in a cylindrical shell. The chiller may be either the *flooded* type, with water circulating through the tubes and refrigerant through the shell (Figure 13.2) or *dry expansion*, with the reverse arrangement

(Figure 13.3). The shell can be made in one piece or can be constructed with bolted removable ends, called *heads*. In the latter case mechanical cleaning and replacement of individual tubes is possible. This construction is more expensive, however. Flooded chillers are generally used on the larger systems.

Figure 13.4. Reciprocating compressor construction. (*Courtesy:* Dunham-Bush, Inc.)

13.4. TYPES OF COMPRESSORS

Positive displacement compressors function by reducing the volume of gas in the confined space, thereby raising its pressure. *Reciprocating*, *rotary*, and *screw* compressors are positive displacement types. *Centrifugal* compressors function by increasing the kinetic energy (velocity) of the gas, which is then converted to an increased pressure by reducing the velocity.

13.5. RECIPROCATING COMPRESSOR

This is the most widely used type, available in sizes from fractional horsepower and tonnage up to a few hundred tons. Construction is similar to the reciprocating engine of a vehicle, with *pistons*, *cylinders*, *valves*, *connecting rods*, and *crankshaft* (Figure 13.4). The suction and discharge valves are usually a thin *plate* or *reed* that will open and close easily and quickly.

Open compressors have an exposed shaft to which the electric motor or other driver is attached externally. *Hermetic* compressors are manufactured with both compressor and motor sealed in a casing (Figure 13.5). In this way there is no possibility of refrigerant loss from leaking around the shaft. The motor is cooled by refrigerant in a hermetic compressor. Most modern open compressors use *mechanical seals*, rather than *packing seals*, to reduce refrigerant leakage. These seals are similar to those used in pumps, as discussed in Chapter 11.

13.6. ROTARY COMPRESSOR

This type has a *rotor* eccentric to the casing; as the rotor turns it reduces the gas volume and increases its pressure (Figure 13.6). Advantages of this compressor are that it has few parts, is of simple construction, and can be relatively quiet and vibration free. Small rotary compressors are

Figure 13.5. Cut-away view of hermetic reciprocating compressor. (*Courtesy:* Dunham-Bush, Inc.)

Figure 13.6. Sectional view of rotary compressor. (Reprinted with permission from the *1979 Equipment ASHRAE Handbook & Products Directory*.)

often used in household refrigerators and window air conditioners.

13.7. SCREW (HELICAL ROTARY) COMPRESSOR

Two meshing helical shaped *screws* rotate and compress the gas as the volume between the screws decreases toward the discharge end. This type of compressor has become popular in recent years due to its reliability, efficiency, and cost. It is generally used in the larger size ranges of positive displacement compressors, in capacities up to about 1000 tons. A screw compressor is shown in Figure 13.7.

13.8. CENTRIFUGAL COMPRESSOR

This type of compressor has vaned *impellers* rotating inside a casing, similar to a centrifugal

Figure 13.7. Cut-away view of screw compressor. (*Courtesy:* Dunham-Bush, Inc.)

Figure 13.8. Hermetic centrifugal refrigeration water chiller. (*Courtesy:* Machinery & Systems Division, Carrier Corp., Syracuse, N.Y.)

pump. The impellers increase the velocity of the gas, which is then converted into a pressure increase by decreasing the velocity. The nature of the centrifugal compressor makes it suitable for very large capacities, up to 10,000 tons; the impellers can be rotated at speeds up to 20,000 RPM, enabling it to handle large quantities of refrigerant. Hermetic centrifugal compressors as well as open compressors are available. Figure 13.8 shows a complete *hermetic centrifugal refrigeration water chiller*, with compress or, condenser, and evaporator.

13.9. CAPACITY CONTROL OF COMPRESSORS

The capacity of a compressor must be regulated to meet the load demand. Control is usually from a signal received from a thermostat or pressurestat (see Chapter 14). In a small reciprocating compressor, capacity is often changed simply by starting and stopping the compressor.

In larger multicylinder compressors a number of steps of capacity can be achieved. In one method the refrigerant gas is *bypassed* around the compressor when less capacity is called for. This method requires a relatively high power input for low capacity. A more efficient load reduction method is accomplished by holding the suction valve open when a reduction in capacity is called for. The cylinder is then simply idling, and a significant reduction in power input results. Mechanical devices called *unloaders*, automatically controlled from a load signal, are used to open the suction valves.

A reliable method of reducing centrifugal compressor capacity is to use *inlet guide vanes*. This is a set of adjustable vanes in the compressor suction that are gradually closed to reduce the volume of refrigerant gas compressed, thus reducing the capacity. The use of inlet guide vanes lessens a problem of centrifugal compressor operation called *surging*. If the gas flow rate is reduced by throttling with a butterfly-type discharge damper, a point will be reached where instability occurs in which the gas is constantly surging back and forth through the compressor. This is a very serious event that could damage the machine. Inlet guide vanes avoid this by curving the flow direction of the gas in an efficient manner that permits capacity reduction down to about 15% of full load without surging.

For centrifugal compressors that are driven by variable speed prime movers, speed reduction is a convenient method of capacity reduction. Both inlet guide vane and speed control are relatively efficient methods of capacity control, the power input decreasing considerably with capacity. Below about 50% capacity, however, the efficiency falls off rapidly. This is one reason why it is desirable to use multiple centrifugal machines in an installation, if practical.

13.10. PRIME MOVERS

Compressors can be driven by electric motors, reciprocating engines, or by steam or gas tur-

bines. Electric motors are most commonly used because of the convenience and simplicity. However, on very large installations, particularly with centrifugal compressors, steam or gas turbines are often used. The high rotating speed of the turbine often matches that of the compressor, whereas expensive speed-increasing gears may be needed when motors are used. The relative energy costs of electricity, steam, or gas often determine which prime mover will be used. In the Middle East, natural gas from the well (which might otherwise be wasted) is often used in gas turbines that drive large centrifugal machines.

13.11. CONDENSERS

The condenser rejects from the system the energy gained in the evaporator and the compressor. Atmospheric air or water are the two most convenient heat sinks to which the heat can be rejected.

In the *air cooled condenser* (Figure 13.9) the refrigerant circulates through a coil and air flows across the outside of the tubing. The air motion may be caused by natural convection effects when the air is heated, or the condenser can include a fan to increase the air flow rate, resulting in greater capacity. Air-cooled condensers are nor-

Figure 13.9. Air cooled condenser. (*Courtesy:* Dunham-Bush, Inc.)

mally installed outdoors. They are available in sizes up to about 50 tons.

Water-cooled condensers are usually of shell and tube construction, similar to shell and tube evaporators. Water from lakes, rivers, or wells is sometimes used when available. Usually, however, natural sources of water are not sufficient, and the water must be recirculated through a *cooling tower.*

Evaporative condensers (Figure 13.10) reject heat to the atmosphere as do air cooled condensers, but by spraying water on the coils some heat is transferred to the water as well as the air, increasing the capacity of the condenser. A pump, piping, spray nozzles, and collection sump are required for the water circulating system. Fans are used to force the air through the unit. Evaporative condensers can be installed indoors as well as outdoors, by using ductwork to discharge the exhaust air outside.

The capacity of condensers must be controlled to maintain the condensing pressure within certain limits. Higher condensing pressures result in more power use, and extremely high pressures can damage the equipment. On the other hand, if the pressure is too low, the flow control device will not operate satisfactorily. An automatic valve regulating water flow rate is a convenient way of controlling capacity of water-cooled condensers. Air-cooled condensers are often controlled by reducing air flow across the coils, through use of dampers or cycling the fan. The control is usually in response to a change in condensing pressure.

Proper water treatment is important for maintaining the capacity of water-cooled condensers. Manufacturers rate water-cooled condenser and chiller capacity on the basis of a water *fouling factor*—a number that represents the thermal resistance of the water film on the tubes. A value of 0.0005 is considered clean water, and ratings are often based on this value. The water treatment should prevent formation of scale that will increase the thermal resistance, resulting in a decrease in refrigeration capacity and an increase in energy required.

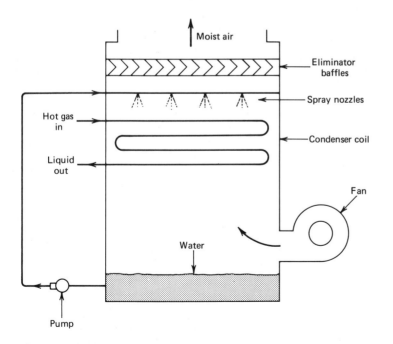

Figure 13.10. Evaporative condenser.

13.12. FLOW CONTROL DEVICES

The restricting device that causes the pressure drop of the refrigerant also regulates the refrigerant flow according to the load. Some of the devices available are the *capillary tube*, *thermostatic expansion valve*, and the *low side float valve*. The first two are used with dry expansion evaporators; the low side float valve is used in flooded chiller evaporators.

The *capillary tube* is a very small diameter tube of considerable length, which thus causes the required pressure drop. It is used often in small units (e.g., domestic refrigerators and window air conditioners) because of its low cost and simplicity.

The *thermostatic expansion valve* (TEV), shown in Figure 13.11, is widely used in dry expansion systems. The small opening between the valve seat and disc results in the required pressure drop. It also does an excellent job of regulating flow according to the need. The operation of a TEV is shown in Figure 13.12. A bulb filled with a fluid is strapped to the suction line, and thus senses the suction gas temperature. This bulb is connected to the valve by a tube in a manner so that the pressure of the fluid in the bulb tends to open the valve more, against a closing spring pressure. If the load in the system increases, the refrigerant in the evaporator picks

Figure 13.11. Cut-away view of thermostatic expansion valve, internally equalized type. (*Courtesy:* Sporlan Valve Co.)

Figure 13.12. Operation of internally equalized thermostatic expansion valve.

up more heat and the suction gas temperature rises. The pressure of the fluid in the bulb increases as its temperature rises, and it opens the valve more. This increases the refrigerant flow needed to handle the increased load. The reverse of all these events occurs when the refrigeration load decreases.

It is important that the refrigerant vapor leaving the evaporator be a few degrees above the saturation temperature (called *superheat*) to ensure that no liquid enters the compressor, which might result in its damage. This is achieved by adjusting the spring pressure to a value that prevents the bulb pressure from opening the valve more unless the gas leaving the evaporator is superheated.

The *internally equalized* TEV has a port connecting the underside of the diaphragm chamber to the valve outlet (Figures 13.11 and 13.12). This neutralizes the effect of any change in evaporator pressure on the balance between spring and bulb pressure. If there is a larger pressure drop in the evaporator, however, this would result in a reduction in superheat. This problem is solved by using an *externally equalized* valve which has a connection to the evaporator outlet rather than the inlet.

A *low side float valve*, as shown in Figure 13.8, is a flow control device that is used with flooded chillers. If too much liquid refrigerant accumulates because flow is not adequate, the float rises and a connecting linkage opens the valve, allowing more flow.

13.13. SAFETY CONTROLS

All refrigeration systems include a number of safety control devices to protect the equipment. The devices required for each system must be determined in each case according to the need. A brief listing of some of the available safety control devices follows.

A **high pressure cut-out** stops the compressor when the refrigerant discharge pressure exceeds a safe limit.

A **low pressure cut-out** stops the compressor when the refrigerant suction pressure is below a safe limit. Usually this is intended as a temperature safety device. The pressure setting on the device corresponds to a temperature at which water freeze-up might occur.

A **low temperature cut-out** senses refrigerant temperature on the low side directly, and serves to protect against freeze-up.

A **low oil pressure cut-out** stops the compressor when lubricating oil pressure is inadequate.

A **flow switch** will stop the compressor when chilled water (or condenser water) flow is inadequate.

When the compressor stops in response to a thermostat, refrigerant may continue to flow to the evaporator due to a vapor pressure difference between the condenser and evaporator. It is not desirable to have the evaporator filled with liquid refrigerant during shutdown because this increases the likelihood of liquid entering the compressor. It also increases the amount of refrigerant absorbed in the crank case oil, thus reducing the lubricating effectiveness of the oil. The problem is solved by using *pump down control*. Instead of having a thermostat control the compressor operation, it controls a *solenoid valve* in the liquid line. This cuts off flow to the compressor. The compressor continues to operate for a time, pumping out the refrigerant from the evaporator. The compressor stops when its low pressure cut-out setting is reached.

13.14. PACKAGED REFRIGERATION EQUIPMENT

Compressors, condensers, evaporators, and accessories are each available separately from manufacturers for selection, purchase, and installation. However, these components may also be available already assembled (packaged) in the factory. There are a number of advantages of using packaged equipment. The components are selected and matched in capacity by the manufacturer, so that they will perform properly together. Installation costs are reduced, as each component does not have to be installed and aligned separately. Controls and interconnecting piping are factory installed, further reducing field costs. The assembled equipment is usually factory tested, reducing the likelihood of operating problems that would have to be corrected on the job. Packaged equipment is available in various combinations, some of which are mentioned below.

Condensing Units The package of compressor and condenser with interconnecting piping and controls is called a *condensing unit* (Figure 13.13). Both water- and air-cooled condensing units are available. Air-cooled units are installed outdoors. When the air cooled condensing unit is located outdoors and the air handling unit and evaporator coil are located indoors, the air conditioning system is called a *split system*. This arrangement is popular for residential air conditioning systems.

Figure 13.14. Packaged chiller. (*Courtesy:* Dunham-Bush, Inc.)

Compressor-Chiller Unit This unit consists of compressor, water chiller, interconnecting piping, and controls. It is often used with a remote air-cooled condenser.

Packaged Chiller This unit, shown in Figure 13.14, contains the complete refrigeration package: compressor, condenser, water chiller, piping and controls, ready to operate when put in place and when external connections are made.

13.15. SELECTION

Refrigeration equipment is selected from manufacturers' ratings after performance requirements are determined. Although the compressor, condenser, and evaporator can be chosen separately, one of the packaged combinations is often used, so we will limit our explanation to these. In any case, selection of individual components is a similar process.

Air-Cooled Condensing Unit The following data are needed for selection of an air-cooled condensing unit:

1. Refrigeration capacity required (load)
2. Condenser ambient temperature
3. Saturated suction gas temperature

Figure 13.13. Water cooled condensing unit. (*Courtesy:* Dunham-Bush, Inc.)

TABLE 13.1. AIR COOLED CONDENSING UNIT RATINGS

CAPACITY DATA* (60 HZ.)**
CONDENSING UNITS-R22

Model	Suction Temp °F	Ambient Temperature °F												EER @ ARI Base Rating Cond
		90°F		95°F		100°F		105°F		110°F		115°F		
		Tons	K.W.	Tons	K.W.	Tons	K.W.	Tons	K.W.	Tons	K.W.	Tons	K.W.	
RCU-008S	30	5.8	7.4	5.6	7.5	5.3	7.6	5.1	7.7	4.7	7.8	4.6	7.9	
	35	6.6	7.8	6.5	8.0	6.1	8.1	5.7	8.2	5.5	8.3	5.3	8.5	11.2
	40	7.4	8.2	7.0	8.4	6.8	8.6	6.6	8.7	6.3	8.9	5.9	9.2	
	45	8.0	8.6	7.9	8.9	7.6	9.1	7.4	9.3	7.0	9.6	6.7	9.8	
RCU-008SS†	30	6.4	7.0	6.2	7.1	5.8	7.2	5.6	7.4	5.5	7.6	5.2	7.7	
	35	7.0	7.4	6.8	7.6	6.6	7.8	6.4	7.9	6.2	8.0	5.8	8.2	12.4
	40	7.7	7.7	7.5	7.9	7.0	8.1	6.9	8.3	6.7	8.5	6.5	8.7	
	45	8.4	8.1	8.0	8.4	7.8	8.6	7.6	8.8	7.4	8.9	7.0	9.1	
RCU-010SS†	30	7.7	8.4	7.5	8.6	7.3	8.7	6.9	8.9	6.7	9.1	6.4	9.3	
	35	8.5	8.9	8.3	9.1	8.0	9.2	7.8	9.4	7.5	9.6	7.3	9.8	12.2
	40	9.2	9.3	9.0	9.5	8.8	9.8	8.6	10.0	8.3	10.2	8.0	10.4	
	45	10.0	9.8	9.8	10.1	9.6	10.3	9.4	10.5	9.0	10.8	8.8	11.0	
RCU-010T	30	8.5	11.5	8.2	11.7	7.9	11.9	7.5	12.0	7.2	12.1	7.0	12.2	
	35	9.4	12.4	9.1	12.6	8.7	12.8	8.4	13.0	8.0	13.1	7.8	13.2	9.7
	40	10.3	13.2	10.0	13.4	9.6	13.6	9.3	13.8	8.9	13.9	8.6	14.1	
	45	11.3	14.0	10.9	14.2	10.5	14.6	10.2	14.7	9.9	14.8	9.5	15.1	

The load is a result of the cooling load calculations. The condenser ambient temperature is usually the outdoor design temperature in summer. Sometimes the condenser is located where the ambient temperature may be even greater than design temperature, and this should be considered. The compressor saturated suction temperature will be equal to the evaporator temperature minus an allowance. This allowance accounts for the pressure drop in the suction line, expressed as an "equivalent temperature drop," usually 2 F. Table 13.1 is an example of air cooled condensing unit ratings.

EXAMPLE 13.1

The refrigeration load for the air conditioning system of a branch of the Big Bank in Tulsa, Oklahoma, is 12 tons. The system uses refriger-

TABLE 13.1 *(Continued)*.

CAPACITY DATA* (60 HZ.)**
CONDENSING UNITS-R22

Model	Suction Temp °F	Ambient Temperature °F												EER @ ARI Base Rating Cond
		90°F		95°F		100°F		105°F		110°F		115°F		
		Tons	K.W.	Tons	K.W.	Tons	K.W.	Tons	K.W.	Tons	K.W.	Tons	K.W.	
RCU-015SS†	30	10.9	12.6	10.6	12.8	10.2	13.0	9.7	13.4	9.5	13.6	9.4	13.8	11.8
	35	12.1	13.2	11.8	13.7	11.4	13.9	11.1	14.2	10.8	14.6	10.5	14.8	
	40	13.3	14.1	12.8	14.2	12.4	14.6	12.0	14.9	11.9	15.4	11.4	15.6	
	45	**14.5**	**14.8**	14.5	15.2	13.8	15.4	13.3	16.0	12.9	16.2	12.5	16.4	
RCU-015T	30	11.2	14.8	10.8	15.1	10.2	15.4	9.7	15.5	9.2	15.7	8.8	15.9	10.3
	35	12.7	15.7	12.1	16.1	11.6	16.4	11.1	16.7	10.7	17.0	10.1	17.2	
	40	14.1	16.8	13.5	17.2	13.0	17.7	12.4	17.9	11.9	18.2	11.3	18.6	
	45	**15.5**	**18.0**	14.9	18.4	14.3	18.8	13.8	19.2	13.3	19.6	12.8	19.9	
RCU-020T	30	14.5	16.8	14.1	17.1	13.5	17.5	13.0	17.9	12.4	18.3	11.9	18.7	11.6
	35	16.1	17.8	15.6	18.1	15.0	18.5	14.5	19.0	14.0	19.2	13.4	19.6	
	40	17.4	18.5	16.9	19.0	16.3	19.5	15.8	20.0	15.4	20.3	14.6	20.4	
	45	**19.0**	**19.7**	18.5	21.0	17.8	20.6	17.3	21.0	16.8	21.4	16.4	21.9	
RCU-020SS†	30	15.7	19.1	15.2	19.5	14.9	19.9	14.3	20.2	14.0	20.5	13.4	21.0	11.1
	35	17.4	20.0	16.8	20.5	16.4	20.9	15.8	21.4	15.5	21.9	14.9	22.4	
	40	19.0	21.3	18.5	21.7	17.9	22.4	17.4	22.8	17.1	23.2	16.3	23.7	
	45	**20.7**	**22.4**	20.1	22.8	19.6	23.5	19.0	24.1	18.5	24.5	17.8	25.0	

NOTES: ARI Base rating conditions 90° ambient, 45° suction temperature. These are indicated by boldface type.
† All models with the suffix 'SS' denote single D/B-Metric accessible Hermetic compressors.
* For capacity ratings at 85°F ambient temperature multiply the ratings of 90°F ambient by 1.03 x Tons and .97 x KW.
** For 50 hertz capacity ratings, derate above table by .85 mutliplier.
(Courtesy of Dunham-Bush, Inc.)

ant R-22, evaporating at 42 F. Select an air-cooled condensing unit.

Solution. A unit will be selected from Table 13.1. Required capacity = 12 tons. Ambient temperature = 98 F (Table A.6). Allow a friction loss in the suction line equivalent to 2 F. There-fore, the saturated suction temperature = 42 − 2 = 40 F.

From Table 13.1, the unit selected is a Model RCU-0155SS air-cooled condensing unit. Capac-ity = 12.4 tons at saturated suction temperature = 40 F, ambient temperature = 100 F. Power input = 14.6 KW. (The capacity at 98 F will be slightly higher than at 100 F ambient, as noted from the table.)

Packaged Water Chiller The following data are needed for selection of a packaged water chiller:

1. Refrigeration load

2. Condenser water temperature leaving unit

3. Condenser water temperature rise

4. Chilled water temperature leaving unit

5. Chilled water temperature drop

6. Fouling factor

The load is determined from cooling load calculations. Condenser water leaving temperature is usually about 5–15 F above ambient wet bulb temperature if a cooling tower is used. The water temperature rise is usually selected between 8–15 F. The leaving chilled water temperature will depend on the cooling coil selection for the air handling equipment. Chilled water temperature ranges of 8–12 F are common.

No exact figures on temperature changes are given because many choices are possible. The designer must frequently try different combina-

TABLE 13.2. PACKAGED WATER CHILLER RATINGS

| Model | Lvg. Chilled Water Temp. °F | Condenser Entering Water Temp. °F | | | | | | | | | | | | | | |
| | | 75° | | | 80° | | | 85° | | | 90° | | | 95° | | |
		Cap. Tons	KW	Cond. GPM	Cap. Tons	KW	Cond. GPM	Cap. Tons	KW	Cond. GPM	Cap. Tons	KW	Cond. GPM	Cap. Tons	KW	Cond. GPM
PCW010T	42	10.3	8.2	30.0	10.1	8.5	29.7	9.9	8.8	30.3	9.7	9.2	29.6	9.5	9.6	28.9
	44	10.7	8.3	30.9	10.5	8.6	30.7	**10.3**	**9.0**	**30.5**	10.0	9.3	29.4	9.9	9.7	29.4
	45	10.9	8.3	31.8	10.7	8.7	31.3	10.5	9.0	31.0	10.2	9.3	30.0	10.1	9.7	29.4
	46	11.1	8.9	32.0	10.9	8.7	31.8	10.7	9.1	31.5	10.4	9.4	30.8	10.2	9.8	29.7
	48	11.5	8.5	33.1	11.3	8.8	32.8	11.1	9.3	32.6	10.8	9.6	31.6	10.6	9.9	30.9
	50	11.9	8.6	34.5	11.7	9.0	34.0	11.5	9.4	33.6	11.2	9.7	32.7	10.9	10.0	32.0
PCW015T	42	15.3	13.0	43.5	14.9	13.4	42.4	14.5	13.8	41.7	14.1	14.2	40.5	13.6	14.6	39.3
	44	15.9	13.1	45.0	15.5	13.5	44.0	**15.0**	**14.0**	**43.4**	14.7	14.3	41.1	14.2	14.9	40.7
	45	16.2	13.2	45.8	15.8	13.6	44.8	15.4	14.0	44.1	15.0	14.5	42.6	14.5	15.0	41.2
	46	16.5	13.3	46.3	16.1	13.7	45.6	15.7	14.1	44.8	15.3	14.6	43.4	14.8	15.1	42.0
	48	17.1	13.4	47.3	16.7	13.8	46.6	16.3	14.2	46.2	15.8	14.7	44.7	15.4	15.3	43.6
	50	17.6	13.6	48.1	17.3	14.0	47.8	16.9	14.4	47.2	16.4	14.9	45.7	15.9	15.5	44.7
PCW020T	42	19.2	15.8	52.9	18.6	16.3	52.3	18.0	16.8	51.6	17.4	17.3	49.8	16.7	17.8	48.5
	44	19.9	16.0	54.9	19.3	16.5	53.7	**18.6**	**17.0**	**52.6**	18.0	17.5	50.7	17.4	18.0	49.6
	45	20.3	16.1	56.1	19.7	16.6	54.9	19.0	17.1	53.6	18.4	17.6	51.9	17.7	18.1	50.6
	46	20.6	16.1	56.5	20.0	16.7	55.3	19.3	17.2	54.1	18.7	17.7	52.6	18.0	18.2	51.7
	48	21.3	16.3	57.4	20.7	16.8	56.8	20.0	17.3	56.2	19.4	17.8	54.8	18.7	18.3	53.9
	50	22.0	16.4	58.9	21.3	16.9	58.6	20.6	17.4	58.4	20.0	17.9	57.0	19.3	18.5	56.2

TABLE 13.2 (Continued).

Model	Lvg. Chilled Water Temp. °F	Condenser Entering Water Temp. °F														
		75°			80°			85°			90°			95°		
		Cap. Tons	KW	Cond. GPM	Cap. Tons	KW	Cond. GPM	Cap. Tons	KW	Cond. GPM	Cap. Tons	KW	Cond. GPM	Cap. Tons	KW	Cond. GPM
PCW025T	42	24.1	19.7	67.9	23.0	20.3	66.5	22.9	20.9	65.1	22.3	21.4	63.1	21.6	22.0	61.7
	44	25.0	19.8	70.7	24.4	20.5	69.3	**23.7**	**21.0**	**68.2**	23.0	21.7	66.0	22.4	22.3	64.9
	45	25.5	19.9	71.8	24.8	20.6	70.5	24.1	21.2	69.1	23.4	21.8	67.2	22.8	22.4	65.9
	46	25.9	20.0	72.8	25.2	20.7	71.7	24.5	21.5	70.4	23.8	21.9	68.5	23.1	22.6	66.9
	48	26.7	20.2	75.1	26.0	20.9	73.8	25.3	21.7	72.5	24.7	22.2	70.6	24.0	22.9	69.1
	50	27.5	20.4	78.6	26.8	21.1	77.1	26.1	21.9	75.2	25.5	22.5	73.1	24.8	23.2	71.6
PCW030T	42	28.3	23.2	80.0	27.5	23.9	79.6	26.6	24.6	79.2	25.9	25.3	77.0	25.1	26.0	75.5
	44	29.3	23.4	83.2	28.4	24.2	82.6	**27.6**	**25.0**	**82.5**	26.7	25.7	79.8	26.0	26.4	78.3
	45	29.8	23.6	85.3	29.0	24.3	84.4	28.1	25.1	84.0	27.3	25.9	81.6	26.5	26.6	79.9
	46	30.3	23.7	86.3	29.7	24.5	85.2	28.5	25.2	85.5	27.8	26.1	83.1	26.9	26.9	81.5
	48	31.2	23.9	88.2	30.5	24.8	87.9	29.6	25.6	87.6	28.7	26.5	85.2	27.8	27.3	83.6
	50	32.1	24.2	88.7	31.3	25.1	90.2	30.6	26.0	88.5	29.7	26.9	86.1	28.7	27.8	85.8
PCW040T	42	41.6	35.0	116.1	40.0	36.0	113.8	38.9	37.2	111.3	37.5	38.4	109.3	36.4	39.6	108.4
	44	43.0	35.4	118.8	41.7	36.5	117.1	**40.4**	**37.6**	**115.4**	39.0	38.9	112.8	38.0	40.2	112.0
	45	43.7	35.7	120.9	42.4	36.8	119.2	41.1	37.9	117.4	39.7	39.2	115.2	38.6	40.4	114.1
	46	44.4	36.0	123.1	43.1	37.1	121.1	41.8	38.0	118.8	40.4	39.4	116.6	39.2	40.6	115.5
	48	45.9	36.6	127.0	44.6	37.8	124.8	43.3	39.0	122.5	41.9	40.2	120.1	40.6	41.3	118.8
	50	47.3	37.1	131.1	46.0	38.4	128.9	44.8	39.6	126.3	43.4	40.8	123.8	42.0	42.0	122.6

* Boldface type indicates ARI rating condition.

NOTES:
1. Ratings are based on 10°F chilled water temperature range. Ratings are applicable for 6ᵛ to 14ᵛ range.
2. Condenser water flow rate data are based on tower water with a 10° rise. See Selection Procedure (page 6) for note regarding city water application.
3. Ratings are based on .0005 fouling factor in the chiller and condenser. For .001 condenser fouling factor multiply capacity shown in ratings by .98 and kW by 1.03. For other fouling factor ratings consult factory.
4. 50 Hz. units are full capacity except PCW040T which is 5/6 capacity.
5. Direct interpolation for conditions between ratings is permissible. Do not extrapolate.
(Courtesy of Dunham-Bush, Inc.)

tions of values, selecting the equipment each time, to find which will result in the best choice. Computer programs for equipment selection provided by the manufacturer are very useful for this task. Table 13.2 is an example of packaged water chiller ratings.

EXAMPLE 13.2

A package water chiller is required for the air conditioning system of the Royal Arms Apartments. The load is 27 tons. Chilled water is cooled from 55 F to 45 F. Condenser water enters at 85 F and leaves at 95 F. The condenser and chilled water fouling factors are 0.0005. Select a suitable unit.

Solution. A unit will be selected from Table 13.2. The fouling factor is a number that describes the cleanliness of the water. The size of the condenser required will depend on this. Table 13.2 is based on a water fouling factor of 0.0005, as noted, so no correction for this will be necessary. If the fouling factor is different, tables from the manufacturer show corrections to the selection. The required conditions are:

Capacity = 27 tons

Leaving chilled water temperature = 45 F

Chilled water temperature drop = 10 F

Entering condenser water temperature = 85 F

Condenser water temperature rise = 10 F

The unit chosen for these requirements is a Model PCW 030T water chiller, which has a capacity of 28.1 tons at required conditions. Power input is 25.1 KW.

Note that Table 13.2 indicates both a reduction in refrigeration capacity and an increase in power required, if the fouling factor of the condenser increases to 0.001. There is a loss of 2% in refrigeration and an increase of 3% in power, resulting in a net increase of 5% in energy use for a given capacity. This points up the importance of maintaining a clean condenser to conserve energy.

Additional information about the equipment is usually required, such as dimensions and weight, and water pressure drop through the chiller and condenser. This can be found in the manufacturer's catalog.

13.16. ENERGY EFFICIENCY

When selecting refrigeration equipment, it will often be found that more than one unit will have the capacity needed. In this case it is useful to know which will give the "best" performance. The most desirable choice is the unit that would produce the most refrigeration with the lowest power input. This can be measured by a performance factor called the *Coefficient of Performance* (COP). The COP is defined as:

$$COP = \frac{refrigeration\ capacity}{equivalent\ power\ input\ to\ compressor}$$

The higher the COP of a refrigeration unit, the less power is required for a given refrigeration requirement. The COP is thus a useful figure in comparing equipment to minimize energy consumption. This will be discussed in more detail in Chapter 15, together with another efficiency measure called the *Energy Efficiency Ratio* (EER).

EXAMPLE 13.3

Determine the coefficient of performance for the package chiller of Example 13.2.

Solution

$$Capacity = 28.1\ tons \times 12,000\ \frac{BTU/hr}{ton}$$

$$= 337,200\ BTU/hr$$

$$Power\ input = 25.1\ KW \times 3410\ \frac{BTU/hr}{KW}$$

$$= 85,590\ BTU/hr$$

Using the equation for the COP

$$COP = \frac{refrigeration\ capacity}{equivalent\ power\ input}$$

$$= \frac{337,200\ BTU/hr}{85,590\ BTU/hr}$$

$$= 3.94$$

The COP found could be compared with values obtained for other possible selections, to see if an improved performance is possible without sacrificing other benefits.

13.17. INSTALLATION OF REFRIGERATION CHILLERS

The procedures for installing specific refrigeration equipment are furnished by the manufacturer. It is not the intent here to either repeat or supersede such instructions. Each manufacturer and each piece of equipment has individual features that require detailed installation instructions. These instructions are often very lengthy. If they were repeated here, they would soon be forgotten by the student who does not regularly carry out these procedures. Therefore, we will discuss here some general points of installation practice that apply to most situations.

1. Check machine for damage or refrigerant leaks (leak detecting devices and their use are described in refrigeration service manuals).

2. Locate chillers and condensers with removable tubes to provide adequate clearance on one end to allow removal of the tube bundle, and on the other for removal of water box heads.

3. Allow clearance on all sides of equipment for comfortable maintenance (3 or 4 ft minimum).

4. Allow adequate clearance in front of control panels for operation and good visibility.

5. Provide vibration isolation supports under compressors and prime movers. Rubber, cork, and springs are some types available. Consult the manufacturer for the proper choice.

6. Install anchor bolts in floor or base and anchor machine. If there is any doubt about whether the floor is adequately strong for the machine or whether a special base is needed, a structural engineer must be consulted.

7. Make water, electrical, and control connections so as not to block access to the machine.

13.18. COOLING TOWERS

Operation When water cooled condensers are used in the refrigeration plant, a steady supply of cooling water must be made available. For reasons already explained, natural sources of water are usually limited. In this case we must arrange to cool the heated water after it passes through the condenser, and then return it to the condenser.

The *cooling tower* is the equipment that accomplishes this. It transfers heat from the condenser water to the atmospheric air (Figure 13.15). Most of the heat transfer is accomplished by the evaporation of a small percentage of the condensing water into the atmosphere. The heat required for evaporation is taken from the bulk of condenser water, thus cooling it. Water from the condenser is pumped to the top of the cooling tower and sprayed down into the tower. The tower has internal baffles called *fill*, which break up the water into finer droplets when the water splashes onto the fill. This improves the heat transfer. The cooled water collects in a basin and is then recirculated to the condenser.

In addition to the water lost due to the evaporative cooling, there are two other causes of

Figure 13.15. Induced draft cooling tower. (*Courtesy:* The Marley Cooling Tower Co.)

water loss. *Drift* loss results from wind carrying water away with the air. *Blowdown* loss results from draining off and discarding a small portion of the water from the basin. This must be done at regular intervals in order to prevent a continual accumulation of minerals that would otherwise occur from the evaporation and drift losses. The losses require provision for *makeup* water. This is done by providing a make-up water supply to the basin, controlled by a float valve level.

Types and Construction The *atmospheric tower* is a type of tower where the air circulation results from air being warmed in the tower and thereby rising from natural convection. The amount of air that will circulate from this effect is quite limited, and atmospheric towers are not often used today. *Mechanical draft* towers use fans to create a high air flow rate. The *induced draft* fan type has the fan located at the tower outlet, whereas the *forced draft* fan type blows the air through (Figure 13.16).

When the air and water move in opposite directions, the tower is called a *counterflow* type. When the air and water move at right angles to each other the tower is called a *crossflow* type. Figure 13.17 shows this difference. There is not necessarily an operating advantage in practice of one type over another. However, sometimes a crossflow tower will be lower in height (although

Figure 13.16. Forced and induced draft fan arrangements for cooling tower.

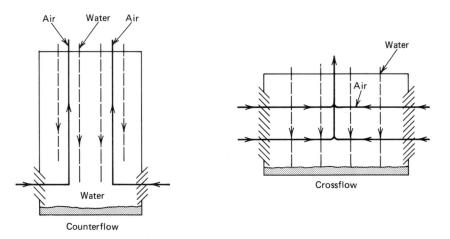

Figure 13.17. Counterflow versus crossflow of air and water in cooling tower.

Figure 13.18. Diagram illustrating refrigeration by absorption.

bigger in length or width) than a counterflow tower for the same capacity. Lower height may be preferable when installed on a roof.

The tower *siding* may be wood, galvanized steel, or plastic. The structural framework may be wood or steel. Redwood is ideal because water will not cause its deterioration. The *fill* may be wood, metal, or plastic. A row of baffles called *eliminators* are provided near the tower outlet to catch and prevent excessive loss of water droplets.

The capacity of a cooling tower depends on the rate of water evaporation. This rate decreases with higher water vapor content (humidity) in the ambient air. Therefore, the higher the ambient wet bulb temperature, the less the capacity of the tower.

Absorption Refrigeration System

Absorption refrigeration machines are often used for large air conditioning systems. The absence of a compressor usually has the advantages of less vibration, noise, and weight than with a vapor compression machine.

13.19. PRINCIPLES

The absorption system uses the principle that some gases will be absorbed by certain other substances. There are many pairs of substances that have this affinity for one another. We are all aware of how table salt absorbs water vapor from the air, thus making it difficult to pour. Another combination is *lithium bromide* (LiBr) and water; lithium bromide will absorb large quantities of water. Because this pair is used in many absorption systems, we will refer to them in our explanation.

Consider a tank partially filled with a *concentrated* liquid solution of lithium bromide (concentrated means that it contains very little water) as shown in Figure 13.18. The space above the liquid is evacuated of any gas, as much as possible, leaving a very low pressure. Water is then sprayed into the tank. Because of the low pressure some of the water will evaporate, requiring heat to do so. A coil circulating water is located under the evaporating sprays. This water furnishes the heat needed for the evaporating spray, and is thereby chilled. The temperature at which the spray water evaporates will depend on the pressure in the tank, according to the saturation pressure-temperature relations of water.

EXAMPLE 13.4

The pressure maintained in the evaporator of a LiBr-water absorption refrigeration machine is 0.147 psia. What is the refrigerant evaporating temperature?

Solution. Water is the refrigerant. From Table A.3, the evaporating (saturation) temperature of water at 0.147 psia is 45 F.

The lithium bromide absorbs the water as both solutions make contact. The lithium bromide eventually absorbs all of the water it can hold, however, and no longer is effective. The water vapor quantity will build up in the tank, raising its pressure, and therefore increasing its evaporating temperature above useful refrigeration temperatures.

In order to have a practical absorption refrigeration system that will operate continuously, the diluted solution of lithium bromide must be reconcentrated and used again. We will now explain how the actual system functions, as shown in Figure 13.19. Typical operating temperatures and pressures are indicated on the diagram.

The *evaporator* operation is as described previously. Spray water (the refrigerant) evaporates

in a tank where the pressure is very low, thus extracting heat from water circulating in a coil. The water that is chilled in the coil is distributed to air conditioning equipment as required. The spray water does not all evaporate, so the liquid water is recirculated by the *refrigerant pump*.

To prevent the pressure from building up in the evaporator the water vapor must be absorbed by lithium bromide. A concentrated solution is stored in a tank called the *absorber*. This solution is sprayed into the absorber and recirculated by the *absorber pump*. The lithium bromide absorbs and draws water vapor from the evaporator space and a low pressure is maintained there. However, the solution gradually becomes too dilute to absorb enough water. To solve this problem, diluted solution is pumped to the *concentrator* (also called *generator*) by a *concentrator pump*. Here it is heated to a temperature that will evaporate some of the water, which has a lower boiling point than the lithium bromide. The reconcentrated solution is then returned to

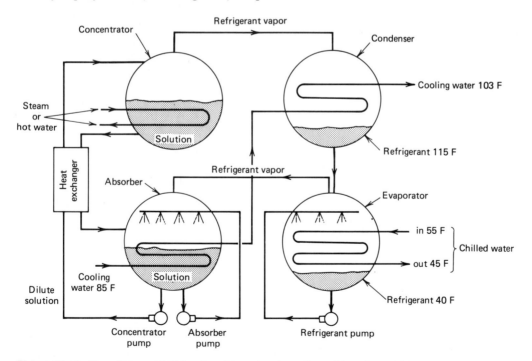

Figure 13.19. Flow diagram of lithium bromide-water absorption refrigeration system.

the absorber. Steam or hot water is used as a source of heat in the concentrator.

The water vapor from the concentrator flows to the *condenser*, where it is condensed to a liquid by giving up heat to water from a cooling tower or natural body of water. The condensed water is then returned to the evaporator, completing the cycle.

Two refinements to this cycle, which improve the system's efficiency, and are shown in Figure 13.19, require explanation. The absorption process generates heat that would raise the temperature of the absorbing solution, making it less effective. The heat is removed by cooling water, which is circulated through a coil in the absorber. The same cooling water is then used in the condenser, as shown.

The second refinement to the cycle is the inclusion of a heat exchanger between the absorber and concentrator. The solution from the absorber is preheated by hot solution returning from the concentrator, thereby saving some of the heat needed in the concentrator.

Another pair of fluids often used in absorption systems is ammonia and water. In this case water is the absorbent and ammonia is the refrigerant. Because of its volatility some water boils off with the ammonia in the generator of the aqua-ammonia absorption system. This requires additional equipment (a *rectifier*) to separate the ammonia from the water. Another disadvantage of this system is that it operates at much higher pressures in the generator (about 300 psia) compared with about 30 psia for the LiBr system.

Small capacity lithium bromide-water absorption units (3–25 tons) with direct-fired generators are also available. They are popular in areas where natural gas is plentiful and inexpensive.

13.20. CONSTRUCTION AND PERFORMANCE

Absorption refrigeration machines are not actually constructed with four separate vessels as was shown in Figure 13.19. (The sketch was made that way for clarity.) To economize on construction costs the four parts are built into two or even one shell, as shown in Figure 13.20.

The machine is completely factory assembled, including *evaporator*, *absorber*, *concentrator*, *condenser*, and *solution pumps*, interconnecting piping, and electric controls. The pumps are hermetic, to prevent any leaks into the system. The machine is carefully evacuated in the factory of air down to an extremely low pressure.

The heat required in the concentrator is furnished either by low pressure steam or hot water. Temperatures around 240 F result in peak efficiency. A steam use rate of 18–19 lb of steam per ton of refrigeration is typical at this temperature. Selection procedures for an absorption chiller are similar to those for reciprocating or centrifugal chillers, and therefore will not be discussed in further detail. Heat source temperatures, load, chilled water, and condensing water temperatures are the factors required to select the proper machine from manufacturers' tables.

The coefficient of performance (COP) of absorption machines is much lower than systems using mechanical compression refrigeration. A COP of 0.65 is typical for large absorption equipment. This corresponds to a heat input rate of about 18,000 BTU/hr per ton of refrigeration. A large compressor driven water chiller may have a COP of 3.5 or higher. That is, it uses only one-fifth of the energy of an absorption machine. This would seem to be an unacceptable waste of energy, but the COP does not show the whole situation. The energy used to drive the compressor is usually electricity generated by a thermal electric utility. Only about one-third of the heat from the fuel in the power plant is converted to electric energy (this is a limitation imposed by the Second law of Thermodynamics, explained in Chapter 15.) Thus, the energy use advantage of the compression refrigeration system is greatly reduced. Further factors make the absorption machine desirable under certain circumstances. The low pressure steam used for the energy

Figure 13.20. Absorption refrigeration machine. (*Courtesy:* Machinery & Systems Division, Carrier Corp., Syracuse, N.Y.)

source is frequently otherwise wasted heat from a process or from a utility company. The total energy balance between the absorption or compression machine may then be equal or may even be in favor of the former.

The COP of the absorption system may be improved considerably if a two-stage generator (concentrator) is used. Heat from the vapor coming from the first generator is used to provide further vaporization of liquid from the absorber. A two-stage machine available from Japan raises the COP to about 1.0, a 50% improvement in energy efficiency.

13.21. SPECIAL APPLICATIONS

The absorption machine has considerable promise for refrigeration in conjunction with solar energy as a heat source. In this arrangement, hot water is heated by solar collection panels and then used in the concentrator as the heat source.

The energy input rate required is high at the water temperatures solar heaters usually can produce (170–200 F), but because there is no depletable fuel used, this is not important.

A popular and efficient combination of refrigeration sources for air conditioning is the centrifugal-absorption combination. In this arrangement a high pressure steam turbine is used to drive a centrifugal refrigeration machine, and the low pressure exhaust steam from the turbine is then used as a heat source in an absorption machine. The overall steam rate per ton by using the steam twice can be very attractive, about 13 lb/ton, as compared with 16 lb/ton for the turbine driven machine alone.

It should be noted that the cooling tower required for an absorption machine will be considerably larger than that needed for a vapor compression cycle machine, due to the larger quantity of heat that must be rejected from the absorber and condenser combined.

The choice of whether to use an absorption or vapor compression machine (or combination) for

an installation is largely a matter of economics, which is a function of relative fuel costs.

13.22. CAPACITY CONTROL

Two modern methods of modulating refrigeration capacity to meet load demands are used with absorption machines. Both of them use a controller that senses and maintains a constant leaving chilled water temperature. With *heat source control* the controller will operate a valve that controls the steam or hot water flow to the concentrator, thereby changing the machine capacity. *Solution modulation* control uses a controller that mixes the absorber solution to vary its concentration, which affects the machine capacity. Safety controls are provided with the machine, and will not be discussed.

13.23. CRYSTALLIZATION

This is an important phenomenon that needs to be understood, especially by the operating engineer. If the lithium bromide solution becomes too concentrated, it changes from a liquid to a solid (crystal) form. The maximum concentration possible decreases as the solution temperature decreases. Therefore, if a solution is already near its maximum concentration and if its temperature is then lowered, it will *crystallize* (solidify).

This is a serious problem in absorption systems, because if it occurs, the crystallized LiBr blocks the piping and the machine stops working. There are three factors that can result in a drop in temperature of the solution:

1. Power failure
2. Condensing water temperature too low
3. Air leakage into the system

We will not explain here how each can cause this effect. Information can be found in manufacturers' manuals. If crystallization occurs, it is necessary to heat the piping where the blockage has occurred. Automatic methods of doing this can be provided by the manufacturer.

13.24. INSTALLATION

Detailed instructions for installation of absorption refrigeration machines are provided by each manufacturer. Some general procedures will be discussed here.

1. Investigate possible installation on an upper floor or penthouse. The light weight and lack of serious noise and vibration make this a feasible alternative to a basement. If the boiler is also located in the same space, this eliminates the need for a boiler stack and much of the piping; both are major cost items in high-rise buildings.
2. Provide rubber isolation pads under the machine.
3. Allow ample clearance for tube removal and for service access as needed on all sides.
4. Install external piping (to boiler, condenser) and electrical and control connections so as not to block access to the machine.

The Heat Pump

13.25. PRINCIPLES

The *heat pump* is a refrigeration system that can be used for both cooling or heating. To those not familiar with refrigeration cycles the heat pump appears to be a mysterious device that operates on some unusual principle; it is not. The heat pump is usually a vapor compression refrigeration machine, which is basically no different in operation or components from that described previously. (An absorption machine can also be used as a heat pump, but this is unusual.)

(a)

(b)

Figure 13.21. The heat pump cycle. (a) Summer cycle (cooling).
(b) Winter cycle (heating).

Normally the purpose of a refrigeration machine is to absorb heat (in the evaporator) from a cooling load. The heat that is rejected in the condenser is thrown away to the atmosphere or a body of water. There is no reason why this heat could not be used to satisfy a heating load. When this is done, the machine is a heat "pump." The refrigeration effect, which is still occurring, may or may not be utilized, depending on need.

Figure 13.21 shows how a heat pump performs in both summer and winter. In summer, with the refrigerant flowing in the direction shown, the room coil serves as the evaporator and the room air is cooled to produce summer cooling. In winter the direction of refrigerant flow is *reversed* after leaving the compressor, so that the room coil serves as the condenser, and the outdoor coil as the evaporator. The room air passing over the room coil therefore receives the heat rejected in the condenser. The heat pump now acts as a heating unit. The heat absorbed in the outdoor coil is the refrigeration effect, but of course it serves no useful purpose in this case.

The reversal of refrigerant flow to switch between heating-cooling is accomplished with a *reversing valve*. This has four ports, two of which are open at any one time to allow flow in the direction chosen. The heat pump is sometimes called a "reverse cycle" air conditioner. This is a misleading name. The cycle is the vapor compression refrigeration cycle. In the explanation given, only the refrigerant flow direction and function of the coils are reversed.

Heat pumps are often supplied as unitary equipment, with all the components assembled as a package by the manufacturer, including the air handling unit. Another arrangement, used more on larger equipment, without reversing refrigerant flow, is to reverse air flow. The evaporator and condenser coils function the same in winter and summer. However, the duct arrangement is made so that room air is circulated to the evaporator coil in summer and to the condenser in winter. The opposite is done for outside air.

Therefore, in winter the room air, passing over the condenser coil, is heated.

One clear advantage of a heat pump is that it can provide heating or cooling from one machine, without any great modifications. In many cases this means that it would have a lower first cost than using separate heating and cooling equipment.

Another advantage that is not apparent without further investigation is that it may have a lower operating cost than separate conventional heating and cooling systems, especially when electric resistance heating would be otherwise used. Although this has always caused considerable interest in the heat pump, high energy costs and shortages have created even more intense interest.

13.26. ENERGY EFFICIENCY

Consideration of how the Energy Balance principle (Chapter 3) applies to the heat pump will show how it may provide heating with a relatively small expenditure of energy. Referring to Figure 13.22, the total energy into the system equals the total energy out:

$$Q_c = Q_p + Q_e$$

where

Q_c = heat rejected from condenser
Q_e = heat absorbed in evaporator
Q_p = heat equivalent of compressor power input

The significance of this equation is that the useful heating Q_c is greater than the energy needed to drive the compressor Q_p by the amount Q_e, which does not require any energy expenditure. Contrast this with any direct heating system, either electrical or by burning a fuel to generate steam or hot water. In these cases, of course, the energy expended is at least equal to the useful heating. A relative measure of the

Figure 13.22. Energy usé in the heat pump.

performance of the heat pump is the heating coefficient of performance, defined as:

$$\text{COP}_h = \frac{Q_c}{Q_p}$$

$$= \frac{\text{heat rejected from condenser}}{\text{heat equivalent of compressor power}}$$

Note that the coefficient of performance of a heat pump does not have the same meaning as when the unit is used for refrigeration. The heat pump COP_h is useful in illustrating the advantage of heating by using electrical energy to drive a heat pump compressor rather than using the electricity directly in resistance heaters.

For packaged heat pumps that include evaporator and condenser fans, the COP is often defined to include these auxiliary power inputs. This lowers the value of the COP slightly, of course.

13.27. SELECTION OF HEAT PUMPS — THE BALANCE POINT

As the evaporating temperature decreases, the heating capacity (the heat rejected from condenser) of a heat pump also decreases, because less heat is absorbed in the evaporator. Since the evaporator is the outdoor coil, a drop in outdoor temperature causes a decrease in the heating capacity.

For typical heating cooling load requirements, if the heat pump is sized to handle the maximum cooling load, its heating capacity will be inadequate below outdoor temperatures often encountered in many climates. For residential applications, an outside temperature of about 30 F is a typical temperature at which the heating capacity of the unit will just match the load. This is called the *balance point*.

At temperatures below the balance point, supplementary heating must be furnished. This is often accomplished by using one or more electric resistance heaters. As the outdoor temperature decreases, the amount of supplementary heat required increases, since both the load is increasing and the heat pump capacity is decreasing. Controls are arranged to activate the resistance heaters in steps as the outdoor temperature drops.

The supplementary heater should always be located downstream of the indoor coil in the ductwork so that the heat does not affect the condensing temperature.

Ratings of some small split system packaged heat pumps suitable for residential or similar applications are shown in Table 13.3. The ratings in the table are based on a space temperature of 70 F for heating and 80 F DB, 67 F WB for cooling, a standard ARI test condition.

When selecting a heat pump, standard practice is to select a unit that will satisfy the design

TABLE 13.3. RESIDENTIAL TYPE PACKAGED HEAT PUMP RATINGS

performance
data

Fedders Flexhermetic Heat Pump Condensing Units are matched to Fedders Evaporator Blower Package Units for complete split system applications. Select performance requirements and matched components below.

FLEX. HEAT PUMP MODEL NO.	MATCHING EVAP. BLOWER MODEL NO.	OUTSIDE TEMP. db °F	CAPACITIES BTU/HR. COOL	CAPACITIES BTU/HR. SENS.	TOTAL WATTS COOL	PSIG PRESSURES SUCTION	PSIG PRESSURES DISCHG.	OUTSIDE TEMP. db °F HEAT	CAPACITY BTU/HR. HEAT	TOTAL WATTS HEAT	PSIG PRESSURES SUCTION	PSIG PRESSURES DISCHG.
CFH024A3	CFB024A2	115	20,000	14,500	4,450	80	388	60	31,000	4,000	68	310
		105	21,500	15,000	4,200	77	351	50	27,000	3,800	59	280
		*95	23,000	16,500	4,000	74	313	*45	25,000	3,600	52	263
		85	24,000	17,000	3,800	69.5	276	40	23,000	3,500	50	250
		75	26,000	18,000	3,550	65	235	30	19,500	3,300	40	225
								*20	16,000	3,150	31.5	200
								10	12,500	2,950	21	165
								0	9,000	2,800	11	135
CFH030A3	CFB036A2	115	24,000	18,000	4,800	82	387	60	35,000	4,000	74	380
		105	26,000	19,000	4,650	79	350	50	31,000	3,750	60	320
		*95	27,000	20,000	4,300	76	312	*45	29,000	3,600	53	287
		85	30,000	21,000	4,100	71.5	274	40	27,000	3,500	46	260
		75	32,000	22,000	3,750	67	234	30	23,000	3,300	37	220
								*20	18,500	3,150	30	189
								10	14,500	2,950	24	170
								0	10,000	2,800	20	159

UPFLOW

FLEX. HEAT PUMP MODEL NO.	MATCHING EVAP. BLOWER MODEL NO.	OUTSIDE TEMP. db °F	CAPACITIES BTU/HR. COOL	CAPACITIES BTU/HR. SENS.	TOTAL WATTS COOL	PSIG PRESSURES SUCTION	PSIG PRESSURES DISCHG.	OUTSIDE TEMP. db °F HEAT	CAPACITY BTU/HR. HEAT	TOTAL WATTS HEAT	PSIG PRESSURES SUCTION	PSIG PRESSURES DISCHG.
CFH036D7A	CFB036A2	115	30,000	21,000	6,250	76	400	60	41,000	5,300	63	409
		105	32,900	23,000	5,850	73	360	50	38,000	5,100	54	367
		*95	35,000	24,500	5,500	70	320	*45	35,000	4,900	50	350
		85	36,800	25,750	5,150	66	285	40	32,000	4,700	45	330
		75	38,200	26,750	4,700	62	240	30	26,000	4,300	37	295
								*20	21,000	3,900	28	260
								10	15,000	3,500	20	225
								0	10,000	3,100	10	185
CFH042D7A	CFB048A2	115	36,000	25,200	6,620	81	370	60	48,000	5,400	65	302
		105	39,000	27,300	6,320	78	335	50	44,900	5,100	55	260
		*95	42,000	29,400	6,000	75	330	*45	42,000	4,900	50	250
		85	44,000	30,800	5,650	71	265	40	37,000	4,100	45	238
		75	46,000	32,200	5,220	66	225	30	31,000	3,600	37	216
								*20	25,000	3,200	30	200
								10	22,000	3,000	22	182
								0	20,000	2,800	17	170
CFH048D3A	CFB048A2	115	39,000	27,300	7,950	79	415	60	52,000	5,900	62	318
		105	42,000	29,400	7,500	76	375	50	48,000	5,600	52	290
		*95	45,000	31,500	7,100	73	335	*45	45,000	5,500	47	280
		85	47,000	32,900	6,700	68	295	40	41,000	5,300	42	262
		75	49,000	34,300	6,150	64	250	30	34,000	5,000	33	240
								*20	28,000	4,700	25	220
								10	24,000	4,500	18	210
								0	20,000	4,300	12	204
CFH048D8A	CFB048A2	115	39,000	27,300	7,900	79	380	60	52,000	5,900	62	318
		105	42,000	29,400	7,450	76	350	50	48,000	5,600	52	290
		*95	45,000	31,500	7,100	73	310	*45	45,000	5,500	47	280
		85	47,000	32,900	6,650	68	270	40	41,000	5,300	42	262
		75	49,000	34,300	6,100	64	230	30	34,000	5,000	33	240
								*20	28,000	4,700	25	220
								10	24,000	4,500	18	210
								0	20,000	4,300	12	204
CFH060D3A	CFB060A2	115	48,000	31,500	9,600	79	390	60	66,000	7,600	60	358
		105	53,000	37,100	9,100	76	348	50	62,000	7,200	51	320
		*95	56,000	39,200	8,600	73	310	*45	57,000	6,900	45	300
		85	59,000	41,300	8,100	68	270	40	51,000	6,500	42	282
		75	61,000	42,700	7,500	64	230	30	42,000	5,800	34	250
								*20	34,000	5,200	25	225
								10	27,000	4,500	18	204
								0	21,000	3,900	10	188
CFH060D8A	CFB060A2	115	48,000	31,500	8,700	79	400	60	66,000	7,200	60	358
		105	53,000	37,100	8,650	76	360	50	62,000	6,800	51	320
		*95	56,000	39,200	8,200	73	322	*45	57,000	6,500	45	300
		85	59,000	41,300	7,750	68	284	40	51,000	6,100	42	282
		75	61,000	42,700	7,200	64	240	30	42,000	5,400	34	250
								*20	34,000	4,800	25	225
								10	27,000	4,100	18	204
								0	21,000	3,500	10	188

* A.R.I. test conditions: Inside temp. 80° F db; 67° F wb (cooling), 70° F db (heating)

Reprinted with the permission of Fedders Corp.

cooling load. The heating capacity of the unit is then plotted at different outdoor temperatures, and a line is drawn showing these heating capacities. Another line is drawn showing the building heating load at different outdoor temperatures (this is a straight line). The intersection of these two lines is the balance point. The supplementary heating is then sized to provide the extra heating capacity below the balance point, if any is required. Example 13.5 illustrates the selection of a heat pump and the determination of the balance point.

EXAMPLE 13.5

Select a heat pump for a residence in Memphis, Tennessee. The design cooling load is 44,000 BTU/hr and the design heating load is 41,000 BTU/hr. The inside design temperatures are 70 F in winter and 80 F in summer. What is the balance point? What is the size of the required supplementary resistance heaters?

Solution. Using Table A.9, the summer and winter outdoor design temperatures are 95 F and 18 F.

Using Table 13.3, a Model CFH048 unit is selected based on the design cooling load. The capacity at 95 F is 45,000 BTU/hr.

Refer to Figure 13.23. Using the columns of heating capacity versus outside heating temperatures, a curve is drawn through these points.

A straight line is drawn from the design heating load point, at 41,000 BTU/hr and 18 F, to 0 BTU/hr (no load) and 70 F. This line represents the building heating load at different outdoor temperatures.

The intersection of these two lines is about 28 F. This is the balance point. The heat pump heating capacity just matches the required building load at an outdoor temperature of 28 F.

The heat pump heating capacity at the balance point is 33,000 BTU/hr. The design heating load is 44,000 BTU/hr. Supplementary heaters

Figure 13.23. Determination of heat pump balance point for Example 13.5.

are required below 28 F. At the design temperature of 18 F, the supplementary heat required is:

$$\text{Supplementary heat} = \text{heating load}$$
$$- \text{heating capacity}$$
$$= 41{,}000 - 27{,}000$$
$$= 14{,}000 \ \text{BTU/hr}$$

$$14{,}000 \ \text{BTU/hr} \times \frac{1 \ \text{kW}}{3410 \ \text{BTU/hr}} = 4.1 \ \text{kW}$$

It is possible to reduce or eliminate supplementary heating by oversizing the heat pump in the cooling cycle. The larger unit will also provide more heating, as can be seen in Table 13.3. This is usually undesirable, however. The oversized unit will cycle too often in the cooling cycle, resulting in uncomfortable conditions, and a shortened life for the compressor and controls. For these reasons heat pumps should always be sized for the cooling load, rather than the heating load.

Example 13.6 illustrates the savings in energy and operating cost by using a heat pump instead of electric resistance heating.

EXAMPLE 13.6

For the heat pump used in Example 13.5, compare the amount of power saved at an outdoor temperature of 30 F by using a heat pump instead of electrical resistance heating. What is the heating COP of the heat pump at this temperature, compared to resistance heaters?

Solution. From Table 13.3, the heating capacity of the Model CFH048 heat pump at 30 F is 34,000 BTU/hr, requiring 5000 W of power for the compressor.

If electrical resistance heating is used, the equivalent to 34,000 BTU/hr is

$$\text{Resistance heating} = 34{,}000 \text{ BTU/hr}$$

$$\times \frac{1 \text{ W}}{3.41 \text{ BTU/hr}}$$

$$= 9970 \text{ W}$$

The savings is 9970 − 5000 = 4970 W.

The COP_h for the heat pump at 30 F is (see Section 13.26)

$$COP_h = \frac{Q_c}{Q_p} = \frac{34{,}000 \text{ BTU/hr}}{5000 \text{ W} \times \dfrac{3.41 \text{ BTU/hr}}{1 \text{ W}}} = 1.99$$

The COP of electric resistance heating = 1.0. That is, the amount of heat output is the same as the amount of electric energy input (see Equation 15.11).

The relationship of the two COP_hs is

$$\frac{\text{Heat pump } COP_h}{\text{Elect. resist. } COP_h} = \frac{1.99}{1.0} \approx 2:1$$

To express this in words, at 30 F the heat pump requires about one-half the power input that electrical resistance heating does to produce the same heat output.

At higher outdoor temperatures the heat pump will perform even better (be more energy efficient). However, at lower outdoor temperatures it will not perform as well.

Typical actual heat pump COP_h values when used for heating range from 1.5–3.0. (This applies to unitary air-to-air heat pumps when operating at an outside air temperature of 47 F.) This means that the heat pump is producing 1.5–3.0 times more heat output for the same energy input than by using electrical resistance heating.

13.28. SOLAR ENERGY — HEAT PUMP APPLICATION

An effective use of the heat pump is in combination with solar energy. The coefficient of performance of the heat pump decreases with a decrease in evaporator temperature (see Chapter 15). With the conventional application of the heat pump, this temperature is lower than ambient air temperature, which results in a low COP in winter. However, a solar energy collector can be used to supply water at a much higher temperature than normally available in winter for the evaporator, say 75–100 F. This moderate temperature can be achieved with a relatively inexpensive collector, and will result in a low heat pump energy use. This arrangement is called the *solar assisted heat pump*, described in Chapter 18.

13.29. REFRIGERANTS

Although many other substances have been used as refrigerants in vapor compression systems, those that are called the *fluorinated hydrocarbons* have become universally adopted in refrigeration equipment for air conditioning. These substances all have important properties for this application

—they are nontoxic, nonflammable, nonexplosive, and noncorrosive. Each has different pressure-temperature boiling point characteristics, so that a suitable refrigerant can be chosen for the application and equipment available. All refrigerants are identified by a standard code numbering system. R-12 and R-22 are fluorinated hydrocarbons whose properties make them very suitable for reciprocating compressors used for air conditioning. R-11, with a relatively high specific volume, is often used in centrifugal compressors, which handle large volume flow rates.

There is concern about fluorinated hydrocarbons escaping and reacting with ozone in the upper layers of the earth's atmosphere, which could deplete the ozone. The ozone filters out ultraviolet rays from the sun, which otherwise have harmful health effects. Steps have already been taken to limit or ban the use of the fluorinated hydrocarbons in aerosol spray devices.

13.30. WATER TREATMENT

Water used in condensers, water chillers, and boilers requires proper chemical treatment. Minerals that exist naturally in water can precipitate as solids and form *scale* that deposits on surfaces, reducing heat transfer. The water can have an acidic character that will cause *corrosion* of metals. *Biological growths* can occur that may cause deterioration of wood or coat surfaces and reduce heat transfer. This is a common problem in cooling towers, where the water is exposed to the atmosphere. The most bizarre example of contamination is "Legionnaires disease," a virus that apparently has been traced in some cases to stagnant water in cooling tower basins. Water treatment in chilled water systems is usually a minor problem, because those systems are closed, but it should not be neglected. A firm that specializes in water treatment should be called in to set up a treatment plan when planning an air conditioning system.

13.31. ENERGY CONSERVATION IN REFRIGERATION

Some methods to consider for conserving energy with refrigeration systems are:

1. Use refrigeration compressors that reduce power requirements as load decreases. For reciprocating compressors this would involve use of cylinder unloaders. For centrifugal compressors this would involve use of speed control or inlet guide vanes.
2. Select and operate equipment with highest evaporating (or chilled water) temperature and lowest condensing temperature consistent with maintaining satisfactory space conditions and satisfactory equipment performance.
3. Use condenser heat for heating needs by recovering heat (Chapter 15).
4. Use multiple equipment on larger projects so that each operates close to full load more often.
5. Use some form of total energy system (such as combined steam turbine centrifugal-absorption machines).

REVIEW QUESTIONS

1. Describe with a sketch the vapor compression refrigeration system.
2. What are the three types of positive displacement compressors?
3. Explain the difference between an open and hermetic compressor.
4. Describe two methods of controlling capacity of reciprocating compressors and of centrifugal compressors.
5. What are the three types of condensers and their features?
6. Describe three types of refrigerant flow devices.

7. What are the causes of water loss in a cooling tower?

8. Describe with a sketch the lithium bromide-water absorption refrigeration system.

9. Describe with a sketch the air-to-air heat pump.

10. Explain what is meant by the term *balance point* of a heat pump.

11. Describe a heat pump defrost cycle.

PROBLEMS

13.1. Select an air cooled condensing unit to handle a load of 15 tons of refrigeration for a bowling alley in Richmond, Virginia. Evaporating temperature is 45 F. Find the COP of the unit.

13.2. Select a package water chiller for a load of 21 tons. Chilled water is cooled from 55 to 44 F. Condenser water enters at 90 F and leaves at 100 F. Chiller and condenser fouling factors are 0.0005. Determine the COP of the unit.

13.3. Find the capacity, KW required, and COP of the unit selected in Problem 13.2 if the condenser fouling factor is 0.001.

13.4. Select a heat pump for a home in Los Angeles, California, for a design cooling load of 46,000 BTU/hr and design heating load of 42,000 BTU/hr. Determine the heating COP.

13.5. Determine the design cooling capacity and COP of the heat pump selected in Problem 13.4.

13.6. A heat pump is to be selected for summer and winter air conditioning of a home in Charleston, South Carolina. The design heating load is 32,000 BTU/hr and the design cooling load is 40,000 BTU/hr. Select a heat pump adequate to handle the summer load. What size electric booster heater, if any, is required in winter?

CHAPTER 14

Automatic Controls

The automatic controls for an HVAC system can be compared in importance and function with the brain and nervous system of a human. Without these, the body, regardless of how physically healthy it is, would be a lifeless mass.

The HVAC controls must be designed and installed to fit the system and must function properly. If not, the air conditioning system will not produce satisfactory conditions.

OBJECTIVES

After studying this chapter you will be able to:

1. Explain the purposes of automatic controls.

2. Identify and describe the elements of each control systems with the aid of a diagram.

3. Explain closed and open loop controls.

4. Explain the types of control action.

5. Describe how electric, electronic, and pneumatic controllers function.

6. Explain the types of valve flow characteristics and damper arrangements.

7. Sketch control diagrams and describe the operation of some basic control systems.

14.1. UNDERSTANDING AUTOMATIC CONTROLS

If asked to name the part of the HVAC system that they found most difficult to understand and work with, probably most designers, contractors, and service people would list the automatic control system. It is true that modern controls are often somewhat involved and that some of the devices used are complex. The major reason few people know how to deal with controls, however, is that they do not understand the basic principles and how to apply them. This is because most information focuses on hardware and how it is connected. This by itself often will not enable the service technician to determine the cause of a malfunctioning system. Nor will it enable the designer to plan the controls to suit the type of HVAC system. In this chapter we will emphasize the principles of control systems. These principles can be applied to electric, electronic, or pneumatic controls with equal case. Although hardware will be discussed, we will not "lose sight of the woods for the trees."

14.2. PURPOSES OF CONTROLS

The controls can serve four different functions:

Maintain Design Conditions   *Controls maintain design conditions* (temperature, humidity) in the space. The heating and cooling capacity of the HVAC system is selected at the design load conditions. Whenever the load (heat gain or loss) is less than the design value, the system capacity is too large. If it produces its full output, the spaces will be overheated or overcooled. The controls must regulate the heating or cooling output of the system to match that of the load. The load varies mainly from changes in outdoor temperature, solar radiation, occupancy, and lights being switched on and off. Although controlling and maintaining conditions in the space are the primary function of the automatic controls, they serve other functions, as follows.

Reduce Human Labor Needed   Controls reduce the amount of human labor needed to operate and maintain the system, thus reducing labor costs and the chances of errors. For exam-

ple, the controls may be designed to open an outside air damper to provide fresh air, rather than have operating personnel do this manually.

Minimize Energy Use and Costs   Controls minimize energy use and costs. One of the most important considerations in planning and operating a control system is based on its ability to minimize the use of energy at all times. For example, the controls may automatically change the amount of outside air introduced to the building so that free cooling is obtained from this air when suitable. Many of the energy conserving control applications are discussed in Chapter 15, but some of these features will be discussed here.

Controls that serve the purposes already described—maintaining space conditions, reducing labor, or conserving energy—are called *operating controls*. Often the same controls are providing all of these functions. Some controls serve a different purpose.

Keep Equipment Operation at Safe Levels   Controls keep operation of equipment at safe levels, thus preventing damage to property or injury to people. This type of control is called a *safety control*. It usually functions as a limiting device, to limit the values of temperature, pressure, or similar variables in the equipment. Safety controls have been discussed to some extent as part of the coverage of equipment in other chapters. Although some further references to them will be made here, our emphasis will be on the operating controls. Safety controls are of the utmost importance, however, and are part of the overall automatic control system.

14.3. THE CONTROL SYSTEM

One way to study a control system is to see it as a collection of many control devices. This leads to little understanding; a better approach is to recognize that every control system has similar elements. This is true regardless of how large and

Figure 14.1. Functional block diagram for an open loop control system.

complex the system is, or whether it is pneumatic, electric, or electronic.

Every control system has the following elements:

1. A *controlled variable*. This is a condition that is to be controlled, such as temperature, humidity, or pressure.

2. A *controller*. This is a device that senses a signal from a change in the controlled variable and then transmits an action to a controlled device to correct the changed condition. Thermostats, humidistats, and pressure-stats are examples of controllers. Note that a controller has two functions—to sense a signal, and to transmit an action based on the signal.

3. A *source of energy*. This provides the means of power to transmit the action from the controller to the controlled device. Two examples are compressed air (pneumatic) and electrical energy sources.

4. A *controlled device*. This is a device which, when receiving the action from the controller, regulates the flow or other effect of a control agent. Examples of controlled devices are a valve, damper, or the motor driving an oil

burner, pump, fan, or refrigeration compressor.

5. A *control agent*. This is the medium regulated by the controlled device. Examples are water flowing through a valve, air through a damper, or the electric current of a motor.

6. A *process plant*. The regulation of the control agent changes the output or operation of equipment called the process plant. Examples are a cooling or heating coil, oil burner, fan, pump, compressor, and duct. The change in performance of the process plant changes the condition of the controlled variable, thus completing the desired action.

The sequence of action is shown by a *functional block control diagram* (Figure 14.1).

An example of a simple control system will help to identify these terms and how the control sequence functions.

EXAMPLE 14.1

A student is studying in a room that has a hot water heating convector with a manual valve. The outdoor temperature drops very suddenly. Sketch a functional block control system dia-

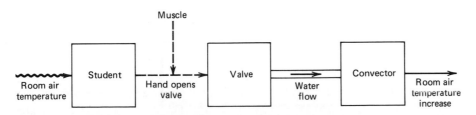

Figure 14.2. Functional block control diagram for Example 14.1.

gram, identify the elements, and explain the sequence of control action.

Solution.   The control diagram is shown in Figure 14.2.

1. The controlled variable is the room air temperature. A change (drop) in this temperature results from the sudden increase in heat loss to the outdoors.
2. The temperature controller (thermostat) is the student. His body senses the change in the control variable. He then transmits an action —turns the valve handle.
3. The energy source for transmitting the action is the human muscle power.
4. The controlled device is the manual valve.
5. The control agent (medium) is the hot water. Opening the valve increases water flow rate through it.
6. The process plant is the convector. Increased water flow rate results in more heat output from the convector to the room, causing a correction in the controlled variable condition —the room air temperature rises.

Example 14.1 was of a manual control system, which is usually not very reliable or accurate. The same events occur in an automatic control system. For example, a thermostat senses the changed room air temperature and transmits an action, for instance closing an electric circuit. An electric power source then opens an automatic, electrically operated valve on the convector.

14.4. CLOSED LOOP (FEEDBACK) AND OPEN LOOP CONTROL SYSTEMS

In Example 14.1 no mention was made of what might occur after the valve was opened. The room temperature might rise to an uncomfortably high level. It might happen quickly if the valve is opened too much, or gradually, if the outdoor temperature increases. Students could resolve this by closing the valve. However, they might have little time left for studying. We will now discuss how this problem is resolved, which leads to an important concept in controls.

In the manual control system described the controller (student) sensed the results of the corrective action (the rise in room temperature) and when this went beyond a satisfactory condition, the student took an opposite corrective action. This is an example of feedback.

Feedback is the transmission of information about the results of an action back to the sensor.

The result of the information being fed back is that the room air temperature changes were continually being corrected by the student. This is an example of a *closed-loop* control system. The functional control diagram for a closed loop system would be as shown in Figure 14.3 for the example of a room thermostat controlling the convector.

The control diagram is the same as before, except that a feedback signal has been added. Note that the system now forms a closed loop. That is, it will respond continuously to the feedback signal, always correcting the value of the

Figure 14.3. Functional block diagram for a closed loop (feedback) control system.

controlled variable. This is the essence of most automatic controls. The sensor in the thermostat reacts to any change in the room temperature (increase or decrease) and sets in motion the corrective action.

A control system without feedback and its effect is called an *open-loop* control system. In the example of the student who acted as a controller, if he had decided to open the valve because he expected colder weather, and not because he felt cold, this would be a case of an open loop, because there is no feedback causing the action. The weakness of this type of control is apparent. The student does not know whether he is opening the valve a proper amount, and does not provide any corrective action unless he returns to using his sensory feeling (feedback). Open-loop controls are used in automatic control systems in certain instances, however, as will be explained later.

14.5. ENERGY SOURCES

Automatic control systems can be classified according to the source of energy they used:

Electric / Electronic   Electric energy is used to actuate the controlled devices. The controller regulates the amount of energy transmitted to the controlled device. If the controller has sensing and transmitting elements that are electronic, the system is called electric/electronic, or electronic for convenience. Otherwise the system is called electric. This is the only basic difference between electric and electronic control systems. The power is always transmitted electrically.

Electric control systems are often used on small installations because they are inexpensive, simple, and easy to install.

Line voltage control systems use electricity at the voltage from the power supply, usually 110 volts. *Low voltage* control systems transform the power supply to low voltages for control use. The choice depends on cost, convenience, and safety.

Pneumatic   Compressed air is used as the source of energy to actuate the controlled devices. The controller regulates the air pressure transmitted to the controlled device.

Pneumatic control systems are often used in large installations. Air compressors are required, and copper or plastic tubing is used to transmit the air. Pneumatic systems are popular because the controlled devices easily lend themselves to modulating action (Section 14.7).

Self-Powered   No external source of energy is used. Power to actuate the controlled devices comes from the medium being controlled. This is usually accomplished by an enclosed fluid that will change pressure in response to a temperature change. A common example is the thermal expansion valve (TEV) refrigerant flow control. Fluid in a bulb changes pressure in response to the temperature it senses. The pressure actuates the valve (Chapter 13). Self-powered controls are practical in certain applications, but are generally not used for the whole control system.

Combinations of electric, electronic, pneumatic, and self-powered controls may be used in one control system when desirable.

14.6. COMPONENT CONTROL DIAGRAM

Although the functional block control diagram is helpful in understanding the operation of the control loop, it is also useful to prepare a *component control diagram* showing the connections between components of the control system and HVAC system. Figure 14.4 is an example of the simple control system described earlier, where a valve to a convector is controlled by a room thermostat.

A dashed line is used to represent the control action, which may come from one of a number of energy souces. The component control diagram is drawn the same regardless of type of energy source, however, which simplifies reading

Figure 14.4. Component control diagram. Room thermostat controls position of valve.

and understanding the diagram. An actual diagram showing all the wiring for an electric system would look much more complicated. Although it is necessary to have detailed diagrams for each installation, they do not aid in understanding, but on the contrary, make it more difficult. For that reason we will use component control diagrams instead of detailed wiring or compressed air piping diagrams in our explanations.

14.7. TYPES OF CONTROL ACTION

There are different types of action that the controller can impart to the controlled device, classified as follows:

Two-Position Action   This is also commonly called "on-off" action. For instance, if the controlled device is a motor, it may be started or stopped (on-off) by the two-position controller. A control valve that moves only to a fully open

or closed position is another example of two-position action.

Differential is a term of importance in two-position control action. It refers to the range of controlled variable values at which action takes place. In the system there are two differentials:

Controller differential is the range set on the control device of the variable values at which it transmits action to the controlled device. For example, if a thermostat is set to move to one position at 70 F and the other position at 72 F, it is said to have a differential of 2 F.

Operating differential is the range that actually occurs in the value of the controlled variable. This differential will often be greater than the controller differential setting because there is a lag in response of the controlled device and medium. For example, when a thermostat causes a hot water heating valve to close, the convector keeps heating the room for a short time due to the hot water still in the unit.

EXAMPLE 14.2

A heating thermostat is set to start an oil burner on a furnace at a room temperature of 70 F. The thermostat has a 2 F differential setting. The response lag is 1 F in either direction. Between what values does the room temperature vary? What is the opening differential?

Solution.   Figure 14.5 illustrates the solution. The thermostat high position (off) setting is

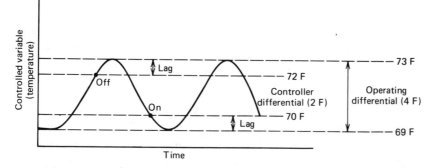

Figure 14.5. Two-position control action, as in Example 14.2.

70 + 2 = 72 F. With a lag of 1 F, however, the room temperature will rise to 72 + 1 = 73 F. Similarly, the room temperature will fall to 70 − 1 = 69 F. The operating differential is therefore 73 − 69 = 4 F.

If the controller differential is set too small, the heating or cooling equipment may cycle on and off too rapidly, a situation called *cycling* or *hunting*.

Timed Two-Position Control   If the operating differential is too great, it may cause uncomfortable conditions. This can be reduced by building *anticipation* into the controller. For example, a small heater may be contained in the thermostat. As soon as there is a signal calling for heating, the heater warms the thermostat faster than the room air would. As a result the thermostat reaches its high setting earlier than it would otherwise, shutting off the controlled device sooner and reducing overheating. In effect the operating differential has been reduced.

Floating Action   In floating action the controlled device is still operated by a two-position type controller. The controlled device is constructed so that it moves gradually between full open and closed. The signal from the power source moves the operating part of the controlled device in one direction. There is a neutral zone (also called *dead zone*) in which no signal is transmitted, leaving the controlled device "float-

ing" in an intermediate position until a new signal is received.

Proportional Action   In proportional action the strength of the signal from the controller varies in proportion to the amount of change in the controlled variable. The controlled device in turn moves proportionally to the signal strength, taking a fixed intermediate position at a point relative to the change in the variable. This type of action can provide much finer response to load changes than the two position types described previously, because the response is proportional to the needs, not an all-or-nothing response. Proportional type controllers and control devices are both required. For example, a modulating hot water valve would partially open or close at a position corresponding to the strength of a signal calling for increased or decreased heating.

There are some important terms used in proportional control that need to be defined.

The *set point* is the desired value of the controlled variable at which the controller is set to maintain.

The *control point* is the actual value of the controlled variable which the controller is maintaining at any given time.

Offset is the difference between the set point and the control point. It is also called *drift* or *deviation*.

Throttling range is the amount of change in the controlled variable required to move the controlled device from one extreme limit of travel to

Figure 14.6. Proportional action control.

the other (full open to full closed). The term *proportional band* of a controller also means its throttling range.

The relationships among these terms are shown graphically in Figure 14.6.

The *sensitivity* of a controller is the relationship between changes in value of the control energy and the controlled variable. For example, a pneumatic controller might have a sensitivity of 1 psi per degree F. This means that a change of 1 F in the controlled variable will change the transmitted supply pressure to the controlled device by 1 psi. The sensitivity of most controllers can be adjusted in the field to provide best control. Generally speaking, the sensitivity should be set to the maximum possible that does not cause hunting—large and continuous changes in the controlled variable.

Proportional Plus Reset Action   This type of control combines proportional action with a *reset* feature. When an offset occurs, the control point is changed automatically back toward the set point. That is, the amount of offset is reduced. The reset is accomplished by using floating action with proportional action. With proportional plus reset control the desirability of control action proportional to the load changes is achieved without the disadvantage of large offset. Figure 14.7 shows how the controlled variable behaves with proportional plus reset; it should be compared with the previous diagrams.

Proportional plus reset action is also called *proportional-integral* (PI) action.

Stability and Hunting   It may seem from the discussions that proportional plus reset action is always the most desirable. This is definitely not so. First, the proper type of control action depends on the job to be accomplished. For instance, for starting or stopping equipment a two-position control is the only suitable one.

An important feature of a control system is its ability to maintain the control variable at a reasonably steady value. This is called stability. Under certain conditions reset may cause instability, as will be explained now.

The speed of response of a reset type device is usually not very fast, due to its type of construction. Consider what happens when the controlled variable is changing very rapidly, perhaps due to sudden and frequent load changes (e.g., continual opening and closing of outside doors with a thermostat located in the room). The controller will be signaling rapidly for control action. Unfortunately, if proportional plus reset action is used, it may not be able to respond quickly enough. The controlled variable will swing widely in value and the control system will become unstable. These wide and rapid swings are called hunting. Proportional action with reset control is not desirable in HVAC systems where the controlled variable changes rapidly.

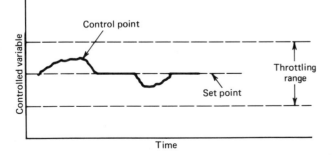

Figure 14.7. Proportional plus reset action control.

An example of a good application of proportional plus reset action is the chilled water temperature controller on a large water chiller. Close control of chilled water temperature (small offset) is desirable, because this results in good control of space temperature and humidity. Therefore, reset action is desirable, but a small offset could cause hunting. However, the air conditioning load usually changes slowly in a large building. Furthermore, the large mass of chilled water also reduces the rate of change of water temperature. Therefore, proportional plus reset action can be used without causing instability.

14.8. CONTROLLERS

As mentioned previously, the controller serves two functions: to *sense* the controlled variable signal and to *transmit* an action to the controlled device as a result of the signal. The variables most often requiring control in HVAC systems are temperature, humidity, pressure, and flow.

Temperature controllers are also called *thermostats*. Numerous types of sensing elements for thermostats are available.

A *bimetal element sensor* is made of two attached strips of different metals. The metals change lengths at different rates when their temperature changes, forcing a bending of the element (Figure 14.8). The bimetal strip may be straight or arranged in other shapes (Figure

Figure 14.9. Remote thermostat with fluid filled bulb-type sensor. (*Courtesy:* Honeywell, Inc.)

14.11). The bimetallic sensing element is used often in *room thermostats*.

Another type of sensor uses a bulb filled with a fluid (Figure 14.9). Changes in temperature cause the fluid pressure to change, and this pressure acts to move a diaphragm or bellows.

This sensor is usually inserted in a duct or pipe. When the sensing bulb is attached directly to the control element, it is called an *immersion thermostat*. When a long capillary tube connects the sensor to the control element, it is called a *remote thermostat*. This allows location of the control in a more convenient, accessible place than might otherwise be possible.

Another type of temperature sensor is called a *resistance element*. This is a thin wire whose electrical resistance changes with temperature. It is applicable to both room type and remote thermostats.

Figure 14.8. Bimetal temperature sensor — bends with temperature change.

Humidity controllers are also called *humidistats*. One type of humidity sensing element uses two different materials attached together that absorb water vapor at different rates, thus bending or moving, much like a bimetal temperature sensor.

Pressure controllers are also called *pressurestats*. The sensing element is often an open tube connected directly to the fluid where pressure is to be controlled. The fluid pressure may act on a diaphragm or bellows, or a mechanical type linkage.

Flow controllers often use pressure as a sensing signal. The velocity of the fluid where flow is to be controlled is converted to a static pressure by a sensing element such as a pitot tube, and this signal is used to control flow.

After the signal is sensed by the controller sensing element it must be transmitted by another part of the controller. Often the signal is also amplified in order to be strong enough to operate the controlled device. The transmitting element may be electric, electronic, or pneumatic.

An electric transmitter may consist simply of two electric contacts that are connected to the controlled device as shown in the thermostat in Figure 14.10. When the bimetal element bends from temperature change, it closes or opens an electrical circuit thus operating the controlled device.

Figure 14.11. Thermostat with closed electric contacts in mercury filled tube — sensor is spiral shaped bimetal element.

A *mercury switch* is often used instead of open electric contacts. A glass tube filled with liquid mercury has two electrodes inserted in it (Figure 14.11). The sensing mechanism tips the tube so that the mercury either completes or breaks the electrical circuit through the electrodes. The mercury switch has the advantage over open contacts of being enclosed, and therefore is not subject to dirt, dust, or moisture, which could increase the resistance of the electrical contacts.

Bimetal elements usually move slowly, resulting in a slow closing and opening of the electrical contacts. This may cause pitting of the contacts, increasing electrical resistance so that the signal is not transmitted properly. It may also result in bounce or chatter of the contacts, which may result in damage to electrical equipment in the circuit, since the circuit opens and closes many more times than normal.

This problem is resolved by causing *snap action* of the contacts, where the contacts open and close quickly. There are various means of achieving this. One way is to use a magnet that pulls the contacts quickly. A mercury switch also acts relatively fast, and is therefore considered snap action.

One type of electric/electronic controller uses a resistance sensing element. The transmitting element is called a Wheatstone *bridge*, which is an electric circuit arranged to deliver a voltage proportional to the signal. This voltage is very

Figure 14.10. Thermostat with open electric contacts to transmit signal.

Figure 14.12. Solenoid-type electrical relay.

small and is therefore afterward amplified. This type of device is suitable for proportional control.

A *relay* is an auxiliary device that is often used with controllers and in other parts of a control circuit. An electrical relay is a device that closes or opens one electrical circuit when a signal is received from another electric circuit. It may be used with controllers when the signal circuit is at a low voltage and the controlled device is to be operated with a high voltage. One type of relay uses a *solenoid* (Figure 14.12). A coil in the low voltage circuit acts as a magnet when electrically energized. The magnetized iron core pulls a contact armature, closing contacts in the high voltage circuit.

Solid state relays use semiconductors to transmit the signal from the control circuit to the operating circuit. They have the advantages of no moving parts, compactness, increased reliability, and rapid action. Solid state devices have become very popular for many applications in control systems.

A pneumatic transmitting element adjusts the air pressure that is supplied to the controlled device. In one type the signal from the sensor moves a flapper that covers the opening to a branch of the tube carrying the control air (Figure 14.13). As the flapper moves away from the opening, some air bleeds out and the pressure in the main line decreases. This reduces the pressure transmitted to the controlled device, causing a changed action. The opposite happens when the signal moves the flap toward the opening. This arrangement is called a *bleed* type controller. That is, some of the control air is bled off from the control circuit. A nonbleed type arrangement is also often used. It has the advantage of not wasting compressed air.

Pneumatic controllers have the desirable feature of being inherently proportional type devices. The amount of air pressure, which varies with the flapper position, can vary the position of the controlled device.

Special Purpose Thermostats A *limiting thermostat* has a built-in maximum or minimum setting of the set point. For example, it may be constructed so that the maximum heating set point is 74 F. If the occupant sets the control temperature at 80 F, it will still control at 74 F. The obvious use of this is to conserve energy.

A *day-night thermostat* is actually two thermostats in one, with two different set points. At night or on weekends the control temperature is set back to conserve energy. It is usually controlled by a time clock.

A *summer-winter thermostat* is a dual-temperature thermostat like a day-night type. It may be

Figure 14.13. Operation of pneumatic thermostat (bleed-type).

controlled manually or by the outdoor temperature.

A *master-submaster* thermostat arrangement is where one (master) thermostat controls and changes the set point of another (submaster) thermostat. This might be used to have an outdoor thermostat reset the control point of a thermostat controlling the hot water temperature in a heating system.

A *dead band* thermostat has a wide differential band (e.g., 8–10 F) within which the thermostat does not call for heating or cooling. This may result in significant energy savings in some applications.

14.9. CONTROLLED DEVICES

Valves, dampers, relays, and motors are examples of controlled devices in HVAC systems.

Control valves may be either two or three-way devices. Three-way valves are either of the *mixing* or *diverting* type (Figure 14.14). A mixing valve has two inlets and one outlet. A diverting valve has one inlet and two outlets.

Two-way valves are used to vary flow rate to the heating or cooling equipment by throttling.

Figure 14.14. Three-way valves.

Mixing and diverting valves can also be used to vary flow rates through the unit, as shown in Figure 14.15, while still maintaining the same total flow rate.

Mixing and diverting valves can also be used to control capacity by varying water temperature instead of quantity (Figure 14.16). In this application supply water from the boiler and return water are mixed to provide water at the desired temperature.

The capacity of a heating or cooling coil can be changed either by varying the water flow rate or the temperature. However, the output does not change as much with flow rate variation as it does with water temperature. For this reason water temperature control is often preferred. On the other hand, flow rate control with a two-way valve is usually less expensive, and normally is used on room terminal units.

Figure 14.15. Use of three-way valves to control flow rate.

Figure 14.16. Use of three-way valves to control supply water temperature.

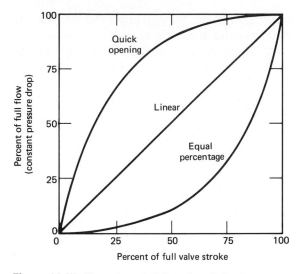

Figure 14.17. Flow characteristics of control valves.

Figure 14.18. Two-position solenoid electric valve.

closed position in response to the signal (Figure 14.18).

Modulating electric valves use a motor as a valve operator that moves the valve stem gradually in response to the signal. The operator for pneumatic valves is either a diaphragm, bellows, or piston that responds to pressure and moves the valve stem (Figure 14.19).

Automatic dampers are used as controlled devices for varying air flow, mixing air, or for bypassing (diverting) air. As with valves, the purpose is usually to vary heating or cooling capacity of the equipment. Except for very small

Valves have different characteristics concerning how the flow varies with valve stroke; this depends on the shape of the valve opening. Valves are classified in three groups: *quick opening*, *linear*, and *equal percentage*. The difference in performance is shown in Figure 14.17. The equal percentage valve is usually best for automatic control of water flow rate in coils because more variation in flow rate can be achieved for a given movement of valve stroke than can be achieved with the other types. This results in better modulation of heating or cooling capacity, since considerable throttling of the water flow rate is required to reduce capacity. When a smaller range of throttling capacity is required, a linear flow valve is adequate.

A quick opening valve is used when almost full flow must occur even with a small change in the controlled variable. It is used, for example, with an outside air preheat coil. When heat is called for, the valve should open wide to prevent freezing of the steam in the coil.

Two-position electrically operated valves use a solenoid to move the valve stem to an open or

Figure 14.19. Modulating pneumatic valve. (*Courtesy:* Honeywell, Inc.)

Parallel   Opposed

Figure 14.20. Multiblade damper arrangements.

sizes they are of multiblade construction in order to achieve better control of the air flow rate. Two arrangements are available; *parallel* or *opposed* blade (Figure 14.20). The opposed blade arrangement will give better modulation of air flow rate. Parallel blade dampers should be used only for two-position (open-closed) control.

Electric motors are used for modulating dampers in electric control systems and pistons are used as damper operators in pneumatic systems.

There are other auxiliary devices used in control systems that will not be discussed here. Although they are of practical importance, a description of them will not add to our understanding of control principles.

14.10. CHOICE OF CONTROL SYSTEMS

There are countless choices and arrangements of controls for HVAC systems. This is why our approach in this chapter up to now has not been merely to describe control systems, but instead has focused on principles. We will now look at some examples of how controls are used.

There is a choice of from where to control the HVAC system. Control can be provided at the heating/cooling source, the pump or fan, or terminal units. For example, the burner or compressor can be started-stopped or modulated. Control can also be provided by varying air or water flow rates using dampers or valves. On many systems a combination of these is used.

In addition to selecting which part of the HVAC system is to be controlled, there are choices as to what controlled variable to use for control—the space, the medium, or outdoor air. Of course the space air is the final temperature being controlled, and it would therefore seem obvious that control should be here, using a room thermostat. On many systems, however, additional control is provided from thermostats sensing outdoor air or the cooling/heating fluid medium.

It is the desire to provide better control and to conserve energy that often determines the choices made. Examples will be described later.

Safety controls are also required and are part of the control system. The student should note that systems have safety controls that are not fully discussed here.

14.11. CONTROL FROM SPACE TEMPERATURE

Control of Burner or Compressor   A simple control arrangement is to have a room thermostat control an oil or gas burner or a refrigeration compressor. In Figure 14.21 a room thermostat T starts and stops a refrigeration compressor motor M of an air conditioning unit. This is the type of control used on a window unit, with the room thermostat mounted on the unit. Larger

Figure 14.21. Space control of refrigeration compressor motor.

Figure 14.22. Space control of HW boiler burner motor or gas valve.

systems usually have more complex controls, to provide better control and to conserve energy.

A similar arrangement for a hot water boiler is shown in Figure 14.22. The room thermostat T starts and stops the oil burner motor in response to room air temperature. In a gas-fired boiler, the thermostat opens or closes a valve in the gas supply line. Not shown, but always required, is a high limit thermostat, a safety control that shuts off the burner when the water rises above a set temperature.

A warm air furnace would use a similar control.

Control of Flow Rate Through Valves A room thermostat may be used to vary the flow of hot water, chilled water, or refrigerant to a terminal unit or coil in a duct (Figure 14.23) using

automatic valves, thus varying the heating or cooling output.

Control of Volume Dampers The room thermostat varies the supply air quantity by controlling a modulating damper D, as seen in Figure 14.24. This control is used in variable air volume (VAV) systems.

Control of Mixing Dampers A room thermostat varies the proportions of hot and cold air from two ducts (Figure 14.25). Two sets of dampers move together so that one closes as the other opens on call from the thermostat. This arrangement is used in both dual duct and multizone systems (Chapter 12).

Control of Face and Bypass Dampers The quantity of air flowing over the cooling coil or bypassed around the coil is varied by the opposing motion of the two dampers (Figure 14.26).

14.12. CONTROL FROM OUTDOOR AIR

Although the outdoor air temperature rather than space temperature could be used to control space temperature, this is seldom done, because it does not provide feedback, as explained previously. However, it is used in combination for certain purposes that will be explained.

Figure 14.23. Space control of water flow rate through terminal unit or coil in duct.

Figure 14.24. Space control of air flow rate through damper.

Control of Outside and Return Air Proportions

Controls on larger systems are often used to vary amounts of outside air, from a minimum fresh air requirement to all outside air. This is done so that the outside air can be used for cooling when suitable. Control may be provided from a mixed air thermostat that adjusts the outside and return air dampers to provide cool outside air when required, as seen in Figure 14.27.

The minimum outside air damper is open during the coldest weather. When outdoor air temperature rises, mixed air thermostat T_1 gradually opens maximum outside air dampers and closes return air dampers, to provide outside air in the range of 50–60 F. Thermostat T_1 operates through a high limit thermostat T_2. When the outdoor air temperature rises to a level at which it has no cooling effect (near room temperature), thermostat T_2 takes over and closes the maximum outside air damper.

An arrangement that will offer even better energy conservation uses an *enthalpy controller* (Figure 14.28). The controller senses wet bulb temperature and therefore enthalpy of the outside and return air streams and sets the air proportion so that outside air is used for cooling whenever its enthalpy, not its temperature, is lower than that of the return air. There are days when the humidity and therefore enthalpy of the outside air may be low enough so that it is useful for cooling, even though its temperature does not indicate this.

Whether using temperature or enthalpy sensing, the outside air (OA) and return air (RA) dampers are modulated to provide cooling from the outside air whenever its temperature is low enough. Controlling these dampers saves operating the refrigeration equipment and also prevents the introduction of excess outside air at high temperatures, both conserving energy. For these reasons this system is called *economizer* control.

Outdoor Temperature Reset

A control arrangement that is sometimes included as part of

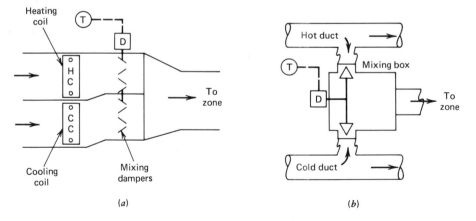

Figure 14.25. Space control of mixing dampers for multizone and dual duct systems. (*a*) Multizone unit. (*b*) Dual duct and mixing box.

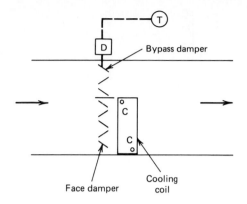

Figure 14.26. Space control of face and bypass dampers.

the control system is to have an outdoor thermostat *reset* (change) the temperature at which a variable is controlled. For example, it may reset the water temperature in a boiler (Figure 14.29). An immersion thermostat T_2 controls the boiler water temperature at its set point through the burner motor. As the outdoor temperature rises, outdoor thermostat T_1 resets the control point of T_2 lower. In this way the hot water supply temperature is inversely proportional to the outside temperature, and overheating in mild weather is reduced. This of course also results in energy conservation. The system usually also includes

additional control from space temperature, such as water flow rate.

The duct heating system in Figure 14.30 operates in a similar manner. Duct thermostat T_1 controls the air supply temperature to the space through the automatic valve. Outdoor thermostat T_1 resets the control point of T_2, to reduce overheating. Further space control might be furnished through variable volume or other means.

14.13. CONTROL FROM HEATING/COOLING MEDIUM

It is often advantageous to control equipment from controllers sensing the conditions in the heating or cooling medium. In the examples described previously the immersion and duct thermostats are controlled from water and supply air temperatures. Another example is shown in Figure 14.31.

A water chiller has a refrigeration compressor whose capacity is controlled by an immersion thermostat in the chilled water supply line. The thermostat modulates the compressor capacity to maintain a constant chilled water temperature. This is often done on large chilled water HVAC systems, with separate space control of water flow rate. In this application medium control is

Figure 14.27. Outdoor temperature control of outside and return air dampers for energy conservation.

Figure 14.29. Outdoor reset of water temperature.

Figure 14.28. Enthalpy control of outside and return air dampers for energy conservation.

Figure 14.30. Outdoor reset of supply air temperature.

Figure 14.31. Control of chilled water temperature.

Figure 14.32. Space temperature and humidity control of cooling and heating coils.

useful because it is desirable to keep the chilled water temperature at a constant value in order to ensure proper dehumidification. Another reason for medium control is that faster response may be achieved by controlling the supply air or water temperature. The methods by which compressor capacity is modulated are discussed in Chapter 13.

14.14. HUMIDITY CONTROL

For humidification in heating systems a space humidistat controls a humidifier. Steam or water spray humidifiers located in the ductwork are used. The cooling coils are often used for both cooling and dehumidifying in a cooling system. In this case some form of reheating after cooling

is used. One arrangement is shown in Figure 14.32.

The cooling coil is controlled by the room thermostat as long as room humidity is below the humidistat setting. When the humidity rises above the control point, the humidistat takes control of the cooling coil, calling for cooling. If the room temperature becomes too low, the thermostat operates the reheat coil. The psychrometrics of this process are explained in Chapter 7.

14.15. COMPLETE CONTROL SYSTEMS

One of the individual temperature control arrangements described previously may serve as the complete control system in a simple heating or cooling system. Often, however, combinations

Figure 14.33. Hot water heating control system example.

of space, outdoor, and medium control are used. This may be done because controls of temperature, humidity, and ventilation are all required, or to provide closer control. It may also be done to conserve energy. Two examples of possible control system arrangements will be described.

A hot water heating control system with individually controlled rooms or zones is shown in Figure 14.33. The controls operate as follows:

1. The immersion (medium) thermostat T_1 controls the hot water supply temperature through the burner operation.

2. The outdoor thermostat T_2 resets the control point of thermostat T_1 as the outdoor temperature varies.

3. Room thermostats T_3 control the terminal unit valves to maintain desired space temperatures.

4. The outdoor thermostat shuts off the pump when the outdoor temperature rises to a value that requires no building heating.

This control system provides good temperature control and also conserves energy. By reducing water supply temperature on mild winter days overheating of rooms is avoided, providing greater comfort and less energy use. By stopping the pump automatically when no heating is needed further energy is conserved.

An example of a control system for a single zone year-round air conditioning system is shown in Figure 14.34. The HVAC system provides summer and winter space temperature control and ventilation, but no humidity control. It operates as follows:

1. The enthalpy controller (or a temperature controller) positions the return air and maximum outside air dampers so that maximum free cooling is achieved during the cooling season. During the heating season minimum outside air is used.

2. The air discharge thermostat T_1 controls the cooling coil (summer) or heating coil (winter) to regulate discharge air temperature.

3. The room thermostat T_2 acts as a master controller to the submaster T_1. In response to a change in room temperature the room thermostat resets the control point of the discharge thermostat.

Using the room thermostat to control the discharge thermostat provides faster response of the system to changes in room temperature. The use of enthalpy control to conserve energy has been explained previously. A summer-winter room thermostat might be used to control at two different room temperatures.

These examples are given to illustrate how controls can be combined. There are hundreds of other arrangements. In each case, however, they are composed of the basic elements and with a little patience the student can analyze their operation.

Figure 14.34. Year-round air conditioning control system example.

Direct Digital Control   As we have learned, the conventional control arrangement is to have a controller sense a signal and then send an action directly to a controlled device. The use of digital computers in HVAC control systems has changed this control sequence. The scheme using computers is called direct digital control (DDC).

We will outline the basic operation and a few advantages of DDC, avoiding the complex terminology and structure of DDC systems. In DDC, conventional controllers are not used. A sensor transmits the sensed signal (e.g., temperature) to the computer. (The signal from the sensor is usually conditioned by intermediate devices so that it is in a form that the computer can understand.) The computer then sends out a signal to operate the controlled device (e.g., a valve) as needed. This signal is also usually conditioned so that it is suitable for the controlled device. This sequence is complex even in our simplified explanation, but achieves some important advantages.

Control changes (e.g., set points) can be made at one central point (the computer) instead of having to be done at each controller. This results in improved conditions and reduced operation and maintenance work.

Energy conservation strategies are easily handled in the computer program, rather than relying on many pieces of hardware and their connections that can get out of calibration and breakdown.

REVIEW QUESTIONS

1. Explain the purposes of automatic control.
2. Sketch a functional block control diagram for an open loop and for a closed loop system. Label and describe basic elements.
3. What are the two most common energy sources for control systems?
4. With the aid of a sketch, describe two-position control action. Explain controller differential and operating differential and why they are different.
5. With the aid of a sketch, describe proportional control action. Explain the terms set point, control point, offset, and throttling range.
6. With the aid of a sketch, describe proportional plus reset control action.
7. Describe three types of thermostat sensors.
8. With the aid of a sketch, describe a mixing valve and a diverting valve.
9. Sketch the piping connections of a mixing valve and a diverting valve to control flow rate through a coil.
10. With the aid of a sketch, describe the three types of flow characteristics of control valves. What are the applications of each type?
11. With the aid of a sketch, describe the two types of multiblade damper arrangements. What are the proper applications of each type?
12. With the aid of a sketch, describe the control of a boiler burner from space temperature.
13. With the aid of a sketch, describe the control of a multizone unit mixing dampers from space temperature.
14. With the aid of a sketch, describe temperature economizer control of outside and return air dampers.
15. With the aid of a sketch, describe outdoor reset control of HW boiler supply temperature.
16. With the aid of a sketch, describe space temperature and humidity control of a cooling and heating coil.
17. With the aid of a sketch, describe a hot water heating system control with individual room temperature control and outdoor reset of supply water temperature.
18. With the aid of a sketch, describe a year-round single zone air conditioning system

that controls space temperature with economizer control for energy conservation.

19. Explain the term "snap action," its purpose,
and two ways of accomplishing it.

20. Explain what a "dead band" thermostat is.
What is its purpose?

PROBLEMS

14.1. A warm air heating system has a room
thermostat that controls the furnace oil
burner motor. Draw a functional block
control diagram and identify the elements.

14.2. A three-way mixing valve on a water chiller
is controlled by an immersion thermostat
to maintain constant chilled water temperature. Draw a functional block control
diagram and identify the elements.

14.3. A room thermostat controls a two-way
modulating valve to a hot water heating
coil in a central air conditioning unit. A
duct thermostat in the supply airstream
maintains a minimum discharge air temperature. Draw a component control diagram.

14.4. An air conditioning system has a DX cooling coil with face and bypass dampers. A
room thermostat opens a two-position
solenoid valve at minimum load and then
modulates the face and bypass dampers.
Draw a component control diagram.

14.5. A cooling thermostat is set to control flow
through a fan coil unit at a room temperature of 78 F. The thermostat has a 2 F
differential setting. The response lag is
2 F. Between what values does the room
temperature vary? What is the operating
differential?

CHAPTER *15*

Energy Utilization and Conservation

*I*n the past little attention was usually given to conserving the energy used by HVAC systems because of the relatively low cost of fuel. Sharply rising fuel prices and concerns of shortages have changed this situation. Costs for energy used in building operations have become such a significant expense that it is necessary that they be kept to a minimum level. This requires a thorough analysis of energy use and conservation in HVAC design, installation, and operation.

In this chapter we will explain procedures for analyzing energy use, energy conservation in design, and efficiency in operation. Some of the energy recovery equipment that has become popular will be described. The use of computers as a tool in analyzing these problems will be discussed.

Some of the energy material covered in other chapters may be repeated here. This is done intentionally so that all of the information is presented together. On the other hand, our presentation can cover only a small part of this subject. The scope of techniques for energy conservation is vast, and new ideas are constantly developing. It should also be noted that energy use and conservation in HVAC systems are closely related to energy use in lighting and other building systems. Our discussion of these subjects will also be limited.

The intention of this chapter is to give the student an idea of how to approach the problem of energy conservation in an organized manner, to indicate what factors should be considered, and how to find this information.

OBJECTIVES

After studying this chapter you will be able to:

1. Calculate thermal energy conservation values as specified in energy codes.
2. Determine the efficiency, COP, or energy efficiency ratio (EER) of energy conversion equipment.
3. Determine the seasonal heating requirements and fuel costs for a building using the degree day method.
4. Describe energy recovery equipment used in air conditioning.
5. Suggest energy conservation procedures in construction, design, installation, and maintenance.
6. Describe uses of computers in the HVAC field.

15.1. STANDARDS AND CODES

It must be recognized that use of energy studies and conservation techniques is no longer optional on the part of the HVAC designer, contractor, or operating personnel. Standards have been developed that have already been adopted as part of almost all state building codes. No one can successfully practice in the HVAC industry without a reasonable knowledge of this subject. Although there is more than one standard being developed, those of ASHRAE that are being used as a basis for many codes, are perhaps best known. *ASHRAE Standard* 90-75 is titled *Energy Conservation in New Building Design. Standard 100P* deals with energy conservation in existing buildings. Some of the information in these standards and in state codes will be referred to here.

Many energy conservation techniques apply to both new and existing buildings, but some have meaning only in one case. Students should keep this in mind when studying this chapter.

TABLE 15.1. MAXIMUM OVERALL COEFFICIENT OF HEAT TRANFER U_0 FOR BUILDINGS

Degree Days	Overall Wall	Roof
5000	0.30	0.08
6000	0.28	0.08
7000	0.26	0.07

Some energy conservation standards for HVAC systems are detailed *prescriptions*; that is, they set down rules for limiting values of thermal resistance, design temperatures, and other criteria. We have already indicated some of the design criteria for temperature, humidity, infiltration, and ventilation in Chapters 3 and 7. Table 15.1 lists a typical standard for heat transfer values of building elements for heating.

The degree day is a measure of the severity of the heating season; it will be defined later in this chapter.

In Table 15.1 the overall wall U is defined as

$$U_{ow} = \frac{U_w \times A_w + U_g \times A_g + U_d \times A_d}{A_{ow}} \quad (15.1)$$

where

U_{ow} = overall wall U-value, BTU/hr-ft^2-F

U_w, U_g, U_d = U-value for exposed walls, windows, doors, BTU/hr-ft^2-F

A_w, A_g, A_d = area of exposed walls, windows, doors, ft^2

A_{ow} = total area of exposed walls, windows, doors, ft^2

EXAMPLE 15.1

A building has the following specifications:

$A_w = 4000$ ft^2, $U_w = 0.15$ BTU/hr-ft^2-F

$A_g = 1000$ ft^2, $U_g = 0.69$

$A_d = 100$ ft^2, $U_d = 0.40$

Degrees days = 6000

Determine the overall wall U, and compare it to the suggested standard.

Solution. Equation 15.1 will be used

$$U_{ow} = \frac{0.15 \times 4000 + 0.69 \times 2000 + 0.40 \times 100}{6100}$$

$$= 0.33 \text{ BTU/hr-ft}^2\text{-F}$$

$$\text{Standard } U_{ow} = 0.28$$

The wall in Example 15.1 does not meet the standard specified, and construction presumably would not be permitted. One solution would be to change the architectural design to reduce the amount of glass, or to change materials to reduce U-values, or both.

This approach may restrict the architect and engineer unnecessarily. For this reason an alternate type of standard called an *energy budget* has been developed, which simply sets a maximum value on energy consumption in a building. For example, a total design (maximum) energy use of 35 BTU/hr per square foot of floor area for new residences and 70 BTU/hr for new commercial buildings has been suggested as a reasonable standard for many climates. This standard usually includes energy for HVAC, lighting, and other building operations. There are many commercial buildings operating that have a design consumption of 150 BTU/hr per square foot, or higher.

Another variation on energy use standards is to specify the maximum amount that should be used for a whole year; for example, 55,000 BTU/hr per square foot per year has been suggested for residences in some climates.

A limitation of relying solely on an energy budget type standard is the difficulty in predicting the energy consumption of a building. If a building is designed to meet a certain standard, and the owner finds it does not, the resolution of the conflict may not be clear. Is the fault with

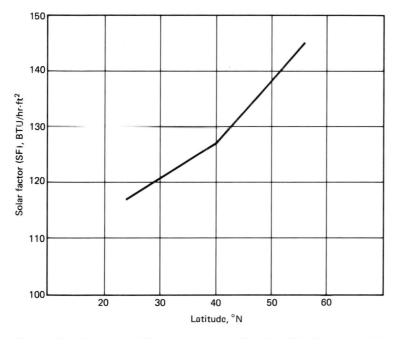

Figure 15.1. Solar factor (SF) values for use in Equation 15.2. (Reprinted with permission from the 1975 ASHRAE Standard 90-75, Energy Conservation in New Building Design.)

TABLE 15.2. VALUES OF TD_w FOR USE
WITH EQUATION 15.2

Wall weight, lb / ft²	0 – 25	26 – 40	41 – 70	Over 70
TD_w, F	44	37	30	23

the architect, engineer, or operating procedures?
Because methods and data for determining an
accurate energy use may themselves often be in
question, this approach may not be adequate by
itself.

There are various opinions on the relative
value of the different types of standards. The
author believes they can be made to compliment
each other, rather than be contradictory.

Table 15.1 is typical of an excerpt from State
Energy Codes, but is not an actual one. The
student should become familiar with the one in
his or her locality.

For summer HVAC energy use, a factor called
the OTTV (overall thermal transfer value) has
been developed and adopted in many codes. It is
defined as follows

$$OTTV = (U_w \times A_w \times TD_w + A_g \times SF \times SC$$
$$+ U_g \times A_g \times TD)/A \qquad (15.2)$$

where

OTTV = overall thermal transfer value, BTU/hr-ft²

U_w, U_g = U-values for exposed wall, glass, BTU/hr-ft²-F

A_w, A_g = area of exposed wall, glass, ft²

SF = solar factor, BTU/hr-ft² (Figure 15.1)

SC = shade coefficient (Table 6.6)

TD_w = wall temperature difference, F (Table 15.2).

Figure 15.2. Recommended overall thermal transfer values (OTTV) for walls, nonresidential buildings. (Reprinted with permission from the 1975 ASHRAE Standard 90-75, Energy Conservation in New Building Design.)

TD = design inside-outside temperature difference, F

A = total exposed area of walls and glass, ft^2

Figure 15.1 lists the SF values for use with Equation 15.2.

Maximum recommended allowable values of the OTTV for nonresidential buildings are shown in Figure 15.2.

EXAMPLE 15.2

A building has the following specifications:

$A_w = 8000$ ft^2, $U_w = 0.20$ BTU/hr-ft^2-F

$A_g = 3000$ ft^2, $U_g = 0.65$

Design TD = 15 F. Location = 36°N latitude

Wall weight = 35 lb/ft^2. SC = 0.61

Determine the OTTV value, and compare it with the maximum allowable.

Solution. Equation 15.2 will be used. From Table 15.2, TD$_w$ = 37 F. From Figure 15.1, SF = 125.

OTTV = $(0.20 \times 8000 \times 37 + 3000 \times 125$

$\times 0.61 + 0.65 \times 3000 \times 15)/11{,}000$

= 28.9 BTU/hr-ft^2

From Figure 15.2, the maximum OTTV = 32.4. The building meets the standards.

15.2. SOURCES OF ENERGY

The energy sources that are available for HVAC system operation can be classified in two groups —*depletable* (nonrenewable) and *nondepletable* (renewable). Depletable energy is stored in fossil fuels: coal, oil, and gas. These sources are quite limited in supply in relation to the rate that they are being used, and therefore will be depleted within a relatively short time (estimates range from 20–100 years for gas and oil to 1000 years for coal). Nondepletable energy sources have an expected long life in terms of human existence, so that they can be considered everlasting. Examples are the sun's radiation, nuclear energy, wind, ocean currents and tides, and geothermal energy (hot water or steam from deep underground). The nondepletable sources are often spread out or diffuse and therefore are very difficult to utilize. However, two principles should always be considered when selecting energy sources for HVAC systems:

1. It is always desirable to minimize the use of energy derived from depletable sources. These sources will eventually be depleted and they are expensive.

2. It is desirable to use energy from nondepletable sources. Of course this energy should still be used efficiently. Even though the energy is sometimes free of cost, the cost of equipment to utilize it increases as more energy is used.

15.3. PRINCIPLES OF ENERGY UTILIZATION

To be able to evaluate possible energy conservation techniques, it is important to understand some principles of energy use and availability. These principles are largely based on the First and Second Laws of Thermodynamics, discussed in Chapter 2, which should be reviewed. The implications of these laws for energy conservation will be emphasized here.

The stored energy in a fossil fuel is released through combustion, in the form of heat. The heat may be used directly if needed or the heat energy may be converted into work or power if needed. (Power is the rate of doing work.)

When the energy in the fuel is released, it is theoretically possible to use all of the equivalent heat released, as seen in Figure 15.3. However,

Figure 15.3. Maximum energy available from furnace or boiler.

Figure 15.4. Typical actual energy available from furnace or boiler.

some of the heat energy released is always wasted. It would be too expensive to build equipment that would have no losses. Figure 15.4 is an example of a more realistic situation.

For each 100 BTU of energy released from the fuel in the boiler about 20 BTUs are lost in the form of the hot combustion gases in the flue. (There are other small losses not shown in the example, such as radiation of heat from the surface of the equipment.) Some of the major efforts in energy conservation are directed to recovering this wasted heat.

Fuel is often burned to heat air, water, or steam for a direct use such as heating of buildings. Another major use, however, is to generate heat that is then converted into power. An example is the generation of electrical power, as shown in Figure 15.5.

Combustion of fuel in the boiler heats steam that is used to drive a steam turbine. The heat in the steam has been converted into mechanical power in the turbine. The turbine drives an electrical generator, creating electrical power. In this case, however, there is a very severe limit on how much of the heat energy can be converted into work, according to the Second Law of Thermodynamics. Under the best practical circumstances only about 30 BTU of the 80 BTU delivered to the turbine can be converted into work. The remainder (50 BTU) must be rejected in the form of heat. Of course this heat may still have a use.

These facts lead to further recommendations concerning energy conservation.

1. It is generally unnecessary and often wasteful to convert heat into power and then back into heat again. An example of this is the generation of electricity at a power plant, as above, and then delivering it to buildings, where through electric resistance heaters, it is converted back to heat. As seen in the examples only 30 BTUs of the original 100 BTUs in the fuel is used, whereas 80 BTUs would be available if the heat was used directly. There are sometimes other advantages to using electric-

Figure 15.5. Typical actual energy available from power plant.

ity directly for heating (convenience, control), but the energy conservation aspects should always be considered.

2. If heat is converted into power, the remainder of the heat that cannot be converted into power should be used for heating, and not wasted, if practical. This is the basis for total energy systems, to be discussed later.

Some other effects related to energy conservation, also explained in Chapter 2 are summarized here:

3. Friction causes a loss of useful energy, and should be minimized, especially in pipe and duct flow.

4. Rapid expansion of fluids causes loss of useful energy, and should be minimized. An example where this may occur is in pressure reducing devices.

5. Mixing of fluids at different temperatures causes a loss of useful energy. This occurs in dual duct and three pipe systems, and should be minimized.

15.4. MEASURING ENERGY UTILIZATION IN POWER PRODUCING EQUIPMENT (EFFICIENCY)

In order to decide whether or not equipment is utilizing energy efficiently there must be standards for measuring its performance. Some of these will be discussed here. Figure 15.6 shows a schematic arrangement of a power producing

device such as a turbine or engine which receives heat and converts some of the heat into power.

Heat (Q_1) flows from the heat source (usually a fuel) at a high temperature to the engine which converts some of this heat to the power (P) which it produces. The remainder of the heat (Q_2) is rejected to a heat sink at a lower temperature. This is usually water or the atmosphere. The values of Q_1, Q_2, and P are related to each other, according to the Energy Equation, as

$$P = Q_1 - Q_2 \qquad (15.3)$$

The efficiency (E_p) of an engine in percent is defined as

$$E_p = \frac{\text{power output}}{\text{equivalent heat input}} \times 100 = \frac{P}{Q_1} \times 100$$
$$(15.4)$$

The power and heat must be expressed in the same units.

Equation 15.4 is a useful measure of energy conservation because we always wish to get the maximum conversion of heat Q_1 to power P, and as small an amount of waste heat Q_2 as possible.

EXAMPLE 15.3

A Diesel engine-generator is being used to generate electricity in a hospital. The operating engineer wants to know how efficiently it is operating. Over a period of one hour the engineer

Figure 15.6. Energy flow for an engine.

measures 5 gallons of fuel oil consumed and 50 KWH of electricity produced. The fuel oil has a heating value of 140,000 BTU/gal. What is the efficiency of the engine-generator?

Solution. Using Equation 15.5, after converting terms into the same units

$$Q_1 = 5 \frac{\text{gal}}{\text{hr}} \times 140,000 \frac{\text{BTU}}{\text{gal}} \frac{1 \text{ KWH}}{3410 \text{ BTU}}$$

$$= 205 \text{ KWH}$$

$$E_p = \frac{P}{Q_1} \times 100 = \frac{50 \text{ KWH}}{205 \text{ KWH}} \times 100$$

$$= 24.4\%$$

The engine in Example 15.3 is converting 24% of the energy in the fuel into electrical power. The remainder is converted into heat which usually is wasted to heat sinks, in this case to both the combustion gases going out the stack and the engine cooling water. Much of this wasted energy could be recovered if heating is needed. A heat exchanger might be used to heat water from the hot combustion gases. This is an example of heat recovery equipment.

EXAMPLE 15.4

An energy consultant called in to investigate using the heat wasted in the engine in Example 15.3 estimates that a heat exchanger will recover 50% of the wasted heat. How much heat can be made available?

Solution. The wasted heat, as seen from Figure 15.6, is Q_2. Using Equation 15.3, solving for Q_2,

$$Q_2 = Q_1 - P = 205 - 50 = 155 \text{ KW}$$

$$= 155 \text{ KW} \times 3410 \frac{\text{BTU/hr}}{\text{KW}}$$

$$= 528,550 \text{ BTU/hr}$$

of which 50% can be recovered, or 264,300 BTU/hr.

The definition of efficiency given previously is correct in relation to the engine efficiency, but in HVAC applications heat, when used, is just as useful a form of energy as power. An overall efficiency could therefore be defined when heat is also used, as

$$E_0 = \frac{\text{useful energy}}{\text{energy input}} \times 100 \qquad (15.5)$$

The same units must be used for both energy terms.

EXAMPLE 15.5

If the recommendations of the energy consultant in Example 15.5 are followed, what is the overall efficiency of energy use?

Solution. The electrical energy and the recovered heat are both useful.

$$\text{Useful energy} = P + 0.5 \, Q_2$$

$$= 50 \text{ KW} + 264,300$$

$$\times \frac{1 \text{ KW}}{3410 \text{ BTU hr}}$$

$$= 127.5 \text{ KW}$$

$$E_0 = \frac{127.5 \text{ KW}}{205 \text{ KW}} \times 100 = 62.2\%$$

Notice the tremendous energy conservation achieved here—more than doubling the useful energy of the total energy consumed, as compared to before.

EXAMPLE 15.6

If the heat recovery device in Example 15.5 were not installed, the building management would have to purchase steam for the heat needed, at a cost of $6.00 per thousand pounds. The amount of recovered heat is needed for 1500 hours per year. How much will be saved on the energy bill if the device is installed?

Solution. Using a figure of 1000 BTU/lb for the latent heat given up by the steam, the savings are

$$\text{Savings} = 264,300 \frac{\text{BTU}}{\text{hr}} \times \frac{1 \text{ lb}}{1000 \text{ BTU}}$$
$$\times \frac{\$6.00}{1000 \text{ lb}}$$
$$= 1500 \text{ hr} = \$2380$$

Although the actual efficiency of an engine expresses how much energy is useful, it does not show what the best possible efficiency could be. There is a limiting maximum value of the efficiency of any engine that converts heat into power, as can be shown from the Second Law of Thermodynamics. This is expressed by the following equation

$$E_m = \frac{T_1 - T_2}{T_1} \times 100 \qquad (15.6)$$

where

E_m = maximum efficiency of an engine
T_1 = temperature at which heat is received from heat source
T_2 = temperature at which heat is rejected to heat sink

In this equation temperatures must be expressed in absolute units.

EXAMPLE 15.7

The combustion gases driving an engine are at 2000 F and they are exhausted to the stack at 600 F. What is the maximum efficiency the engine could have?

Solution. Using Equation 15.6, with the temperatures expressed in Rankine (abolute units)

$$T_1 = 2000 \text{ F} + 460 = 2460 \text{ R}$$
$$T_2 = 600 \text{ F} + 460 = 1060 \text{ R}$$
$$E_m = \frac{T_1 - T_2}{T_2} \times 100 = \frac{1400}{2460} \times 100 = 57\%$$

Example 15.7 shows that the maximum possible efficiency of an engine is always much less than 100%. This is important to realize, because in some cases it may not be worthwhile to seek very small possible improvements in efficiency.

15.5. MEASURING ENERGY CONSERVATION IN COOLING EQUIPMENT — THE COP AND EER

Refrigeration and air conditioning equipment consumes rather than produces power—the opposite of engines. Efficiency, as described in the previous section, is a measure of the relative power output of an engine, and therefore is not meaningful when applied to power consuming equipment. The *Coefficient of Performance* (COP) is a useful measure devised to measure and compare performance of air conditioning and refrigeration equipment Figure 15.7 shows the energy flows in a refrigeration machine.

Heat (Q_1) flows from the heat source at a low temperature, cooling it. This is the useful refrigeration effect. Power (P) is required in a compressor. Heat (Q_2) flows to the heat sink. According to the Energy Equation

$$Q_2 = Q_1 + P \qquad (15.7)$$

The COP is defined as

$$\text{COP} = \frac{\text{useful cooling capacity}}{\text{equivalent power input}} = \frac{Q_1}{P} \quad (15.8)$$

Both terms must be expressed in the same energy units. The COP is a good measure of energy

Figure 15.7. Energy flow for a refrigeration machine.

conservation because we always want to get the maximum amount of cooling Q_1 with the minimum value of power input P.

EXAMPLE 15.8

What is the COP of a model PCW-030T (Table 13.2) refrigeration water chiller when producing chilled water at 44 F and with entering condensing water temperature of 85 F?

Solution.   From Table 13.2 the capacity is listed as 27.6 tons with 25.0 KW required motor power. Using Equation 15.8, after converting to the same units for both terms

$$Q_1 = 27.6 \text{ tons} \times \frac{12,000 \text{ BTU/hr}}{1 \text{ ton}}$$

$$\times \frac{1 \text{ KW}}{3410 \text{ BTU/hr}}$$

$$= 96.8 \text{ KW}$$

$$\text{COP} = \frac{Q_1}{P} = \frac{96.8 \text{ KW}}{25.0 \text{ KW}} = 3.9$$

Another measure of energy conservation quite similar to the COP is the *Energy Efficiency Ratio* (EER). It is defined as

$$\text{EER} = \frac{\text{useful cooling capacity in BTU/hr}}{\text{power input in watts}} = \frac{Q_1}{P}$$
$$(15.9)$$

The EER has exactly the same two terms as the COP, only they are expressed in different units, as noted. The EER has been developed

because it is easier for the consumer to use and understand. It has already become a legal requirement for manufacturers to label the EER values for certain types of air conditioning equipment. These EER values are measured at a standard set of temperature conditions so that equipment is comparable.

EXAMPLE 15.9

A consumer shopping for a window air conditioning unit sees a 9200 BTU/hr capacity unit in a store, with a labeled EER rating of 6.2. Another unit does not have an EER label, but a sales brochure lists the cooling capacity at 9200 BTU/hr and the power input at 1200 watts, at standard conditions. Which unit will be the most efficient in energy use?

Solution.   Using Equation 15.9 to compute the EER of the second unit

$$\text{EER} = \frac{9200 \text{ BTU/hr}}{1200 \text{ W}} = 7.7$$

The second unit has a significantly better energy use efficiency. The power consumption of the first unit at full load would be

$$P = \frac{Q_1}{\text{EER}} = \frac{9200 \text{ BTU/hr}}{6.2 \text{ BTU/hr/W}} = 1480 \text{ W}$$

and would therefore be less desirable. Of course the consumer should find out why the second unit does not have the required label!

Any compressor driven cooling device has a maximum possible COP, similar to the maximum efficiency of an engine. This is expressed by the following equation

$$COP_m = \frac{T_1}{T_2 - T_1} \qquad (15.10)$$

where

COP_m = maximum possible COP for a cooling device

T_1 = temperature at which heat is absorbed from cooling load

T_2 = temperature at which heat is rejected to heat sink

The temperatures must be in absolute units in the equation.

EXAMPLE 15.10

A refrigerating unit operates at an evaporating temperature of 40 F and a condensing temperature of 100 F. What is the maximum coefficient of performance it could have?

Solution. Using Equation 15.10, with temperatures in Rankine units

$$COP_m = \frac{T_1}{T_2 - T_1} = \frac{500}{60} = 8.3$$

15.6. MEASURING ENERGY CONSERVATION IN THE HEAT PUMP

When refrigeration equipment is used as a heat pump for heating (Chapter 13), a modified form of the COP is used as a measure of efficient energy use, as follows:

$$COP_h = \frac{\text{useful heat output}}{\text{heat equivalent of power input}} = \frac{Q_2}{P}$$
$$(15.11)$$

The maximum possible COP for a heat pump when used for heating is expressed by Equation 15.12

$$COP_{hm} = \frac{T_2}{T_2 - T_1} \qquad (15.12)$$

where T_2 and T_1 have the same meaning as in Equation 15.10 and are, as before, in absolute units.

EXAMPLE 15.11

A contractor is offered a heat pump that is claimed to have a heating capacity of 800,000 BTU/hr with a compressor requiring 25 KW, while operating between 30 F and 100 F. The price is right, so the contractor buys it. Was this a good decision?

Solution. The claimed COP should be compared with the maximum possible COP.

$$COP_h = \frac{800,000 \text{ BTU/hr}}{25 \text{ KW} \times 3410 \text{ BTU/hr/KW}} = 9.4$$

$$COP_{hm} = \frac{T_2}{T_2 - T_1} = \frac{560}{70} = 8.0$$

The claimed performance is better than the maximum possible under the conditions, so the contractor is being misled. Actually any real COP would be much lower than 8.0 due to energy losses from friction and other causes.

15.7. MEASURING ENERGY CONSERVATION IN HEATING EQUIPMENT

For equipment that converts energy directly to heat (boilers, furnaces), efficiency (E) is defined as a measure of energy conservation as follows

$$E = \frac{\text{useful heat output}}{\text{equivalent heat input}} \times 100 \quad (15.13)$$

As discussed previously, unlike engines and refrigeration equipment, there is no theoretical limit below 100% on efficiency in heating equipment. Practically, however, efficiency never approaches 100%, except for electric resistance heating. There are always some energy losses, and a reasonable balance must be made between equipment cost and efficiency. The term heating coefficient of performance (COP_h), similar to that for the heat pump, is also used to express efficiency of heating equipment.

EXAMPLE 15.12

A packaged gas-fired steam boiler has a catalog rated output of 2 million BTU/hr and a listed fuel consumption of 2510 ft^3/hr of natural gas with a heating value of 1000 BTU/ft^3. What is the boiler full load efficiency?

Solution. The heat input of the boiler will be calculated, and then Equation 15.13 will be used.

$$Input = 2510 \text{ ft}^3 \text{ hr} \times 1000 \text{ BTU/lb}$$
$$= 2,510,000 \text{ BTU/hr}$$
$$E = \frac{2,000,000}{2,510,000} \times 100 = 79.7\%$$

15.8. MEASURING ENERGY CONSERVATION IN PUMPS AND FANS

Efficiency of energy use in pumps and fans is defined as

$$E = \frac{\text{power output}}{\text{power input}} \times 100 \qquad (15.14)$$

Both terms must be in the same units. If the power input is taken as the power at the fan or pump shaft, it is called brake horsepower (BHP), and the efficiency is that of the pump or fan alone. If the power input is that required at the motor, the combined efficiency of the motor and pump or fan is found. The power output of a pump was given in Equation 11.1.

EXAMPLE 15.13

An operating engineer wants to determine the efficiency of a condenser water pump and motor. The pump flow rate is 260 GPM, with a total head of 40 ft. The electrical meters show the pump motor is drawing 4.2 KW. What is the pump-motor efficiency?

Solution. Finding the power output (Equation 11.1)

$$WHP = \frac{260 \times 40}{3960} = 2.63 \text{ HP}$$
$$= 2.63 \text{ HP} \times 0.746 \text{ KW/HP} = 1.96 \text{ KW}$$

Using Equation 15.14

$$E = \frac{1.96 \text{ KW}}{4.2 \text{ KW}} \times 100 = 46.7\%$$

After determining the energy use efficiency of equipment by one of the above methods, specific recommendations for equipment improvement can be made, if needed. Although many of these recommendations have been discussed previously with each type of equipment, they will be referred to again later in this chapter. Before any steps are taken, however, an energy analysis of the whole system should be made.

15.9. MEASURING ENERGY USE IN EXISTING BUILDING HVAC SYSTEMS

Before individual energy conservation steps can be decided upon, an analysis of the energy consumption of the system should be carried out. In existing buildings this is best done by an actual hourly measurement of energy used throughout the system for the whole year. This is often not fully possible because of the lack of suitable instrumentation. It is recommended that instruments be added where practical and that this procedure be carried out.

The most complete procedure in analyzing energy use in an existing building includes measuring energy consumption of each piece of equipment, flow rates, temperatures, pressures, noting types of systems, controls, and their operations. Other interrelated system operations would also be analyzed (e.g., lighting). The physical conditions of the building would be noted. The results will suggest what conservation steps should be implemented.

The amount of detail to which the analysis is carried out largely depends on how much conservation gain is expected and the cost of the analysis itself. It is not possible to describe one single energy analysis procedure, because each case will be different. A list of some energy conserving measures that may result from studies will be listed later.

If it is not possible to measure the actual energy consumption in an existing building, a reasonably accurate estimate can be made by an energy simulation of the system operation, as is done in the design stages of new buildings. This will be explained in the next section.

If the cost of a detailed energy analysis is prohibitive, a "walk-through" survey and recommendations by an experienced energy consultant may suffice, but are not usually as effective.

15.10. MEASURING ENERGY USE IN NEW BUILDING HVAC SYSTEMS

A thorough energy conservation study requires determination of how much energy a building uses. In new building design this requires predicting energy consumption. The most complete analysis would consist of the following steps, in order:

1. Calculation of building heating and cooling loads on an hourly basis for the whole year. A complete heat balance is carried out, taking into account heat storage, internal gains, and changing climatic conditions.

2. Simulation of the system performance. This involves describing the performance of each component of the system by equations, and then solving these equations using the values of loads calculated above. Load requirements of boilers and refrigeration equipment are determined. Duct and piping flow rates and pressure drops are calculated, from which pump and fan power requirements are determined. Seasonal changes are accounted for, such as proportions of outside air used.

3. The equipment loads are transformed into energy requirements by utilizing efficiency characteristics of the equipment. Using unit fuel costs, the total energy costs can then be calculated.

These steps may be done for every hour of the year or perhaps for a selected number of hours and then summed up for the year.

To utilize it as an energy conservation tool, the analysis is carried out for different building construction features to see which uses less energy. For example, the type and amount of glass may be changed, or varied solar shading devices may be considered. Different types of HVAC systems may also be compared to see which is most efficient.

These procedures usually require the use of computers. The calculations are far too lengthy and complex to attempt them manually. Many computer programs are available that will handle these tasks satisfactorily.

A detailed energy conservation analysis is required in most large buildings because of new code design standards, and also because of client demand. However, for initial approximate energy estimates simpler procedures are available.

These may be adequate for a final energy analysis for small buildings, particularly residences.

15.11. THE DEGREE DAY METHOD

This is a well-established single calculation procedure for estimating energy requirements for heating. It can be done for the whole year or on a monthly basis. It is based on the assumption that a building will be in a heat balance with its surroundings at a certain average daily outside temperature. In the past this temperature was assumed to be 65 F. Over a one-day period at this average temperature, solar and internal heat gains produce heat storage effects that result in no heat requirements, yet a comfortable temperature is maintained.

It is assumed that the heating load is proportional to the temperature difference between 65 F and the average daily outdoor temperature. This leads to the *degree day* concept. For example, if the average daily temperature is 50 F on a given day, that day has 15 degree days (65 F–50 F). An average 35 F day would have 30 degree days. Having twice the number of degree days, the heating load would be expected to be twice as much on the 35 F day as on the 50 F day.

The degree day is therefore based on weather data, and can be calculated from known average daily temperatures, as above. This data can be summed up for a month or a whole year. Tables A.6 and 18.3 list expected degree days for a number of cities.

The degree day can be used directly to estimate the heating energy requirements of a building after the design heating load is determined.

Equation 15.15, resulting from the degree day concept, will give the heating requirements over a time period.

$$Q_0 = \frac{Q}{\text{TD}} \times D \times 24 \qquad (15.15)$$

where

Q_0 = heat energy consumed over period considered, BTU

Q = design heating load, BTU/hr

TD = design temperature difference, F

D = number of degree days per period

EXAMPLE 15.14

A building has a design heating load of 86,000 BTU/hr. The design temperature difference is 70 F. On a day with an average temperature of 30 F, how much heat is required for the whole day?

Solution. The number of degree days on this day is $D = 65 - 30 = 35$. Using Equation 15.14

$$Q_0 = \frac{86,000 \text{ BTU/hr}}{70 \text{ F}} \times 35 \text{ F} \times 24 \text{ hr}$$
$$= 1,030,000 \text{ BTU}$$

Example 15.14 is used to explain how the degree day is used, but the procedure needs to be modified in practical use. In the first place, recent investigation has found that in modern homes the assumption of 65 F outdoor temperature as that at which no heat loss occurs is often too high, because of increased internal heat gains and better insulation. This is accounted for by a correction factor C_D (Figure 15.8). The values shown are typical but may vary considerably due to factors such as type of construction and use.

Another effect that must be considered is the seasonal performance of the heating equipment. This is dependent on its full and part load efficiency, oversizing in relation to the design heating load, and use of energy conservation devices (Chapter 4). A correction factor E is used to account for performance effect. A suggested value of E for fossil fuel fired equipment is 0.55. This does not include possible effects of oversizing or energy conservation devices. For electric resistance heating a value of 1.0 for E should be used, because all of the input energy is converted to useful heat.

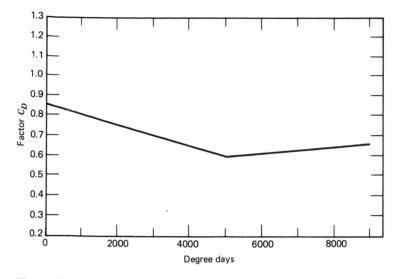

Figure 15.8. Correction factor C_D for degree days. (Reprinted with permission from the *1980 Systems, ASHRAE Handbook & Product Directory*.)

The inclusion of the two correction factors results in a modified degree day equation for calculating the seasonal energy requirements, as follows:

$$Q_0 = \frac{Q}{TD} \times D \times 24 \times C_D \times \frac{1}{E} \quad (15.16)$$

where

Q_0, Q, TD, D are as in Equation 15.15, and
C_D = correction factor for degree days
E = correction factor for heating equipment performance

The design heating load Q_0 should include an allowance for duct or pipe heat losses to unconditioned spaces (see Chapter 3) and additional infiltration allowance from stack effect, if significant. The *ASHRAE Systems Handbook* has further discussion on this and related subjects.

EXAMPLE 15.15

A residence located in Milwaukee, Wisconsin has a design heating load of 62,000 BTU/hr. Inside design temperature is 70 F. Estimate the annual energy consumption.

Solution. From Table A.6, $D = 7863$ degree days, TD $= 70 - (-7) = 77$ F. From Figure 15.8, $C_D = 0.65$. Assuming $E = 0.55$, using equation 15.16

$$Q_0 = \frac{62,000}{77} \times 7640 \times 24 \times 0.65 \times \frac{1}{0.55}$$

$$= 174,500,000 \text{ BTU/yr}$$

Using the heating value of the fuel and its unit cost, the estimated annual energy cost can be found.

EXAMPLE 15.16

Gas with a heating value of 1000 BTU/ft^3 and a cost of \$0.60 per 100 ft^3 is used to heat the house in the previous example. What is the expected average yearly heating cost?

This water is too dirty to circulate through the chilled water hydronic system, however, so a heat exchanger must be included to transfer heat from the chilled water system which is operated in its usual manner. The free cooling operation is effective in the fall and spring when wet bulb temperatures are low and the refrigeration load is light. There are other variations on this arrangement that will not be discussed here.

15.15. LIGHT HEAT RECOVERY

A considerable amount of heat is generated by the high level lighting systems in modern buildings. There are numerous ways of recovering this heat. Room return air can be circulated around special water-cooled lighting fixtures that are used to provide hot water. In addition to furnishing heat for any required uses, removing part of the lighting heat from the room also reduces the cooling load, thus achieving a double energy conservation.

15.16. TOTAL ENERGY SYSTEMS

A project which generates all of its energy needs (power and heat) at the site and which utilizes the maximum part of this energy that is practically available is called a *total energy system* (TES). The term *cogeneration* of power and heat

is also used. There are many types and combinations of these systems. Electrical power may be generated from a gas turbine, reciprocating engine, or steam turbine, any of which drive electrical generators. This equipment always rejects heat.

The aim of a total energy system is to utilize the waste heat rejected from one piece of equipment for heating or to operate other equipment. The decision about whether or not to use a TES relies primarily on energy cost and conservation. When electricity is generated at a utility, about 65% of the energy in the fuel is rejected as heat. The argument for a TES is that if the power is generated at the buildings, it is convenient and perhaps less expensive to utilize the waste heat for required heating needs, rather than purchasing both heating fuel and power. (Of course there are some cases where power generating systems have been planned which heat and distribute hot water or steam long distances.) Detailed cost analyses are necessary before making a decision to use a TES. Sometimes only part of a building's energy needs are selected for total energy use. Two examples will illustrate some of the many possible combinations of total energy systems. In Figure 15.15 a reciprocating internal combustion engine is used to drive a generator for electric power and light. About 70% of input energy from the fuel oil goes to heat in the combustion gases, engine cooling water, and lu-

Figure 15.15. Heat recovery from an engine-generator for hot water.

Figure 15.16. Use of steam waste heat for cooling.

bricating oil. Instead of wasting this energy, most of it can be used for building heating needs.

The stack gases, cooling water, and lubricating oil are circulated through heat exchangers to produce hot water. In this way about 80%, rather than 30%, of the energy input can be used.

An example of a partial total energy system is shown in Figure 15.16. A high pressure steam boiler is used in winter for the building heating system, using steam or hot water from a heat exchanger. In summer the steam is used to drive a steam turbine-centrifugal compressor refrigeration machine combination. Low pressure steam exhausting from the turbine is used to operate an absorption refrigeration machine. Further refinements could be use of condenser water from the refrigeration machines for heating needs (service hot water), as well as extracting heat from the boiler combustion gases.

In this case only the building heating and cooling needs are integrated into a total energy system. Electricity would be purchased and used in a conventional manner.

15.17. ENERGY CONSERVATION METHODS

We have discussed procedures for measuring the efficiency of energy use and have described specific devices for recovering waste heat. Our next task is to determine how to achieve more energy conservation, if it is required. We will consider separately the areas of HVAC system design, installation, and operation. In each case a few specific suggestions will be given. These are not intended as complete checklists, which would be far too lengthy and are not appropriate here.

For convenience and clarity, energy conservation in design will be divided into four categories: building construction, design criteria, system design, and controls.

Methods of achieving conservation in design usually are considered in the planning of new buildings. Some of the methods may be applicable to existing buildings through *retro-fitting*—changes to the existing system. Whether or not a specific method is practical in existing buildings depends on the nature of each case. In some cases the decision is obvious (an existing building could not be turned around to reduce solar heat gain).

15.18. BUILDING CONSTRUCTION

Reduced HVAC energy consumption in building construction usually is a direct result of minimizing heat gains and losses. Some suggestions follow.

1. Use exterior wall and roof materials with high thermal resistance. An R-20 value is not unreasonable for roofs, considering present energy costs.

2. Avoid excessive use of exterior glass (which has a low R-value and high solar heat gain). An exception to this is a residence where there is a great solar intensity during most of the winter and a moderate temperature. Some homes are designed with a large southern facing glass expanse to utilize the resulting winter solar radiation heat.

3. Plan the site for reduced gains or losses. Trees or other objects may reduce solar heat gains and infiltration.

4. Orient the building to minimize solar heat gain in summer (and maximize it in winter).

5. Use internal shading devices and external shading overhangs or side reveals.

6. Use double or even triple window glazing in severe winter climates.

7. Avoid excessive lighting requirements.

8. Use window sash that has a tight seal at the frame. Consider possible use of nonoperable windows. However, this prevents use of natural ventilation in case the mechanical system fails.

EXAMPLE 15.19

A building is being designed that will have 20,000 ft^2 of roof area. The winter design temperature difference is 70 F. Degree days are 6500. The architect asks the consulting engineer to approximate how much gas will be saved in the winter if a roof with an insulation value of R-20 is substituted for a roof with an R-5 value. Gas at 1000 BTU/ft^3 is used for heating.

Solution. The design heat losses for the two types of roof are calculated, and the difference found

$$\text{for R-5 roof } U = \frac{1}{R} = \frac{1}{5} = 0.20 \text{ BTU/hr-ft}^2\text{-F}$$

$$\text{for R-20 roof } U = \frac{1}{20} = 0.05 \text{ BTU/hr-ft}^2\text{-F}$$

R-5:
$$Q = 0.20 \times 20,000 \times 70 = 280,000 \text{ BTU/hr}$$

R-20:
$$Q = 0.05 \times 20,000 \times 70 = \underline{70,000}$$
$$\text{Difference} = \overline{210,000} \text{ BTU/hr}$$

To find the annual savings, the degree days are used. For a rough approximation we will use the unmodified degree day method.

$$\text{BTU saved} = \frac{210,000}{70} \times 6500 \times 24$$

$$= 468,000,000 \text{ BTU/yr}$$

$$\text{Gas saved} = \frac{468,000,000 \text{ BTU/hr}}{1000 \text{ BTU/ft}^3}$$

$$= 468,000 \text{ ft}^3/\text{yr}$$

15.19. DESIGN CRITERIA

HVAC system design values used in the past have often resulted in systems that consume excessive amounts of energy. Some recommended energy conserving design factors are listed here. Many of these are based on ASHRAE Standards.

1. Use a 68–72 F maximum winter indoor design temperature (75 F and over has often been used in the past), except for special occupancy.

2. Use the $97\frac{1}{2}\%$ winter outdoor design temperature (Table A.6). As an example of the savings here, this is 15 F for New York City, instead of 0 F, the recommended value found in practically all old tables. This does not reduce the heat loss directly, of course, because it is proportional to the actual temperature difference. However, it will result in less oversized equipment, and therefore lower energy consumption.

3. Use a 78 F or 80 F summer indoor design temperature (75 F or lower has often been used in the past).

4. Use the $2\frac{1}{2}\%$ summer outdoor design temperature. For Cleveland, Ohio, this is 88 F DB. Many old tables list a temperature of 100 F!

5. Use the summer outdoor wet bulb temperature coinciding with the DB, not the maximum, which usually occurs at a different time. For Dallas, Texas, this would result in a 75 F WB instead of 78 F.

6. Use a design window infiltration rate of no more than 0.5 CFM per foot of crack, for wellfitted windows. Often designers use much larger values based on existing windows.

7. Use minimum recommended design ventilation rates consistent with codes. For systems using recirculated air, ASHRAE Standards allow 33% of listed values.

8. Use cooling load calculation procedures and data that account for the building thermal storage.

9. Use correct procedures and data for calculating duct and piping friction losses, accounting for duct pressure regain if significant.

15.20. SYSTEM DESIGN

The type of HVAC system and equipment affects the consumption of energy. Some suggestions that should be considered are:

1. Examine the possible use of a total energy system compared with conventional systems.

2. Recover waste heat in exhaust air by use of heat recovery devices, to heat or cool ventilation air.

3. Use refrigeration condenser water for heating.

4. Recover waste heat from hot combustion gases with a heat exchanger.

5. Compare the energy consumption of a heat pump with separate systems for heating and cooling.

6. Examine the feasibility of a solar energy collector system for heating service hot water or even for space heating or cooling.

7. Use a variable air volume (VAV) type air conditioning system, if suitable. This is generally the most energy efficient system, because the reduction in air quantity directly lowers power use.

8. Avoid terminal reheat systems, which cool and then reheat air, unless the reheat comes from waste energy.

9. Dual duct and multizone systems or any other types that mix hot and cold air (or water) are inherently energy wasteful, although steps can be taken to minimize energy losses.

10. Choose the highest chilled water (evaporating) temperature and the lowest condensing temperature that results in satisfactory space conditions. This results in minimum refrigeration machine power consumption.

11. Choose the largest satisfactory chilled water and hot water temperature range, thereby reducing water flow rates and pumping power.

12. Choose lowest reasonable friction loss in ducts and piping, reducing fan and pumping power.

13. Select pumps and fans near the most efficient operating point.

14. Use multiple equipment so that at part loads some equipment can be shut down. The equipment operating then will be closer to full load, at which it is often more efficient.

15. For equipment that is to be operated at part load, speed reduction is usually the most efficient means of control.

16. Try to avoid using preheat coils for outside air. This often results in heating and recooling, a ridiculous waste of energy. Instead

design the intake plenum so that the outside and return air mix thoroughly.

17. Consider reducing CHW and HW pump flow rates when load reduces.

15.21. CONTROLS

Considerable energy conservation can be achieved through selection of proper automatic controls. Some of the suggestions have already been discussed in this chapter because they involve both controls and other parts of the system. Further suggestions are:

1. Use enthalpy control for the supply air. Enthalpy control devices sense and measure the total enthalpy of outside and return air, and adjust the dampers to provide the air mixture proportion that will provide the most economical natural cooling, thus reducing or eliminating need for refrigeration at times.
2. Use automatic time switching to start and stop equipment according to need.
3. Use night and weekend automatic temperature set-back for unoccupied spaces.
4. Investigate the possible use of a central computerized control system designed to provide the most efficient operation at all times.

The energy saving features of control systems are discussed further in Chapter 14.

15.22. INSTALLATION

The system must be installed so that it does not waste energy. Some suggestions are:

1. Install the system as designed, unless some apparent energy wasteful decision has been made; if so, verify before a change is made.
2. Provide air-tight duct joints, test for leakage, and seal if necessary.
3. Avoid obstructions in ducts.

4. Make gradual transitions in ducts.
5. Use wide radius turns in ducts and use turning vanes if necessary.
6. Provide ample insulation on ducts, piping, and equipment.
7. Use dampers that have tight closing features.
8. Install air vents at proper locations in water and steam distribution systems.
9. Locate thermostats so that they do not sense an abnormal condition. For example, a cooling thermostat exposed to solar radiation might cause the cooling system to operate unnecessarily.
10. Use low pressure drop valves in water lines, consistent with function. For example, butterfly valves have a lower pressure loss than globe valves.
11. Install ducts and piping with shortest lengths of run and fewest changes of direction.
12. Test and balance the complete HVAC system in accordance with design requirements.

15.23. OPERATION AND MAINTENANCE

No energy conservation design procedures will be successful without follow-up of system operation to achieve and maintain minimum energy consumption. Some suggestions are:

1. Retest and rebalance the system completely at regular scheduled intervals during its lifetime. Often the initial balancing procedures are all a system ever receives, and it gradually loses efficiency. Think of how inefficiently a car would perform if it never had a tune up!
2. Shut down any equipment when not needed. The use of the space must be examined before this decision is made. For example, even though a space is unoccupied, possible freeze ups must be prevented.

3. Set back temperatures when spaces are not occupied. In winter the temperature can often be set back to 55 F.

4. Start equipment shut down or temperature set back shortly before occupants leave.

5. Start equipment as late as possible before occupants arrive, consistent with achieving comfort.

6. Utilize natural precooling at night from outdoor air (enthalpy control may do this automatically).

7. Close outdoor air dampers near the beginning and end of operation each day when equipment is providing heating or cooling.

8. Turn off unneeded lighting.

9. Check that solar shading devices are used properly.

10. Check and replace or clean air filters when resistance reaches design conditions.

11. Clean all heat transfer surfaces regularly (coils, tubes).

12. Set chilled water temperatures (high) and condensing water temperatures (low), consistent with comfort and design values. Condensing water temperatures must not be so low that the refrigeration equipment malfunctions.

13. Adjust burners and draft on furnaces and boilers periodically to ensure complete combustion and minimum excess air.

14. Arrange for reliable water treatment services from a specialist for boiler and condensing water systems. Dirt in the system will reduce heat transfer as well as harm equipment.

15. Check operation of air vents regularly.

16. Limit the maximum electric power demand of the system (utility companies usually charge extra for high demands). This is accomplished by starting each piece of equipment in sequence, with ample time in between. In addition some equipment (e.g., electrically driven refrigeration compressors)

can be furnished with devices that limit their power demand.

17. Check ductwork regularly for leaks that may develop.

18. Check outside windows and doors regularly for abnormal cracks.

19. Perform routine maintenance regularly (lubrication, belt tension, inspection for damage or breakage).

20. Check that objects (books, clothing, furniture) are not obstructing air flow through room terminal units.

21. Check that air distribution outlets have not been obstructed or tampered with.

22. Check that room thermostats or humidistats have not been readjusted by a unauthorized personnel. Provide locks if necessary.

23. Set thermostats at 68 F in winter and 78–80 F in summer, except for special occupancies or uses.

15.24. COMPUTERS IN HVAC SYSTEMS

Computers have a growing application in the design, installation, and operation of HVAC and other building systems. The computer can improve energy conservation efforts, reduce construction and operating costs, and result in a better quality system.

In HVAC system design the computer can be used for the following tasks:

1. Load calculations. The computer can perform heating and cooling load calculations more accurately and at a much faster rate than can be done manually. It enables the designer to consider calculating the loads for varied types of building construction to find which is the minimum.

2. Simulation of system performance and energy estimation (as described previously).

3. Calculation of duct and piping sizes.

4. Selection of equipment. Manufacturers often have computer programs for this task.

A computer can be used by the mechanical contractor to assist installation in the following ways:

1. To maintain an inventory of supplies and equipment so that shortages do not develop.

2. To determine costs of the installation as it proceeds.

3. To plan the installation for the shortest construction time and lowest cost. This would involve determining when equipment would be shipped to the job, when it should be installed, how many workers to use, and similar decisions a contractor must make. The computer can do this more accurately than can be done manually, thus resulting in an installation that costs less and is completed more quickly.

Computers and related equipment can be used in building HVAC operations to sense and control conditions. Some examples are:

1. Space conditions throughout the building (room temperatures, humidity).

Figure 15.17. Central console for controlling HVAC system. (*Courtesy:* Honeywell, Inc.)

2. System conditions (pressures, flow rates).

3. Equipment conditions (power, speed).

The information may be transmitted to a central control console, shown visually on screens, and may be recorded automatically. For situations that may become dangerous a visual or audible alarm signal may be included. A central console is shown in Figure 15.17.

If it is necessary to adjust any unsatisfactory conditions, this can be done manually, or a more advanced computerized control system can be installed to send signals to control devices (valves, dampers) to adjust conditions automatically. This more advanced function of computers will undoubtedly grow in application because it provides the best energy conservation.

15.25. COMPUTER PROGRAMMING

Any individual employed in technical work in the HVAC industry will benefit by some knowledge of computer use. The best way to become familiar with computers is to learn how to program one. *Programming* refers to the instructions that are given to a computer. These instructions must be in a form called the *programming language*. There are a number of computer languages, each having different features and advantages. After preparing the instruction that the computer is to carry out, it must be transmitted to the computer. One way of transmitting data is to put it on punch cards, which are then processed by the equipment. Another form is through typewriter *terminals*, which may also have screens where the programmer can visually check the input to the computer. This is a very convenient and fast method. The program can be transmitted from the terminal to the computer by telephone lines if the distance is great. This also allows a number of terminals to be used with one computer.

The program instructions are carried out by the computer *hardware*, the equipment that does the calculations or other tasks. The results (output) are usually transmitted back to the programmer. They may be in the form of typed output, or may be visually produced on the same terminal screen that is used to show the input.

15.26. PROGRAMMING LANGUAGES

The small desk top personal computers (PCs), technically called microcomputers, are now familiar to the general public because of their extensive and varied uses. PCs are now powerful enough to handle many tasks encountered in HVAC work.

The programming language most popular for PCs is called BASIC. BASIC is simple to learn (a few hours are adequate) and is suitable for HVAC tasks.

There are also commercially available pro-

```
        BUILDING HEATING LOAD
                  &
            ENERGY ANALYSIS
            ***************

                 BY J.SMITH 3-3-87
                 ENERGY ASSOC.
   NOTRUMP & CO.
   ANYTOWN

            TRIAL #1
            ********
    ITEM      U     A    TD    Q
   ----------------------------------
   WALL      .28  4200   73   85850
   WINDOW    .56  1280   73   52330
   ROOF      .17  3200   73   39710
   DOOR      .6    120   73    5260
   EDGE      .55   220   73    8830
                               --------
   HEAT TRANS                   197240
   VENT/INF                      42560
                               --------
   TOTAL LOAD                   239800   BTUH
```

Figure 15.18. Building heating load computer program output.

```
    BUILDING HEATING ENERGY ANALYSIS
    ************************************
            FOR OIL VS. GAS
            ****************

   PROJECT NAME: AZ HEALTH CENTER
   LOCATION: MIDVILLE USA
   ANALYSIS BY: A.SMART
   DATE: 6-12-88

                 TRIAL NO.1
            ------------------

   INSIDE TEMP, F=72
   OUTSIDE TEMP, F=2
   DESIGN TD, F=70
   DEGREE DAYS=5730
   BUILDING HEATING LOAD,BTU/HR=176000

            OIL HEATING RESULTS
            -------------------
   PRICE         $/GAL=.97
   HEATING VALUE    BTU/GAL=141000
   AFUE             %=72
   OIL USED      GAL/YR=3410
   COST          $/YR=3310

            GAS HEATING RESULTS
            -------------------
   PRICE         $/THERM=.69
   HEATING VALUEBTU/CF=1100
   AFUE             %=74
   GAS USED      THERMS=4670
   COST          $/YR=3220

   GAS IS CHEAPER!
```

Figure 15.19. Computer program output for building heating energy analysis, comparing oil versus gas.

grams that can be used with virtually no previous instruction. Programs may be obtained through sources such as manufacturers, government agencies, utility companies, and independent concerns.

Programs are available for heating and cooling load calculations, duct and pipe sizing, equipment selection, and energy analysis. Figures 15.18 and 15.19 show the program output of two programs prepared especially for student use. The first program calculates the building heating load. The second program is a fuel use and cost analysis for oil and gas. Similar programs for cooling loads have also been used successfully by students.

PROBLEMS

15.1. A gas turbine with an efficiency of 26% is used to drive a centrifugal refrigeration compressor. The turbine uses 32,000 ft^3/hr of natural gas. What HP is available to drive the compressor?

15.2. A heat recovery exchanger is used to recover 30% of the wasted heat from the gas turbine in Problem 15.1 to heat hot water from 50 F to 120 F. How many GPM of water can be heated if all the recovered waste heat is used?

15.3. Determine the overall efficiency of the gas turbine heat recovery exchanger used in Problems 1 and 2.

15.4. A gas turbine burns fuel at a temperature of 1500 F. The stack gases are at 500 F. What is the maximum efficiency that could be achieved?

15.5. A refrigeration machine with a cooling capacity of 34 tons of refrigeration has a COP of 3.6. What is the power input to the compresor?

15.6. A refrigeration unit operates at an evaporating temperature of 38 F and a condensing temperature of 95 F. What is the maximum COP it could have at these conditions?

15.7. A wise consumer wants to check if the window unit she purchased is operating as efficiently as claimed. The unit is supposed to have an EER of 7.8 and a capacity of 9500 BTU/hr, at design conditions. A portable watt-meter is clamped on the outlet cord and it reads 1560 watts. Does the unit meet its claims?

15.8. A building being planned will have a design heating load of 840,000 BTU/hr.

A decision is to be made as to which of the following heating systems shall be used:

A. No. 2 oil-fired boiler, 78% efficiency.

B. Heat pump, $COP_h = 3.2$.

C. Electric resistance heaters.

On the basis of minimum energy use, which system should be selected?

15.9. In Problem 15.8, if fuel oil is $0.85/gallon and electricity is $0.07/KWH, which system would be chosen for minimum operating cost?

15.10. A pump with a mechanical efficiency of 68% delivers 87 GPM of water at a total head of 31 ft w. It is driven by a motor with a 92% efficiency. Electricity costs $0.06/KWH. The pump operates 10 hours a day, 50 weeks a year. What is the expected yearly electric billing cost?

15.11. A house in Boston, Massachusetts has a design heating load of 71,000 BTU/hr. The inside design temperature is 72 F. Using the modified degree day method, determine the expected yearly energy consumption if No. 2 oil at $0.85/gal is used as fuel. What would the yearly heating cost be?

15.12. Compare the expected yearly energy consumption found in Problem 15.11 with that of the unmodified degree day method.

15.13. The heating system in a building utilizes 16,000 CFM of ventilation and exhaust air. The inside temperature is 70 F and the average outdoor winter temperature is 32 F. The heating system operates 1600 hours in winter. A heat recovery device with an 80% efficiency is being considered to recover heat in the exhaust air. The building uses No. 6 fuel oil at $0.95/gal. What would be the yearly cost savings if the heat recovery device were installed?

15.14. The following two choices are being considered for the construction of a building's walls and windows

A. $A_w = 7200 \text{ ft}^2$,

$U_w = 0.14 \text{ BTU/hr-ft}^2\text{-F}$

$A_g = 1600 \text{ ft}^2$, $U_g = 1.10$

B. $A_w = 4100 \text{ ft}^2$, $U_w = 0.10$

$A_g = 4700 \text{ ft}^2$, $U_g = 0.62$

Determine which design has the better overall wall heat transfer coefficient.

15.15. A building in Albany, New York, has a roof constructed of built-up roofing, 4 in. gravel agg. concrete, and metal lath and $\frac{3}{4}$ in. plaster ceiling. Does the roof meet the minimum heat transfer coefficient suggested in this book? If not, what would be the R-value of insulation required to meet the standard?

15.16. A building has the following specifications

$A_w = 3400 \text{ ft}^2$, $U_w = 0.24 \text{ BTU/hr-ft}^2\text{-F}$

$A_g = 1200 \text{ ft}^2$, $U_g = 1.06$

Location: Tampa, Florida. Inside design temperature = 78 F.
Wall weight = 45 lb/ft^2. SC = 0.56
Determine the OTTV of the building, and compare it with the maximum allowable suggested in this book.

15.17. Calculate the overall heat transfer coefficient for your home, apartment house, school, and compare it with the standard in your state.

15.18. Calculate the OTTV for your home, apartment house, and school, and compare it with the standard in your state.

15.19. Calculate the design energy budget in BTU/hr-ft^2 for heating and for cooling for your home, apartment house, school, and compare it with the standards suggested in this book (or other standards).

15.20. An energy budget code requires buildings to use a maximum of 75 BTU/hr-ft^2 of energy from depletable sources from all uses in summer. A 35,000 ft^2 building has a design air conditioning load of 95 tons. The air conditioning system requires 1.3 KW of power per ton. Lighting and miscellaneous power sources use 140 KW of power. Determine if this building meets the code.

CHAPTER *16*

Instrumentation, Testing, and Balancing

*T*he design and installation of an air conditioning system may be carried out properly, but if it is not adjusted and balanced to meet design conditions, it will not perform satisfactorily. The subject of this chapter is the instrumentation and procedures used in balancing systems.

The complexity of modern air conditioning systems may make the balancing process quite involved. The student and even some experienced personnel frequently are unaware of the difficulties and requirements of balancing. In the past it was often possible to get by with making a few adjustments and checking if people were comfortable. This is no longer satisfactory on large systems. Organized procedures are required to balance a system so that it will result in comfort in all seasons. Furthermore, the in-

creased need for minimizing energy waste also requires correct balancing techniques. An improperly balanced system will almost certainly use excess energy.

The testing and balancing process is often carried out by the contractor upon completion of the installation but for large systems a whole new profession has grown, with organizations and specialists who do only this work. This has happened not only because of the complexity of the task, but it also means an independent organization verifies that the system is operating correctly.

Above all, skill in balancing requires a sound understanding of air conditioning principles, as described in this book or elsewhere.

OBJECTIVES

After studying this chapter you will be able to:

1. Choose appropriate temperature, pressure, and velocity measuring instruments for balancing HVAC systems.
2. Carry out duct traverse calculations.
3. List the data to be collected for balancing systems.
4. List the procedures for balancing air and water flow systems.

16.1. DEFINITIONS

The phrase *testing, adjusting, and balancing* (TAB) has become popular as a description of what is commonly called "balancing." *Testing* is the process of operating and checking the performance of equipment. *Balancing* is the process of proportioning the correct flow of air and water throughout the system, through mains, branches, equipment, and terminal units. *Adjusting* is the process of regulating and setting variables so that a balanced system is achieved. The variables may be speeds, temperatures, pressures, flow rates, and so forth, and are regulated by adjusting some device that controls a variable. For example, a valve may be regulated to adjust flow rates in the process of balancing. Hereafter we will use the abbreviation TAB to refer to the complete operation.

16.2. INSTRUMENTATION

The success of TAB depends on adequate *instrumentation*. These instruments are used for measuring temperature, pressure, velocity, flow rate, speed, heat flow, and electrical energy. We will discuss types of instruments and their applications. Of course some of these instruments are

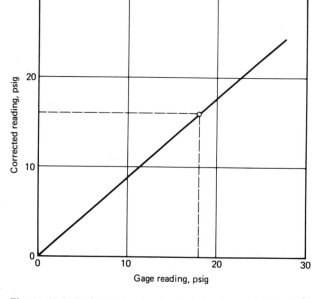

Figure 16.1. Calibration curve for pressure gage in Example 16.1.

also used in routine operating and maintenance work.

Instruments have varied degrees of accuracy. For TAB work, an accuracy within 5% of the true value is more than adequate, and is usually available without exorbitant instrument costs. Instruments should always be *calibrated* before each use. That is, the accuracy of their readings should be checked against those of an instrument or procedure known to be more accurate, and then adjusted. This should be done over the whole range of values, not just at one point. Some instruments have a means of calibration adjustment to correct their settings. With others a *calibration curve* can be made and then the readings corrected by using the curve.

EXAMPLE 16.1

Figure 16.1 is a calibration curve made for a certain pressure gage by comparing its readings with that of a manometer. During balancing a reading of 18 psig on the gage is taken at some station. What is the correct pressure?

Solution. Using Figure 16.1, at a reading of 18 psig on the gage, reading the corrected value, it is 16 psig.

It is important to not increase errors by reading instruments incorrectly. The eye of the reader should be directly in line with both the scale and pointer. If the eye is to the side, above or below, a serious error can be made.

16.3. TEMPERATURE

Temperature measuring instruments can all be called *thermometers*, although one type, the *liquid-in-glass* thermometer, is most commonly known. All temperature measuring devices utilize, as a means of developing a temperature scale, the fact that properties of materials change with changes in their temperature.

The *liquid-in-glass*, or *stem* thermometer utilizes the fact that liquids change volume with change in temperature. Mercury is the liquid most frequently used. A bulb at one end of the thermometer is inserted in the fluid. When reading temperatures of liquids in pipes, a permanent *well* should be installed at points where readings are required, because obviously the pipe cannot be cut into to take readings. The well is an insert that can be filled with oil, into which the thermometer is placed. Thermometers with an accuracy of ± 0.5 F (± 1 C) are suitable for HVAC work.

Liquid-in-glass thermometers can be affected significantly by radiant heat, causing erroneous readings. When there is radiant energy nearby, they can be protected by a suitable shield.

The *resistance* thermometer utilizes the fact that the electrical resistance of a metal wire changes with temperature. The resistance is measured by an electrical device whose indicator reads temperature directly.

When two wires of different metals have their ends joined together, it is found that they will generate a very small voltage. This arrangement is called a *thermocouple*. The voltage generated varies with temperature, and this provides a means for constructing a temperature measuring device. The wires are connected to a *potentiometer*, an electrical device that measures the voltage, and whose scale reads temperature directly (Figure 16.2).

The thermocouple and resistance thermometers have the great advantage of allowing temperature measurements at remote points. Furthermore, a large number of points can be connected to one readout instrument by simple switching. They are often used to provide the system operating engineer with temperature readings from hundreds of locations at one central console.

A *dial-type* thermometer with a metal stem (Figure 16.3) has a *bimetallic* element in the stem that moves with temperature change, actuating

Figure 16.2. Temperature measuring instrument using thermocouple and potentiometer. (*Courtesy:* Alnor Instrument Co.)

Figure 16.3. Dial-type thermometer with bimetal sensing stem. (*Courtesy:* Airserco Manufacturing Co.)

the dial. They are very convenient for quick field checks, but have limited accuracy. One type, the *contact* thermometer, can be placed directly on a pipe to read the water temperature inside, although the reading is only approximate.

16.4. PRESSURE

Manometers are simple and accurate pressure measuring instruments, which utilize the pressure that will lift a column of liquid. Water and mercury are the two most common liquids used. The *U-tube manometer* is the simplest arrangement (Figure 16.4). One end of the manometer is connected to the location where pressure is to be measured, and the other is usually left open to the atmosphere. The reading on the manometer

is therefore *gage pressure*—the difference between the pressure being measured and the atmospheric pressure. Sometimes the difference between the pressures at two locations in a pipe or duct is to be measured. In this case the ends of the manometer are connected to the two locations.

For measuring smaller pressures water is used in a manometer. Mercury is used for measuring larger pressures.

EXAMPLE 16.2

A TAB technician is going to use a U-tube manometer to measure a pressure difference. It is known that the pressure difference is in the range of 3-4 psi. Should the technician use a mercury or water manometer?

Solution. Assuming water is used, the conversion factor (Table A.2) to change to feet of water

Figure 16.4. U-tube manometer. (*Courtesy:* Dwyer Instruments, Inc.)

is

$$p = 4 \text{ psi} \times \frac{2.3 \text{ ft w.}}{1 \text{ psi}} = 9.2 \text{ ft w.}$$

If mercury is used,

$$p = 4 \text{ psi} \times \frac{2.04 \text{ in. Hg}}{1 \text{ psi}} = 8.2 \text{ in. Hg}$$

This means that a manometer more than 9 ft in height would be needed to read these pressures if water were used, a very clumsy arrangement!

Because mercury has a much greater density than water, it should be used.

EXAMPLE 16.3

A technician takes a mercury U-tube manometer reading of 9 in. Hg. One end of the manometer is attached to the location where pressure is to be measured, the other is open to the atmosphere. What is the pressure at the location, in psig?

Solution. Using the proper conversion factor from Table A.2

$$p = 9 \text{ in. Hg} \times \frac{1 \text{ psi}}{2.04 \text{ in. Hg}} = 4.4 \text{ psig}$$

A mercury *barometer* is a single vertical tube manometer that is closed at the top, evacuated, and filled with mercury. The bottom end has an open pool of mercury which is exposed to atmospheric pressure. Therefore, the mercury barometer is used to read atmospheric pressure.

For reading small differences in pressure more accurately, an *inclined manometer*, also called a *draft gage*, is used. By inclining the manometer, the vertical distance can be divided more finely, allowing more accurate readings. The inclined manometer is used often to measure small pressures in air ducts.

The *Bourdon tube* pressure gage is the most commonly used type for installation in pipe lines and on vessels. A hollow metal curved tube changes its shape with pressure. By means of a linkage to a pointer the pressure is read on a scale on the face of the gage. This type of device reads gage pressures (Figure 16.5). A *compound gage* is used to read pressures above or below atmospheric pressure.

The Bourdon tube pressure gage is rugged, reliable, and relatively inexpensive. It is not suit-

Figure 16.5. Pressure gage, Bourdon tube type. (*Courtesy:* Weksler Instruments Corp.)

Figure 16.6. Pressure gage, magnetic field actuated. (*Courtesy:* Dwyer Instruments, Inc.)

able for reading very small pressure differences and gets out of correct calibration easily.

A popular dial-type pressure gage which has a diaphragm that moves with pressure is shown in Figure 16.6. There is no mechanical linkage to the pointer—the transmission of pressure is by use of a magnetic field. It can read pressure differences to accuracies of 0.01 in. w. and is relatively inexpensive. It is quite popular for measuring pressure differences in ducts and across air filters.

16.5. VELOCITY

For very approximate air velocity studies smoke can be used. The smoke, generated by a mechanical device or candle, is injected into the airstream and observed. More practical use of smoke is to

examine the air distribution patterns in rooms and to find air leaks from ducts.

Anemometers and the *Pitot tube* are most frequently used for measuring air velocities. The Pitot tube is used together with manometers to determine velocity. It consists of two concentric

Figure 16.7. Pitot tube. (*Courtesy:* Dwyer Instruments, Inc.)

Figure 16.8. Use of Pitot tube to measure velocity pressure in a duct.

tubes (Figure 16.7). The inner tube has an opening at the tip which is pointed directly at the oncoming airstream and therefore reads *total pressure*. The outer tube has holes in the circumference that are perpendicular to the airstream, and therefore *static pressure* is transmitted through this tube.

By connecting the total pressure outlet to one leg of a manometer and the static pressure outlet to the other leg, as shown in Figure 16.8, the difference between these pressures is read on the manometer. The difference between total and static pressure is called *velocity pressure*.

As was explained in Chapter 8, the velocity pressure is related to velocity for air by Equation 8.7

$$V = 4000\sqrt{H_v}$$

where

$V =$ air velocity, ft/min
$H_v =$ velocity pressure, in. w.

Using the above equation, a table relating velocity (V) to velocity pressure (H_v) can be constructed (Table 16.1). This table is convenient to use in field testing and balancing work when measuring velocities and flow rates, as an example in the next section illustrates.

Figure 16.9. Rotating vane anemometer. (*Courtesy:* Davis Instruments Manufacturing Co.)

Figure 16.10. Deflecting vane anemometer. (*Courtesy:* Bacharach Instrument Co.)

The *rotating vane* anemometer has a propeller that spins as the air flows past it (Figure 16.9). The dial reads total distance in linear feet of air. A stopwatch is used with the device so that the velocity in feet per minute can be found.

The *deflecting vane* anemometer (Figure 16.10) has a vane in a case with an opening. As air flows by, it deflects the vane. There is a linkage to a pointer and scale where the air velocity is read.

The *hot wire* anemometer utilizes the fact that the electrical resistance of a wire changes with temperature. The temperature of the wire in turn depends on the cooling effect of the air velocity flowing past it. The sensor is connected to an electrical readout device which reads air velocity directly. It is a very sensitive instrument that is used to measure low velocities, such as drafts in occupied spaces.

TABLE 16.1. VELOCITY, V (FPM) VS. VELOCITY PRESSURE, H_V (IN. W.G.) FOR STANDARD AIR (.075 lbs / ft^3)

H_V	V	H_V	V	H_V	V
.01	400	.31	2230	.61	3127
.02	566	.32	2260	.62	3153
.03	693	.33	2301	.63	3179
.04	801	.34	2335	.64	3204
.05	895	.35	2369	.65	3229
.06	981	.36	2403	.66	3254
.07	1060	.37	2436	.67	3279
.08	1133	.38	2469	.68	3303
.09	1201	.39	2501	.69	3327
.10	1266	.40	2533	.70	3351
.11	1328	.41	2563	.71	3375
.12	1387	.42	2595	.72	3398
.13	1444	.43	2626	.73	3422
.14	1498	.44	2656	.74	3445
.15	1551	.45	2687	.75	3468
.16	1602	.46	2716	.76	3491
.17	1651	.47	2746	.77	3514
.18	1699	.48	2775	.78	3537
.19	1746	.49	2804	.79	3560
.20	1791	.50	2832	.80	3582
.21	1835	.51	2850	.81	3604
.22	1879	.52	2888	.82	3625
.23	1921	.53	2916	.83	3657
.24	1962	.54	2943	.84	3669
.25	2003	.55	2970	.85	3690
.26	2042	.56	2997	.86	3709
.27	2081	.57	3024	.87	3729
.28	2119	.58	3050	.88	3758
.29	2157	.59	3076	.89	3779
.30	2193	.60	3102	.90	3800

16.6. FLOW RATES

The Pitot tube and types of anemometers described earlier can be used to measure air flow rate by applying the continuity Equation 8.1.

$$VFR = A \times V$$

where

VFR = volume flow rate of fluid
A = cross-sectional area of duct
V = velocity of fluid

The Pitot tube is often used to find air flow rate in a duct. The velocity at any section usually varies across the duct, and therefore an average velocity must be found. This is done by placing the tube at a number of different locations (called a *traverse*), measuring the velocity pressures, calculating velocities, and averaging them. For rectangular ducts, readings at 16 or more locations are taken (Figure 16.11). When using a Pitot tube care must be taken that the probe points directly into the airstream.

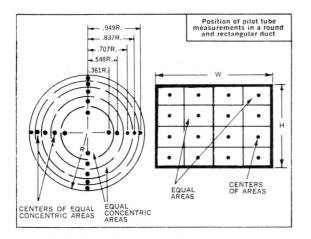

Figure 16.11. Positions of Pitot tube traverse measurements for round and rectangular ducts.

24 in.

.05	.07	.06	.05
.05	.09	.08	.06
.06	.10	.10	.05
.04	.06	.06	.04

12 in.

Figure 16.12. Duct traverse Pitot tube readings for Example 16.4.

EXAMPLE 16.4

Figure 16.12 is a cross section of a duct, listing the velocity pressure traverse readings taken with a Pitot tube. Readings are in in. w.g. What is the air flow rate in CFM through the duct?

Solution. Using Table 16.1, the velocity at each traverse reading corresponding to the velocity pressure are recorded, as follows, and the average velocity is then calculated.

Hv	V
.05	895
.07	1060
.06	981
.05	895
.05	895
.09	1201
.08	1133
.06	981
.06	981
.10	1266
.10	1266
.05	895
.04	801
.06	981
.06	981
.04	801
	16013

$$V = \frac{16013}{16}$$
$$= 1000 \text{ FPM}$$

The flow rate in CFM is now found

$$A = 24 \times \frac{12}{144} = 2 \text{ ft}^2$$

$$\text{VFR} = A \times V = 2 \text{ ft}^2 \times 1000 \text{ ft/min}$$

$$= 2000 \text{ CFM}$$

The rotating vane and deflecting vane anemometers are used mainly for measuring flow rates through air diffusers and grilles. However, a modified version of the continuity equation must be used. The outlet area of the air distribution device differs somewhat from the measured area because the air does not flow through it evenly. The outlet manufacturer will furnish this information. Furthermore, each type of anemometer will read slightly different velocities. The continuity equation is revised when using an anemometer to read:

$$\text{VFR} = A_k \times V_k$$

where

VFR = volume flow rate, CFM
A_k = effective air distribution outlet area, ft^2
V_k = average air velocity through outlet, ft/min

Water flow rates are often measured by using the relationship between velocity and pressure loss in piping. Two devices used are the *orifice plate* (Figure 6.13) and the *Venturi tube*. The pressure loss, which varies with the velocity, is measured with a manometer. Equations are developed for each device so that the velocity and flow rate can be calculated.

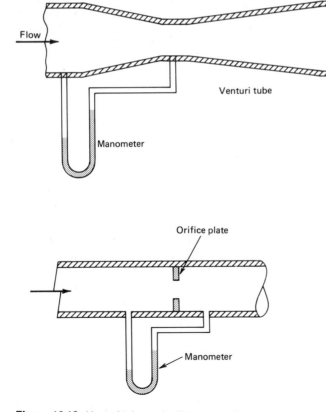

Figure 16.13. Venturi tube and orifice plate for measuring flow rate in pipes.

The orifice plate and Venturi tube must be installed in a straight length of pipe with no turns or obstructions for a considerable distance both upstream and downstream as specified by the supplier or by ASME, to prevent erroneous readings.

The system pump can be used as an instrument for approximate measure of system flow rate. By measuring the head across the pump and referring to the pump characteristic curves, the flow can be found. Similarly, any other device (e.g., chiller, valve) whose pressure-flow characteristics are known, can be used in the same way.

There are other types of liquid flow meters, but they will not be discussed here.

16.7. HEAT FLOW

Instruments have been developed that read heat flow through a surface such as a building wall or pipe. This device utilizes the heat conduction equation (Chapter 3). By measuring temperatures on both sides of the wall and with a known thermal resistance, it determines the heat flow. It can also be used to measure thermal resistance of a material.

A relatively new method of measuring heat flow is to use *infrared photography*. Temperature variations show different shades of brightness on infrared photos (Figure 16.14). This can be used to locate sources of high or low heat flow. It has become popular as an aid in energy conservation by pinpointing high heat losses through building surfaces.

16.8. HUMIDITY

The *sling psychrometer* (Figure 16.15) is the instrument used most often to find the humidity of air in HVAC work. It consists of two liquid-in-glass thermometers, one with a cloth wick wrapped around the sensing bulb. The wick is soaked in water and the apparatus is spun rapidly. The wetted stem thermometer will read

Figure 16.14. Infrared photograph locating excessive heat losses. (*Courtesy:* Thermo Test Division, Fuel Savers, Inc.)

Figure 16.15. Sling psychrometer. (*Courtesy:* Airserco Manufacturing Co.)

the air *wet bulb* temperature and the other thermometer reads the *dry bulb* temperature. The moisture content of the air can then be found by using the psychrometric chart (Chapter 7). There are also instruments called *hygrometers* that read humidity directly. They usually have a material that changes shape with change in humidity, and use a linkage to a pointer on a dial face.

16.9. EQUIPMENT SPEED

Measuring speeds of rotating equipment (fans, pumps, compressors) is often required in TAB work. A chronometric *tachometer* has a shaft with a tip that is placed on the countersunk end of the rotating shaft of the equipment whose speed is to be measured. The dial reads RPM directly. Other types are the electric and vibrating reed tachometers.

The *stroboscope* is a very useful speed measuring device used when physical contact with the rotating equipment is difficult. It has a flashing light whose frequency is adjustable. The light is pointed toward any rotating part of the equipment. When the frequency of flashing is equal to the speed of rotation (or any multiple), the equipment appears to stand still.

16.10. ELECTRICAL ENERGY

The measurement of current, voltage, and electric power use of motors is required in TAB work. *Ammeters*, *voltmeters*, and *wattmeters* are sometimes permanently wired into the system to

Figure 16.16. Clamp-on volt amp ohmmeter. (*Courtesy:* Amprobe Instrument.)

measure these characteristics. For field TAB use, however, the portable *clamp-on* combination voltmeter and ammeter is very convenient (Figure 16.16). The U-shaped jaw can be opened and clipped around an electric wire. The magnetic field created by the current flow operates the instrument.

16.11. TESTING AND BALANCING

Organized procedures for TAB are necessary to avoid an endless, wasted process. If the work is done blindly by simply going around from one outlet to another, good balancing will never be achieved, particularly on larger systems. The procedures for air and water systems will be discussed separately. The work can be divided into two stages in each case, preparatory steps and the main balancing procedure. Air system balancing is usually more complex and lengthy than water system balancing.

16.12. PREPARATION FOR AIR SYSTEM BALANCING

Before starting the system balancing routine, it is advisable to carry out a series of preliminary checks and steps. Otherwise time will be wasted later in carrying out these useful tasks. Some of this work may be the responsibility of the mechanical contractor, and some the responsibility of the TAB technician. This varies from job to job. In any case all of it should be done. The following steps are recommended:

1. Obtain ductwork drawings that can be used to indicate air flow rates and velocities through all ducts and air distribution devices. If either the engineering drawings or shop drawings are suitable, use them. Otherwise prepare simple single line drawings. List all design velocities and flow rates on the drawings. Show the locations of all dampers.

2. Obtain all equipment data from the design specifications and from manufacturers. This

includes any performance information required for testing and balancing. The following list is helpful but is not intended to cover every case:

General. Indoor and outdoor design conditions (winter and summer).

Fans. Type, size, CFM, static pressure, BHP, RPM, motor data. Obtain performance curves if possible.

Filters. Type, CFM, pressure loss, effective area (A_k), effective velocity (V_k) for balancing device to be used, air outlet temperature.

Coils. CFM, air pressure loss, face area, air temperatures and humidity, capacity (both sensible and latent for cooling coils).

Dampers. Types, pressure loss, CFM.

Mixing Boxes, VAV Units, Induction Units. CFM, pressure loss, any special information furnished by manufacturer.

3. Prepare standard TAB report forms for recording data. An apparatus report form and air outlet report form are recommended.

4. Select and obtain the instrumentation most suitable for each task.

5. Decide where all measurements will be taken and check the installation to see if access is available. Arrange for suitable access with the contractor if necessary. For example, access doors in ducts may have to be provided.

6. Check the system to see if all dampers and controls are in correct positions for balancing.

7. Schedule times for the TAB work. One period in summer and one in winter are usually necessary (for a year-round system), and preferably close to outdoor design conditions.

8. Start and operate the system and check that all equipment is performing satisfactorily. This would usually be done under the supervision of the mechanical contractor.

9. It is advisable on large systems, especially high-pressure systems, to check for major leaks by injecting smoke into the system. Any leaks should be closed with sealant or tape.

If the above preliminary procedures are carried out, considerable time and expense can be saved during balancing.

16.13. THE AIR SYSTEM BALANCING PROCESS

1. Check that the preparatory steps have been completed.

2. Measure fan speeds and adjust to design values.

3. Measure total system CFM by one or more of the following methods:

   (a) Pitot tube traverse in the main duct.

   (b) Anemometer readings across coils in the air-handling unit. Readings across the filters or dampers are not acceptable—they will be inaccurate.

   (c) Static pressure reading across the fan and reference to the fan performance curve. If this method is used the static pressure reading must be corrected for the system effect (Chapter 10).

Although any of the above methods can be used, the Pitot tube traverse is more accurate. If the CFM is within ±10% of design, proceed to the next step. If not, first check if there is some major problem such as closed dampers, malfunctioning equipment, incorrect design, and correct. If everything is satisfactory, adjust fan speed to bring CFM to within ±10% of design. Most large fans are furnished with adjustable speed drives.

4. Measure and adjust CFM to within ±10% of design in major branches using branch dampers. Use a Pitot tube traverse.

5. Measure and adjust flow at air outlets as follows:

   (a) Start with outlets farthest from fan, working backward.

(b) If there are a number of outlets on one branch, adjust total CFM to branch to approximately design value.

(c) Measure and adjust flow at each outlet to ±10% of design. Remember to use effective area and effective velocity readings. A deflecting vane-type anemometer should be used. Rotating vane types sometimes do not give satisfactory readings.

(d) Repeat the process at each outlet a second time. After the first set of adjustments it will often be found that some of the outlets do not read correctly. Repeat a third time, if necessary. Accuracy should be at least within ±10% of design. Some installations require ±5%. Record all readings.

6. Check room air distribution for drafts or dead spots. Make minor adjustments if needed. (This may be the contractor's responsibility.)

7. Check and record all performance data on fans, filters, and other equipment listed previously.

8. Measure air DB and WB temperatures before and after coils, preferably at full load. If not at full load, the manufacturer may furnish information so that the predicted full load performance can be determined.

9. Carry out a similar procedure for the return air system. This should be done at the same time as the supply system, not afterward.

10. Send in the bill for your fee!

Some TAB technicians prefer to start at the first outlet on a branch, rather than the last. For small installations some of the steps may be unnecessary. For instance, if the air handling equipment is packaged, much of the testing of individual components is not practical and perhaps not necessary. A straightforward adjustment of air flow at each outlet is the main balancing task.

16.14. PREPARATION FOR WATER SYSTEM BALANCING

1. Obtain or prepare a piping flow diagram showing all equipment. Record flow rates and temperatures on the diagram.

2. Obtain all equipment data from the design specifications and from the manufacturers. This includes any performance information required for testing and balancing. The following list is helpful but is not intended to cover every case:

Pumps. Type, size, GPM, head, BHP, RPM, motor data. Obtain performance curves.

Water Coils. Physical characteristics (rows, circuiting), capacity (sensible and latent), GPM, water pressure loss.

Chillers. Capacity, GPM, water pressure loss, motor data.

Boilers. Capacity, water or steam temperatures, flow rates.

Terminal Units. Type, size, capacity (sensible and latent), GPM, pressure loss, water temperatures, motor data (if any).

Cooling Tower. Type, size, capacity, air DB and WB temperatures, water temperatures, and flow rates.

3. Prepare report forms for recording data.

4. Select and obtain the instrumentation most suitable for each TAB task. Calibrate all instruments.

5. Decide where all measurements will be taken and check the installation to see if access is available.

6. Check the system to see if all valves and controls are in correct positions for balancing.

7. Schedule times for the TAB work. One period in summer and one in winter are usually necessary (for a year-round system), and preferably close to outdoor design conditions.

8. Start and operate the system and check that all equipment is performing satisfactorily.

16.15. THE WATER SYSTEM BALANCING PROCESS

1. Check that the preparatory steps have been completed.

2. Check pump speed with design. Pumps do not usually have adjustable speed drives, so it is unlikely there will be any significant difference between actual and design speeds.

3. Gradually close the pump discharge balancing valve, recording suction and discharge head and motor amps and volts. Do this for a number of settings from valve full open to full closed. Use this information to plot an actual pump performance curve. Correct the pump head for any significant differences in velocity heads entering and leaving (Chapter 8). Use one pressure gage with alternate connections to the suction and discharge to avoid errors between gages. Compare the results with the manufacturer's data. *Note*: Do not operate the pump at shut-off (with the valve completely closed) for any significant length of time, as it may overheat.

4. Adjust system flow to about 110% of design GPM, according to pump curve.

5. If the decision has been made to balance flow rates in mains and branches manually, adjust manual balancing valves and read flow rates on the instruments installed (orifice plates or Venturi tubes) until flows are approximately correct.

   This detailed procedure is sometimes not carried out on water systems, because adequate balancing can be accomplished at the equipment. Furthermore automatic flow control valves are often used at the equipment.

6. Check and balance flow rates through chiller and large coils. This can be done by using instruments or reading pressure loss through coil and using manufacturer's curves of pressure loss versus flow rate. Adjust balancing valves, to within ±10% of design GPM.

7. Check and balance flow rates to terminal units also using pressure loss versus flow rate data, to within ±10% of design GPM.

   An alternate method of checking design flow is to measure water temperature in and out of units. This is not as accurate, however, and is recommended only as an additional check.

8. Repeat the balancing process until no change is found.

9. Measure pump head and motor data and check flow rate.

10. Measure and adjust water flow to cooling tower, using pump curve data or instruments. Check performance of cooling tower by measuring water flow rate and temperatures, air DB and WB temperatures in and out. Calculate capacity. If not done on a design day, the manufacturer can furnish data on predicting full load performance.

11. Carry out performance tests of boilers, chillers, and cooling towers. Details will not be explained here because this is usually done on large systems with the aid of and in the presence of the manufacturer's field engineer. These are called *witness tests*. Measurements taken are similar to those described for other equipment.

On small systems many of the above steps are combined or are not necessary.

16.16. ENERGY CONSERVATION

A special listing of energy conservation methods in TAB work would be repetitious, in view of our description of balancing procedures. Obviously an unbalanced system can be highly wasteful in pump and fan power usage, and in the excess energy required if it is overheating or overcooling. It is shocking to see how much effort and expense are put into the design and installation of some HVAC systems, whereas the TAB procedures are done inadequately.

Even when a system is thoroughly and properly balanced it will inevitably become unbalanced with time. Piping will become rough, settings on dampers and valves may be changed by vibrations or unauthorized tampering, and equipment may get dirty or wear out. Included in all maintenance schedules should be consideration of a thorough rebalancing procedure every few years.

16.17. SOUND MEASUREMENT

Overall sound levels are measured with a sound level meter. Acceptable room noise levels are often specified on the dBA weighted scale (Chapter 10), which most instruments are designed to read.

If sound level readings are required at individual frequencies, a meter called a *sound analyzer* is used. This is useful when trying to discover sources of noise. For example, a centrifugal compressor may produce high-pitched sounds, which would be picked up when the analyzer was set to high frequencies.

REVIEW QUESTIONS

1. Explain the terms testing, adjusting, and balancing
2. List four types of temperature measuring instruments and how they function
3. Explain the difference between a U-tube and an inclined manometer
4. Describe three types of anemometers
5. Sketch and describe a Pitot tube connection to measure velocity
6. List the equipment data to be obtained for air system balancing
7. List the equipment data to be obtained for water system balancing

PROBLEMS

16.1. A mercury manometer is going to be used to test the performance of a pump whose maximum head is 28 ft w. What is the minimum height the manometer should have?

16.2. A rotating vane anemometer is used to measure the air flow rate exiting from a 20 in. × 10 in. duct. The average reading on the instrument after 30 sec is 800 linear ft. Determine the flow rate in CFM.

16.3. A Pitot tube traverse is used to find the air flow rate in a duct. Figure 16.17 indicates the velocity pressure readings at each point in in. w. Find the air flow rate in CFM.

16.4. A contact thermometer is used to make an approximate check of the performance of a hot water convector. The convector has a rated output of 12,000 BTU/hr. It is circulating 4 GPM of water entering at 218 F. What should the thermometer read if it is placed against the outlet pipe?

16.5. List all the locations and probable causes of excess heat loss indicated by the infrared photograph shown in Figure 16.14.

16.6. In testing the discharge head produced by a pump, the contractor uses the gage whose calibration curve is shown in Figure 16.1.

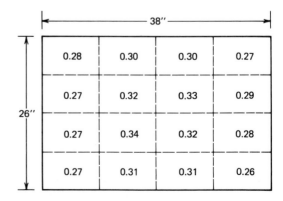

Figure 16.17. Sketch for Problem 16.3.

If the required head is 46 ft w., what should be the reading on the gage?

16.7. An absorption refrigeration machine being tested is supposed to have a pressure in the absorber of 0.12 psia. A mercury manometer is used to check the pressure.

What should be the reading on the manometer in mm of Hg vacuum?

16.8. A 12-in. air diffuser is to be balanced to a flow rate of 700 CFM. The effective outlet area is 0.57 ft^2. What should be the average velocity reading on an anemometer?

CHAPTER *17*

Planning and Designing the HVAC System

In this chapter we will explain the steps followed in the planning and design of two HVAC projects. Many of the individual items of knowledge studied previously will be utilized. Understanding these procedures will not only be useful to those interested primarily in designing

HVAC systems; it will increase the knowledge of the contractor, operator, and service engineer as to how the system functions, thereby enabling them to perform their work more efficiently. Practical problems that occur on an actual job will be faced.

OBJECTIVES

After studying this chapter you will be able to:
1. Design a small hydronic heating system.
2. Design a small air conditioning system.

17.1. PROCEDURES FOR DESIGNING A HYDRONIC SYSTEM

The steps in planning and designing hydronic systems are outlined as follows (numbers follow-

ing each item are the principal chapters where the subject is covered). In addition, recommendations on energy conservation (Chapter 15) should always be considered for each step.

1. Calculate the heating/cooling loads (3, 6).
2. Select the types and location of terminal units (5).
3. Choose the type of system piping arrangement. Plan the distribution to the terminal units (5).
4. Determine the water flow rates and temperatures throughout the system (5).

445

5. Select the sizes of terminal units (5).

6. Determine pipe sizes (8).

7. Plan the piping layout in the building and locate valves (5, 9, 17).

8. Select the pump (11).

9. Select the boiler/chiller (4, 13).

10. Select the compression tank (11).

11. Provide and locate accessories required for proper operation and maintenance: air vents, drains, unions, expansion devices, anchors, supports, insulation (9, 11).

12. Select the control system (14).

13. Prepare final plans and specifications (17).

Although these steps are generally carried out in the order shown, the nature of the project sometimes requires a slight change in sequence. On smaller projects steps are often combined.

We will describe each of these steps in some detail, as they apply to the following project.

PROJECT I. *Design a hot water heating system for the residence whose floor plans are shown in Figure 17.1. The building is located in Des Moines, Iowa, with construction as follows:*

FIRST FLOOR

WALLS: *frame with 4 in. brick veneer, sheathing, R-7 insulation, inside finish. (U = 0.09 BTU / hr-ft²-F).*

Figure 17.1. Floor plan of residence for Project I.

ROOF: *wood frame pitched roof on rafters, R-11 insulation, finished ceiling (U = 0.07 BTU / hr-ft^2-F).*

WINDOWS: *insulating (double) glass with $\frac{1}{4}$ in. air space. Double hung metal sash. Infiltration rate = 0.5 CFM / ft. Window dimensions give width first and height last.*

DOORS: *1 in. wood, 7 ft H × 3 ft W. Infiltration rate = 1.0 CFM / ft.*

BASEMENT

WALLS: *8 in. concrete block below grade.*

WINDOWS: *single glass, pivoted metal casement type. Infiltration rate = 0.8 CFM / ft. (Note that this is a high rate. These windows probably need better weatherstripping.)*

It should be noted that we have arbitrarily chosen a hot water heating system for the project. A warm air system might also have been chosen. The choice depends on costs, type of construction, convenience, and individual tastes, items we will not discuss further.

17.2. CALCULATING THE HEAT LOAD

The steps outlined in Chapter 3 will be followed here as they apply to the project. The data is recorded on a Room Heating Load Calculation form (Figure 17.2).

1. The recommended outdoor design temperature is selected from Table A.9, using the 97.5% level. An indoor design temperature of 72 F is chosen as a comfortable yet energy conserving design value (thermostat temperature may be set lower).

2. The room dimensions are taken from the architectural plans. Dimensions are converted to decimals for calculations. For example, in Bedroom No. 1 the north wall length 9 ft 6 in. is recorded as 9.5 ft. It is more precise to take dimensions from center lines of partitions and walls, rather than inside dimensions. Closet areas are included as part of a room when of significant size. Very small corners or offsets may be ne-

glected. Small hallways usually may be neglected. The decision in each case depends on whether the heat transfer values would be significant.

3. The wall areas are calculated from room plan dimensions and height. For example, for Bedroom No. 1 the exposed wall length is 9.5 + 12 = 21.5 ft. The gross exposed wall area is therefore 21.5 × 8 = 172 ft^2. Allowing for the window openings, the net wall area is 172 − 24 = 148 ft^2.

   The stairs are included as part of the kitchen area. Some designers might neglect this small area.

   The Dining Area is calculated as if it were a separate room because it will be heated by an individual terminal unit.

4. The heat transfer coefficients U are found in Tables A.4 to A.8.

5. Using the conduction heat transfer equation (3.8) the rate of heat transfer through each exposed material surface is calculated. If there is no temperature difference across the material, then there is no heat transfer. The basement is open and it is assumed that it will be at room temperature from heat given off by equipment and piping. The total heat transfer from each room is listed in the appropriate subtotal row. Note that all calculated values are rounded off to three places of accuracy.

6. Length of cracks are found from the plans.

7. The air infiltration rates are as specified (this information is obtained from the window manufacturer).

8. Infiltration heat losses are calculated from Equation 3.9. For corner rooms the infiltration on that side with the largest value is used.

9. The room heat transfer losses and infiltration losses are added to find the room total heat loss (load).

10. Building heating load. Figure 17.3 shows the calculation of the building heating load. The

Project __W. Guthrie Residence__   Location __Des Moines, Iowa__   Indoor DB __72__ F
Engrs. __Enersave Assoc.__   Calc. by __EP 1-22-88__   Chk. by __KM 2-1-88__   Outdoor DB __-5__ F

Room	Bedroom No. 1				Bedroom No. 2				Bedroom No. 3			
Plan Size	9.5' x 12' + 2' x 3'				11' x 12.5'				9.5' x 9.5' + 2' x 4'			
Heat Transfer	U	x A	x TD	= BTU/hr	U	x A	x TD	= BTU/hr	U	x A	x TD	= BTU/hr
Wall	.09	148	77	1030	.09	164	77	1140	.09	60	77	420
Window	.60	24	77	1110	.60	24	77	1110	0.60	16	77	740
Door												
Roof/ceiling	.07	120	77	650	.07	138	77	740	.07	98	77	530
Floor												
Partition												
Subtotal				2790				2990				1690
Infiltration (CFM) 1.1 x A x B x TC =												
Window	1.1	.5	17	77	720	1.1	.5	17	77	720	1.1	.5	20	77	850
Door	1.1					1.1					1.1				
Subtotal				720				720				850
Room Load				3510				3710				2540

Room	Bath				Kitchen				Living Room			
Plan Size	7' x 9' + 4' x 2'				17' x 9'				22' x 13'			
Heat Transfer	U	x A	x TD	= BTU/hr	U	x A	x TD	= BTU/hr	U	x A	x TD	= BTU/hr
Wall	.09	46	77	320	.09	82	77	570	.09	217	77	1500
Window	.60	10	77	460	.60	36	77	1660	.60	45	77	2130
Door					.64	21	77	1030	.64	21	77	1030
Roof/ceiling	.07	55	77	300	.07	153	77	820	.07	286	77	1540
Floor												
Partition												
Subtotal				1080				4080				6200
Infiltration (CFM) 1.1 x A x B x TC =												
Window	1.1	.5	15.5	77	660	1.1	.5	33	77	1400	1.1	.5	57	77	2410
Door	1.1					1.1	1.0	20	77	1690	1.1	1.0	20	77	1690
Subtotal				660				3090				4100
Room Load				1740				7170				10300

Infiltration CFM	Column A	Column B
Crack method air change method	CFM per ft or crack air changes per hr	length of crack, ft room volume/60, ft^3

Figure 17.2. Room heating load calculations for Project I.

Room Heating Load Calculations

Project *W. Guthrie Res.* _____ Location _____ Indoor DB _____ F

Engrs. _____ Calc. by _____ Chk. by _____ Outdoor DB _____ F

Room	*Dining Area*				*Basement*							
Plan Size	*9' x 9'*				*42' x 22' + 11' x 3'*							
Heat Transfer	U	x A	x TD	= BTU/hr	U	x A	x TD	= BTU/hr	U	x A	x TD	= BTU/hr
Wall	.09	106	77	730	.20	670	20	2680				
					.09	220	77	1520				
Window	.60	20	77	920	1.10	48	77	4070				
Door	.64	21	77	1030								
Roof/ceiling	.07	81	77	440								
Floor					.10	957	20	1910				
Partition												
	Subtotal			3120				10180				
Infiltration	(CFM) 1.1 x A x B x TC =				(CFM) 1.1 x A x B x TC =				(CFM) 1.1 x A x B x TC =			
Window	1.1				1.1	.8	30	77	2030	1.1		
Door	1.1	1.0	20	77	1690	1.1				1.1		
	Subtotal			1690				2030				
	Room Load			4810				12210				

Room												
Plan Size												
Heat Transfer	U	x A	x TD	= BTU/hr	U	x A	x TD	= BTU/hr	U	x A	x TD	= BTU/hr
Wall												
Window												
Door												
Roof/ceiling												
Floor												
Partition												
	Subtotal											
Infiltration	(CFM) 1.1 x A x B x TC =				(CFM) 1.1 x A x B x TC =				(CFM) 1.1 x A x B x TC =			
Window	1.1				1.1				1.1			
Door	1.1				1.1				1.1			
	Subtotal											
	Room Load											

Infiltration CFM	Column A	Column B
Crack method air change method	CFM per ft or crack air changes per hr	length of crack, ft room volume/60, ft^3

Figure 17.2 *(Continued).* Room heating load calculations for Project I.

Building Heating Load Calculations

Project __W. Guthrie Residence__

Location __Des Moines, Iowa__   Calc. by EP 2-8-88

Engineers __Enersave Assoc.__   Chk. by KM 2-12-88

	DB, F	W; gr./lb.
Indoor	72	
Outdoor	-5	
Diff.	77	

Heat Transfer	Q =	U	x	A	x	TD	=	BTU/hr
Roof		.07		957		77		5160
Walls (1st floor)		.09		1072		77		7430
(base, below gr.)		.20		670		20		2680
(base, above gr.)		.09		220		77		1520
Windows (1st floor)		.60		175		77		8090
(basement)		1.10		48		77		4070
Doors		.64		63		77		3100
Floor (basement)		.10		957		20		1910

Subtotal 33960

Infiltration Q_S = 1.1 x __116__ CFMx __77__ TC = 9830

Ventilation Q_S = 1.1 x _____ CFMx _____ TC =

Humidific Q_L = 0.68 x _____ CFMx ___ gr./lb.

Building Net Lead

Piping or duct loss ___ %   43790

Duct leakage ___ %

Pickup factor __40__ %   17520

Domestic HW Load

Boiler or furnace gross load   61310

Note: Calculation of infiltration CFM

	N	E	S	W
1st fl. windows	44		47	17
doors	20	20	20	
base. windows	24		24	16
CFM per side	88	20	91	33

½ total CFM (88 + 20 + 91 + 33) x ½ = 116 CFM
Use 116 CFM

Figure 17.3. Building heating load calculations for Project I.

areas and infiltration rates are found directly from building plans, rather than by adding up individual room areas in Figure 17.2.

The estimated infiltration is found by choosing the larger value between the side with the greatest infiltration (91 CFM on the south side) or one-half the total infiltration of all sides (116 CFM). The calculations are shown in Figure 17.3.

Based on knowledge of building conditions and operations, additions are made to the calculated building net load. In this example we have assumed that the piping heat loss is included in the basement load and that night setback of temperature requires 40% excess capacity for pickup (see Chapters 3 and 4).

The final total is the required boiler gross output. A suitable boiler should be selected using this value.

17.3. TYPE AND LOCATION OF TERMINAL UNITS

Baseboard radiation will be used, because it is inexpensive, occupied little space and is attrac-

tive—features that are desirable in a residence. Terminal units will be located under windows to prevent cold down drafts. When there is more than one window in a room, units will be located under each, if possible. The basement will not be heated with terminal units, because it will be "unfinished." Heat from the boiler and pipe surface will be adequate.

17.4. PIPING SYSTEM ARRANGEMENT

For a small, modest home either a series loop system or a one-pipe main, with the advantage of regulating flow to each unit, would be used. The advantages of a two-pipe arrangement do not gain enough benefits in such a small installation to justify the higher cost.

We will use a split series loop piping arrangement. Actually, the temperature drop in a single series loop will probably not be excessive on such a small project, but the split loop is chosen only to make a slightly more complicated design procedure, providing more instructional value for the student. (It is suggested that the student also design other systems for the additional experience.)

Figure 17.4. Schematic piping arrangement for Project I.

The connection of piping to units is now determined. The boiler should be located near the chimney in the basement, so that the vent connection is short. The building shape lends itself to splitting the loop naturally on the north and south sides, starting with Bedrooms 1 and 2.

A schematic piping sketch showing this arrangement is now made (Figure 17.4). Note that there are terminal units under both windows in Bedrooms 1 and 2, for increased comfort. In preparing this diagram it is often helpful to use transparent sketch paper placed over the architectural plans. The designer then sketches the locations of terminal units and connecting piping. These sketches are also used to aid in preparation of the finished drawings.

17.5. FLOW RATES AND TEMPERATURES

The procedures described in Chapter 5 will be followed. Let us try a system water temperature drop of 14 F. The required system flow rate to handle the building heating load (Equation 5.2) is

$$\text{Flow rate} = \frac{Q}{500 \times \text{TC}} = \frac{43,790}{500 \times 14} = 6 \text{ GPM}$$

Because there are two loops, the flow splits between them. We could either determine the quantity to be delivered to each, or we could assume each has exactly a 14 F temperature drop and then calculate the required flow rate in each. Because the two parts of the building have about the same load, it is satisfactory to assume 3 GPM each. If the loads were very different, using the same flow rate in each would result in very unequal temperature drops in each loop. This might occasionally cause problems in selection of a unit.

A supply temperature is now chosen in the recommended range. We will choose 210 F, which is a somewhat arbitrary decision. Higher temper-

atures might reduce the length of radiation slightly, but it is only a small part of the cost of the installation. Furthermore, the higher the temperature, the more severe a burn might occur from contact with exposed pipe. Even lower temperatures might be preferred by some designers for residential systems.

17.6. SELECTION OF TERMINAL UNITS

The size of each terminal unit is selected with the aid of the manufacturer's rating tables. The water flow rate and average temperature for the unit must be known. A flow rate of 3 GPM has already been chosen.

The average temperature for each unit could be determined by calculating the temperature drop through each unit, using Equation 5.2. In this system, however, it will be accurate enough to use the average water temperature for the whole system as if it were that for each unit. This is because the capacity of the baseboard does not vary much with small temperature differences. If the temperature changes were large, this would not be allowable, nor would it be allowable for a type of unit whose capacity varies greatly with water temperature. (An important case is chilled water terminal units.) The system average water temperature is

$$T_{\text{ave}} = 210 - 14/2 = 203 \text{ F}$$

The nearest lower listing in Table 5.1 is 200 F, so this value will be used in selecting each unit. A $\frac{3}{4}$ in. pipe size will be used. The capacity at these conditions is 740 BTU/hr per foot of length. For Bedroom 1 the length of baseboard needed to produce the heat output required is therefore

$$\text{Length} = \frac{3510 \text{ BTU/hr}}{740 \text{ BTU/hr per ft}} = 4.7 \text{ ft}$$

For this particular room the radiation is to be installed under both windows, so two units, each $2\frac{1}{2}$ ft long, or perhaps one 3 ft and the other 2 ft

TABLE 17.1. SELECTION OF TERMINAL UNITS FOR PROJECT I

Room	Unit	BTU/hr	GPM	t_{ave}	BTU/hr per ft	L, ft	Type
Bedroom 1	A	1760	3	200	740	3	Base.
Bedroom 1	B	1760	3	200	740	3	Base.
Bath	C	1740	3	200	740	3	Base.
Kitchen	D	7170	3	200		4	Convector.
Dining	E	4810	3	200	740	7	Base.
Living	F	10,300	3	200	740	14	Base.
Bedroom 3	G	2540	3	200	740	4	Base.
Bedroom 2	H	1860	3	200	740	3	Base.
Bedroom 2	I	1860	3	200	740	3	Base.

long would be used. The selection of the units for other rooms is carried out in a similar manner. The results are shown in Table 17.1.

Sometimes the required length of a unit is greater than the outside wall length available. One solution is to use adjacent inside walls. Of course if there are cabinets or other obstructions, this prevents use of a wall. In such cases a convector or other type of unit with high capacity may solve the problem, as in the kitchen.

17.7. PIPE SIZING

The procedure described in Chapter 8 will be used to find the pipe diameters and system pressure loss.

1. The piping system sketch is shown in Figure 17.4.
2. The flow rate in each line has been determined, and is shown on the sketch.
3, 4. Copper tubing will be used because it is easy to work with. Furthermore, the baseboard radiation is copper tube. A trial size of the main is made. Using Figure 8.15, either a 1 in. or $1\frac{1}{4}$ in. Type L will have a friction loss within the recommended range with a flow rate of 6 GPM (2.8 and 1.0 ft of water per 100 ft, respectively). The 1 in. diameter will be

chosen for a small initial cost saving. However, operating cost will be slightly higher.

5. Using the chosen friction loss rate as a guide (equal friction method), the two loop pipe sizes will be chosen. The flow rate is 3 GPM in each. From Figure 8.15 it is seen that a $\frac{3}{4}$ in. diameter provides the friction loss rate closest to that desired, 3.0 ft of water per 100 ft. The velocity is about 2 FPS, an acceptable value.

We have used the suggested equal friction method for sizing the pipe. Actually in such a simple system the procedure does not have to be followed rigorously. Either a 1 in. or $1\frac{1}{4}$ in. pipe could have been used for the larger line and a $\frac{3}{4}$ in. or $\frac{1}{2}$ in. pipe for the subloops, because balancing will be a simple operation.

17.8. PIPING OR DUCT LAYOUT

The physical layout of the piping, including all changes in direction, valves, and other items in the lines, must be known in order to find the system pressure loss. This information is then used to determine the required pump head. Although the final drawings show the detailed piping layout, it suffices to make freehand sketches to scale for the purpose here. Often the simple sketch made previously is used, with some imagination necessary to visualize the changes in direction.

To plan the piping layout, place transparent sketch paper over each floor plan and sketch the desired location of piping and equipment to be installed on that floor. Some recommended good practice guidelines for locating piping (which also apply to ductwork), are:

1. Piping and ducts are usually run parallel or perpendicular to building walls.

2. Use the minimum number of changes in direction consistent with item 1.

3. The location should minimize the difficulty of installation. For example, avoid locating piping in cramped quarters where the pipe fitter cannot work easily.

4. The location should provide for ease of operation. For instance, a pipe line containing a valve that is frequently used and yet is difficult to reach would be poor planning.

Figure 17.5. HVAC plans for Project I.

Piping that obstructs equipment control panels is another example of a poor layout.

5. The layout should provide for ease of maintenance. Piping and ducts should not be located where they interfere with access to equipment requiring servicing or replacement.

6. The layout should not interfere with normal use of the space. For this reason horizontal pipes are usually run overhead and close to walls. They cannot be located where they block or displace furniture or openings.

7. The location should not physically interfere with or disturb proposed installations of other trades (electrical and plumbing). As an example, ductwork running directly underneath a lighting fixture would block some light. HVAC planners must coordinate their layouts with the other trades to prevent possible interferences. This requires studying their plans.

8. Avoid penetration of structural members (columns or beams). If this seems unavoid-able, permission from the structural engineer must be obtained.

9. The layout should not create any safety hazards. For example, a valve located above a transformer would be poor practice, in case of a valve leak.

10. The layout should conform with aesthetic requirements of the architecture. An obvious example is that piping and ducts must be concealed above hung ceilings or behind walls in many types of spaces.

11. The layout must meet all applicable Code requirements.

In addition to the general recommendations in this list, each project has its own peculiarities that affect the choice of piping and duct location.

The piping layout developed for this project is shown in Figure 17.5. The student should check the plans to see if it follows good practice. Lines are run in the basement overhead. Note that where obstructions are encountered on the occupied floor, the piping drops down to the basement.

TABLE 17.2. PRESSURE DROP CALCULATIONS FOR PROJECT I

Section	Item	D in.	GPM	V FPS	EL ft	No.	TEL, ft	H_f ft w./ 100 ft	H_f ft w.
Main	Pipe	1	6	2.3	43		43	2.8	
	Boiler				11		11		
	Gate				1.0	3	3		
	Check				42	2	84		
	El				2.6	7	18		
	Tee				2.6	1	3		
					Subtotal		$\overline{162}$	$\times$ 2.8 / 100 =	4.5
Loop									
A	Pipe	$\frac{3}{4}$	3	2.0	78		78	3.0	
	Globe				22	1	22		
	El				2.0	18	36		
	Tee				4.0	1	4		
					Subtotal		$\overline{140}$	$\times$ 3.0 / 100 =	4.2
								Total H_f =	8.7

Gate valves shall be located on both sides of any equipment to be serviced (see Figure 4.9). Globe valves or other flow regulating valves have been located in each loop so that flow can be balanced to the design values.

17.9. PUMP SELECTION

The first step in selecting the pump is to calculate the system pressure loss. From inspection of the piping layout it is clear that the circuit with the greatest pressure loss is the loop on the north side of the building. Not only is the total length of piping longer, but there are more changes in direction. The lengths are measured, the fittings are counted, and equivalent lengths are determined. Pressure losses in each section are calculated, then added. The results are recorded in Table 17.2. Students should go through the procedure to see if they obtain the same results. Be sure to include all vertical distances and elbows. When some information is not given, an estimate is often accurate enough. For example, a vertical distance of 2 ft from the horizontal lines in the basement to the terminal units is assumed.

The system friction loss is 8.7 ft w. Because the piping is a closed loop, there is no additional pump pressure needed to lift the water, and therefore the required pump head is also 8.7 ft w. An in-line pump will be used for its low cost and small capacity needed. From Figure 11.4 a No. 102 pump is adequate, developing about 11 ft w. head at 6 GPM, using a $\frac{1}{8}$ HP motor.

17.10. BOILER SELECTION

A boiler will be selected from Figure 4.21. It is assumed that the fuel is natural gas. On an actual project the choice of fuel is made on the basis of cost, convenience, and availability. The required boiler gross output is 73,880 BTU/hr.

The boiler that will satisfy the load is:

Model no. GG100
Net I-B-R output = 69,600 BTU/hr
Gross output = 80,000 BTU/hr
Input = 100,000 BTU/hr
Flue diameter = 6 in.

Note that in order to obtain the boiler gross capacity, we have used the piping and pickup losses known from job conditions in the load calculations. If no specific knowledge of these losses were available, we would have used the industry standard losses (Chapter 4).

The boiler full load efficiency can also be determined:

$$E = \frac{80,000}{100,000} \times 100 = 80\%$$

17.11. COMPRESSION TANK

The size of the compression tank is found by using Equation 11.7. The volume of water in the system must first be determined, using Table 9.2. Lengths are approximated from building plans.

1 in. tube, 43 ft × 0.044
gal/ft = 1.9 gal.
$\frac{3}{4}$ in. tube, 160 ft ×
0.025 gal/ft = 40
Boiler = 12
System volume = $\overline{54\ \text{gal}}$

The value for the boiler water volume is furnished by the manufacturer.

The terms in Equation 11.7 are:

$t = 210\ \text{F}$
$H_a = 14.7\ \text{psi} \times 2.3\ \text{ft w.}/1\ \text{psi} = 34\ \text{ft w.}$
$H_t = 5 \times 2.3 + 4 + 34 = 50\ \text{ft w. abs}$
$H_0 = 45\ \text{psia} \times 2.3 = 104\ \text{ft w. abs}$

In these calculations, a fill pressure of 5 psi was

Figure 17.6. Control diagram for Project I.

used, and an estimated system height of 4 ft above the tank elevation used to find H_t. Boiler relief valve setting is 30 psig. Substituting in the equation:

$$V = \frac{[0.00041(210) - 0.0466]54}{34/50 - 34/104} = 6.0 \text{ gal}$$

17.12. ACCESSORIES

The location and types of valves to be used have already been discussed. Air vents will be installed at high points. This would include points where the horizontal piping from the radiation drops down to the basement. A drain connection with valve will be installed at the return to the boiler. Connections to the boiler and pump will be made with unions so that they can be removed easily. An expansion loop will be provided in the long 1 in. line. The $\frac{3}{4}$-in. lines have enough natural offsets to permit expansion. All piping in the basement will be insulated. Proper support hangers will be furnished for all horizontal piping.

17.13. CONTROLS

The application calls for a simple, inexpensive control system. A room thermostat controls the pump operation in response to a call for heat. A high limit immersion thermostat controls the gas

burner to maintain a set hot water temperature. A control diagram is shown in Figure 17.6. Safety controls would be provided in addition to these operating controls.

The room thermostat is located on an inner wall of the living room. Although direct control of temperature is from this room only, the smallness of the house should result in limited temperature variations in other rooms. A dual thermostat will be used to set back night temperatures and conserve energy.

17.14. PLANS AND SPECIFICATIONS

The final step of the project is to prepare finished *plans* and *specifications* of the heating system. The drawings would be similar to those in Figure 17.5, made to a scale convenient for easy reading. A *plan view* is made for each floor, and *details* of connections to equipment are shown separately.

The specifications will include descriptions of materials and equipment. Some excerpts from a typical specification will be given to familiarize the student with how they are written. Much of the information is prepared with the aid of the manufacturer's equipment specifications.

BOILER   Furnish and install a low pressure hot water heating boiler of the gas-fired, packaged, cast-iron type, designed and constructed in accordance with and approved by ASME, I-B-R, and AGA Stan-

dards. It shall have a net I-B-R rating of at least 69,600 BTU / hr with a piping and pick-up factor of 1.15 and a full load efficiency of no less than 80%.

The boiler shall be preassembled with heat exchanger, air eliminator, combustion chamber, and insulated jacket. It shall have a gas burner and manifold and gas controls as a unit, including automatic gas valve, 24 volt, with pilot, filter and pressure regulator, 100% pilot shut off, main gas valve, pilot burner, and thermocouple. The boiler shall include a combination pressure-temperature gauge, ASME relief valve, boiler drain valve, and draft hood. It shall be as manufactured by _ _ or approved equal.

BASEBOARD RADIATION Furnish and install baseboard heating elements where shown on the plans, of the capacity specified. They shall be of $\frac{3}{4}$ in. nominal copper tubing with aluminum fins. Furnish complete enclosures for wall-to-wall covering, including front and back panels, end caps, corner, and trim. All components shall have shell white baked on finish. Furnish support brackets, expansion cradles and dampers. Baseboard shall be as manufactured by _ _ or approved equal.

Similar specifications would be included for all other items in the system. The specifications must also include legal and contractual statements, covering subjects such as guarantees and liabilities.

17.15. ENERGY USE AND CONSERVATION

The recommendations for energy conservation have been applied in planning the system. This includes the design temperatures chosen, the R-values of construction materials, the use of storm sash, and windows with small infiltration rates. The control system includes night setback of temperature. The pump is operated intermittently rather than continuously.

An estimation of fuel used can be made using the modified degree day method. Applying Equa-

tion 15.16:

$$Q_0 = \frac{Q}{TD} \times D \times 24 \times C_D \times \frac{1}{E}$$
$$= \frac{43,790}{77} \times 6588 \times 24 \times 0.62 \times \frac{1}{0.55}$$
$$= 101,000,000 \text{ BTU}$$

Using natural gas with a heating value of 1000 BTU ft^3 the fuel quantity used is

$$101,000,000/1000 = 10,100 \text{ ft}^3 \text{ per year}$$

17.16. PROCEDURES FOR DESIGNING AN ALL-AIR SYSTEM

The steps in planning and designing all-air systems are outlined as follows (numbers in parentheses are chapter references):

1. Calculate the heating/cooling loads (3, 6)
2. Determine the supply air conditions (7)
3. Choose the type of system (12)
4. Plan the equipment and duct locations (17)
5. Determine duct sizes (8)
6. Determine sizes of air distribution devices (10)
7. Select equipment (5, 10, 12, 13)
8. Provide and locate accessories required for proper operation and maintenance (9, 16)
9. Select the control system (14)
10. Prepare final plans and specifications (17)

These steps are generally carried out in the order shown, but occasionally some change in the sequence is necessary, particularly when information required to complete one step is dependent on data found in a later step. Sometimes steps can be combined. The recommendations on energy conservation should always be considered for each step.

We will describe each of these steps in some detail as they apply to the following project.

PROJECT II. *Design an all-air summer air conditioning system for the department store whose floor plans are shown in Figure 17.7. The building is located in Fort Worth, Texas, with construction and conditions as follows:*

WALLS: *4 in. face brick, 8 in. cinder block, furred gypsum wallboard. Dark colored.*

ROOF: *4 in.-lightweight concrete deck, suspended ceiling.*

PARTITIONS: *4 in. cinder block.*

BUILDING CONSTRUCTION: *medium weight.*

GLASS: *single clear plate glass. No shading.*

LIGHTING: *3.0 W / ft^2, including ballast factor.*

OCCUPANCY: *500 people.*

EQUIPMENT: *16 motors $\frac{1}{4}$ HP each, operating 50% of time (vending machines and similar devices).*

DOOR INFILTRATION: *1 CFM / ft.*
    Only sales area will be air conditioned.

17.17. CALCULATING THE COOLING LOAD

The steps outlined in Chapter 6 will be followed here as they apply to the project. The data are

Figure 17.7. Floor plan of building for Project II.

COMMERCIAL COOLING LOAD CALCULATIONS

Project _Bargain Dept. Store_   Room _Building, peak_   Engr. _Miller & Moroze_
Location _Ft. Worth, Texas_   Lat. _32N_   Calc. by _EP 6/6/80_   Check _FC 6/8/80_

Design Conditions		DB F	WB F	RH %	W gr/lb
	Outdoor	99	74		86
	Room	78		50	71

Daily Range _22_   F. Ave. _88 F_
Day _June 21_   Time _5 PM_

Conduction	Dir.	Color	U	A, ft² Gross	A, ft² Net	CLTD, F Table	CLTD, F Corr.	F_c	RSHG BTU/hr
Glass			1.04		2124	13	16	.87	30,750
Wall (Group D)	SW	D	.22		2198	27	28	.87	11,780
	NE	D	.22		2198	24	29	.87	12,220
	NW	D	.22		354	18	23	.87	1,560
Roof/ceiling		D	.13	27,790		65	70	.87	220,000
Floor									
Partition			.40		2478		12	.87	10,350
Door									

Solar	Dir.	Sh.	SHGF	A	SC	CLF	F_c	
Glass	NW	no	176	2124	.94	.51	.87	155,910

$$K = (U_w A_w + U_g A_g)/L = (.22 \times 4750 + 1.04 \times 2124)/491 = 6.63$$
$$F_c = 1 - 0.02K = 1 - 0.02(6.63) = 0.87$$
$$CFM = \frac{RSHG}{1.1(T_R - T_{SA})} = \frac{999,540}{1.1(78-59)} = 48,000$$
$$RSHR = RSHG/RTHG = \frac{999,540}{1,188,000} = 0.84$$

Lights _83,370_ W × 3.41 × _—_ BF × _/_ CLF × _.87_ F_c → 247,330   RLHG BTU/hr
People _315_ SHG × _500_ n × _/_ CLF × _.87_ F_c → 135,030
_325_ LHG × _500_ n → 162,500
Equipment _____ → 8000
Infiltration 1.1 × _1650_ CFM × _19_ TC → 34,490
0.68 × _1650_ CFM × _15_ gr/lb → 16,830

Subtotal → 867,400   179,330

SA duct gain → 17,400
SA duct leakage _8_ % → 70,000   8910
SA fan gain (draw through) _5_ % → 45,000

Room Heat Gain → 999,800   188,240   1,188,000   RTHG BTU/hr

Ventilation 1.1 × _5000_ CFM × _19_ TC → 104,500
0.68 × _5000_ CFM × _15_ gr/lb → 51,000
SA fan gain (blow through) _____ %
Pump gain
RA duct gain
RA fan gain _2.5_ % → 22,000

Cooling Load → 1,126,000   239,000   1,365,000

Tons → 114

Figure 17.8. Cooling load calculations for Project II.

recorded on the Commercial Cooling Load Calculations form (Figure 17.8).

Step 1. The recommended outdoor design conditions are selected from Table A.9, at the 2.5% level, including the mean coincident wet bulb. An indoor design condition of 78 F DB and 50% RH is chosen as comfortable yet energy conserving. Perhaps even 80 F could have been used, because occupancy is for short periods of time.

Steps 2–3. Dimensions are taken from the architectural plan. Areas are calculated and recorded. The NW wall has a 12 ft height of glass the full length of the building. Doors are glass.

Step 4. Heat transfer coefficients are found from Tables 6.1, 6.3, or A.4–A.8.

Steps 5–7. The peak load time must be determined. Because there is only one air conditioned space, there is no problem of calculating peak loads for both individual rooms and for the whole building. From Table 6.6, at 32°N latitude the peak NW glass solar gain occurs in June (176 BTU/hr). Because the CLTD for the roof will also be maximum at this time of year (Table 6.1), this is the peak month. (If peak solar gain were in the fall, we would have to calculate the solar and conduction gains at that time to see if the total is a peak. This might be true if the glass faced south.)

The peak time of day must now be determined. From inspection of Table 6.8, the peak solar gain through the glass is at 6 PM (CLF = 0.53). At 5 PM, however, the roof heat transfer is slightly higher (65) than at 6 PM (64) as seen from Table 6.1, yet the solar heat gain is almost as great as at 6 PM (CLF = 0.51). A partial calculation of external heat gains will be carried out to determine if 5 PM or 6 PM is the peak time.

All values are found from the appropriate tables. The CLTD values are found using Tables 6.1 and 6.2, with corrections as explained in Chapter 6. For example, the CLTD for the SW wall at 5 PM in June would be found using the following data: weight of wall is in Group D class. Average outdoor temperature on a design day is $99 - 22/2 = 88$ F. Inside design temperature is 78 F; wall color is dark. Using Equation 6.2

$$CLTD_c = (CLTD + LM) \times K + (78 - t_R)$$

$$+ (t_0 - 85)$$

$$= (27 - 2) \times 1 + (78 - 78)$$

$$+ (88 - 85) = 28 \, F$$

The results of the calculations are as follows:

At 6 PM:

Solar	SHGF	A	SC	CLF	BTU/hr
Glass NW	176	× 2124	× 0.94	× 0.53 =	186,240
Conduction	U	A	CLTD		
Glass	1.04	× 2124	× 15	=	33,130
Roof	0.13	× 27790	× 69	=	249,280
NW wall	0.22	× 354	× 27	=	2100
SW wall	0.22	× 2198	× 33	=	15,960
NE wall	0.22	× 2198	× 30	=	14,510
				Sum =	501,220

At 5 PM:

Solar					BTU/hr
Glass NW	176	× 2124	× 0.94	× 0.51 =	179,210
Conduction					
Glass	1.04	× 2124	× 16	=	35,340
Roof	0.13	× 27790	× 70	=	252,890
NW wall	0.22	× 354	× 23	=	1790
SW wall	0.22	× 2198	× 28	=	13,540
NE wall	0.22	× 2198	× 29	=	14,020
				Sum =	496,790

Those items that are not read directly from the table and that may need clarification will be explained.

The CLTD values are corrected for month, latitude, color of surface, inside temperature, and average daily temperature.

All sensible heat gains except infiltration and ventilation are corrected by the factor F_c for heat lost to the surroundings as follows

$$K = \frac{U_w A_w + U_g A_g}{L}$$

$$= \frac{0.22 \times 4750 + 1.04 \times 2124}{491}$$

$$= 6.63$$

$$F_c = 1 - 0.02\ K = 1 - 0.02(6.63) = 0.87$$

The spaces adjacent to the air conditioned area are assumed to be at 90 F. If exhaust air from the air conditioned area is used to ventilate these spaces, this might be a reasonable assumption, otherwise a higher temperature might result.

Lighting   The lighting total is:

$$30\ \text{W}/\text{ft}^2 \times 27{,}790\ \text{ft}^2 = 83{,}370\ \text{W}$$

The ballast factor BF has already been included in the value for the lighting intensity. All the lights are presumed turned on. The cooling system will be shut down as soon as the store closes, so CLF = 1.0.

People   The number of people (500) presumably was determined by the building owner. If this information is not available, Table 6.15 can be used as an estimate. The CLF is 1.0, because the cooling system is shut down when the store closes. Table 6.11 is used to find the heat gains.

   The solar and conduction heat gain is slightly higher at 6 PM, and it would seem that this is therefore the peak time. The infiltration and ventilation gains have not been included, however, and these will be less at 6 PM due to the drop in outdoor temperature. They will very likely decrease by more than the very small dif-

ference found above (this will be verified in the final calculations), and therefore 5 PM in June is the peak time.

Step 8. The items comprising the heat gains can now be calculated, and are recorded in Figure 17.8. The values are found from the following tables:

Item	Table
Design conditions	A.9
SHGF (Sensible Heat Gain Factor)	6.6
SC (Shading Coefficient)	6.7
CLF (Cooling Load Factor), solar	6.8
CLTD (Cooling Load Temperature Difference), glass	6.5
CLTD, wall	6.2, 6.3
CLTD, roof	6.1
U	6.1, 6.3, A.4–A.8
CLF, people	6.12
Ventilation	6.15
People gain	6.11
Motor gain	6.14

The student should confirm all values listed in Figure 17.8.

Equipment   The equipment (motors) estimate has been furnished by the building owner. The resulting heat output, assuming 50% operation, is

$$1000\ \text{BTU}/\text{hr} \times 16 \times 0.5 = 8000\ \text{BTU}/\text{hr}$$

Infiltration   It is unlikely that the air conditioning system can pressurize such a large space significantly, so infiltration should be included. As is often true, the estimate may be quite approximate because it is based on uncertain figures.

   Because there will be heavy traffic, infiltration through the doors should be estimated both by the crack method and by door usage, taking the largest value if there is a great difference. By the crack method

$$1 \text{ CFM/ft} \times 34 \text{ ft} \times 4 = 136 \text{ CFM}$$

It is assumed that rate of entry at peak design time is three persons per minute. The building owner can often aid in furnishing this information. Using a 550 CFM infiltration per door opening gives $550 \times 3 = 1650$ CFM. This value will be used.

The outdoor design temperature must be corrected for time of day. From Table 6.16, the decrease is 2 F, and the resulting air TC = $(99 - 2) - 78 = 19$ F. Note that the outdoor temperature decrease at 6 PM is 4 F. This confirms that the peak load will be at 5 PM. The extra 2 F results in a sensible heat equivalent of $1.1 \times 1650 \times 2 = 3630$ BTU/hr. When the ventilation air heat difference is also included, the result will be a total load greater at 5 PM than at 6 PM. The student should confirm these totals.

The individual room heat gains are calculated and tabulated in the appropriate places in Figure 17.8, and summed up to find the total RSHG and RLHG.

Duct Heat Gains The ductwork will be located above the hung ceiling and therefore the heat gain to it must be included. Equation 3.8 will be used. In order to use the equation, however, we must know the temperature inside and outside the duct and its area, all of which are not accurately determined until later in the design process. Approximate estimates must be made now, and these can be corrected later if they are in gross error. The following assumptions will be made:

1. Total CFM = 48,000. This is based on an experienced guess of 1.6 CFM/ft^2 of floor area. (Many applications have a range of from 0.5-2 CFM/ft^2.)
2. Four ducts will be used, each handling 12,000 CFM. The average CFM in each duct is 6000

CFM. The average duct area will therefore be 4 ft^2 (based on a 1500 FPM velocity). A typical duct size with this cross-sectional area is 32 in. × 18 in. This results in a surface area for each of

$$A = \left(\frac{2(32)}{12} + \frac{2(18)}{12} \right) 140 = 1160 \text{ ft}^2$$

3. The ducts will be insulated to an overall $R = 8.0$. Therefore, $U = 1/8.0 = 0.125$.
4. The duct supply air temperature is assumed as 60 F, and the temperature in the hung ceiling is 90 F.

Applying this information to Equation 3.8

$$Q = U \times A \times TD = 0.125 \times 1160 \times (90 - 60)$$
$$= 4350 \text{ BTU/hr}$$

The total heat gain for the four ducts is

$$Q = 4 \times 4350 = 17,400 \text{ BTU/hr}$$

The temperature rise of the supply air is found from the sensible heat equation

$$TC = \frac{Q}{1.1 \times \text{CFM}} = \frac{4350}{1.1 \times 6000} = 0.7 \text{ F}$$

There will be no significant heat gain to the return air ducts, because their length will be very short.

The heat equivalent of the duct air leakage must be determined, because the ducts are outside the conditioned space. An estimate of 8% of the RSHG will be used.

Fan Heat Gain The temperature rise and heat gain from the fans must be determined. It is assumed that the supply fan SP = 20 in. w.g. A draw-through fan arrangement will be used; therefore, the heat gain will become part of the

RSHG. From Chapter 6, the value is 5%, or 45,000 BTU/hr. The temperature rise is

$$TC = \frac{Q}{1.1 \times CFM} = \frac{45,000}{1.1 \times 48,000} = 0.9\,F$$

It is assumed that the return air fan SP = 1.0 in. w.g., for a heat gain of 2.5%; this is about 25,000 BTU/hr. Assuming 40,000 CFM of return air, the temperature rise is

$$TC = \frac{25,000}{1.1 \times 40,000} = 0.6\,F$$

These values can be checked when the final selection of equipment is made later. If they are grossly in error, these heat gain calculations should be revised.

Step 9. The cooling load can now be determined. The number of occupants at peak time is estimated at 500. Table 6.15 recommends

minimum and maximum ventilation requirements of 7-15 CFM per person for this application. It has been decided to use 10 CFM per person. Ventilation loads are calculated using these values.

The sensible, latent, and total cooling loads are found by adding room gains to ventilation and return air gains. (Any blow-through fan and pump gains would also be added.) There will be no diversity of loads, because all lights are on at all times, and the people occupancy occurs at the same time. No allowance for temperature swing will be taken, because it is decided that the resulting maximum temperature would be unacceptably high for comfort.

Step 10. The required design supply air conditions can now be determined. The solution will be worked out with the aid of a psychrometric chart (Figure 17.9), as explained in Chapter 7.

Figure 17.9. Psychrometric chart analysis of air conditioning processes for Project II.

The supply air CFM is determined. A trial value of the coil leaving air temperature is chosen, 59 F. The required CFM is:

$$CFM = \frac{RSHG}{1.1 \times TC} = \frac{999,540}{1.1 \times (78 - 59)}$$

$$= 48,000 \text{ CFM}$$

The RSHR line is drawn. The RSHR is

$$RSHR = \frac{RSHG}{RTHG} = \frac{999,540}{1,188,000} = 0.84$$

A line with this slope is drawn through the room condition (1). An allowance must be made for the temperature rise from the supply fan and duct heat gain; this is 0.9 + 0.7 = 16 F. Therefore, the supply air temperature to the room is 60.6 F. This point (2) is located on the RSHR line. A horizontal line (2-3) is then drawn to 59 F, accounting for the fan and duct heat gain. This establishes the air condition (3), leaving the coil at 59 F and 57 F WB. The temperature rise in the return air fan is 0.6 F. Therefore, the return air temperature is 78 + 06 = 786 F DB, point (4). The mixing line is now drawn from this point to the outside air condition (5). The proportion of outside air to total air quantity is 5000/48,000 = 0.10. The mixed air condition can now be located at this proportional distance from the return air condition along the mixing line. This is 0.10(97 − 78.6) + 78.6 = 80.4 F DB with a 66.2 F WB.

The effective coil process line (6-3) is now drawn. The required coil contact factor (CF) and bypass factor (BF) can now be determined. The CF is the ratio of the length of line (6-3) to (6-7). Using a scale to measure this, CF = 0.85. The (effective) coil surface temperature (7) is 55.0 F.

The cooling load imposed on the coil can be calculated from the psychrometric chart. Using the enthalpy change (6-3) in Equation 7.15:

$$Q_t = 4.5 \times CFM(h_6 - h_3)$$

$$= 4.5 \times 48,000(30.9 - 24.5)$$

$$= 1,382,400 \text{ BTU/hr}$$

The sensible and latent portions of the cooling coil load can also be found by finding the horizontal and vertical proportions to point (a) of line (6-3) on the enthalpy scale:

$$Q_s = 4.5 \times CFM(h_a - h_3)$$

$$= 4.5 \times 48,000(29.8 - 24.5)$$

$$= 1,145,000 \text{ BTU/hr}$$

$$Q_l = 4.5 \times CFM(h_6 - h_a)$$

$$= 4.5 \times 48,000(30.9 - 29.8)$$

$$= 237,600 \text{ BTU/hr}$$

These figures check closely with the cooling load calculation sums, which would be used to specify the cooling coil requirements. If a coil is available, the trial value of supply air temperature is satisfactory. If not, a new trial value would be selected.

17.18. TYPE OF SYSTEM

An all-air single zone system will be used, with DX refrigerant coils, because there is only one space, and the equipment can all be located in one place, with air distributed directly to the space from there.

17.19. EQUIPMENT AND DUCT LOCATIONS

The suggestions in Section 17.8 on planning the installation should be referred to as a guideline. The shape and size of the building suggest a few rows of ducts running the length or width of the building, above the suspended ceiling. The refrigeration and air handling equipment will be located on the ends of the building. Roof-mounted units will be used because of their low cost and convenience for a one-story building. This also saves usable floor space in the building. Four duct mains will be used, with four air diffusers in each. This seems to provide a reasonably simple duct arrangement but with enough diffusers so that good air distribution will result.

A separate air handling unit will be used for each main. It is possible that two units could be used for the building, but this would leave the

Elevation view of connections to roof units

Figure 17.10. HVAC plan for Project II.

system with only 50% of capacity if one were out of service. Furthermore, 50 tons is about the limit at which many roof-mounted units are manufactured. A sketch of the arrangement is shown in Figure 17.10.

17.20. DUCT SIZES

Because the total supply air quantity is 48,000 CFM, there is 12,000 CFM for each of the four air handling units, if the air is distributed equally among them. In this particular application consideration might be given to supplying a greater proportion of air to the side of the building receiving the solar load through the glass. On the other hand, the load tends to equalize throughout the space by internal heat transfers. We will assume an equal air distribution is satisfactory, taking note that on an actual installation this should be studied further before a decision is made. Normally one checks similar existing installations for their experiences.

The air quantities in each section of duct can now be determined. Because there are 16 air diffusers, each supplies 3000 CFM. The second section of duct has 12,000 − 3000 = 9000 CFM. The remaining sections are found in the same manner.

Low pressure, galvanized steel rectangular ductwork will be used. The ducts will be sized by the equal friction method. A friction loss rate of 0.1 in. w. per 100 ft will be used. Using Figure 8.21, for the first duct section at 12,000 CFM, the velocity is 1800 FPM. This is a reasonable maximum velocity for noise levels in retail stores (Table 8.6). The equivalent round diameter is listed as 35 in.

The available depth above the hung ceiling for the duct is determined from architectural and structural drawings. Assume that the maximum duct depth allowable is 24 in. Using Figure 8.23, a 44 in. × 24 in. rectangular duct is approximately equivalent to a 35 in. round duct. The second duct section has an equivalent 31.5 in. diameter. A 36 in. × 24 in. size is selected. Only

one dimension of the duct is changed each time, if possible, to make a simpler transition fitting. In addition, the duct shape is kept as square as possible. This minimizes the aspect ratio, reducing the amount of sheet metal and insulation used. The other duct sections are sized in the same manner. The results are shown as follows, and are also indicated on Figure 17.10:

Section	CFM	V, FPM	Friction Loss in. w./ 100 ft	Eq. D, in.	Duct Size, in.
AB	12,000	1800	0.1	35	44 × 24
BC	9000	1650	0.1	32	36 × 24
CD	6000	1500	0.1	27	26 × 24
DE	3000	1250	0.1	21	26 × 15

17.21. AIR DISTRIBUTION DEVICES

Each diffuser is covering a radius of approximately 20 ft equally in all directions. The type listed in Table 10.4 will be used. A 24 in. diffuser is the best selection. It has approximately a 19 ft throw radius with an NC-37 noise level at 3000 CFM. Both the NC level and velocity are suitable for retail stores. A 20 in. diffuser has a suitable radius of diffusion but an NC-level of about 47, which is above that recommended.

Return air grilles will be located in the ceiling directly under the air handling units on one end of the building. There will be a total of 40,000 CFM of return air. The remainder of the supply will be used to ventilate the service areas in the rear of the building. The return grilles will be located well above the occupied zone. Therefore, a reasonably high face velocity can be used without resulting in excessive noise. Each of the returns handles 10,000 CFM. Grilles with a face area of 12 ft^2 will be used.

17.22. EQUIPMENT

Packaged roof-mounted units will be used. They will consist of compressors, evaporator coil, air cooled condenser, supply air fan, dampers, and cabinet. Permanent, cleanable, panel-type filters will be used. A gas-fired furnace section can also be furnished if the unit is to be used for heating. A detailed specification will be described later.

The required fan pressure characteristics must be determined. Information on pressure drops through each component is obtained from the manufacturer. This data, plus the friction loss through the ducts are summarized in Table 17.3.

The required supply fan total pressure is 1.79 in. w.g. In the prior calculations a radius-type elbow was assumed. No significant pressure losses occur in the gradual transition pieces. The values for pressure losses in the takeoff and diffuser were obtained from the manufacturers. Note that a filter pressure drop of 0.40 in. w.g. has been used. This allows for dirt build-up, because the clean filter pressure drop is much less. If the latter figure were used, the fan would be under-sized. The operating engineer must be sure to change filters at 0.4 in. w.g., not greater.

A return air fan will be included. A short section of ductwork from the return air grilles is required. The fan pressure requirements would be determined in a manner similar to that for the supply fan.

The selection of the roof-mounted units would be based on the design sensible latent and total loads and conditions. Because there are four units, all of the same size, the requirements for each would be

$$Q_s = \frac{1,126,000}{4} = 282,000 \text{ BTU/hr}$$

$$Q_l = \frac{239,000}{4} = 60,000$$

$$Q_t = \overline{342,000} \text{ BTU/hr}$$
$$= 28.5 \text{ tons}$$

Entering air = 80.4 FDB, 66.2 F WB

Leaving air = 59 F DB, 57 F WB

CFM = 12,000

TABLE 17.3. SUMMARY OF FAN PRESSURE CHARACTERISTICS

Item	CFM	Duct Size	V, FPM	Friction Loss in. w./ 100 ft	L, ft	Loss coeff, C	V.P. in. w.	Pr. loss in. w.
UNIT								
Inlet	12,000							0.32
Filter								0.40
Coil								0.41
Outlet								0.25
DUCT								
AB	12,000	44 × 24	1800	0.1	20			0.02
Elbow						0.20	0.20	0.04
BC	9000	36 × 24			40			0.04
CD	6000	26 × 24			40			0.04
DE	3000	26 × 15			40			0.04
Takeoff								0.08
Diffuser								0.15
						Total system pressure loss =		1.79

A selection from a manufacturer's catalog would be made with this data.

17.23. ACCESSORIES

Fire dampers and smoke detectors might be required in the ductwork, as determined by Codes. Control grids and balancing dampers will be provided in the collar to each diffuser. Return, exhaust, and outside dampers will be furnished. All supply ductwork will be insulated with vapor barrier covered insulation for an overall R-8.

17.24. AUTOMATIC CONTROL SYSTEM

The controls in roof-mounted package units usually come in one of a few fixed choices as a "programmed" arrangement. That is, the controls are completely wired and built into the equipment, and only certain combinations are available. The cooling system controls will be specified as follows:

1. An economizer control will be furnished. When the outdoor air temperature is between 60 F and 78 F, a mixed air thermostat shall modulate outside and return air dampers to provide "free" cooling when called for. When the outdoor air temperature reaches the high limit setting, an outdoor thermostat will position the dampers to minimum outside air. This type of control is similar to that described and shown in Chapter 14.
2. On call for further cooling a two-stage room thermostat shall operate the first stage of the compressor. This may be accomplished either by using a compressor furnished with one step of cylinder unloading, or by furnishing a unit with two compressors. On call for further cooling the second stage of compressor cooling shall operate. The opposite sequence of events will occur when the room temperature is satisfied.

The refrigeration compressor will have a pump down cycle (Chapter 13). A solenoid valve to the evaporator coil will open and close as called for by the room thermostat. When the solenoid valve closes, the compressor will continue to operate until the suction pressure reaches a set value. At this point a pressurestat will stop the compressor. The compressor will have a normal complement of safety controls—high and low pressure cut-outs and oil pressure failure control. The condenser will have a pressure control that maintains adequate head pressure when the outdoor air temperature is low. This will function by cycling the condenser fans. The room thermostats will be mounted on interior columns at shoulder height, away from drafts or heat sources. The control system will be low voltage electric.

17.25. PLANS AND SPECIFICATIONS

The floor plans and any necessary details are prepared, using the previous sketches as a guide. The architectural and structural plans and those of the other trades are used to insure that there are no interferences. The plans would be similar to that shown in Figure 17.10, drawn to scale.

The specifications would be of a form similar to those described for Project I. For example, the specifications for the roof-mounted unit might read as follows:

Furnish and install where shown on the plans four (4) roof-mounted air conditioning units complete with compressors, evaporator, fans, air cooled condenser, cabinet, and accessories. Each unit shall have a total cooling capacity of 342,000 BTU/hr and a sensible cooling capacity of 282,000 BTU/hr with 12,000 CFM of air entering at 80.4 F DB and 66.2 F WB with 99 F DB air entering the condenser.

The supply fan shall have a 1.80 in. w.g. total pressure with 7.5 BHP. The fan motor shall be $7\frac{1}{2}$ HP, 220 V., 3 phase, 60 hz. The pressure drop through the unit components shall not exceed 1.4 in. w.g.

The cabinet shall be of 20 gauge galvanized steel panel construction with an acrylic paint and completely weatherized, using stainless steel fasteners. Access panels shall provide easy access to all components. Panels shall have weatherproof gaskets and chrome-plated handles. The cabinet shall be lined inside with 1 in. of glass fiber insulation and neoprene coated.

The unit shall have two semihermetic compressors (or one compressor with automatic one-stage unloaders), complete with oil pump, suction and discharge service valves, oil level sight glass, suction gas filter, and crankcase heater.

The evaporator coil shall be of copper tube and aluminum fin with the number of rows required to handle the sensible and total cooling capacities specified. The refrigeration equipment shall be completely piped and charged, with thermal expansion valve, distributor, sight glass, and drier, using copper tubing.

The air-cooled condenser shall be complete with coils, fans, and motors. Coils shall be copper tube with aluminum fins.

All electrical components shall be weatherproof, in accordance with UL (Underwriter's Laboratory) requirements. The unit shall have a return air section complete with centrifugal fan and fresh air, return air, and exhaust dampers. Furnish a roof mounting curb. The units shall be completely tested at the factory. The units shall be as manufactured by __ __ or approved equal.

17.26. ENERGY CONSERVATION

The recommended 2.5% outdoor design dry bulb temperature and coincident mean wet bulb have been used. Cooling load calculation procedures that account for energy storage have been used, thus reducing the design load considerably. (The load calculated by one of the older methods not accounting for storage is about 20% greater!). This will result in smaller equipment that will operate closer to full load, and therefore more efficiently.

The roof load is fairly high, and perhaps more insulation should be used. The glass solar load is quite high. Solar heat absorbing glass or internal shading is probably unacceptable to the owners, however, as they want potential customers to be attracted by the visibility of the interior. An outside shading overhang might be considered,

but it is possible that it would have to be too long to be effective; this should be determined. Door infiltration is considerable. Revolving doors would reduce infiltration greatly but this might slow down customer ingress.

The control system has been designed for outside air free cooling, using temperature sensing. Enthalpy sensing would provide even more energy conservation, but is perhaps excessive in cost for the number of hours it would be effective. The weather data could be checked to determine this.

One major energy conservation device that probably should be considered in an air-to-air heat exchanger between ventilation and exhaust air, especially if the system is also used for winter heating. A calculation will show the value of summer operation. Assuming that a heat recovery wheel with 80% efficiency is used, if it recovers both sensible and latent heat, the energy saved at design load is

$$Q = 4.5 \times 5000 \times 0.80(37.7 - 30.1)$$
$$= 136,800 \text{ BTU/hr} = 11.4 \text{ tons}$$

This is not only a savings in energy consumption, but the size of all the cooling equipment could be reduced by 11.4 tons. Assuming an equivalent full load cooling season of 1500 hours, and 1.4 KW per ton required by the compressor and fans, and with an electric power cost of $0.05/KWH, the annual summer savings would be:

$$11.4 \text{ tons} \times 1.4 \text{ KW/ton} \times 1500 \text{ hr}$$
$$\times \$0.05/\text{KWH} = \$1200$$

This is a substantial and continual saving in both fuel and operating cost.

PROBLEMS

17.1. Plan and design a series loop hydronic heating system for the residence shown in Figure 17.1, for construction in your town. Use construction materials and conditions suitable for the climate, as recommended by the energy standards in this book or in your state.

17.2. Design a reverse return heating system for the residence shown in Figure 17.1, as recommended in Problem 17.1.

17.3. Design a split series loop (or reverse return) heating system for the residence described in Problem 3.20, to be built in your town, using materials and conditions recommended by energy conservation standards.

17.4. Plan and design an all-air summer air conditioning system for the factory described in Problem 3.21, to be constructed in your town, using materials and conditions recommended for energy conservation.

17.5. Plan and design an all-air summer air conditioning system for the store described in Example 6.17, to be constructed in your town, using materials and conditions recommended for energy conservation.

17.6. Carry out the designs recommended in Problems 17.1–17.5 for other locations.

CHAPTER

Solar Heating and Cooling Systems

Solar energy systems can be used to provide domestic hot water and hot water or warm air for air conditioning, in place of boilers or furnaces. The advantage of course is that there is no depletable fuel consumed and thus no cost for the heat energy used. The disadvantages have been the equipment cost and problems in collecting this energy. However, sharply increasing energy costs have stimulated the development of practical solar heating systems. Solar heating of hot water is already quite com-mon in Japan, Israel, and Australia. The growth of solar energy for space heating of buildings is also developing rapidly in many countries, and is practical where fuel costs are relatively high.

We will discuss basic principles and features of solar heating equipment and systems here. A relatively simple yet practical method of system design and collector sizing will be explained.

A solar heating system is composed of three essential parts: a collector, a storage system, and a distribution and terminal system.

OBJECTIVES

After studying this chapter, you will be able to:

1. Explain the construction of a flat plate solar collector.

2. Sketch basic solar hot water and space heating systems.

3. Calculate the required size of a solar heat collector and storage tank.

473

4. Perform an economic payback analysis for solar heating.

18.1. SOLAR COLLECTORS

The function of the *collector* is to receive and capture as much of the radiant energy from the sun as is practical. There are three types of collectors: concentrating, flat plate, and evacuated tube.

The *flat plate* collector has a receiving surface that is flat in shape. The collector is stationary and therefore receives the maximum radiation only when it faces the sun directly. At all other times it "sees" the sun at an angle, and its effective surface area is smaller, resulting in collection of less energy. Figure 18.1 demonstrates this. The effective surface is equal only to the projected area facing the sun.

A typical flat plate collector construction for heating water or other liquids is shown in Figure 18.2. The *transparent cover* (either one or two plates) admits solar radiation. Both glass and plastics have been used. The *absorber plate* collects the radiant energy. It is constructed of a flat panel and tubing through which the water is circulated as it is heated. The plate and tubing

may be steel, aluminum, plastic, or copper. It is often coated black to increase energy absorption. The cover plates reduce the heat that would be reradiated out from the hot panel. The solar radiation received has a short wavelength, and passes through the transparent cover plate. The absorber plate is heated by the radiation it absorbs, and in turn heats the water. Some of the energy received is reradiated outward by the absorber plate. This radiation, however, has a longer wavelength, which does not pass outward through the cover plate. The cover also reduces the amount of convected heat lost to the surrounding air. Insulation is provided on the other sides to reduce heat losses there.

A flat plate collector construction for heating air is shown in Figure 18.3. Air is circulated through the space behind the absorber plate and is heated directly. In the example shown the plate has fins to increase the heat transfer surface.

Concentrating or focusing collectors have a concave curved shape surface that concentrates the rays received. The result is a collector that can produce a much higher temperature than the flat plate collector. Concentrating collectors are often designed to move and track the sun, thus they always receive maximum radiation. The flat plate collector, although less efficient, has been favored due to its usually lower cost, but new

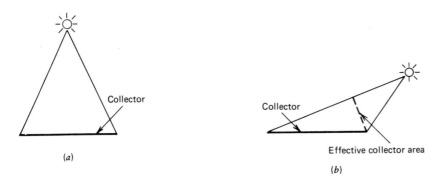

Figure 18.1. Effect of the angle between the collector and sun on solar radiation received. (*a*) Collector directly facing sun — all surface area effective. (*b*) Collector not directly facing sun — effective area smaller than actual surface area.

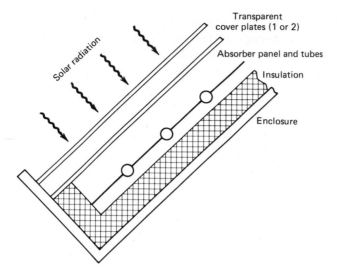

Figure 18.2. Typical flat plate collector for heating liquids.

and improved types of concentrating collectors are being developed.

The *evacuated tube* collector has a tube within a tube construction. The outer tube is transparent and transmits the solar radiation. The inner tube contains the circulating liquid that is heated. The annular space between the tube is evacuated of air. This greatly reduces the convection heat losses from the collector, similar to how a thermos bottle functions. The efficiency of the evacuated tube collector is quite high, as much as double that of the flat plate collector. It can also produce water at very high temperatures (over 300 F). With reduced production costs it should add greatly to the use of solar heating and cooling systems.

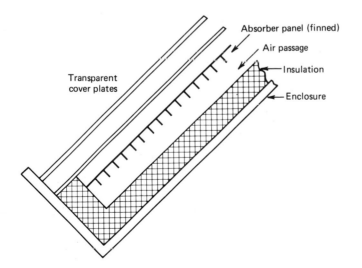

Figure 18.3. Typical flat plate collector for heating air.

Figure 18.7. Solar heating system with air collector and warm air distribution.

very popular use of a solar heating system, because it can then also be used for direct heating when the heating load is sufficiently low, and supplemented or replaced by the heat pump when necessary.

18.4. SOLAR COOLING SYSTEMS

Although solar energy systems produce only heat, any cooling system that receives its operating energy from a heat source can be driven by using the solar system heat. One method is to use the solar heated hot water to operate an absorption refrigeration system. Because the efficiency of flat plate solar collectors decreases greatly as the required temperature increases, the cost of the solar system may be quite high to produce the water temperatures required to operate absorption machines (about 190 F).

A solar heating system can be used as a heat source of a Rankine power cycle turbine which in turn is used to drive a conventional refrigeration compressor. The Rankine cycle is the same as that in a steam turbine. When used with a solar heating system, however, a fluid is used that will vaporize at a temperature available from the solar collector. The cooling system for the building shown in the frontispiece illustration uses this method.

18.5. SOLAR RADIATION ENERGY

In order to determine the capacity of solar heating equipment, the quantity of solar energy that is received by the collector must be found. The solar radiation received just outside the Earth's atmosphere is approximately 429 BTU/hr-ft^2 on a surface normal to (faces directly) the sun. It is about 14 BTU/hr-ft^2 higher in winter and lower in summer, due to the slightly elliptical orbit of the Earth around the sun. The solar energy received at the Earth's surface, called *insolation* (I), is much less than the above value, due to the absorbing and scattering effect of the atmosphere. That part of the solar radiation which reaches the Earth's surface and is not affected by the atmosphere is called *direct* radiation. It ranges from 0 to about 350 BTU/hr-ft^2 on a surface normal to the sun, depending on the length of path through the atmosphere and its condition. A portion of the scattered radiation also reaches the Earth's surface. This is called *diffuse* radiation. Radiation may also be *reflected* back from the ground onto surfaces.

In order to design solar energy collectors, it is necessary to find that part of the energy which is received on surfaces that are not normal to the sun. Collectors are usually fixed in position and therefore cannot collect the full insolation on a surface that is always normal to the sun's rays.

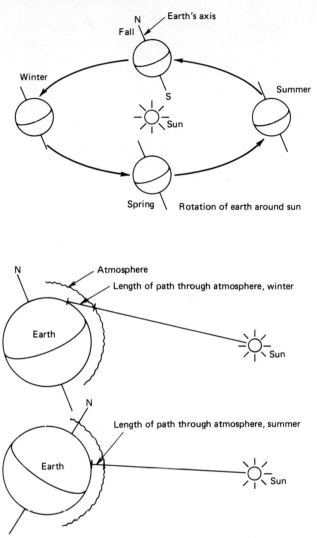

Figure 18.8. Effect of tilt of Earth's axis on solar radiation received.

There is an hourly variation in the angle between the collector surface and the sun due to the rotation of the earth. There is also a daily variation because in winter the Earth's axis is tilted further from the sun than in summer. This results in the sun being low on the horizon in winter, and high in summer. These variations

also change according to the distance from the equator (latitude).

Figure 18.8 illustrates how the tilt of the Earth's axis relative to the sun (called *declination*) affects the amount of solar radiation received. When the Earth is tilted away from the sun, as in winter, the length of path of the

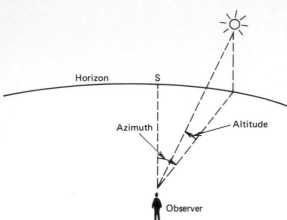

Figure 18.9. Illustration of the altitude and azimuth of the sun.

radiation through the atmosphere is longer, resulting in less radiation reaching the Earth's surface, despite the fact that the Earth is slightly closer to the sun in the winter.

The change in declination results in the sun appearing higher in the sky in the summer than in the winter. This affects the desirable angle that a collector should be tilted at in order to maximize the radiation received. Some general recommendations will be given later. The elevation of the sun above the horizon is called its *altitude* and its horizontal position relative to south is called its *azimuth*. Both are expressed in degrees. Figure 18.9 illustrates this.

18.6. INSOLATION TABLES

The solar insolation values for clear days have been determined and reported in tables. Table 18.1 is an example, shown for latitudes of 28°N to 48°N. *Solar time* is used; this is the time as read on a sun dial facing true south. Equations are available for converting solar time to local time, if necessary. The table shows insolation values monthly and hourly, for flat surfaces—those normal to the sun, horizontal (flat on the

ground), and tilted at various elevations above the horizontal. In all cases the surfaces are facing south. The insolation values listed are the total of direct and diffuse radiation except for the values listed for a normal surface, which includes only direct insolation. Interpolation between values can be used when accuracy requires it.

It is possible for radiation to be reflected from the ground onto a tilted surface. Reflected radiation is not included in the table, because it varies with local conditions.

EXAMPLE 18.1

Find the solar insolation at 10 AM and 12 N solar time on January 21 in Phoenix, Arizona, for surfaces tilted as follows: horizontal, 32°, 42°, and 90°, facing south.

Solution. Phoenix is at 33°N latitude, so the 32°N latitude table may be used, without significant error. Reading from the table, the results are

Time	Insolation in BTU / hr-ft²-F At Tilt Angles Listed			
	horiz.	32°	42°	90°
10 AM	167	256	269	221
12 N	209	308	321	253

The insolation tables also show the total daily insolation for each tilt angle, which is simply the sum of the hourly values for a day.

18.7. CLEARNESS FACTOR

The natural clearness of the atmosphere is not the same at all locations. It varies with the natural amount of water vapor and dust in the air. For example, the clearness in the southeastern part of the United States is less than in the Rocky Mountain area. In order to allow for this,

TABLE 18.1. CLEAR DAY INSOLATION VALUES

SOLAR POSITION AND INSOLATION, 24°N LATITUDE

DATE	AM	PM	ALT	AZM	NORMAL	HORIZ.	14	24	34	54	90
JAN 21	7	5	4.8	65.6	71	10	17	21	25	28	31
	8	4	16.9	58.3	239	83	110	126	137	145	127
	9	3	27.9	48.8	288	151	188	207	221	228	176
	10	2	37.2	36.1	308	204	246	268	282	287	207
	11	1	43.6	19.6	317	237	283	306	319	324	226
	12		46.0	0.0	320	249	296	319	332	335	232
	SURFACE DAILY TOTALS				2766	1622	1984	2174	2300	2360	1766
FEB 21	7	5	9.3	74.6	158	35	44	49	53	56	46
	8	4	22.3	67.2	263	116	135	145	150	151	102
	9	3	34.4	57.6	298	187	213	225	230	228	141
	10	2	45.1	44.2	314	241	273	286	291	287	168
	11	1	53.0	25.0	321	276	310	324	328	323	185
	12		56.0	0.0	324	288	323	337	341	335	191
	SURFACE DAILY TOTALS				3036	1998	2276	2396	2446	2424	1476
MAR 21	7	5	13.7	83.8	194	60	63	64	62	59	27
	8	4	27.2	76.8	267	141	150	152	149	142	64
	9	3	40.2	67.9	295	212	226	229	225	214	95
	10	2	52.3	54.8	309	266	285	288	283	270	120
	11	1	61.9	33.4	315	300	322	326	320	305	135
	12		66.0	0.0	317	312	334	339	333	317	140
	SURFACE DAILY TOTALS				3078	2270	2428	2456	2412	2238	1022
APR 21	6	6	4.7	100.6	40	7	5	4	4	3	2
	7	5	18.3	94.9	203	83	77	70	62	51	10
	8	4	32.0	89.0	256	160	157	149	137	122	16
	9	3	45.6	81.9	280	227	227	220	206	186	41
	10	2	59.0	71.8	292	278	282	275	259	237	61
	11	1	71.1	51.6	298	310	316	309	293	269	74
	12		77.6	0.0	299	321	328	321	305	280	79
	SURFACE DAILY TOTALS				3036	2454	2458	2374	2228	2016	483
MAY 21	6	6	8.0	108.4	86	22	15	10	9	4	5
	7	5	21.2	103.2	203	98	85	73	59	44	12
	8	4	34.6	98.5	248	171	159	145	127	106	15
	9	3	48.3	93.6	269	233	224	210	190	165	16
	10	2	62.0	87.7	280	281	275	261	239	211	22
	11	1	75.5	76.9	286	311	307	293	270	240	34
	12		86.0	0.0	288	322	317	304	281	250	37
	SURFACE DAILY TOTALS				3036	2556	2447	2286	2072	1800	246
JUN 21	6	6	9.3	111.6	97	29	20	12	12	11	7
	7	5	22.3	106.8	201	103	87	73	53	41	13
	8	4	35.5	102.6	242	173	158	142	122	99	16
	9	3	49.0	98.7	263	234	221	204	182	155	18
	10	2	62.6	95.0	274	280	269	253	229	199	18
	11	1	76.3	90.8	279	309	300	283	259	227	19
	12		89.4	0.0	281	319	310	294	269	236	22
	SURFACE DAILY TOTALS				2994	2574	2422	2230	1952	1700	204

DATE	AM	PM	ALT	AZM	NORMAL	HORIZ.	14	24	34	54	90
JUL 21	6	6	8.2	109.0	81	23	16	11	10	9	6
	7	5	21.4	103.8	195	98	85	73	59	44	13
	8	4	34.8	99.2	239	169	157	143	125	104	16
	9	3	48.4	94.5	261	231	221	207	187	161	18
	10	2	62.1	89.0	272	278	270	256	235	206	21
	11	1	75.7	79.2	278	307	302	287	265	235	32
	12		86.6	0.0	280	317	312	298	275	245	36
	SURFACE DAILY TOTALS				2932	2526	2412	2250	2036	1766	246
AUG 21	6	6	5.0	101.3	35	7	5	4	4	4	2
	7	5	18.5	95.6	186	82	76	69	60	50	11
	8	4	32.2	89.7	241	158	154	146	134	118	16
	9	3	45.9	82.9	265	223	222	214	200	181	39
	10	2	59.3	73.0	278	273	275	268	252	230	58
	11	1	71.6	53.2	284	304	309	301	285	261	71
	12		78.3	0.0	286	315	320	313	296	272	75
	SURFACE DAILY TOTALS				2864	2408	2402	2316	2168	1958	470
SEP 21	7	5	13.7	83.8	173	57	60	60	59	56	26
	8	4	27.2	76.8	248	136	144	146	143	136	62
	9	3	40.2	67.9	278	205	218	221	217	206	93
	10	2	52.3	54.8	292	258	275	278	273	261	116
	11	1	61.9	33.4	299	291	311	315	309	295	131
	12		66.0	0.0	301	302	323	327	321	306	136
	SURFACE DAILY TOTALS				2878	2194	2342	2366	2322	2212	992
OCT 21	7	5	9.1	74.1	138	32	40	45	48	50	42
	8	4	22.0	66.7	247	111	129	139	144	145	99
	9	3	34.1	57.1	284	180	206	217	223	221	138
	10	2	44.7	43.8	301	234	265	277	282	279	165
	11	1	52.5	24.7	309	268	301	315	319	314	182
	12		55.5	0.0	311	279	314	328	332	327	188
	SURFACE DAILY TOTALS				2868	1928	2198	2314	2364	2346	1442
NOV 21	7	5	4.9	65.8	67	10	16	20	24	27	29
	8	4	17.0	58.4	232	82	108	123	135	142	124
	9	3	28.0	48.9	282	150	186	205	217	224	172
	10	2	37.3	36.3	303	203	244	265	278	283	204
	11	1	43.8	19.7	312	236	280	302	316	320	222
	12		46.2	0.0	315	247	293	315	328	332	228
	SURFACE DAILY TOTALS				2706	1610	1962	2146	2268	2324	1730
DEC 21	7	5	3.2	62.6	30	3	7	11	11	12	14
	8	4	14.9	55.3	225	71	99	116	129	139	130
	9	3	25.5	46.0	281	137	176	198	214	223	184
	10	2	34.3	33.7	304	189	234	258	275	283	217
	11	1	40.4	18.2	314	221	270	295	312	320	236
	12		42.6	0.0	317	232	282	308	325	332	243
	SURFACE DAILY TOTALS				2624	1474	1852	2058	2204	2286	1808

Column group: **BTUH/SQ. FT. TOTAL INSOLATION ON SURFACES** — NORMAL, HORIZ., and SOUTH FACING SURFACE ANGLE WITH HORIZ. (14, 24, 34, 54, 90).

TABLE 18.1 (*Continued*).

SOLAR POSITION AND INSOLATION, 32°N LATITUDE

January – June

DATE	AM	PM	ALT	AZM	NORMAL	HORIZ.	22	32	42	52	90
JAN 21	7	5	1.4	65.2	1	0	0	0	0	1	1
	8	4	12.5	56.5	203	56	93	106	116	123	115
	9	3	22.5	46.0	269	118	175	193	206	212	181
	10	2	30.6	33.1	295	167	235	256	269	274	221
	11	1	36.1	17.5	306	198	273	295	308	312	245
	12		38.0	0.0	310	209	285	308	321	324	253
	SURFACE DAILY TOTALS				2458	1288	1839	2008	2118	2166	1779
FEB 21	7	5	7.1	73.5	121	22	34	37	40	42	38
	8	4	19.0	64.4	247	95	127	136	140	141	108
	9	3	29.9	53.4	288	161	206	217	222	220	158
	10	2	39.1	39.4	306	212	266	278	283	279	193
	11	1	45.6	21.4	315	244	304	317	321	315	214
	12		48.0	0.0	317	255	316	330	334	328	222
	SURFACE DAILY TOTALS				2872	1724	2188	2300	2345	2322	1644
MAR 21	7	5	12.7	81.9	185	54	60	60	59	56	32
	8	4	25.1	73.0	260	129	146	147	144	137	78
	9	3	36.8	62.1	290	194	222	224	220	209	119
	10	2	47.3	47.5	304	245	280	283	278	265	150
	11	1	55.0	26.8	311	277	317	321	315	300	170
	12		58.0	0.0	313	287	329	333	327	312	177
	SURFACE DAILY TOTALS				3012	2084	2378	2403	2358	2246	1276
APR 21	6	6	6.1	99.9	66	14	9	6	6	5	3
	7	5	18.8	92.2	206	86	78	71	62	51	10
	8	4	31.5	84.0	255	158	156	148	136	120	35
	9	3	43.9	74.2	278	220	225	217	203	183	68
	10	2	55.7	60.3	290	267	279	272	256	234	95
	11	1	65.4	37.5	295	297	313	306	290	265	112
	12		69.6	0.0	297	307	325	318	301	276	118
	SURFACE DAILY TOTALS				3076	2390	2444	2356	2206	1994	764
MAY 21	6	6	10.4	107.2	119	36	21	13	13	12	13
	7	5	22.8	100.1	211	107	88	75	60	44	15
	8	4	35.4	92.9	250	175	159	145	127	105	33
	9	3	48.1	84.7	269	233	223	209	188	163	56
	10	2	60.6	73.3	280	277	273	259	237	208	72
	11	1	72.0	51.9	285	305	305	290	268	237	77
	12		78.0	0.0	286	315	315	301	278	247	77
	SURFACE DAILY TOTALS				3112	2582	2954	2284	2064	1788	469
JUN 21	6	6	12.2	110.2	131	45	26	16	15	14	9
	7	5	24.3	103.4	210	115	91	76	59	41	14
	8	4	36.9	96.8	245	180	159	143	122	99	16
	9	3	49.6	89.4	264	236	221	204	181	153	19
	10	2	62.2	79.7	274	279	268	251	227	197	41
	11	1	74.2	60.9	279	306	299	282	257	224	56
	12		81.5	0.0	280	315	309	292	267	234	60
	SURFACE DAILY TOTALS				3084	2634	2436	2234	1990	1690	370

July – December

DATE	AM	PM	ALT	AZM	NORMAL	HORIZ.	22	32	42	52	90
JUL 21	6	6	10.7	107.7	113	37	22	14	13	12	8
	7	5	23.1	100.6	203	107	87	75	60	44	14
	8	4	35.7	93.6	241	174	158	143	125	104	16
	9	3	48.4	85.5	261	231	220	205	185	159	31
	10	2	60.9	74.3	271	274	269	254	232	204	54
	11	1	72.4	53.3	277	302	300	285	262	232	69
	12		78.6	0.0	279	311	310	296	273	242	74
	SURFACE DAILY TOTALS				3012	2558	2422	2250	2030	1754	458
AUG 21	6	6	6.5	100.5	59	14	9	7	6	6	4
	7	5	19.1	92.8	190	85	77	69	60	50	12
	8	4	31.8	84.7	240	156	152	144	132	116	33
	9	3	44.3	75.0	263	216	220	212	197	178	65
	10	2	56.1	61.3	276	262	272	264	249	226	91
	11	1	66.0	38.4	282	292	305	298	281	257	107
	12		70.3	0.0	284	302	317	309	292	268	113
	SURFACE DAILY TOTALS				2902	2352	2388	2296	2144	1934	736
SEP 21	7	5	12.7	81.9	163	51	56	56	55	52	30
	8	4	25.1	73.0	240	124	140	141	138	131	75
	9	3	36.8	62.1	272	188	213	215	211	201	114
	10	2	47.3	47.5	287	237	270	273	268	255	145
	11	1	55.0	26.8	294	268	306	309	303	289	164
	12		58.0	0.0	296	278	318	321	315	300	171
	SURFACE DAILY TOTALS				2808	2014	2288	2308	2264	2154	1226
OCT 21	7	5	6.8	73.1	99	19	29	32	34	36	32
	8	4	18.7	64.0	229	90	120	128	133	134	104
	9	3	29.5	53.0	273	155	198	208	213	212	153
	10	2	38.7	39.1	293	204	257	269	273	270	188
	11	1	45.1	21.1	302	236	294	307	311	306	209
	12		47.5	0.0	304	247	306	320	324	318	217
	SURFACE DAILY TOTALS				2696	1654	2100	2208	2252	2232	1588
NOV 21	7	5	1.5	65.4	2	0	0	0	1	1	1
	8	4	12.7	56.6	196	55	91	104	113	119	111
	9	3	22.6	46.1	263	118	173	190	202	208	176
	10	2	30.8	33.2	289	166	233	252	265	270	217
	11	1	36.2	17.6	301	197	270	291	303	307	241
	12		38.2	0.0	304	207	282	304	316	320	249
	SURFACE DAILY TOTALS				2406	1280	1816	1980	2084	2130	1742
DEC 21	8	4	10.3	53.8	176	41	77	90	101	108	90
	9	3	19.8	43.6	257	102	161	180	195	204	183
	10	2	27.6	31.2	288	150	221	244	259	267	226
	11	1	32.7	16.4	301	180	258	282	298	305	251
	12		34.6	0.0	304	190	271	295	311	318	259
	SURFACE DAILY TOTALS				2348	1136	1704	1888	2016	2086	1794

BTUH/SQ. FT. TOTAL INSOLATION ON SURFACES — SOUTH FACING SURFACE ANGLE WITH HORIZ. (22, 32, 42, 52, 90)

TABLE 18.1 *(Continued).*

SOLAR POSITION AND INSOLATION, 40°N LATITUDE

DATE	AM	PM	ALT	AZM	NORMAL	HORIZ	30	40	50	60	90
JAN 21	8	4	8.1	55.3	142	23	56	74	81	85	84
	9	3	16.8	44.0	239	83	155	171	182	187	171
	10	2	23.8	30.9	274	127	218	237	246	250	223
	11	1	28.4	16.0	289	154	257	277	290	293	253
	12		30.0	0.0	294	164	270	291	303	306	263
	SURFACE DAILY TOTALS				2182	948	1660	1810	1906	1944	1725
FEB 21	7	5	4.8	72.7	69	10	19	21	23	24	22
	8	4	15.4	62.2	224	73	114	122	125	127	107
	9	3	25.0	50.2	274	132	195	205	209	208	167
	10	2	32.8	35.9	295	178	256	267	271	267	210
	11	1	38.1	18.9	305	206	295	306	310	304	236
	12		40.0	0.0	308	216	306	319	323	317	245
	SURFACE DAILY TOTALS				2640	1414	2060	2162	2202	2176	1730
MAR 21	7	5	11.4	80.2	171	46	55	55	54	51	35
	8	4	22.5	69.6	250	114	140	141	138	131	89
	9	3	32.8	57.3	282	173	215	217	213	202	138
	10	2	41.6	41.9	297	218	273	276	271	258	176
	11	1	47.7	22.6	305	247	310	313	307	293	200
	12		50.0	0.0	307	257	322	326	320	305	208
	SURFACE DAILY TOTALS				2916	1852	2308	2330	2284	2174	1484
APR 21	6	6	7.4	98.9	89	20	11	8	7	7	4
	7	5	18.9	89.5	206	87	77	70	61	50	12
	8	4	30.3	79.3	252	152	153	145	133	117	53
	9	3	41.3	67.2	274	207	221	213	199	179	93
	10	2	51.2	51.4	286	250	275	267	252	229	126
	11	1	58.7	29.2	292	277	308	301	285	260	147
	12		61.6	0.0	293	287	320	313	296	271	154
	SURFACE DAILY TOTALS				3092	2274	2412	2320	2168	1956	1022
MAY 21	5	7	1.9	114.7	1	0	0	0	0	0	0
	6	6	12.7	105.6	144	49	25	15	13	13	9
	7	5	24.0	96.6	216	114	89	76	60	44	13
	8	4	35.4	87.2	250	175	158	144	125	104	25
	9	3	46.8	76.0	267	227	221	206	186	160	60
	10	2	57.5	60.9	277	267	275	255	233	205	89
	11	1	66.2	37.1	283	293	301	287	264	234	108
	12		70.0	0.0	284	301	312	297	274	243	114
	SURFACE DAILY TOTALS				3160	2552	2442	2264	2040	1760	724
JUN 21	5	7	4.2	117.3	22	4	3	2	2	2	1
	6	6	14.8	108.4	155	60	30	18	17	16	10
	7	5	26.0	99.7	216	123	92	77	59	41	14
	8	4	37.4	90.7	246	182	159	142	121	97	16
	9	3	48.8	80.2	263	233	219	202	179	151	47
	10	2	59.8	65.8	272	272	266	248	224	194	74
	11	1	69.2	41.9	277	296	296	278	253	221	92
	12		73.5	0.0	279	304	306	289	263	230	98
	SURFACE DAILY TOTALS				3180	2648	2454	2224	1974	1670	610

BTUH/SQ. FT. TOTAL INSOLATION ON SURFACES — SOUTH FACING SURFACE ANGLE WITH HORIZ.

DATE	AM	PM	ALT	AZM	NORMAL	HORIZ	30	40	50	60	90
JUL 21	5	7	2.3	115.2	2	0	0	0	0	0	0
	6	6	13.1	106.1	138	50	26	17	15	14	9
	7	5	24.3	97.2	208	114	89	75	60	44	14
	8	4	35.8	87.8	241	174	157	142	124	102	24
	9	3	47.2	76.7	259	225	218	203	182	157	58
	10	2	57.9	61.7	269	265	266	251	229	200	86
	11	1	66.7	37.9	275	290	296	281	258	228	104
	12		70.6	0.0	276	298	307	292	269	238	111
	SURFACE DAILY TOTALS				3062	2534	2409	2230	2006	1728	702
AUG 21	6	6	7.9	99.5	81	21	12	9	8	7	5
	7	5	19.3	90.0	191	87	76	69	60	49	12
	8	4	30.7	79.9	237	150	150	141	129	113	50
	9	3	41.8	67.9	260	205	216	207	193	173	89
	10	2	51.7	52.1	272	246	267	259	244	221	120
	11	1	59.3	29.7	278	273	300	292	276	252	140
	12		62.3	0.0	280	282	311	303	287	262	147
	SURFACE DAILY TOTALS				2916	2244	2354	2258	2104	1894	978
SEP 21	7	5	11.4	80.2	149	43	51	51	49	47	32
	8	4	22.5	69.6	230	109	133	134	131	124	84
	9	3	32.8	57.3	263	167	206	208	203	193	132
	10	2	41.6	41.9	280	211	260	265	260	247	168
	11	1	47.7	22.6	287	239	298	301	295	281	192
	12		50.0	0.0	290	249	310	313	307	292	200
	SURFACE DAILY TOTALS				2708	1788	2210	2228	2182	2074	1416
OCT 21	7	5	4.5	72.3	48	7	14	15	17	17	16
	8	4	15.0	61.9	204	68	106	113	117	118	100
	9	3	24.5	49.8	257	126	185	195	200	198	160
	10	2	32.4	35.6	280	170	245	257	261	257	203
	11	1	37.6	18.7	291	199	283	295	299	294	229
	12		39.5	0.0	294	208	295	308	312	306	238
	SURFACE DAILY TOTALS				2454	1348	1962	2060	2098	2074	1654
NOV 21	8	4	8.2	55.4	136	28	63	72	78	82	81
	9	3	17.0	44.1	232	82	152	167	178	183	167
	10	2	24.0	31.0	268	126	215	233	245	249	219
	11	1	28.6	16.1	283	153	254	273	285	288	248
	12		30.2	0.0	288	163	267	287	298	301	258
	SURFACE DAILY TOTALS				2128	942	1636	1778	1870	1908	1688
DEC	8	4	5.5	53.0	89	14	39	45	50	54	56
	9	3	14.0	41.9	217	65	135	157	164	171	153
	10	2	20.7	29.4	261	107	200	221	235	242	221
	11	1	25.0	15.2	280	134	239	262	276	283	252
	12		26.6	0.0	285	143	253	275	290	296	263
	SURFACE DAILY TOTALS				1978	782	1480	1634	1740	1795	1696

TABLE 18.1 (Continued).

SOLAR POSITION AND INSOLATION, 48°N LATITUDE

DATE	SOLAR TIME AM	SOLAR TIME PM	SOLAR POSITION ALT	SOLAR POSITION AZM	BTUH/SQ. FT. TOTAL INSOLATION ON SURFACES — NORMAL	HORIZ	SOUTH FACING SURFACE ANGLE WITH HORIZ. 38	48	58	68	90
JAN 21	8	4	3.5	54.6	37	4	17	19	21	22	22
	9	3	11.0	42.6	185	46	120	132	140	145	139
	10	2	16.9	29.4	239	83	190	206	216	220	206
	11	1	20.7	15.1	261	107	231	249	260	263	243
	12		22.0	0.0	267	115	245	264	275	278	255
	SURFACE DAILY TOTALS				1710	596	1360	1478	1550	1578	1478
FEB 21	7	5	2.4	72.2	12	1	4	4	4	4	4
	8	4	11.6	60.5	188	49	95	102	105	106	96
	9	3	19.7	47.7	251	100	178	187	191	190	167
	10	2	26.2	33.3	278	139	240	251	255	251	217
	11	1	30.5	17.2	290	165	278	290	294	288	247
	12		32.0	0.0	293	173	291	304	307	301	258
	SURFACE DAILY TOTALS				2330	1080	1880	1972	2024	1978	1720
MAR 21	7	5	10.0	78.7	153	37	49	49	47	45	35
	8	4	19.5	66.8	236	96	131	132	129	122	96
	9	3	28.2	53.4	270	147	205	207	203	193	152
	10	2	35.4	37.8	287	187	263	266	261	248	195
	11	1	40.3	19.8	295	212	300	303	297	283	223
	12		42.0	0.0	298	220	312	315	309	294	232
	SURFACE DAILY TOTALS				2780	1578	2208	2228	2182	2074	1632
APR 21	6	6	8.6	97.8	108	27	13	9	8	7	5
	7	5	18.6	86.7	205	85	76	69	59	48	21
	8	4	28.5	74.9	247	142	149	141	129	113	69
	9	3	37.8	61.2	268	191	216	208	194	174	115
	10	2	45.8	44.6	280	228	268	260	245	223	152
	11	1	51.5	24.0	286	252	301	294	278	254	177
	12		53.6	0.0	288	260	313	305	289	264	185
	SURFACE DAILY TOTALS				3076	2106	2358	2266	2114	1902	1262
MAY 21	5	7	5.2	114.3	41	9	4	4	3	3	2
	6	6	14.7	103.7	162	61	27	16	15	13	10
	7	5	24.6	93.0	219	118	89	75	60	43	13
	8	4	34.7	81.6	248	171	156	142	123	101	45
	9	3	44.3	68.3	264	217	217	202	182	156	86
	10	2	53.0	53.3	274	252	252	240	220	200	120
	11	1	59.5	28.6	279	274	298	281	258	228	141
	12		62.0	0.0	280	281	306	283	258	225	149
	SURFACE DAILY TOTALS				3254	2482	2418	2234	2010	1728	982
JUN 21	5	7	7.9	116.5	77	21	9	9	8	7	7
	6	6	17.2	106.2	172	74	33	19	18	16	12
	7	5	27.0	95.8	220	129	93	77	59	39	15
	8	4	37.1	84.6	246	181	157	140	119	95	35
	9	3	46.9	71.6	261	225	216	198	175	147	74
	10	2	55.8	54.8	269	259	262	244	220	189	105
	11	1	62.7	31.2	274	280	291	273	248	216	126
	12		65.5	0.0	275	287	301	283	258	225	133
	SURFACE DAILY TOTALS				3312	2626	2420	2204	1950	1644	874

DATE	SOLAR TIME AM	SOLAR TIME PM	SOLAR POSITION ALT	SOLAR POSITION AZM	BTUH/SQ. FT. TOTAL INSOLATION ON SURFACES — NORMAL	HORIZ	SOUTH FACING SURFACE ANGLE WITH HORIZ. 38	48	58	68	90
JUL 21	5	7	5.7	114.7	43	10	5	5	4	4	3
	6	6	15.2	104.1	156	62	28	18	16	15	11
	7	5	25.1	93.5	211	118	89	75	59	42	14
	8	4	35.1	82.1	240	171	154	140	121	99	43
	9	3	44.8	68.8	256	215	214	199	178	153	83
	10	2	53.5	51.9	266	250	261	246	224	195	116
	11	1	60.1	29.0	271	272	291	276	253	223	137
	12		62.6	0.0	272	279	301	286	263	232	144
	SURFACE DAILY TOTALS				3158	2474	2386	2200	1974	1694	956
AUG 21	6	6	9.1	98.3	99	28	14	10	9	8	6
	7	5	19.1	87.2	190	85	75	67	58	47	20
	8	4	29.0	75.4	232	141	145	137	125	109	65
	9	3	38.4	61.8	254	189	210	201	187	168	110
	10	2	46.4	45.1	266	225	260	252	237	214	146
	11	1	52.2	24.3	272	248	293	285	268	244	169
	12		54.3	0.0	274	256	304	296	279	255	177
	SURFACE DAILY TOTALS				2898	2086	2300	2200	2046	1836	1208
SEP 21	7	5	10.0	78.7	131	35	44	44	43	40	31
	8	4	19.5	66.8	215	92	124	124	121	115	90
	9	3	28.2	53.4	251	142	196	197	193	183	143
	10	2	35.4	37.8	269	181	252	254	248	236	185
	11	1	40.3	19.8	278	205	287	289	284	269	212
	12		42.0	0.0	280	213	299	302	296	281	221
	SURFACE DAILY TOTALS				2568	1522	2102	2118	2070	1966	1546
OCT 21	7	5	2.0	71.9	4	0	1	1	1	1	1
	8	4	11.2	60.2	165	44	86	91	95	95	87
	9	3	19.3	47.4	233	94	167	176	180	178	157
	10	2	25.7	33.1	262	133	228	239	242	239	207
	11	1	30.0	17.1	274	157	266	277	281	276	237
	12		31.5	0.0	278	166	276	291	294	288	247
	SURFACE DAILY TOTALS				2154	1022	1774	1860	1890	1866	1626
NOV 21	8	4	3.6	54.7	36	4	17	19	21	22	22
	9	3	11.2	42.7	179	46	117	129	137	141	135
	10	2	17.1	29.5	233	83	186	202	212	215	201
	11	1	20.9	15.1	255	107	227	245	255	258	238
	12		22.2	0.0	261	115	241	259	270	272	250
	SURFACE DAILY TOTALS				1668	598	1336	1448	1518	1544	1442
DEC 21	9	3	8.0	40.9	140	27	87	98	105	110	109
	10	2	13.6	28.2	214	63	164	180	192	197	190
	11	1	17.3	14.4	242	86	207	226	239	244	231
	12		18.6	0.0	250	94	222	241	254	260	244
	SURFACE DAILY TOTALS				1444	446	1136	1250	1326	1364	1304

Reprinted with permission from the 1974 *ASHRAE Transactions*, Vol. 80, Part II.

Figure 18.10. Clearness factors for the continental United States. (*Note:* Dashed lines show clearness factors; solid lines are average annual insolation in Langleys/day [Langleys/day × 2.69 = BTU/hr]).

corrections to the insolation values may be required. The values shown in Table 18.1 are based on a *clearness factor* (CF) of 1.0. Figure 18.10 is a map that shows clearness factors (also called clearness numbers) for the continental United States. Note that the values may differ for summer and winter in some localities.

EXAMPLE 18.2

Find the total daily insolation for a clear day on March 21 in Denver, Colorado, for a south facing surface tilted at a 50° elevation above the horizontal.

Solution. Denver is at 40°N latitude. From Table 18.1 the total daily insolation on March 21

for a surface at a 50° tilt is 2284 BTU/ft². From Figure 18.10, however, it is noted that CF = 1.05 for that part of Colorado. Correcting the value from the table

$$I = 2284 \times 1.05 = 2398 \text{ BTU/ft}^2 \text{ per day}$$

The clearness factor does not account for the effects of industrial air pollution. When smog is present, it may reduce insolation. The solar heating system designer must determine this from personal knowledge of the area.

18.8. ORIENTATION AND TILT ANGLES

In order to maximize the total daily amount of radiation collected, a flat plate collector should be *oriented south*. This is why Table 18.1 lists insolation only at this orientation. If, for some reason, a collector is not oriented south, corrections can be made to the insolation values from data available in the literature. However, for an orientation as much as 15° away from south, the loss in radiation is negligible for practical purposes.

There may be special circumstances where it is desirable to orient the collector other than south. Some local climates have a regular pattern of fog or clouds all morning or afternoon, for example. In that case an orientation slightly southeast or southwest might be better. Interference from nearby objects, of course, might dictate the collector orientation.

The optimum *tilt angles* above the horizontal for collecting the maximum daily total radiation have been determined. Because the elevation of the sun changes seasonally, the angle depends on the use of the solar system. Recommended values are:

Service	Optimum Tilt Angle
Heating only	Latitude angle +10°
Heating and cooling	Latitude angle
Cooling only	Latitude angle −10°

Small variations in the above tilts will not affect the results significantly. Some authorities recommend a tilt of latitude +15° for heating.

18.9. SUNSHINE HOURS

Table 18.1 lists insolation for clear days, without clouds. In order to determine the economic feasibility of installing a solar heating system, the amount of solar energy available monthly and seasonally must be found. Therefore, an estimate must be made of the amount of time the sun is shining. Table 18.2 lists the mean percentage of sunshine hours expected for various locations, on a monthly and annual basis. The total monthly values found from Table 18.1 must be corrected for this.

EXAMPLE 18.3

Determine the monthly solar insolation in March for the surface described in Example 18.2.

Solution. From Example 18.2 the daily insolation is 2398 BTU/ft^2 per day on a clear day. From Table 18.2 the mean percentage of sunshine hours in March in Denver is 65. Applying this factor, the monthly solar insolation in March is

2398 BTU/ft^2 per day × 31 days × 0.65

= 48,320 BTU/ft^2

A condition that is not accounted for in this correction for cloudy periods is the effect of very short periods of sunshine. On days when the sun appears and disappears often, there may not be enough sustained direct radiation to bring the collector up to any significant operating temperature. Data on this effect are not yet readily available. However, the procedures described here are relatively conservative, thus balancing this effect.

The method we have used does not allow for diffuse radiation on cloudy days. But it is questionable if this radiation would usually be great enough to operate or "drive" a typical flat plate collector. Below a certain insolation value a collector will not absorb any net amount of heat.

Under special local conditions ground reflectance of radiation in the collector surface should be included. For example, if it is known that the area surrounding the collector will always have snow in winter, the resulting reflected

TABLE 18.2. MEAN PERCENTAGE OF POSSIBLE SUNSHINE

State and Station	Years	Jan.	Feb.	Mar.	Apr.	May	June	July	Aug.	Sept.	Oct.	Nov.	Dec.	Annual
Ala. Birmingham	56	43	49	56	63	66	67	62	65	66	67	58	44	59
Montgomery	49	51	53	61	69	73	72	66	69	69	71	64	48	64
Alaska, Anchorage	19	39	46	56	58	50	51	45	39	35	32	33	29	45
Fairbanks	20	34	50	61	68	55	53	45	35	31	28	38	29	44
Juneau	14	30	32	39	37	34	35	28	30	25	18	21	18	30
Nome	29	44	46	48	53	51	48	32	26	34	35	36	30	41
Ariz. Phoenix	64	76	79	83	88	93	94	84	84	89	88	84	77	85
Yuma	52	83	87	91	94	97	98	92	91	93	93	90	83	91
Ark. Little Rock	66	44	53	57	62	67	72	71	73	71	74	58	47	62
Calif. Eureka	49	40	44	50	53	54	56	51	46	52	48	42	39	49
Fresno	55	46	63	72	83	89	94	97	97	93	87	73	47	78
Los Angeles	63	70	69	70	67	68	69	80	81	80	76	79	72	73
Red Bluff	39	50	60	65	75	79	86	95	94	89	77	64	50	75
Sacramento	48	44	57	67	76	82	90	96	95	92	82	65	44	77
San Diego	68	68	67	68	66	60	60	67	70	70	70	76	71	68
San Francisco	64	53	57	63	69	70	75	68	63	70	70	62	54	66
Colo. Denver	64	67	67	65	63	61	69	68	68	71	71	67	65	67
Grand Junction	57	58	62	64	67	71	79	76	72	77	74	64	58	69
Conn. Hartford	48	46	55	56	54	57	60	62	60	57	55	46	46	56
D. C. Washington	66	46	53	56	57	61	64	64	62	64	61	54	47	58
Fla. Apalachicola	26	59	62	62	71	77	70	64	63	62	74	66	53	65
Jacksonville	60	58	59	66	71	71	63	62	61	58	58	61	53	62
Key West	45	68	75	78	78	76	63	65	65	65	65	69	53	71
Miami Beach	48	66	72	73	73	68	62	65	67	62	62	65	65	67
Tampa	63	63	67	71	74	75	66	61	64	64	67	67	61	68
Ga. Atlanta	65	48	53	57	65	68	68	62	63	65	67	60	47	60
Hawaii, Hilo	9	48	42	41	34	31	41	44	38	42	41	34	36	39
Honolulu	53	62	64	60	62	64	66	67	70	70	68	63	60	65
Lihue	9	48	48	48	46	51	60	58	59	67	58	51	49	54
Idaho, Boise	20	40	48	59	67	68	75	89	86	81	66	46	37	66
Pocatello	21	37	47	58	64	66	72	82	81	78	66	48	36	64
Ill. Cairo	30	46	53	59	65	71	77	82	79	75	73	56	46	65
Chicago	66	44	49	53	56	63	69	73	70	65	61	47	41	59
Springfield	59	47	51	54	58	64	69	76	72	73	64	53	45	60
Ind. Evansville	48	42	49	55	61	67	73	78	76	73	67	52	42	64
N. Mex. Albuquerque	28	70	72	72	76	79	84	76	75	81	80	79	70	76
Roswell	47	69	72	75	77	76	80	76	75	74	74	74	69	74
N. Y. Albany	63	43	51	53	53	57	62	63	61	58	54	39	38	53
Binghamton	63	31	39	41	44	50	56	54	51	47	43	29	26	44
Buffalo	49	32	41	49	51	59	67	70	67	60	51	31	28	53
Canton	43	37	47	50	48	54	61	63	61	54	45	30	31	49
New York	83	49	56	57	59	62	65	66	64	64	61	53	50	59
Syracuse	49	31	38	45	50	58	64	67	63	56	47	29	26	50
N. C. Asheville	57	48	53	56	61	64	63	59	59	62	64	59	48	58
Raleigh	61	50	56	59	64	67	65	62	62	63	64	62	52	61
N. Dak. Bismarck	65	52	58	56	57	58	61	73	69	62	59	49	48	59
Devils Lake	55	53	60	59	60	59	62	71	67	59	56	44	45	58
Fargo	62	47	60	63	64	62	63	73	69	60	57	39	46	59
Williston	46	41	49	56	61	63	67	78	81	74	68	46	48	60
Ohio, Cincinnati	69	41	46	52	56	62	69	72	65	68	60	46	39	57
Cleveland	65	29	36	45	52	61	67	71	68	62	54	32	25	50
Columbus	65	36	44	49	54	63	68	71	68	66	60	44	35	55
Okla. Oklahoma City	62	57	60	63	64	65	74	78	78	74	68	64	57	68
Oreg. Baker	46	41	49	56	61	63	67	83	81	74	62	46	37	60
Portland	69	27	34	41	49	52	55	70	65	55	42	28	23	48
Roseburg	29	24	32	40	51	57	59	79	77	68	42	28	18	51
Pa. Harrisburg	60	43	52	55	57	61	65	68	64	62	58	47	43	57
Philadelphia	66	45	56	57	58	61	65	65	61	62	61	53	49	57
Pittsburgh	63	32	39	45	50	57	62	64	61	62	54	39	30	51
R. I. Block Island	48	45	54	47	56	58	60	62	62	60	59	50	44	56
S. C. Charleston	61	58	60	65	72	73	70	66	66	67	68	60	57	66
Columbia	55	53	57	62	68	69	68	63	65	64	68	64	51	63
S. Dak. Huron	62	55	62	60	62	65	68	76	72	66	61	52	49	63
Rapid City	53	58	62	63	62	61	66	73	73	69	66	58	54	64
Tenn. Knoxville	62	42	49	53	59	64	66	64	59	64	64	53	41	57
Memphis	55	44	51	57	64	68	74	73	74	70	69	58	45	64
Nashville	63	42	47	54	60	65	69	69	68	69	65	55	42	59
Tex. Abilene	14	64	68	73	66	73	86	83	85	73	71	72	66	73
Amarillo	54	71	71	75	75	75	82	81	81	79	76	76	70	76
Austin	33	46	50	57	60	62	72	76	79	70	70	57	49	63

487

TABLE 18.2 *(Continued).*

City													
Ft. Wayne	48	38	44	51	55	62	69	74	69	64	58	41	38	57
Indianapolis	63	41	47	49	55	62	68	74	70	68	64	48	39	59
Iowa. Des Moines	66	56	56	56	59	62	66	75	70	64	64	53	48	62
Dubuque	54	48	52	52	58	60	63	73	67	61	55	44	40	57
Sioux City	52	55	58	58	59	63	67	75	72	67	65	53	50	63
Kans. Concordia	52	60	60	62	63	65	73	79	76	72	70	64	58	67
Dodge City	70	67	66	68	68	68	74	78	78	76	75	70	67	71
Wichita	46	61	63	64	64	66	73	80	77	73	69	67	59	69
Ky. Louisville	59	41	47	52	57	64	68	72	69	68	64	51	39	59
La. New Orleans	69	49	50	57	63	66	64	58	60	64	70	60	46	59
Shreveport	18	48	54	58	60	69	78	79	80	77	65	60	69
Maine, Eastport	58	45	51	52	51	53	57	55	57	54	50	37	40	50
Mass. Boston	67	47	56	57	56	59	64	64	61	58	48	48	48	57
Mich. Alpena	45	29	43	52	56	59	64	70	64	52	44	24	22	51
Detroit	69	34	42	48	52	58	65	69	66	61	54	35	29	53
Grand Rapids	56	26	37	48	54	60	66	72	67	58	50	31	22	49
Marquette	55	31	40	47	52	53	56	63	57	47	38	24	24	47
S. Ste. Marie	60	28	44	50	54	54	59	63	58	45	36	21	22	47
Minn. Duluth	49	47	55	60	58	58	60	68	63	53	47	36	40	55
Minneapolis	45	49	54	55	57	60	64	72	69	60	54	40	40	56
Miss. Vicksburg	66	46	50	57	64	69	73	69	72	74	71	60	45	64
Mo. Kansas City	69	55	57	59	60	64	70	76	70	67	59	52	65
St. Louis	68	48	49	56	59	64	68	72	68	67	65	54	44	61
Springfield	45	48	54	57	60	63	69	77	71	65	61	58	48	63
Mont. Havre	55	49	58	61	63	63	65	78	75	64	57	48	46	62
Helena	65	46	55	58	60	59	63	77	74	63	57	48	43	60
Kalispell	50	28	40	49	57	58	60	77	73	61	50	28	20	53
Nebr. Lincoln	55	57	59	60	60	63	69	76	71	67	66	59	55	64
North Platte	53	63	63	64	62	64	72	78	74	72	70	62	58	68
Nev. Ely	21	61	64	68	65	67	79	81	81	79	73	67	62	72
Las Vegas	19	74	77	78	81	85	91	84	86	92	84	83	75	82
Reno	51	59	64	69	75	77	82	90	89	86	76	68	56	76
Winnemucca	53	52	60	64	70	76	83	90	90	86	75	62	53	74
N. H. Concord	44	48	53	55	53	51	56	57	58	55	50	43	43	52
N. J. Atlantic City	62	51	57	58	59	62	65	67	66	65	54	58	52	60
Brownsville	37	44	49	51	57	65	73	78	78	67	70	54	44	61
Del Rio	36	53	55	61	63	60	66	80	75	69	66	58	52	63
El Paso	53	74	77	81	85	87	87	78	78	80	82	80	73	80
Ft. Worth	33	56	57	65	66	67	75	78	78	74	70	63	58	68
Galveston	66	50	50	55	61	69	76	71	72	70	74	62	49	63
San Antonio	57	48	51	56	58	60	69	75	74	69	67	55	49	62
Utah, Salt Lake City	22	48	53	61	68	73	82	84	82	73	56	49	69
Vt. Burlington	54	34	43	48	47	53	59	62	59	51	43	25	24	46
Va. Norfolk	60	50	57	60	63	67	66	66	63	64	60	51	62
Richmond	56	49	55	59	63	67	65	62	63	64	58	50	61
Wash. North Head	44	28	37	42	48	48	48	50	46	48	41	31	27	41
Seattle	26	27	34	42	48	53	62	56	53	36	28	24	45
Spokane	62	26	41	53	64	68	82	79	68	53	28	22	58
Tatoosh Island	49	26	36	39	45	47	48	44	47	38	26	23	40
Walla Walla	44	24	35	51	63	67	72	86	84	72	59	33	20	60
Yakima	18	34	49	62	70	72	74	86	86	74	61	38	29	65
W. Va. Elkins	55	33	37	42	47	55	55	53	56	55	51	41	33	48
Parkersburg	62	30	36	42	49	56	60	63	60	53	37	29	48
Wis. Green Bay	57	44	51	55	56	58	64	70	65	58	52	40	55
Madison	59	44	49	52	53	58	64	70	66	60	56	41	38	56
Milwaukee	59	44	48	53	56	60	65	73	67	62	56	44	39	57
Wyo. Cheyenne	63	65	66	64	61	59	68	69	69	65	63	66
Lander	57	66	70	71	66	65	74	76	75	72	67	61	62	69
Sheridan	52	56	61	62	61	61	67	74	71	67	60	53	52	64
Yellowstone Park	35	39	51	55	57	56	63	71	65	57	45	38	56
P. R. San Juan	57	64	69	71	66	59	62	65	67	61	63	63	65	65

Based on period of record through December 1959, except in a few instances.

These charts and tabulation derived from "Normals, Means, and Extremes" table in U.S. Weather Bureau publication *Local Climatological Data.*

* Reprinted from ASHRAE TRANSACTIONS 1974, Volume 80, PART II, by permission of the American Society of Heating, Refrigerating and Air-Conditioning Engineers, Inc."

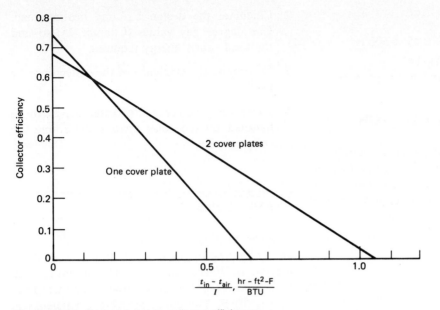

Figure 18.11. Typical flat plate collector efficiency.

radiation should be calculated. We will not discuss this subject here.

the amount of insolation received. This data can be plotted as shown in Figure 18.11.

18.10. COLLECTOR PERFORMANCE

The energy actually collected by a solar collector will be considerably less than the amount received. Some of the energy will be lost back to the surroundings, because the collector is at higher temperature. The efficiency of a collector is defined as the ratio of (heat) energy output to energy input. The efficiency depends on the physical construction of the collector. Increasing the number of cover plates raises the efficiency. The materials and configuration of the absorber plate, coating, and insulation also affect the efficiency. Manufacturers furnish specific data on their collectors.

For a given collector the efficiency will vary with the temperature difference between the collector and the surrounding ambient air, and with

EXAMPLE 18.4

Find the amount of heat collected per hour by a 500 ft² flat plate collector of the type shown in Figure 18.11, with two cover plates, at a time when the solar insolation received is 200 BTU/hr-ft² and the surrounding air temperature is 40 F. The inlet water temperature is 110 F.

Solution. Calculating the value used on the horizontal axis of Figure 18.11,

$$\frac{t_{in} - t_{air}}{I} = \frac{110 - 40}{200} = 0.35$$

From the figure, the efficiency is read on the curve for two cover plates as 0.45. The heat collected is

$$200 \text{ BTU/hr-ft}^2 \times 0.45 \times 500 \text{ ft}^2$$
$$= 45{,}000 \text{ BTU/hr}$$

18.11. SIZING THE COLLECTOR

The main task in designing a solar heating system is to determine the size of the collector to be used. The information we have developed can be used to do this in a number of ways, one of which we will describe here. In this method the percentage of total yearly building heating load that the solar heating system shall handle is first decided on, and based on this the collector size can be determined. Experience has shown that for many typical applications a range of 50–75% of the total yearly load is a good range to assign to the task of the solar heating system, to result in an economic combination of owning and operating costs. The steps in sizing the collector are as follows:

1. Calculate the total solar insolation received for the year.

2. Calculate the design building heating load. Use degree day values (Chapter 15) to find the total yearly energy required.

3. Determine the efficiency of the collector to be used.

4. Based on a percentage of total energy to be handled by the solar system, determine the collector size.

EXAMPLE 18.5

A solar heating system to handle 65% of the yearly heating requirements is to be installed for a home in Washington, D.C. The design heating load is 41,000 BTU/hr and the indoor temperature is 70 F. The collector average efficiency is 42%. Collector orientation is south and the tilt is 50°. Determine the required collector size. Assume CF = 1.0.

Solution.   The insolation received is calculated as explained previously. Table 18.3 shows the results of heat collected.

TABLE 18.3. RESULTS OF HEAT COLLECTION

	Insolation			Collected Energy		
	$\dfrac{\text{BTU}}{\text{day-ft}^2}$	% Sun	$\dfrac{\text{BTU}}{\text{day-ft}^2}$ average	Coll. eff.	$\dfrac{\text{BTU}}{\text{day-ft}^2}$	$\dfrac{\text{BTU}}{\text{month-ft}^2}$
Sept.	2182	68	1484	.42	623	18,690
Oct.	2098	60	1259	.42	529	16,400
Nov.	1870	46	860	.42	361	10,830
Dec.	1740	39	678	.42	285	8,840
Jan.	1906	41	781	.42	328	10,170
Feb.	2202	46	1013	.42	425	11,900
Mar.	2284	52	1188	.42	499	15,470
Apr.	2168	56	1214	.42	510	15,300
May	2040	62	1265	.42	531	16,460
				Collected Yearly Total	=	124,060 BTU / ft²

The total yearly heating requirement, using the degree day method, is

$$\frac{41,000 \text{ BTU/hr}}{(70 - 19)} \times 24 \text{ hr} \times 4224 \text{ degree days}$$

$$= 81,500,000 \text{ BTU}$$

The size of the solar collector to handle 65% of the heat required is

$$\text{Collector area} = \frac{81,500,000 \text{ BTU}}{124,060 \text{ BTU/ft}^2} \times 0.65$$

$$= 427 \text{ ft}^2$$

In the above procedure we assume an average yearly collector efficiency. For a more precise calculation the varying collector efficiency would be determined from manufacturer's curves for the water and ambient temperatures, and insolation values expected each month.

18.12. ECONOMIC ANALYSIS

When considering the installation of a solar heating system, either in a new building, or retrofitting an existing system, an analysis must be made to determine if the fuel cost savings will offset the increased capital equipment expenditures. The first step in doing this is to simply find the annual fuel costs savings from that part of the heating requirements handled by the solar system.

EXAMPLE 18.6

Determine the expected yearly energy and fuel cost savings for the home in Example 18.5 if the solar heating system described is installed. Fuel oil at $.90 a gallon is normally used, with a heating value of 145,000 BTU/gal. The auxiliary heating system has an average efficiency of 60%.

Solution. The solar system handles 65% of the yearly requirements. The net yearly depletable energy saved is therefore

$$81,500,000 \text{ BTU} \times 0.65 = 52,975,000 \text{ BTU}$$

The amount of fuel oil saved is

$$52,975,000 \text{ BTU} \times \frac{1 \text{ gal}}{145,000 \text{ BTU}} \times \frac{1}{0.60}$$

$$= 609 \text{ gal}$$

The yearly fuel cost saving is

$$609 \text{ gal} \times \$0.90/\text{gal} = \$548$$

In order to decide whether the annual savings are justified, we must determine how long it will take to recover our initial investment. If this period is reasonable, then we are willing to make the investment, because thereafter the annual savings will in a sense be profit. There are numerous economic procedures for estimating the justification of the investment. One popular procedure is called the *present worth* method. This approach will tell us how much of an initial investment is justified if we wish to have it paid back in a given number of years. The following equations are used to determine this:

$$P = \frac{S(1 + i_e)^t - 1}{i_e(1 + i_e)^t} \quad (18.1)$$

$$i_e = \frac{1 + i}{1 + j} - 1 \quad (18.2)$$

where

P = initial investment justifiable (present worth)
S = initial yearly savings, $
t = desired pay back period, years
i = annual interest rate for borrowing money, decimals
j = annual rate of fuel price increase, decimals
i_e = effective interest rate, from Equation 18.2

An important characteristic of this method is that it accounts directly for rising fuel prices.

TABLE 18.4. AVERAGE MONTHLY DEGREE DAYS FOR SELECTED LOCATIONS

Location	July	August	September	October	November	December	January	February	March	April	May	June	Yearly Total
Alabama, Birmingham	0	0	6	93	363	555	592	462	363	108	9	0	2551
Alaska, Anchorage	245	291	516	930	1284	1572	1631	1316	1293	879	592	315	10864
Arizona, Tucson	0	0	0	25	231	406	471	344	242	75	6	0	1800
Arkansas, Little Rock	0	0	9	127	465	716	756	577	434	126	9	0	3219
California, San Francisco	81	78	60	143	306	462	508	395	363	279	214	126	3015
Colorado, Denver	6	9	117	428	819	1035	1132	938	887	558	288	66	6283
Connecticut, Bridgeport	0	0	66	307	615	986	1079	966	853	510	208	27	5617
Delaware, Wilmington	0	0	51	270	588	927	980	874	735	387	112	6	4930
District of Columbia, Washington	0	0	33	217	519	834	871	762	626	288	74	0	4224
Florida, Tallahassee	0	0	0	28	198	360	375	286	202	86	0	0	1485
Georgia, Atlanta	0	0	18	124	417	648	636	518	428	147	25	0	2961
Hawaii, Honolulu	0	0	0	0	0	0	0	0	0	0	0	0	0
Idaho, Boise	0	0	132	415	792	1017	1113	854	722	438	245	81	5809
Illinois, Chicago	0	12	117	381	807	1166	1265	1086	939	534	260	72	6639
Indiana, Indianapolis	0	0	90	316	723	1051	1113	949	809	432	177	39	5699
Iowa, Sioux City	0	9	108	369	867	1240	1435	1198	989	483	214	39	6951
Kansas, Wichita	0	0	33	229	618	905	1023	804	645	270	87	6	4620
Kentucky, Louisville	0	0	54	248	609	890	930	818	682	315	105	9	4660
Louisiana, Shreveport	0	0	0	47	297	477	552	426	304	81	0	0	2184
Maine, Caribou	78	115	336	682	1044	1535	1690	1470	1308	858	468	183	9767
Maryland, Baltimore	0	0	48	264	585	905	936	820	679	327	90	0	4654
Massachusetts , Boston	0	9	60	316	603	983	1088	972	846	513	208	36	5634
Michigan, Lansing	6	22	138	431	813	1163	1262	1142	1011	579	273	69	6909
Minnesota, Minneapolis	22	31	189	505	1014	1454	1631	1380	1166	621	288	81	8382
Mississippi, Jackson	0	0	0	65	315	502	546	414	310	87	0	0	2239
Missouri, Kansas City	0	0	39	220	612	905	1032	818	682	294	109	0	4711
Montana, Billings	6	15	186	487	897	1135	1296	1100	970	570	285	102	7049
Nebraska, Lincoln	0	6	75	301	726	1066	1237	1016	834	402	171	30	5864
Nevada, Las Vegas	0	0	0	78	387	617	688	487	335	111	6	0	2709

TABLE 18.4 *(Continued)*.

Location													
New Hampshire, Concord	6	50	177	505	822	1240	1358	1184	1032	636	298	75	7333
New Jersey, Atlantic City	0	0	39	251	549	880	936	848	741	420	133	15	4812
New Mexico, Albuquerque	0	0	12	229	642	868	930	703	595	288	81	0	4348
New York, Syracuse	6	28	132	415	744	1153	1271	1140	1004	570	248	45	6756
North Carolina, Charlotte	0	0	6	124	438	691	691	582	481	156	22	0	3191
North Dakota, Bismarck	34	28	222	577	1083	1463	1708	1442	1203	645	329	117	8851
Ohio, Cleveland	9	25	105	384	738	1088	1159	1047	918	552	260	66	6351
Oklahoma, Stillwater	0	0	15	164	498	766	868	664	527	189	34	0	3725
Oregon, Pendleton	0	0	111	350	711	884	1017	773	617	396	205	63	5127
Pennsylvania, Pittsburgh	0	9	105	375	726	1063	1119	1002	874	480	195	39	5987
Rhode Island, Providence	0	16	96	372	660	1023	1110	988	868	534	236	51	5954
South Carolina, Charleston	0	0	0	59	282	471	487	389	291	54	0	0	2033
South Dakota, Rapid City	22	12	165	481	897	1172	1333	1145	1051	615	326	126	7345
Tennessee, Memphis	0	0	18	130	447	698	729	585	456	147	22	0	3232
Texas, Dallas	0	0	0	62	321	524	601	440	319	90	6	0	2363
Utah, Salt Lake City	0	0	81	419	849	1082	1172	910	763	459	233	84	6052
Vermont, Burlington	28	65	207	539	891	1349	1513	1333	1187	714	353	90	8269
Virginia, Norfolk	0	0	0	136	408	698	738	655	533	216	37	0	3421
Washington, Spokane	9	25	168	493	879	1082	1231	980	834	531	288	135	6655
West Virginia, Charleston	0	0	63	254	591	865	880	770	648	300	96	9	4476
Wisconsin, Milwaukee	43	47	174	471	876	1252	1376	1193	1054	642	372	135	7635
Wyoming, Casper	6	16	192	524	942	1169	1290	1084	1020	657	381	129	7410
Alberta, Calgary	109	186	402	719	1110	1389	1575	1379	1268	798	477	291	9703
British Columbia, Vancouver	81	87	219	456	657	787	862	723	676	501	310	156	5515
Manitoba, Winnipeg	38	71	322	683	1251	1757	2008	1719	1465	813	405	147	10579
New Brunswick, Fredericton	78	68	234	592	915	1392	1541	1379	1172	753	406	141	8671
Nova Scotia, Halifax	58	51	180	457	710	1074	1213	1122	1030	742	487	237	7361
Ontario, Ottawa	25	81	222	567	936	1469	1624	1441	1231	708	341	90	8735
Quebec, Montreal	9	43	165	521	882	1392	1566	1381	1175	684	316	69	8203
Saskatchewan, Regina	78	93	360	741	1284	1711	1965	1687	1473	804	409	201	10806

Source. Heating, Ventilating, and Air Conditioning, F.C. McQuiston and J. D. Parker, Copyright © 1977, John Wiley & Sons. Reprinted by permission.

EXAMPLE 18.7

Mr. and Mrs. U. R. Wise, owners of the house in Example 18.5, want to recover their investment on a retrofit solar heating system addition in 10 years. They can borrow money at 6% under a special government energy conservation program. Fuel costs rise at an annual rate of 12%. What should be the maximum initial investment?

Solution. Applying the present worth method equations

$$i_e = \frac{1 + 0.06}{1 + 0.12} - 1 = -0.054$$

$$P = \frac{S(1 + i_e)^t - 1}{i_e(1 + i_e)^t}$$

$$= 548\frac{(1 - 0.054)^{10} - 1}{-0.054(1 - 0.054)^{10}} = \$7530$$

If the initial cost of adding the solar heating system is $7530 or less, the owners would be wise to install it.

18.13. STORAGE SYSTEM SIZING

Studies have been made which indicate that a storage tank size of 1.5–2.5 gallons of water per square foot of collection area is usually the optimum economic tank size, regardless of climate.

This provides at the most about one day's heat storage.

EXAMPLE 18.8

A storage tank of 2.0 gal/ft^2 of collector area is used for the solar heating system for the home in Example 18.5. How long can the system operate on storage on a typical March day?

Solution. The heat required for a day in March is first determined. To do this, the monthly degree day data is required. Table 18.4 presents this information for some cities.

First, the BTU/deg day are determined

$$\frac{41,000 \text{ BTU/hr}}{(70 - 19)\text{F}} \times 24 \text{ hr}$$

$$= 19,294 \text{ BTU/deg day}$$

There are 626 degree days in March. The average degree days in a day are

$$\frac{626}{31} = 20.2 \text{ deg days}$$

The daily heat required is

$$19,294 \text{ BTU/deg day} \times 20.2 \text{ deg days}$$

$$= 389,740 \text{ BTU}$$

The storage tank volume is

$$2.0 \text{ gal/ft}^2 \times 427 \text{ ft}^2 = 854 \text{ gal} \times 8.3 \text{ lb/gal}$$

$$= 7088 \text{ lb}$$

We will assume the water is heated from 90 F–140 F in the tank. The heat available is therefore

$$Q = m \times c \times \text{TC}$$

$$= 7088 \text{ lb} \times 1 \text{ BTU/lb-F} \times 50 \text{ F}$$

$$= 354,440 \text{ BTU}$$

The tank has slightly less than one day's storage to handle the heating requirements on a March day. Extended periods of cloudiness and conditions in colder months would require use of the conventional auxiliary heat source.

18.14. APPROXIMATE SYSTEM DESIGN DATA

The U.S. Department of Energy has furnished information based on typical conditions that can

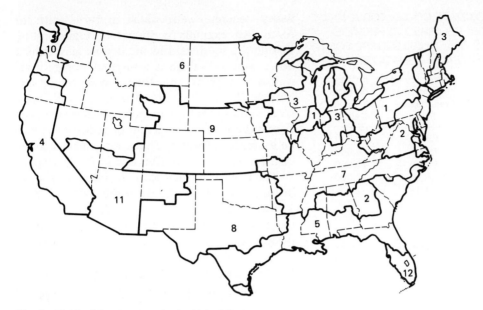

Figure 18.12. Climatic zones in the United States.

be used for approximate solar heating and hot water system sizing and economic analysis. Figure 18.12 shows climatic zones in the United States. This map is used in conjunction with Tables 18.5 and 18.6 for preliminary estimates of collector and storage tank sizes and energy sav-

ings for a 1500 ft² home and 10,000 ft² building. Note that the figures are not grossly different from those resulting from our calculations.

18.15. PASSIVE SOLAR HEATING SYSTEMS

A solar heating system that utilizes the building structure to provide some or all of the heating is called a *passive* system. An *active* system uses collectors and ducts or piping. Those that combine both are called *hybrid* systems.

In a simple passive system design, the building is constructed with a large southward facing glass window. This arrangement is called a *direct* passive system because the sun directly heats the interior materials, which then heat the room air. If the system is properly designed, heat which is stored in the walls and floor in the day will continue to heat the room at night. Heavy mass materials are desirable for this purpose. Sometimes the sides of the building not facing south are built into a hillside.

In an *indirect* passive solar heating system, the sun heats a thermal storage element (e.g., a

TABLE 18.5. APPROXIMATE COLLECTOR AND STORAGE TANK SIZES REQUIRED TO PROVIDE THE HEATING AND HOT WATER NEEDS OF A 1500 SQUARE FOOT HOME

Climatic Zone	Percent of Energy Supplied by Solar	Collector Area, Square Feet	Storage Tank Capacity, Gallons
1	71	800	1,500
2	72	500	750
3	66	800	1,500
4	73	300	500
5	75	200	280
6	70	750	1,500
7	70	500	750
8	71	200	280
9	72	600	1,000
10	58	500	750
11	85	200	280
12*	85	45	80

TABLE 18.6. APPROXIMATE COLLECTOR AND STORAGE TANK SIZES REQUIRED TO PROVIDE THE HEATING NEEDS OF A 10,000 SQUARE FOOT BUILDING LOCATED IN EACH CLIMATIC ZONE

Climatic Zone	Percent of Energy Supplied by Solar	Collector Area, Square Feet	Storage Tank Capacity, Gallons
1	74	5,330	10,000
2	71	3,330	5,000
3	68	5,330	10,000
4	73	2,000	4,000
5	75	1,210	2,000
6	72	5,000	7,500
7	71	3,330	5,000
8	74	1,330	2,000
9	75	4,000	6,000
10	60	3,330	5,000
11	77	1,000	1,500
12	*		

heavy concrete wall) which in turn heats the room. An example is shown in Figure 18.13. Behind the southward facing double glass (this is not a window!) there is a thermal storage inner wall with openings into the room at the top and bottom (Figure 18.13*a*). The inner wall has a large mass so that it can store considerable heat. It may be made of concrete or rocks, or it may be a tankful of water. This arrangement is often called a *Trombe* wall after one of its developers.

The low winter sun heats the inner wall which in turn heats the air between the wall and glass. The warmed air rises and enters the room through the top opening. Cool air from the room flows through the bottom opening to replace the warm air and a convection current is created (Figure 18.13*b*). Sometimes a fan is used to increase the flow of air. At night the warmed wall will still

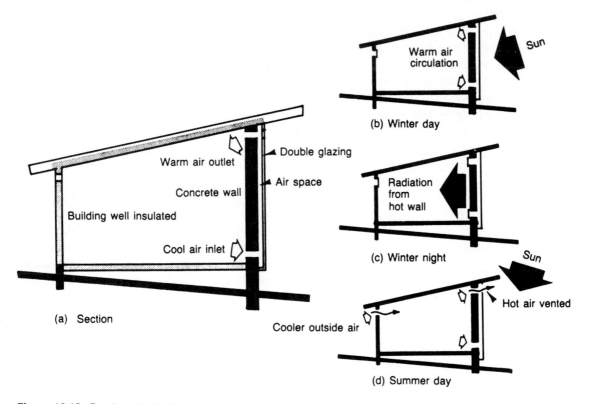

(a) Section

(b) Winter day

(c) Winter night

(d) Summer day

Figure 18.13. Passive solar heating arrangement (called "Trombe" wall).

heat the room, and the openings in the wall are closed to prevent circulation (Figure 18.13c). Furthermore, by providing vent openings under the roof, cool air from the north side can be circulated in summer (Figure 18.13d).

There are many other ingenious passive solar heating systems that have been developed in conjunction with the architectural design. For example, greenhouses which have been integrated as part of the building and roofs with water heat storage ponds.

PROBLEMS

18.1. Find the solar insolation on a south facing surface at 12° N, 3 PM, and the daily total on January 21 in Los Angeles, California, for a surface tilted at 32°, 42°, and normal to the sun.

18.2. What is the optimum collector tilt angle to collect the maximum daily solar insolation in November for (a) Boston, Massachusetts and (b) your town?

18.3. Find the total daily clear day insolation on February 21 for a collector at a tilt angle recommended for domestic hot water heating in a home in Savannah, Georgia.

18.4. Find the average daily insolation in January for a surface tilted at the recommended angle for heating in (a) Dallas, Texas and (b) Detroit, Michigan.

18.5. Determine the efficiency of a flat plate collector for a solar heating system, of the type shown in Figure 18.11, with a single cover plate, at 1 PM on a clear February day in Springfield, Illinois. The outdoor temperature is 30 F and the inlet water temperature is 80 F. What is the

heat output of a 500 ft^2 collector at this time?

18.6. If two cover plates are used on the collector in Problem 18.5, what is the heat output at the time listed?

18.7. Determine the collector size for a home in Kansas City, Missouri, with a design heating load of 45,000 BTU/hr, to handle 70% of the yearly heating requirements. The average collector efficiency is 44%.

18.8. Determine the collector size for (a) your home and (b) your school, for a solar heating system to handle 70% of the yearly heating requirements.

18.9. Determine the expected yearly fuel and cost savings for the solar heating systems in Problems 18.7 and 18.8. Average furnace efficiency is 60%. Use prevailing fuel costs.

18.10. Determine a suitable storage tank size for the solar heating systems in Problems 18.7 and 18.8. Find the amount of heat stored and compare it with that required on an average February day. The water temperature draw down in the tank is 55 F.

18.11. Compare the preliminary estimate sizes of collector and storage tank and energy savings from Tables 18.5 and 18.6 to those found in your calculations in Problems 18.7 through 18.10.

18.12. For the building described in Problems 18.7 and 18.8, determine how much it is worth investing in a solar heating system added to the present heating system. Special loans are available at 5%. Assume a pay back period of 8 years. Repeat the problem for a pay back period of 15 years.

Bibliography

The following sources provide reference material related to subjects contained in this book, as well as various codes and standards.

1. Air Conditioning and Refrigeration Institute (ARI), Arlington, VA.
2. Air Movement and Control Association, Inc. (AMCA), Arlington Heights, IL.
3. American Gas Association (AGA), Arlington, VA.
4. American Society of Heating, Refrigerating and Air-Conditioning Engineers (ASHRAE), Atlanta, GA.
   Standard 90-80, 1980. Energy Conservation in New Building Design
   Equipment Handbook, 1983
   Systems Handbook, 1984
   Fundamentals Handbook, 1985
   Load Calculation Manual, 1985
   Applications Handbook, 1987
   Cooling and Heating Load Calculation Manual, 1979. Prepared by ASHRAE for U.S. Department of Housing and Urban Development.
5. American Society of Mechanical Engineers (ASME), New York, NY.
6. Hydronics Institute, Berkeley Heights, NJ.
7. Sheet Metal and Air Conditioning Contractors Association, Inc. (SMACNA), Vienna, VA.

Appendix

TABLE A.1. ABBREVIATIONS AND SYMBOLS

A	area	c	specific heat
BF	bypass factor	D	diameter
BF	ballast factor	D	degree days
BHP	brake horsepower	DB	dry bulb temperature
BTU	British termal unit	DP	dew point temperature
C	degrees Celsius	d	density
C	loss coefficient	d.a.	dry air
C	thermal conductance	dB	decibel
C_D	correction factor for degree days	E	edge factor
C_F	correction factor for equipment part load efficiency	E	efficiency
		EER	energy efficiency ratio
CF	contact factor	E.L.	equivalent length
CFM	ft^3/min	EST	effective surface temperature
CHW	chilled water	ET*	effective temperature
CLF	cooling load factor	ETD	equivalent temperature difference
CLTD	cooling load temperature difference	F	degrees Fahrenheit
$CLTD_c$	cooling load temperature difference, corrected	F	force
		F_c	correction factor for heat transfer to surroundings
COP	coefficient of performance, cooling		
COP_h	coefficient of performance, heating	f	friction factor

501

TABLE A.1 *(Continued).*

f	correction factor for ceiling ventilation	p	pressure
ft	feet	p_{abs}	absolute pressure
GPM	gallons per minute	p_{atm}	atmospheric pressure
g	gravitational constant	p_g	gage pressure
g	grams	p_{vac}	vacuum pressure
gal	gallons	psf	lb/ft^2
gr	grains	psi	$lb/in.^2$
H	total enthalpy	psia	$lb/in.^2$ absolute
H	head pressure	psig	$lb/in.^2$ gage
H_e	elevation head	Q	heat; heat transfer rate
H_f	friction loss	Q_l	latent heat
H_p	pump or fan head	Q_s	sensible heat
H_s	static head	Q_t	total heat
$\mathbf{H}_t$	total head	R	degrees Rankine
H_v	velocity head	R	thermal resistance
Hg	mercury	R	gas constant
Hz	frequency	R	recovery factor
HP	horsepower	RH	relative humidity
HW	hot water	RLHG	room latent heat gain
h	specific enthalpy	RSHG	room sensible heat gain
h_f	specific enthalpy of saturated liquid	RSHR	room sensible heat ratio
h_{fg}	latent heat of vaporization	RTHG	room total heat gain
h_g	specific enthalpy of saturated vapor	SC	shade coefficient
hr	hours	SHGF	solar heat gain factor
in.	inch	SHR	sensible heat ratio
J	Joule	SPR	static pressure regain
K	unit length conductance	s.g.	specific gravity
K	correction factor for color of surface	sec	seconds
K	degrees Kelvin	T	temperature, absolute
KW	kilowatt	TC	temperature change
k	thermal conductivity	TD	temperature difference
kg	kilogram	t	temperature
L	length	U	overall heat transfer coefficient
LM	correction factor for latitude and month	V	volume
lb	pounds	V	velocity
ME	mechanical efficiency	VFR	volume flow rate
m	mass	v	specific volume
m	meters	W	work
min	minutes	W	watts
N	newton	W	humidity ratio
N	speed, revolutions per minute	W'	humidity ratio, gr/lb d.a.
NC	noise criteria	WB	wet bulb temperature
NPSH	net positive suction head	WHP	water horsepower
n	number of people	w	weight
OTTV	overall thermal transfer value	w.	water
P	power	w.g.	water gage
Pa	Pascal		

TABLE A.2. UNIT EQUIVALENTS (CONVERSION FACTORS)

To change from one set of units to another, multiply known quantity and unit by the ratio of unit equivalents that results in the desired units

LENGTH
    U.S.: 12 in. = 1 ft = 0.333 yd
    metric: 1 m = 100 cm
                 = 1000 mm = 10^{-3}
             km = 10^{6} microns
    U.S.– metric: 1 ft = 0.30 m
    SI unit is the m

AREA
    U.S.: 144 in.^{2} = 1 ft^2
    U.S.– metric: 1 ft^2 = 0.093 m^2
    SI unit is the m^2

VOLUME
    U.S.: 1728 in.^{3} = 1 ft^3 = 7.48 gal.
    U.S.– metric: 1 ft^3 = 0.0283 m^3
    SI unit is the m^3

MASS
    U.S.: 1 lb = 16 oz = 7000 gr
    metric: 1 kg = 1000 g
    U.S.– metric: 2.2 lb = 1 kg
    SI unit is the kg

FORCE
    U.S.– metric: 1 lb = 4.45 N
    SI unit is the N

VELOCITY
    U.S.: 1 ft / sec = 0.68 mi / hr
    SI unit is the m / sec

DENSITY
    U.S.– metric: 1 lb / ft^3 = 16.0 kg / m^3
    SI unit is the kg / m^3

PRESSURE
    U.S.: 1 psi = 2.3 ft w. = 2.04 in. Hg
    metric: 1 atm = 101,300 N / m^3
            1 mm Hg = 133.3 Pa
    U.S.– metric: 14.7 psi = 1 atm
    SI unit is the N / m^2 (Pa)

TEMPERATURE
    U.S.: F = R − 460
    metric: C = K − 273
    U.S.– metric: F = (9 / 5) C + 32
                 C = 5 / 9(F − 32)
    SI unit is the K

ENERGY
    U.S.: 1 Btu = 778 ft-lb
    metric: 1 J = 1 W-sec = 0.239 cal
    U.S.– metric: 1 Btu = 1055 J = 252 cal
    SI unit is the J

POWER (RATE OF ENERGY)
    2545 Btu / hr = 1 hp = 0.746 kW
                 = 33,000 ft-lb / min
    3410 Btu / hr = 1 kW
    1 ton of refrigeration = 12,000 Btu / hr
                 = 4.72 HP = 3.52 kW
    SI unit is the W

SPECIFIC HEAT
    U.S.– metric: 1 Btu / lb-F = 1 cal / gm C
                 = 4.2 kJ / kg C

HEAT TRANSFER COEFFICIENT U
    U.S.– metric: 1 Btu / hr-ft^2-F = 5.68 W / m^2 C

VOLUME FLOW RATE
    U.S.– metric: 1 CFM = 1.70 m^3 / hr

**APPROXIMATE EQUIVALENTS FOR
    WATER ONLY (AT 60°F)**
    Density: 8.33 lb = 1 gal
            62.4 lb = 1 ft^3
    Flow rate: 1 GPM = 500 lb / hr

TABLE A.3. PROPERTIES OF SATURATED STEAM AND SATURATED WATER

Temperature Table

Temp F	Abs press, psi	Specific vol, cu ft / lb Sat liquid v_f	Specific vol, cu ft / lb Sat vapor v_g	Enthalpy, Btu / lb Sat liquid h_f	Enthalpy, Btu / lb Evap h_{fg}	Enthalpy, Btu / lb Sat vapor h_g
32	0.08854	0.01602	3306	0.00	1075.8	1075.8
35	0.09995	0.01602	2947	3.02	1074.1	1077.1
40	0.12170	0.01602	2444	8.05	1071.3	1079.3
45	0.14752	0.01602	2036.4	13.06	1068.4	1081.5
50	0.17811	0.01603	1703.2	18.07	1065.6	1083.7
55	0.2141	0.01603	1430.7	23.07	1062.7	1085.8
60	0.2563	0.01604	1206.7	28.06	1059.9	1088.0
65	0.3056	0.01605	1021.4	33.05	1057.1	1090.2
70	0.3631	0.01606	867.9	38.04	1054.3	1092.3
75	0.4298	0.01607	740.0	43.03	1051.5	1094.5
80	0.5069	0.01608	633.1	48.02	1048.6	1096.6
85	0.5959	0.01609	543.5	53.00	1045.8	1098.8
90	0.6982	0.01610	468.0	57.99	1042.9	1100.9
95	0.8153	0.01612	404.3	62.98	1040.1	1103.1
100	0.9492	0.01613	350.4	67.97	1037.2	1105.2
105	1.1016	0.01615	304.5	72.95	1034.3	1107.3
110	1.2748	0.01617	265.4	77.94	1031.6	1109.5
115	1.4709	0.01618	231.9	82.93	1028.7	1111.6
120	1.6924	0.01620	203.27	87.92	1025.8	1113.7
125	1.9420	0.01622	178.61	92.91	1022.9	1115.8
130	2.2225	0.01625	157.34	97.90	1020.0	1117.9
135	2.5370	0.01627	138.95	102.9	1017.0	1119.9
140	2.8886	0.01629	123.01	107.9	1014.1	1122.0
145	3.281	0.01632	109.15	112.9	1011.2	1124.1
150	3.718	0.01634	97.07	117.9	1008.2	1126.1
155	4.203	0.01637	86.52	122.9	1005.2	1128.1
160	4.741	0.01639	77.29	127.9	1002.3	1130.2
165	5.335	0.01642	69.19	132.9	999.3	1132.2
170	5.992	0.01645	62.06	137.9	996.3	1134.2
175	6.715	0.01648	55.78	142.9	993.3	1136.2
180	7.510	0.01651	50.23	147.9	990.2	1138.1
185	8.383	0.01654	45.31	152.9	987.2	1140.1
190	9.339	0.01657	40.96	157.9	984.1	1142.0
200	11.526	0.01663	33.64	168.0	977.9	1145.9
212	14.696	0.01672	26.80	180.0	970.4	1150.4
220	17.186	0.01677	23.15	188.1	965.2	1153.4
240	24.969	0.01692	16.323	208.3	952.2	1160.5
260	35.429	0.01709	11.763	228.6	938.7	1167.3
280	49.203	0.01726	8.645	249.1	924.7	1173.8
300	67.013	0.01745	6.466	269.6	910.1	1179.7
350	134.63	0.01799	3.342	321.6	870.7	1192.3
400	247.31	0.01864	1.8633	375.0	826.0	1201.0
450	422.6	0.0194	1.0993	430.1	774.5	1204.6
500	680.8	0.0204	0.6749	487.8	713.9	1201.7
550	1045.2	0.0218	0.4240	549.3	640.8	1190.0

Pressure Table

Abs press, psi	Temp, F	Specific vol, cu ft / lb Sat liquid v_f	Specific vol, cu ft / lb Sat vapor v_g	Enthalpy, Btu / lb Sat liquid h_f	Enthalpy, Btu / lb Evap h_{fg}	Enthalpy, Btu / lb Sat vapor h_g
0.50	79.58	0.01608	641.4	47.6	1048.8	1096.4
1.0	101.74	0.01614	333.6	69.7	1036.3	1106.0
2.0	126.08	0.01623	173.73	94.0	1022.2	1116.2
3.0	141.48	0.01630	118.71	109.4	1013.2	1122.6
4.0	152.97	0.01636	90.63	120.9	1006.4	1127.3
5.0	162.24	0.01640	73.52	130.1	1001.0	1131.1
6.0	170.06	0.01645	61.98	138.0	996.2	1134.2
7.0	176.85	0.01649	53.64	144.8	992.1	1136.9
8.0	182.86	0.01653	47.34	150.8	988.5	1139.3
9.0	188.28	0.01656	42.40	156.2	985.2	1141.4
10	193.21	0.01659	38.42	161.2	982.1	1143.3
14.7	212.00	0.01672	26.80	180.0	970.4	1150.4
20	227.96	0.01683	20.089	196.2	960.1	1156.3
25	240.07	0.01692	16.303	208.5	952.1	1160.6
30	250.33	0.01701	13.746	218.8	945.3	1164.1
40	267.25	0.01715	10.498	236.0	933.7	1169.7
50	281.01	0.01727	8.515	250.1	924.0	1174.1
60	292.71	0.01738	7.175	262.1	915.5	1177.6
70	302.92	0.01748	6.206	272.6	907.9	1180.6
80	312.03	0.01757	5.472	282.0	901.1	1183.1
90	320.27	0.01766	4.896	290.6	894.7	1185.3
100	327.81	0.01774	4.432	298.4	888.8	1187.2
110	334.77	0.01782	4.049	305.7	883.2	1188.9
120	341.25	0.01789	3.728	312.4	877.9	1190.4
130	347.32	0.01796	3.455	318.8	872.9	1191.7
140	353.02	0.01802	3.220	324.8	868.2	1193.0
150	358.42	0.01809	3.015	330.5	863.6	1194.1
160	363.53	0.01815	2.834	335.9	859.2	1195.1
170	368.41	0.01822	2.675	341.1	854.9	1196.0
180	373.06	0.01827	2.532	346.1	850.8	1196.9
190	377.51	0.01833	2.404	350.8	846.8	1197.6
200	381.79	0.01839	2.288	355.4	843.0	1198.4
250	400.95	0.01865	1.8438	376.0	825.1	1201.1
300	417.33	0.01890	1.5433	393.8	809.0	1202.8
350	431.72	0.01913	1.3260	409.7	794.2	1203.9
400	444.59	0.0193	1.1613	424.0	780.5	1204.5
450	456.28	0.0195	1.0320	437.2	767.4	1204.6
500	467.01	0.0197	0.9278	449.4	755.0	1204.4
600	486.21	0.0201	0.7698	471.6	731.6	1203.2
700	503.10	0.0205	0.6554	491.5	709.7	1201.2
800	518.23	0.0209	0.5687	509.7	688.9	1198.6
900	531.98	0.0212	0.5006	526.6	668.8	1195.4
1000	544.61	0.0216	0.4456	542.4	649.4	1191.8
1200	567.22	0.0223	0.3619	571.7	611.7	1183.4
1500	596.23	0.0235	0.2760	611.6	556.3	1167.9

TABLE A.4. THERMAL RESISTANCE R OF BUILDING AND INSULATING MATERIALS
(hr · ft^2-F / BTU)

Description	Density lb/ft^3	Resistance (R) Per Inch	Per Listed Thickness
BUILDING BOARD			
Boards, Panels, Subflooring, Sheathing			
Woodboard Panel Products			
Asbestos-cement board	120	0.25	--
Asbestos-cement board..........................0.125 in.	120	--	0.03
Asbestos-cement board.........................0.25 in.	120	--	0.06
Gypsum or plaster board.......................0.375 in.	50	--	0.32
Gypsum or plaster board........................0.5 in.	50	--	0.45
Gypsum or plaster board......................0.625 in.	50	--	0.56
Plywood (Douglas Fir)..............................	34	1.25	--
Plywood (Douglas Fir)..........................0.25 in.	34	--	0.31
Plywood (Douglas Fir).........................0.375 in.	34	--	0.47
Plywood (Douglas Fir)..........................0.5 in.	34	--	0.62
Plywood (Douglas Fir)........................0.625 in.	34	--	0.77
Plywood or wood panels.........................0.75 in.	34	--	0.93
Vegetable Fiber Board			
Sheathing, regular density.................0.5 in.	18	--	1.32
....0.78125 in.	18	--	2.06
Sheathing intermediate density.............0.5 in.	22	--	1.22
Nail-base sheathing........................0.5 in.	25	--	1.14
Shingle backer...........................0.375 in.	18	--	0.94
Shingle backer..........................0.3125 in.	18	--	0.78
Sound deadening board.....................0.5 in.	15	--	1.35
Tile and lay-in panels, plain or			
acoustic..	18	2.50	--
....0.5 in.	18	--	1.25
....0.75 in.	18	--	1.89
Laminated paperboard.............................	30	2.00	--
Homogeneous board from			
repulped paper..................................	30	2.00	--
Hardboard			
Medium density.................................	50	1.37	--
High density, service temp. service			
underlay......................................	55	1.22	--
High density, std. tempered.......................	63	1.00	--
Particleboard			
Low density....................................	37	1.85	--
Medium density.................................	50	1.06	--
High density...................................	62.5	0.85	--
Underlayment.............................0.625 in.	40	--	0.82
Wood subfloor...............................0.75 in.		--	0.94
BUILDING MEMBRANE			
Vapor--permeable felt............................	--	--	0.06
Vapor--seal, 2 layers of mopped			
15-lb felt....................................	--	--	0.12
Vapor--seal, plastic film.........................	--	--	Negl.
FINISH FLOORING MATERIALS			
Carpet and fibrous pad............................	--	--	2.08
Carpet and rubber pad.............................	--	--	1.23
Cork tile................................0.125 in.	--	--	0.28
Terrazzo......................................1 in.	--	--	0.08
Tile--asphalt, linoleum, vinyl, rubber..............	--	--	0.05
vinyl asbestos................................			
cermaic.......................................			
Wood, hardwood finish........................0.75 in.			0.68
INSULATING MATERIALS			
Blanket and Batt			
Mineral Fiber, fibrous form processed			
from rock, slag or glass			
approx. 2-2.75 in.............................	0.3-2.0	--	7
approx. 3-3.5 in.............................	0.3-2.0	--	11
approx. 3.50-6.5.............................	0.3-2.0	--	19
approx. 6-7 in...............................	0.3-2.0		22
approx. 8.5 in...............................	0.3-2.0		30

TABLE A.4 *(Continued).*

Description	Density lb/ft^3	Resistance (R) Per Inch	Per Listed Thickness
Board and Slabs			
Cellular glass................................	8.5	2.63	--
Glass fiber, organic bonded...................	4-9	4.00	--
Expanded rubber (rigid).......................	4.5	4.55	--
Expanded polystyrene extruded			
Cut cell surface...........................	1.8	4.00	--
Expanded polystyrene extruded			
Smooth skin surface........................	2.2	5.00	--
Expanded polystyrene extruded			
Smooth skin surface........................	3.5	5.26	--
Expanded polystyrene, molded beads............	1.0	3.57	--
Expanded polyurethane (R-11 exp.).............	1.5	6.25	--
(Thickness 1 in. or greater)..............	2.5		
Mineral fiber with resin binder..............	15	3.45	--
Mineral fiberboard, wet felted			
Core or roof insulation...................	16-17	2.94	--
Acoustical tile...........................	18	2.86	--
Acoustical tile...........................	21	2.70	--
Mineral fiberboard, wet molded			
Acoustical tile...........................	23	2.38	--
Wood or cane fiberboard			
Acoustical tile...............0.5 in.	--	--	1.25
Acoustical tile...............0.75 in.	--	--	1.89
Interior finish (plank, tile)................	15	2.86	--
Wood shredded (cemented in			
preformed slabs)..........................	22	1.67	--
LOOSE FILL			
Cellulosic insulation (milled paper or			
wood pulp)................................	2.3-3.2	3.13-3.70	--
Sawdust or shavings..........................	8.0-15.0	2.22	--
Wood fiber, softwoods........................	2.0-3.5	3.33	--
Perlite, expanded............................	5.0-8.0	2.70	--
Mineral fiber (rock, slag or glass)			
approx. 3.75-5 in........................	0.6-2.0		11
approx. 6.5-8.75 in......................	0.6-2.0		19
approx. 7.5-10 in........................	0.6-2.0		22
approx. 10.25-13.75......................	0.6-2.0		30
Vermiculite, exfoliated......................	7.0-8.2	2.13	--
	4.0-6.0	2.27	--
Roof Insulation			
Preformed, for use above deck			
Different roof insulations are available in			
different thicknesses to provide the design			
C values listed. Consult individual manu-			1.39
facturers for actual thickness of their			to
material..................................			8.33
MASONRY MATERIALS			
Concretes			
Cement mortar................................	116	0.20	--
Gypsum-fiber concrete 87.5% gypsum,			
12.5% wood chips..........................	51	0.60	--
Lightweight aggregates including expanded	120	0.19	--
shale, clay or slate; expanded slags;	100	0.28	--
cirders; pumice; vermiculite; also	80	0.40	--
cellular concretes	60	0.59	--
	40	0.86	--
	30	1.11	--
	20	1.43	
Perlite, expanded............................	40	1.08	
	30	1.41	
	20	2.00	
Sand and gravel or stone aggregate			
(oven dried)..............................	140	0.11	
Sand and gravel or stone aggregate			
(not dried)...............................	140	0.08	
Stucco.......................................	116	0.20	

TABLE A.4 *(Continued).*

Description	Density lb/ft³	Resistance (R) Per Inch	Resistance (R) Per Listed Thickness
MASONRY UNITS			
Brick, common...	120	0.20	--
Brick, face...	130	0.11	--
Clay tile, hollow:			
1 cell deep..................................3 in.	--	--	0.80
1 cell deep..................................4 in.	--	--	1.11
2 cells deep.................................6 in.	--	--	1.52
2 cells deep.................................8 in.	--	--	1.85
2 cells deep................................10 in.	--	--	2.22
3 cells deep................................12 in.	--	--	2.50
Concrete blocks, three oval core:			
Sand and gravel aggregate....................4 in.	--	--	0.71
....................8 in.	--	--	1.11
...................12 in.	--	--	1.28
Cinder aggregate............................3 in.	--	--	0.86
............................4 in.	--	--	1.11
............................8 in.	--	--	1.72
...........................12 in.	--	--	1.89
Lightweight aggregate........................3 in.	--	--	1.27
(expanded shale, clay, slate................4 in.	--	--	1.50
or slag; pumice)............................8 in.	--	--	2.00
...........................12 in.	--	--	2.27
Concrete blocks, rectangular core.			
Sand and gravel aggregate			
2 core, 8 in. 36 lb............................	--	--	1.04
Same with filled cores.........................	--	--	1.93
Lightweight aggregate (expanded shale,			
clay, slate or slag, pumice):			
3 core, 6 in. 19 lb............................	--	--	1.65
Same with filled cores.........................	--	--	2.99
2 core, 8 in. 24 lb............................	--	--	2.18
Same with filled cores.........................	--	--	5.03
3 core, 12 in. 38 lb...........................	--	--	2.48
Same with filled cores.........................	--	--	5.82
Stone, lime or sand................................	--	0.08	--
Gypsum partition tile:			
3 x 12 x 30 in. solid..........................	--	--	1.26
3 x 12 x 30 in. 4-cell.........................	--	--	1.35
4 x 12 x 30 in. 3-cell.........................	--	--	1.67
PLASTERING MATERIALS			
Cement plaster, sand aggregate.....................	116	0.20	--
Sand aggregate........................0.375 in.	--	--	0.08
Sand aggregate.........................0.75 in.	--	--	0.15
Gypsum plaster:			
Lightweight aggregate...................0.5 in.	45	--	0.32
Lightweight aggregate.................0.625 in.	45	--	0.39
Lightweight agg. on metal lath........0.75 in.	--	--	0.47
Perlite aggregate.............................	45	0.67	--
Sand aggregate................................	105	0.18	--
Sand aggregate........................0.5 in.	105	--	0.09
Sand aggregate......................0.625 in.	105	--	0.11
Sand aggregate on metal lath.........0.75 in.	--	--	0.13
Vermiculite aggregate.........................	45	0.59	--
ROOFING			
Asbestos-cement shingles...........................	120	--	0.21
Asphalt roll roofing...............................	70	--	0.15
Asphalt shingles...................................	70	--	0.44
Built-up roofing........................0.375 in.	70	--	0.33
Slate.....................................0.5 in.	--	--	0.05
Wood shingles, plain and plastic film faced.........	--	--	0.94

TABLE A.4 *(Continued).*

Description	Density lb/ft³	Resistance (R) Per Inch	Resistance (R) Per Listed Thickness
SIDING MATERIALS (On Flat Surface)			
Shingles			
Asbestos-cement........................	120	--	0.21
Wood, 16 in., 7.5 exposure...........	--	--	0.87
Wood, double, 16 in., 12 in. exposure....	--	--	1.19
Wood, plus insul. backer board, 0.3125 in....	--	--	1.40
Siding			
Asbestos-cement, 0.25 in., lapped........	--	--	0.21
Asphalt roll siding....................	--	--	0.15
Asphalt insulting siding (0.5 in. bed.)....	--	--	1.46
Wood drop, 1 x 8 in...................	--	--	0.79
Wood, bevel, 0.5 x 8 in., lapped........	--	--	0.81
Wood, bevel, 0.75 x 10 in., lapped......	--	--	1.05
Wood, plywood, 0.375 in., lapped........	--	--	0.59
Wood, medium density siding, 0.4375 in....	40	0.67	
Aluminum or Steel, over sheathing			
Hollow-backed........................	--	--	0.61
Insulating-board backed nominal			
0.375 in............................	--	--	1.82
Insulating-board backed nominal			
0.375 in., foil backed...............			2.96
Architectural glass....................	--	--	0.10
WOODS			
Maple, oak, and similar hardwoods.......	45	0.91	--
Fir, pine, and similar softwoods........	32	1.25	--
Fir, pine, and similar softwoods....0.75 in.	32	--	0.94
...........................1.5 in.		--	1.89
...........................2.5 in.		--	3.12
...........................3.5 in.		--	4.35

TABLE A.5. THERMAL RESISTANCE R OF SURFACE AIR FILMS AND AIR SPACES (hr · ft²-F / BTU)

Surface air films

	Direction of Heat Flow	R-Value
STILL AIR (interior surfaces)		
Horizontal	Upward	0.61
Sloping - 45 degree	Upward	0.62
Vertical	Horizontal	0.68
Sloping - 45 degree	Downward	0.76
Horizontal	Downward	0.92
MOVING AIR (exterior surfaces)		
15 mph Wind (Winter)	Any	0.17
7.5 mph Wind (Summer)	Any	0.25

Air spaces

Position of Air Space	Direction of Heat Flow	Thickness of Air Space 1/2"	3/4"	1½"	3½"
Horizontal	Up	0.84	0.87	0.89	0.93
45° Slope	Up	0.90	0.94	0.91	0.96
Vertical	Horizontal	0.91	1.01	1.02	1.01
Horizontal	Down	0.92	1.02	1.14	1.21
45° Slope	Down	0.92	1.02	1.09	1.05

TABLE A.6. TYPICAL BUILDING ROOF AND WALL CONSTRUCTION CROSS-SECTIONS

SECTIONS

	R_i	U_r
Asphaltic shingles ½" plywood sheathing 2 x trusses @ 16"-24" o.c. Batt or loose fill insulation Vapor Barrier	11 19 22 30	0.08 0.05 0.04 0.03
Built-up roof ½" Plywood deck 2 x rafters @ 16"-24" o.c. Batt or loose fill Insulation ½" Drywall ceiling Vapor Barrier	11 19 22 30	0.07 0.05 0.04 0.03
Built-up roof Rigid insulation Metal deck Bar joists @ 16"-24" o.c. lay-in ceiling	No ceiling 5.5 8 11 15 Ceiling 5.5 8 11 15	0.14 0.11 0.08 0.06 0.10 0.08 0.06 0.05
Metal roof Steel purlins Batt insulation lay-in ceiling Vapor Barrier	No ceiling 3.5 8 11 19 Ceiling 3.5 8 11 19	0.23 0.11 0.06 0.05 0.13 0.08 0.07 0.04

WALL SECTION

	R_i	U_w
Metal siding Insulation Metal studs or girts Vapor Barrier OR Insulated metal Sandwich panel	3.5 8 11 19	0.23 0.11 0.08 0.05
Metal siding Insulation Metal studs or girts ½" drywall Vapor Barrier OR Insulated metal Sandwich panel	3.5 8 11 19	0.17 0.10 0.07 0.05
Brick veneer Air space Sheathing Insulation Metal studs @ 16"-24" o.c. ½" drywall Vapor Barrier	8 11 19	0.09 0.07 0.05

509

TABLE A.6 (Continued).

WALL SECTIONS	R_i	U_w
Wood or metal siding, Sheathing, Insulation, 2 x 4 studs @ 16"-24" o c, ½" Drywall, Vapor Barrier **OR** Rigid insulation	8 11 14	0.09 0.08 0.07
2 x 6 studs, same construction as above	8 11 19	0.09 0.07 0.05
Face brick, Air space, Sheathing, Insulation, 2 x 4 studs @ 16"-24" o c, ½" Drywall, Vapor Barrier **OR** Rigid Insulation	8 11 19	0.09 0.07 0.06
2 x 6 studs, same construction as above	8 11 19	0.08 0.07 0.05

WALL SECTIONS	DENSITY	R_i	U_w
Face brick, Air space, 8" Conc block, Core insulation	Conc Block 80# cu ft	no insulation core insulation	0.23 0.12
	Conc Block 120# cu ft	no insulation core insulation	0.27 0.16
Face brick, Air space, Rigid insulation, 8" Conc block, Core insulation	Conc Block 80# cu ft	Core Insulation and R_i 3 5.5 8 11	0.09 0.07 0.06 0.05
	Conc Block 120# cu ft	Core Insulation and R_i 3 5.5 8 11	0.11 0.09 0.07 0.06
Face brick, Air space, 8" Conc block, Core insulation, Insulation, ½" drywall, Vapor Barrier	Conc Block 80# cu ft	Core Insulation and R_i 3.5 (1 x 2 @ 16" o c) 5.5 8 (2 x 3 @ 16" o c)	0.08 0.07 0.06
Rigid Insulation, Vapor Barrier	Conc Block 120# cu ft	Core Insulation and R_i 3.5 (1 x 2 @ 16" o c) 5.5 8 (2 x 3 @ 16" o c)	0.10 0.08 0.07

TABLE A. 6(Continued).

WALL SECTIONS

WALL SECTIONS	DENSITY	W	R_i	U_w
Concrete	80#/cu. ft.	6"	—	0.31
		8"	—	0.25
		12"	—	0.18
	120#/cu. ft.	6"	—	0.50
		8"	—	0.42
		12"	—	0.32
Concrete / Insulation / Metal furring/studs @ 16" to 24" o.c. / ½" drywall / Vapor Barrier — OR — Rigid insulation / Vapor Barrier	80#/cu. ft.	6"	3	0.15
			5.5	0.11
			8	0.09
			11	0.07
		8"	3	0.13
			5.5	0.10
			8	0.08
			11	0.06
		12"	3	0.11
			5.5	0.09
			8	0.07
			11	0.06
	120#/cu. ft.	6"	3	0.18
			5.5	0.13
			8	0.10
			11	0.07
		8"	3	0.17
			5.5	0.12
			8	0.09
			11	0.07
		12"	3	0.15
			5.5	0.11
			8	0.09
			11	0.07

WALL SECTIONS

WALL SECTIONS	DENSITY	W	R_i	U_w
Concrete block	Conc. Block 80#/cu. ft	6"	—	37
		8"	—	34
		12"	—	29
	Conc. Block 120#/cu. ft	6"	—	47
		8"	—	43
		12"	—	38
Concrete block / Core Insulation	Conc. Block 80#/cu. ft	6"	—	18
		8"	—	14
		12"	—	10
	Conc. Block 120#/cu. ft	6"	—	26
		8"	—	22
		12"	—	17
Concrete block / Core Insulation / Batt or rigid Insulation / ½" drywall / Vapor Barrier	Conc. Block 80#/cu. ft	6"	3.5	11
			5.5	09
			8	07
		8"	3.5	09
			5.5	08
			8	05
		12"	3.5	07
			5.5	06
			8	05
	Conc. Block 120#/cu. ft	6"	3.5	13
			5.5	10
			8	08
		8"	3.5	12
			5.5	09
			8	06
		12"	3.5	10
			5.5	03
			8	06

TABLE A.7. OVERALL HEAT TRANSFER COEFFICIENT *U* FOR BUILDING
CONSTRUCTION COMPONENTS

Construction	U-Value in Btu/hr·ft²·°F	
	Summer	Winter
WALLS		
Frame with wood siding, sheathing, and inside finish		
No insulation	.22	.23
R-7 insulation (2 in.-2½ in.)	.09	.09
R-11 insulation (3 in.-3½ in.)	.07	.07
Frame with 4 in. brick or stone veneer, sheathing, and inside finish		
No insulation	.24	.24
R-7 insulation	.09	.09
R-11 insulation	.07	.07
Frame with 1 in. stucco, sheathing, and inside finish		
No insulation	.29	.29
R-7 insulation	.10	.10
R-11 insulation	.07	.07
Masonry:		
8 in. concrete block, no finish	.49	.51
12 in. concrete block, no finish	.45	.47
Masonry (8 in. concrete block):		
Inside finish:		
furred gypsum wallboard (½ in.); no insulation	.29	.30
furred, foil backed gypsum wallboard (½ in.); no insulation	.29	.30
1 in. polystyrene insulation board (R-5), and ½ in. gypsum wallboard	.13	.13
Masonry (8 in. cinder block or hollow clay tile):		
Inside finish:		
furred gypsum wallboard (½ in.); no insulation	.25	.25
furred, foil-backed gypsum wallboard (½ in.); no insulation	.17	.17
1 in. polystyrene insulation board (R-5), and ½ in. gypsum wallboard	.12	.12
Masonry (4 in. face brick and 8 in. cinder block or 8 in. hollow clay tile):		
Inside finish:		
furred gypsum wallboard (½ in.); no insulation	.22	.22
furred, foil-backed gypsum wallboard (½ in.); no insulation	.15	.16
1 in. polystyrene insulation board (R-5), and ½ in. gypsum wallboard	.12	.12
Masonry (12 in. hollow clay tile or 12 in. cinder block):		
Inside finish:		
furred gypsum wallboard (½ in.); no insulation	.24	.24
furred, foil-backed gypsum wallboard (½ in.); no insulation	.16	.17
1 in. polystyrene insulation board (R-5), and ½ in. gypsum wallboard	.12	.12
Masonry (4 in. face brick, 4 in. common brick):		
Inside finish:		
furred gypsum wallboard (½ in.); no insulation	.28	.28
furred, foil-backed gypsum wallboard (½ in.); no insulation	.18	.19
1 in. polystyrene insulation board (R-5), and ½ in. gypsum wallboard	.13	.13
Masonry (8 in. concrete or 8 in. stone):		
Inside finish:		
furred gypsum wallboard (½in.); no insulation	.33	.34
furred, foil-backed gypsum wallboard (½ in); no insulation	.21	.21
1 in. polystyrene insulation board (R-5), and ½ in. gypsum wallboard	.14	.14
Metal with vinyl inside finish, R-7 (3 in. glass fiber batt)	.14	.14
PARTITIONS		
Frame (½ in. gypsum wallboard one side only):		
No insulation	.55	.55
Frame (½ in. gypsum wallboard each side):		
No insulation	.31	.31
R-11 insulation	.08	.08
Masonry (4 in. cinder block):		
No insulation, no finish	.40	.40
No insulation, one side furred gypsum wallboard (½ in.)	.26	.26
No insulation, both sides furred gypsum wallboard (½ in.)	.19	.19
One side 1 in. polystyrene insulation board (R-5), and ½ in. gypsum wallboard	.13	.13

TABLE A.7 *(Continued).*

Construction	U-Value in Btu/hr·ft²·°F	
	Summer	Winter
CEILING-FLOOR		
Frame (asphalt tile floor, 5/8 in. plywood, 25/32 in. wood subfloor, finished ceiling):		
Heat flow up	.23	.23
Heat flow down	.20	.19
Concrete (asphalt tile floor, 4 in. concrete deck, air space, finished ceiling):		
Heat flow up	.34	.33
Heat flow down	.26	.25
ROOF (flat roof, no finished ceiling)		
Steel deck:		
No insulation	.64	.86
1 in. insulation (R-2.78)	.23	.25
2 in. insulation (R-5.56)	.15	.16
1 in. Wood deck:		
No insulation	.40	.48
1 in. insulation (R-2.78)	.19	.21
2 in. insulation (R-5.56)	.12	.13
2.5 in. Wood deck:		
No insulation	.25	.26
1 in. insulation (R-2.78)	.15	.16
2 in. insulation (R-5.56)	.10	.11
4 in. Wood deck:		
No insulation	.17	.18
1 in. insulation (R-2.78)	.12	.12
2 in. insulation (R-5.56)	.09	.09
ROOF-CEILING (flat roof, finished ceiling)		
Steel deck:		
No insulation	.33	.40
1 in. insulation (R-2.78)	.17	.19
2 in. insulation (R-5.56)	.12	.13
1 in. Wood deck:		
No insulation	.26	.29
1 in. insulation (R-2.78)	.15	.16
2 in. insulation (R-5.56)	.11	.11
2.5 in. Wood deck:		
No insulation	.18	.20
1 in. insulation (R-2.78)	.12	.13
2 in. insulation (R-5.56)	.09	.10
4 in. Wood deck:		
No insulation	.14	.15
1 in. insulation (R-2.78)	.10	.10
2 in. insulation (R-5.56)	.08	.08
4 in. Light weight concrete deck:		
No insulation	.14	.15
6 in. Light weight concrete deck:		
No insulation	.10	.11
8 in. Light weight concrete deck:		
No insulation	.08	.09
2 in. Heavy weight concrete deck:		
No insulation	.32	.38
1 in. insulation (R-2.78)	.17	.19
2 in. insulation (R-5.56)	.11	.12
4 in. Heavy weight concrete deck:		
No insulation	.30	.36
1 in. insulation (R-2.78)	.16	.18
2 in. insulation (R-5.56)	.11	.12
6 in. Heavy weight concrete deck:		
No insulation	.28	.33
1 in. insulation (R-2.78)	.16	.17
2 in. insulation (R-5.56)	.11	.12

TABLE A.7 *(Continued).*

Construction	U-Value in Btu/hr·ft²·°F	
	Summer	Winter
ROOF CEILING (wood frame pitched roof, finished ceiling on rafters)		
No insulation	.28	.29
R-19 insulation (5½ in.-6½ in.)	.05	.05
ROOF-ATTIC-CEILING (attic with natural ventilation)		
No insulation	.15	.29
R-19 insulation (5½ in.-6½ in.)	.04	.05
FLOORS		
Floor over unconditioned space, no ceiling		
Wood frame:		
No insulation	.33	.27
R-7 insulation (2 in.-2½ in.)	.09	.08
Concrete deck:		
No insulation	.59	.43
R-7 insulation	.10	.09
DOORS		
Solid wood:		
1 in. thick	.61	.64
1½ in. thick	.47	.49
2 in. thick	.42	.43
Steel:		
1½ in. thick, mineral fiber core	.58	.59
1½ in. thick, polystyrene core	.46	.47
1½ in. thick, urethane foam core	.39	.40

TABLE A.8. OVERALL HEAT TRANSFER COEFFICIENT U FOR GLASS (BTU / hr · ft²-F)

VERTICAL PANELS (EXTERIOR WINDOWS, SLIDING GLASS DOORS AND PARTITIONS) -- FLAT GLASS, GLASS BLOCK, AND PLASTIC SHEET

Description	Exterior Winter	Exterior Summer	Interior Summer
Flat Glass			
single glass	1.10	1.04	0.73
insulating glass--double			
1/4" air space[a]	0.58	0.61	0.49
1/2" air space[b]	0.49	0.56	0.46
1/2" air space, low emmittance coating[c]			
e = 0.20	0.32	0.38	0.32
e = 0.40	0.38	0.45	0.38
e = 0.60	0.43	0.51	0.42
insulating glass--triple[d]			
1/4" air spaces[a]	0.39	0.44	0.38
1/2" air spaces[b]	0.31	0.39	0.30
storm windows			
1" to 4" air space[a]	0.50	0.50	0.44
Plastic Sheet			
single			
1/8" thick	1.06	0.98	--
1/4" thick	0.96	0.89	--
1/2" thick	0.81	0.76	--
insulating unit--double			
1/4" air space[a]	0.55	0.56	--
1/2" air space[b]	0.43	0.45	--
Glass Block[e]			
6 x 6 x 4 in. thick	0.60	0.57	0.46
8 x 8 x 4 in. thick	0.56	0.54	0.44
--with cavity divider	0.48	0.46	0.38
12 x 12 x 4 in. thick	0.52	0.50	0.41
--with cavity divider	0.44	0.42	0.36
12 x 12 x 2 in. thick	0.60	0.57	0.46

HORIZONTAL PANELS (SKYLIGHTS) -- FLAT GLASS, GLASS BLOCK, AND PLASTIC DOMES

Description	Exterior Winter	Exterior Summer	Interior Summer
Flat Glass			
single glass	1.23	0.83	0.96
insulating glass--double			
1/4" air space[a]	0.65	0.54	0.59
1/2" air space[b]	0.59	0.49	0.56
1/2" air space, low emittance coating[c]			
e = 0.20	0.48	0.36	0.39
e = 0.40	0.52	0.42	0.45
e = 0.60	0.56	0.46	0.50
Glass Block[e]			
11 x 11 x 3 in. thick with cavity divider	0.53	0.35	0.44
12 x 12 x 4 in. thick with cavity divider	0.51	0.34	0.42
Plastic Domes[f]			
single-walled	1.15	0.80	--
double-walled	0.70	0.46	--

ADJUSTMENT FACTORS FOR VERTICAL AND HORIZONTAL PANELS

Description	Single Glass	Double or Triple Glass	Storm Windows
Windows			
All Glass	1.00	1.00	1.00
Wood Sash--80% Glass	0.90	0.95	0.90
Wood Sash--60% Glass	0.80	0.85	0.80
Metal Sash--80% Glass	1.00	1.20[g]	1.20[g]
Sliding Glass Doors			
Wood Frame	0.95	1.00	--
Metal Frame	1.00	1.10[g]	--

[a] 1/8" glass
[b] 1/4" glass
[c] coating on either glass surface facing air space
[d] window design: 1/4" glass - 1/8" glass - 1/4" glass
[e] nominal dimensions
[f] based on area of opening
[g] with thermal break

TABLE A.9. OUTDOOR DESIGN CONDITIONS

Location, USA	LAT Deg.	Winter		Summer			
		Degree Days	97.5% DB	2.5% Coinc. DB	WB	Daily DB Range	2.5% WB
ALABAMA							
Birmingham	33	2550	21	94	75	21	77
Montgomery	32	2290	25	95	76	21	79
ALASKA							
Anchorage	61	10860	−18	68	58	15	59
Fairbanks	64	14280	−47	78	60	24	62
ARIZONA							
Phoenix	33	1770	34	107	71	27	75
Tucson	32	1800	32	102	66	26	71
ARKANSAS							
Little Rock	34	3220	20	96	77	22	79
CALIFORNIA							
Los Angeles	34	1350	40	89	70	20	71
Sacramento	38	2420	32	98	70	36	71
San Diego	32	1460	44	80	69	12	70
San Francisco	37	3000	40	71	62	14	62
COLORADO							
Denver	39	6280	1	91	59	28	63
CONNECTICUT							
Hartford	41	6240	7	88	73	22	75
DELAWARE							
Wilmington	39	4930	14	89	74	20	76
D.C.							
Washington	38	4220	17	91	74	18	77
FLORIDA							
Miami	25	210	47	90	77	15	79
Tampa	28	680	40	91	77	17	79
GEORGIA							
Atlanta	33	2960	22	92	74	19	76
Savannah	32	1820	27	93	77	20	79
HAWAII							
Honolulu	21	0	63	86	73	12	75
IDAHO							
Boise	43	5810	10	94	64	31	66
ILLINOIS							
Chicago	41	5880	2	91	74	15	77
Springfield	39	5430	2	92	74	21	77
INDIANA							
Indianapolis	39	5700	2	90	74	22	76
South Bend	41	6440	1	89	73	22	75
IOWA							
Des Moines	41	6590	−5	91	74	23	77
Dubuque	42	7380	−7	88	73	22	75

TABLE A.9 *(Continued).*

Location, USA	LAT Deg.	Winter		Summer			
		Degree Days	97.5% DB	2.5% Coinc. DB	WB	Daily DB Range	2.5% WB
KANSAS							
Topeka	39	5180	4	96	75	24	78
Wichita	37	4620	7	98	73	23	76
KENTUCKY							
Lexington	38	4680	8	91	73	22	76
Louisville	38	4660	10	93	74	23	77
LOUISIANA							
Baton Rouge	30	1560	29	93	77	19	80
New Orleans	30	1250	33	92	78	16	80
MAINE							
Portland	43	7510	−1	84	71	22	72
MARYLAND							
Baltimore	39	4110	13	91	75	21	77
MASSACHUSETTS							
Boston	42	5630	9	88	71	16	74
Pittsfield	42	7580	−3	84	70	23	72
MICHIGAN							
Detroit	42	6230	6	88	72	20	74
Lansing	42	6910	1	87	72	24	74
MINNESOTA							
Minneapolis	44	8380	−12	89	73	22	75
MISSISSIPPI							
Jackson	32	2240	25	95	76	21	78
MISSOURI							
Kansas City	39	4710	6	96	74	20	77
St. Louis	38	4480	8	94	75	18	77
MONTANA							
Billings	45	7050	−10	91	64	31	66
NEBRASKA							
Omaha	41	6610	−3	91	75	22	77
NEVADA							
Las Vegas	36	2710	28	106	65	30	70
Reno	39	6330	11	93	60	45	62
NEW HAMPSHIRE							
Concord	43	7380	−3	87	70	26	73
NEW JERSEY							
Newark	40	4590	14	91	73	20	76
Trenton	40	4980	14	88	74	19	76
NEW MEXICO							
Albuquerque	35	4350	16	94	61	27	65
NEW YORK							
Albany	42	6880	1	88	72	20	74
Buffalo	43	7060	6	85	70	21	73
NYC	40	4870	15	89	73	17	75

TABLE A.9 (Continued).

Location, USA	LAT Deg.	Winter		Summer			
		Degree Days	97.5% DB	2.5% Coinc. DB	WB	Daily DB Range	2.5% WB
NORTH CAROLINA							
Charlotte	35	3190	22	93	74	20	76
Raleigh	35	3390	20	92	75	20	77
NORTH DAKOTA							
Bismark	46	8850	−19	91	68	27	71
OHIO							
Cincinnati	39	4410	6	90	72	21	75
Cleveland	41	6350	5	88	72	22	74
OKLAHOMA							
Okla. City	35	3725	13	97	74	23	77
Tulsa	36	3860	13	98	75	22	78
OREGON							
Portland	45	4640	24	86	67	21	67
PENNSYLVANIA							
Philadelphia	39	4490	14	90	74	21	76
Pittsburgh	40	5050	7	88	71	19	73
PUERTO RICO							
San Juan	18		68	88		11	80
RHODE ISLAND							
Providence	41	5950	9	86	72	19	74
SOUTH CAROLINA							
Charleston	32	1790	28	92	78	13	80
SOUTH DAKOTA							
Sioux Falls	43	7840	−11	91	72	24	75
TENNESSEE							
Memphis	35	3020	18	95	76	21	79
Nashville	36	3580	14	94	74	21	77
TEXAS							
Dallas	32	2360	22	100	75	20	78
Ft. Worth	32	2410	22	99	74	22	77
Houston	29	1280	33	95	77	18	79
UTAH							
Salt Lake City	40	6050	8	95	62	32	65
VERMONT							
Burlington	44	8270	−7	85	70	23	72
VIRGINIA							
Richmond	37	3870	17	92	76	21	78
Roanoke	37	4150	16	91	72	23	74
WASHINGTON							
Seattle	47	4420	27	82	66	19	67
Spokane	47	6660	2	90	63	28	64
WEST VIRGINIA							
Charleston	38	4480	11	90	73	20	75

TABLE A.9 *(Continued).*

Location, USA	LAT Deg.	Winter		Summer			
		Degree Days	97.5% DB	2.5% Coinc. DB	WB	Daily DB Range	2.5% WB
WISCONSIN							
Milwaukee	43	7640	− 4	87	73	21	74
WYOMING							
Cheyenne	41	7380	− 1	86	58	30	62

Location, Canada							
ALBERTA							
Edmonton	53	10270	− 25	82	65	23	66
BRITISH COLUMBIA							
Vancouver	49	5520	19	77	66-	17	67
MANITOBA							
Winnipeg	49	10680	− 27	86	71	22	73
NOVA SCOTIA							
Halifax	44	7360	5	76	65	16	67
ONTARIO							
Ottawa	45	8740	− 13	87	71	21	73
Toronto	43	6830	− 1	87	72	20	74
QUEBEC							
Montreal	45	7900	− 10	85	72	17	74

Location, Other Countries				2.5% DB			
ARGENTINA							
Buenos Aires	35s		34	89		22	76
AUSTRALIA							
Melbourne	38s		38	91		21	69
AUSTRIA							
Vienna	48		11	86		16	69
BRAZIL							
Rio de Janeiro	23s		60	92		11	79
CHINA							
Shanghai	31		26	92		16	81
COLOMBIA							
Bogotá	5		46	70		19	59
CUBA							
Havana	23		62	91		14	81
EGYPT							
Cairo	30		46	100		26	75
ENGLAND							
London	51		26	79		16	66
FRANCE							
Paris	49		25	86		21	68

TABLE A.9 *(Continued).*

Location, Other Countries	LAT Deg.	Winter Degree Days	Winter 97.5% DB	Summer 2.5% DB	Summer Daily DB Range	Summer 2.5% WB
GERMANY						
Berlin	52		12	81	19	67
INDIA						
New Delhi	29		41	107	26	82
IRAN						
Tehran	36		24	100	27	74
ISRAEL						
Tel Aviv	32		41	93	16	73
ITALY						
Rome	42		33	92	24	73
JAPAN						
Tokyo	36		28	89	14	80
MEXICO						
Mexico City	19		39	81	25	60
NIGERIA						
Lagos	6		71	91	12	82
POLAND						
Warsaw	52		8	81	19	70
SAUDIA ARABIA						
Jedda	21		60	103	22	84
SOUTH AFRICA						
Capetown	34s		42	90	21	71
SOVIET UNION						
Moscow	56		−6	81	21	67
SPAIN						
Madrid	40		28	91	25	69

Room Heating Load Calculations

P. _____ of _____ PP.

Project _____ Location _____ Indoor DB _____ F

Engrs. _____ Calc. by _____ Chk. by _____ Outdoor DB _____ F

Room															
Plan Size															
Heat Transfer	U	x	A	x	TD	=	BTU/hr	U	x	A	x	TD	=	BTU/hr	
Wall															
Window															
Door															
Roof/ceiling															
Floor															
Partition															
Subtotal															

Infiltration (CFM): 1.1 x A x B x TC =
Window 1.1
Door 1.1
Subtotal
Room Load

(Second identical block repeats below.)

Infiltration CFM	Column A	Column B
Crack method / air change method	CFM per ft or crack / air changes per hr	length of crack, ft / room volume/60, ft^3

Figure A.1. Room heating load calculations form.

521

Project _____

Location _____ Calc. by _____

Engineers _____ Chk. by _____

	DB, F	W; gr./lb.
Indoor		
Outdoor		
Diff.		

Heat Transfer	Q =	U	x	A	x	TD	=	BTU/hr
Roof								
Walls								
Windows								
Doors								
Floor								

Subtotal _____

Infiltration Q_S = 1.1 x _____ CFMx _____ TC = _____

Ventilation Q_S = 1.1 x _____ CFMx _____ TC = _____

Humidific Q_L = 0.68 x _____ CFMx ____ gr./lb. _____

Building Net Load _____

Piping or duct loss _____ %

Duct leakage _____ %

Pickup factor _____ %

Domestic HW Load _____

Boiler or furnace gross load _____

Notes:

Figure A.2. Building heating load calculations form.

COMMERCIAL COOLING LOAD CALCULATIONS

Project _____   Room _____   Engr. _____

Location _____   Lat. _____   Calc. by _____   Check _____

Design Conditions		DB F	WB F	RH %	W gr/lb
	Outdoor				
	Room				

Daily Range _____ F. Ave. _____
Day _____ Time _____

Conduction	Dir.	Color	U	A, ft²		CLTD, F		F_c	RSHG BTU/hr
				Gross	Net	Table	Corr.		
Glass									
Wall (Group)									
Roof/ceiling									
Floor									
Partition									
Door									

Solar	Dir.	Sh.	SHGF	A	SC	CLF	F_c
Glass							

$$K = (U_w A_w + U_g A_g)/L =$$

$$F_c = 1 - 0.02K =$$

$$CFM = \frac{RSHG}{1.1\,(T_R - T_{SA})} =$$

$$RSHR = RSHG/RTHG =$$

Lights _____ W × 3.41 × _____ BF × _____ CLF × _____ F_c
People _____ SHG × _____ n × _____ CLF × _____ F_c
_____ LHG × _____ n
Equipment _____
Infiltration   1.1 × _____ CFM × _____ TC
            0.68 × _____ CFM × _____ gr/lb

RLHG BTU/hr

SA duct gain
SA duct leakage _____ %
SA fan gain (draw through) _____ %

Subtotal

Ventilation 1.1 × _____ CFM × _____ TC
            0.68 × _____ CFM × _____ gr/lb
SA fan gain (blow through) _____ %
Pump gain
RA duct gain
RA fan gain _____ %

Room Heat Gain

RTHG BTU/hr

Cooling Load

Tons

Figure A.3. Commercial cooling load calculation form.

RESIDENTIAL COOLING
LOAD CALCULATIONS

Project _____   Outdoor _____ F. Inside _____ F. Outdoor Daily Range _____ F.

Location _____   Engr. _____ Calc. _____ Check _____

Room Name																	
Plan Size																	
	U	A	ETD	BTU/hr	U	A	ETD	BTU/hr	U	A	ETD	BTU/hr	U	A	ETD	BTU/hr	
Wall gross/net																	
Roof/ceiling																	
Floor																	
Partition																	
Door																	
	Dir.		CLF		Dir.		CLF		Dir.		CLF		Dir.		CLF		
Windows																	
Infiltration																	
People																	
Appliances																	
RSHG																	

Room Name																	
Plan Size																	
	U	A	ETD	BTU/hr	U	A	ETD	BTU/hr	U	A	ETD	BTU/hr	U	A	ETD	BTU/hr	
Wall gross/net																	
Roof/ceiling																	
Floor																	
Partition																	
Door																	
	Dir.		CLF		Dir.		CLF		Dir.		CLF		Dir.		CLF		
Windows																	
Infiltration																	
People																	
Appliances																	
RSHG																	

Building Total   NOTES:

Calc. RSHG   = _____

Duct gain   % = _____

Duct leak   % = _____

Total RSHG   = _____

Cool. Load =   × RSHG = _____

Figure A.4. Residential cooling load calculation form.

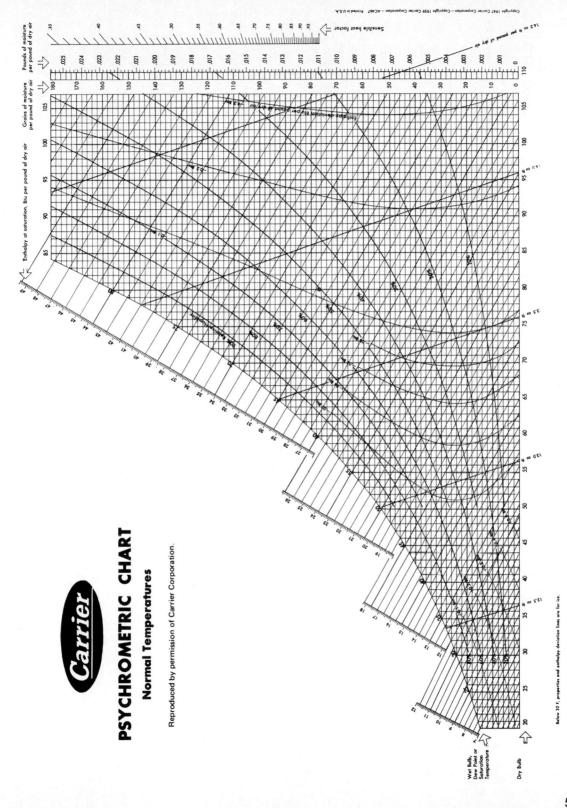

Copyright 1947 Carrier Corporation — Copyright 1959 Carrier Corporation — AC-647   Printed in U.S.A.

PSYCHROMETRIC CHART

Normal Temperatures

Reproduced by permission of Carrier Corporation.

Figure A.5. Psychrometric chart, U.S. units.

Figure A.6. Psychrometric chart, SI units.

Index